PRODUCT PLATFORM AND PRODUCT FAMILY DESIGN

PRODUCT PLATFORM AND PRODUCT FAMILY DESIGN
Methods and Applications

Edited by

Timothy W. Simpson[1], Zahed Siddique[2], and Jianxin (Roger) Jiao[3]
[1]The Pennsylvania State University, University Park, Pennsylvania; [2]The University of Oklahoma, Norman, Oklahoma; [3]Nanyang Technological University, Singapore

Timothy W. Simpson
The Pennsylvania State University
329 Leonhard Building
University Park, PA 16802 U.S.A.

Zahed Siddique
School of Aerospace and Mechanical Engineering
University of Oklahoma
865 ASP Avenue
Norman, OK 73019 U.S.A.

Jianxin (Roger) Jiao
School of Mechanical and Aeropace Engineering
Nanyang Technology University
Nanyang Avenue 50
Singapore 639798

Product Platform and Product Family Design

Library of Congress Control Number: 2005932994

ISBN-10: 0-387-25721-7   e-ISBN-10: 0-387-29197-0
ISBN-13: 9780387257211   e-ISBN-13: 9780387291970

Printed on acid-free paper.

9 8 7 6 5 4 3 2 1   SPIN 11388074

springeronline.com

Contents

Contributing Authors

Janet K. Allen
The Georgia Institute of Technology, Atlanta, Georgia

Daniel Bowman
Pittiglio, Rabin, Todd & McGrath (PRTM), Waltham, Massachusetts

Jordan J. Cox
Brigham Young University, Provo, Utah

Olivier L. de Weck
Massachusetts Institute of Technology, Cambridge, Massachusetts

Ryan Fellini
The University of Michigan, Ann Arbor, Michigan

Sebastian Fixson
The University of Michigan, Ann Arbor, Michigan

Kikuo Fujita
Osaka University, Osaka, Japan

Johannes I. M. Halman
University of Twente, The Netherlands

Adrian P. Hofer
Hofer & Partner, Wollerau, Switzerland

Tobias Holmqvist
Chalmers University of Technology, Göteburg, Sweden

Katja Höltta-Otto
MIT Center for Innovation in Product Development, Cambridge,
Massachusetts and Helsinki University of Technology, Espoo, Finland

Jianxin (Roger) Jiao
Nanyang Technological University, Singapore

Harshavardhan Karandikar
ABB Corporate Research Center, Ladenburg, Germany

Harrison M. Kim
University of Illinois Urbana-Champaign, Urbana, Illinois

Michael Kokkolaras
The University of Michigan, Ann Arbor, Michigan

Tucker M. Marion
The Pennsylvania State University, University Park, Pennsylvania

Farrokh Mistree
The Georgia Institute of Technology, Atlanta, Georgia

Manojkumar Natarajan
University of Oklahoma, Norman, Oklahoma

Srinivas Nidamarthi
ABB Corporate Research Center, Ladenburg, Germany

Kevin Otto
Robust Systems and Strategy, LLC, Watertown, Massachusetts

Panos Y. Papalambros
The University of Michigan, Ann Arbor, Michigan

Jaeil Park
The Pennsylvania State University, University Park, Pennsylvania

Magnus Persson
Chalmers University of Technology, Göteburg, Sweden

Shaligram Pokharel
Nanyang Technological University, Singapore

Gregory M. Roach
Brigham Young University Idaho, Rexburg, Idaho

David W. Rosen
The Georgia Institute of Technology, Atlanta, Georgia

Steven B. Shooter
Bucknell University, Lewisburg, Pennsylvania

Zahed Siddique
University of Oklahoma, Norman, Oklahoma

Timothy W. Simpson
The Pennsylvania State University, University Park, Pennsylvania

Henri J. Thevenot
The Pennsylvania State University, University Park, Pennsylvania

Karin Uller
Infotiv, Göteburg, Sweden

Wim van Vuuren
KPMG Advisory Services, Malta

Christopher B. Williams
The Georgia Institute of Technology, Atlanta, Georgia

Lianfeng Zhang
Nanyang Technological University, Singapore

Yiyang Zhang
Nanyang Technological University, Singapore

Preface

   To compete in today's global marketplace, many companies are utilizing product families to increase variety, improve customer satisfaction, shorten lead-times, and reduce costs. The key to a successful product family is the platform from which it is derived. In the past decade, there has been a flurry of activity to develop methods and tools to facilitate platform-based product family development, and this book showcases the efforts of more than thirty experts in academia and industry who are working to bridge the gap between (i) planning and managing families of products and (ii) designing and manufacturing them. Front-end issues related to platform-driven product development, platform planning, platform selection and evaluation, platform leveraging, and product family positioning are discussed along with methods for optimizing product platforms and product families. Back-end issues related to the realization of product families, including techniques for estimating production costs, planning process platforms, and commonalizing shapes to facilitate manufacturing are also presented. Industrial applications are also included to demonstrate how platform-based product development can impact product definition, product design, and process design.

Acknowledgments

This book would not have been possible were it not for the individual authors who contributed to this book—we would like to thank them for their patience and hard work. We are also indebted to the external reviewers and many authors who helped us review chapters to ensure a consistent and high level of quality. We would also like to thank the anonymous reviewers who supported our book proposal, making this whole endeavor possible. Finally, a special thanks is extended to Mrs. Erin Peterson in the Department of Mechanical & Nuclear Engineering at The Pennsylvania State University for her editorial assistance on many of these chapters.

Dr. Simpson wishes to acknowledge the National Science Foundation for supporting his time and effort under CAREER Grant No. DMI-0133923 and ITR Grant No. II3-0325402. Any opinions, findings, and conclusions or recommendations presented in this book are those of the editors and individual authors and do not necessarily reflect the views of the National Science Foundation. Dr. Siddique would like to acknowledge the School of Aerospace and Mechanical Engineering at University of Oklahoma for supporting his time and effort. Dr. Jiao would like to acknowledge the School of Mechanical and Aerospace Engineering at Nanyang Technological University, Singapore for supporting his time and effort. He also extends his gratitude to support from the Singapore NTU-Gintic Collaborative Research Scheme under U01-A-130B.

Chapter 1

PLATFORM-BASED PRODUCT FAMILY DEVELOPMENT
Introduction and Overview

Timothy W. Simpson[1], Zahed Siddique[2], and Jianxin (Roger) Jiao[3]

[1]*Departments of Mechanical & Nuclear Engineering and Industrial & Manufacturing Engineering, The Pennsylvania State University, University Park, PA 16802;* [2]*School of Aerospace and Mechanical Engineering, University of Oklahoma, Norman, OK 73019;* [3]*School of Mechanical and Aerospace Engineering, Nanyang Technological University, Singapore 639798*

1. PRODUCT VARIETY AND CUSTOMIZATION

Nearly a century ago, Ford Motor Company was producing Model T's in, as Henry Ford has been quoted, "any color you want—so long as it's black". Today, customers can select from more than 3.8 million different varieties of Ford cars based on model type, exterior and interior paint color, and packages and options listed on http://www.fordvehicles.com/. And that does not even include the staggering array of choices available with Ford's minivans, trucks, and sport utility vehicles, or any of the models offered under Ford Motor Company's "global family of brands", namely, Lincoln, Mercury, Mazda, Volvo, Jaguar, Land Rover, or Aston Martin. Ford is not alone as nearly every automotive manufacturer produces a wide variety of vehicles so that nearly every customer can find one that meets his/her specific needs. And it is not only in the automotive industry—consumers can purchase a nearly endless variety of goods and services: bicycles, motorcycles, appliances, computers, audio and video equipment, clothes, food and beverage, pharmaceuticals, software, banking and financial services, telecommunications services, and travel services.

Consequently, many companies struggle to provide as much variety for the market as possible with as little variety between products as possible. "New products must be different from what is already in the market and must meet customer needs more completely," says Pine (1993a), who attributes the increasing attention on product variety and customer demand to the saturation of the market and the need to improve customer satisfaction. Sanderson and Uzumeri (1997, p. 3) state that, "The emergence of global markets has fundamentally altered competition as many firms have known it" with the resulting market dynamics "forcing the compression of product development times and expansion of product variety." Findings from studies of the automotive industry (Alford, et al., 2000; MacDuffie, et al., 1996; Womack, et al., 1990) and empirical surveys of manufacturing firms (Chinnaiah, et al., 1998; Duray, et al., 2000) confirm these trends, as does evidence from Europe's "customer-driven market" (Wortmann, et al., 1997).

Since many companies typically design new products one at a time, Meyer and Lehnerd (1997, p. 2) have found that the focus on individual customers and products results in "a failure to embrace commonality, compatibility, standardization, or modularization among different products or product lines." Mather (1995, p. 378) finds that "Rarely does the full spectrum of product offerings get reviewed at one time to ensure it is optimal for the business." Erens (1997, p. 2) notes that "If sales engineers and designers focus on individual customer requirements, they feel that sharing components compromises the quality of their products." The end result is a "mushrooming" or diversification of products and parts that can overwhelm customers (Huffman and Kahn, 1998; Mather, 1995; Stalk and Webber, 1993); Nissan, for example, reportedly had 87 different varieties of steering wheels for one of their cars (Chandler and Williams, 1993). While offering a wide variety of products has both positive and negative effects (Anderson and Pine, 1997; Galsworth, 1994; Ho and Tang, 1998), the proliferation of product variety can incur substantial costs within a company (Child, et al., 1991; Ishii, et al., 1995a; Lancaster, 1990). "The imperative today," write Anderson and Pine (1997, p. 3), "is to understand and fulfill each individual customer's increasingly diverse wants and needs—while meeting the co-equal imperative for achieving low cost."

In the past decade, there has been a flurry of research activity in the engineering design community to develop methods and tools to facilitate product platform and product family design to provide cost-effective product variety and customization. In the next section, we discuss definitions, approaches, and examples of product platform and product family design to provide a foundation for the chapters that follow in this book. In Section 3, we discuss how the chapters in this book are organized to provide academia and industry with a collection of the state-of-the-art methods and tools for platform-based product family development from the engineering design community.

2.   DEFINITIONS, APPROACHES, AND EXAMPLES

2.1   Defining product platforms and product families

Many companies these days are developing product platforms and designing families of products based on these platforms to provide sufficient variety for the market while maintaining the necessary economies of scale and scope within their manufacturing and production processes. In general terms, a *product family* is a group of related products that is derived from a product platform to satisfy a variety of market niches. Meanwhile, a *product platform* can be either narrowly or broadly defined as:

- "a set of common components, modules, or parts from which a stream of derivative products can be efficiently developed and launched" (Meyer and Lehnerd, 1997, p. 7)
- "a collection of the common elements, especially the underlying core technology, implemented across a range of products" (McGrath, 1995, p. 39)
- "the collection of assets [i.e., components, processes, knowedge, people and relationships] that are shared by a set of products" (Robertson and Ulrich, 1998, p. 20)

A review of the literature suggests that product platforms have been defined diversely, ranging from being general and abstract (for example, Robertson and Ulrich, 1998) to being industry and product specific (for example, Sanderson and Uzumeri, 1995). Moreover, the meaning of platform differs in scope: some definitions and descriptions focus primarily on the product/artifact itself (Meyer and Utterack, 1993) while others try to explore the platform concept in terms of a firm's value chain (Sawhney, 1998). Additional definitions for platforms and families are given throughout this book, reflecting both industry- and application-specific perspectives with which the product platform and ensuing family of products are defined. Defining the product platform within a company is perhaps one of the most challenging aspects of product family design (see Chapter 2).

Regardless of the specific definition used, product platforms can offer a multitude of benefits when applied successfully. As Robertson and Ulrich (1998, p. 20) point out, "By sharing components and production processes across a platform of products, companies can develop differentiated products efficiently, increase the flexibility and responsiveness of their manufacturing processes, and take market share away from competitors that develop only one product at a time." Other benefits include reduced development time and system complexity, reduced development and production costs, and improved ability to upgrade products. A product platform can also facilitate customization by enabling a variety of products to be quickly and easily developed to satisfy the needs and requirements of distinct market niches (Pine, 1993a). Platforms also promote better learning across products and can reduce testing and certification of complex products such as aircraft (Sabbagh, 1996), spacecraft (Caffrey, et al., 2002), and aircraft engines (Rothwell and Gardiner, 1990). Additional and more specific benefits can be found in many chapters throughout the book.

For instance, platforms in the automotive industry enable greater flexibility between plants and increase plant usage—sharing underbodies between models can yield a 50% reduction in capital investment, especially in welding equipment—and can reduce product lead times by as much as 30% (Muffatto, 1999). In the 1990's, automotive manufacturers that employed a platform-based product development approach gained a 5.1 percent market share per year while those that did not lost 2.2 percent (Cusumano and Nobeoka, 1998). In the late 1990's, Volkswagen saved an estimated $1.5 billion per year in development and capital costs using platforms, and they produced three of the six automotive platforms that successfully achieved production volumes over one million in 1999 (Bremmer, 1999; Bremmer, 2000). Their platform consists of the floor group, drive system, running gear, along with the unseen part of the cockpit as shown in Figure 1-1 and is shared across 19 models marketed under its four brands: Volkswagen, Audi, Seat, and Skoda.

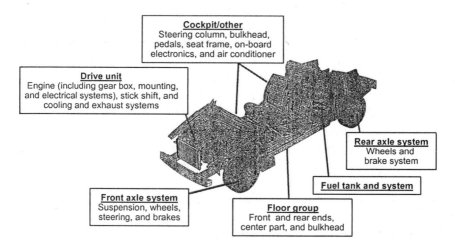

Figure 1-1. Volkswagen's platform definition; adapted from (Wilhelm, 1997).

While many researchers espouse the benefits of platforms, there are potential drawbacks and downsides to platform-based product development (see Chapter 3). For instance, despite the success of Volkswagen's platform strategy, it has been criticized for creating cars that are too similar

(Anonymous, 2002; Miller, 2002) and has suffered from its own success in platforming: lower-end models are cannibalizing sales of the higher-end models in the Europe and the U.S. The Audi TT also had unexpected technical difficulties at high speeds due to problems with the rear wheel down force, and the problems were attributed to the utilization of the aforementioned A-platform (de Weck, et al., 2003). Too much commonality can adversely impact a brand's image. For example, in the late 1980s, engineers at Chrysler were accused of having "fallen asleep at the typewriter with our finger stuck on the K key" (Lutz, 1998, p. 17) due to over-usage of the K-car platform and lack of distinctive new products. Platform-based approaches can also impose additional costs on product development. The fixed costs of developing a product platform can be enormous—Ulrich and Eppinger (2004) found that developing a product platform can cost two to ten times more than a single product—and sharing components across low-end and high-end products can increase unit variable costs due to over-designed low-end products (Fisher, et al., 1999; Gupta and Krishnan, 1998a). In the automotive industry, Muffato (1999) found that up to 80% of total vehicle development cost is spent on platform development (including engine and transmission); others argue that platform development accounts for only 60% of these costs (Sundgren, 1999). Krishnan and Gupta (2001) develop a mathematical model to examine some of the costs of platform-based product development and find that platforms are inappropriate for extreme market diversity or high levels of non-platform scale economies.

Therefore, the key to a successful product family lies in properly balancing the inherent tradeoff between commonality and distinctiveness: designers must balance the commonality of the platform with the individual performance (i.e., distinctiveness) of each product in the family (see Part I). As a result, designing a product platform and corresponding family of products embodies all of the challenges of product design while adding the complexity of coordinating the design of multiple products in an effort to increase commonality across the set of products without compromising their distinctiveness (see Part II). Successful approaches to product family design are discussed next along with several industry examples (see Part IV also).

2.2   Approaches to product family design

There are two basic approaches to product family design (Simpson, et al., 2001a). The first is a *top-down (proactive platform) approach* wherein a company strategically manages and develops a family of products based on a product platform and its derivatives. For instance, Sony has strategically managed the development of its Walkman® products using carefully designed product platforms and derivatives (Sanderson and Uzumeri, 1997).

Similarly, Kodak's product platform-based response to Fuji's introduction of the QuickSnap® single-use camera in 1987 enabled them to develop products faster and more cheaply, allowing them to regain market share and eventually overtake Fuji (Wheelwright and Clark, 1995).

The second is a *bottom-up (reactive redesign) approach* wherein a company redesigns or consolidates a group of distinct products to standardize components to improve economies of scale. For example, after working with individual customers to develop 100+ lighting control products, Lutron redesigns its product line around 15-20 standard components that can be configured into the same 100+ models from which customers could choose (Pessina and Renner, 1998). Black & Decker (Lehnerd, 1987) and John Deere (Shirley, 1990) have benefited from similar redesign efforts to reduce variety in their motor and valve lines, respectively.

The prominent approach to platform-based product development, be it top-down or bottom-up, is through the development of a *Module-Based Product Family* wherein product family members are instantiated by adding, substituting, and/or removing one or more functional modules from the platform. An alternative approach is through the development of a *Scale-Based Product Family* wherein one or more scaling variables are used to "stretch" or "shrink" the platform in one or more dimensions to satisfy a variety of market niches. We note that module- and scale-based product family design are also referred to by many as *configurable* and *parametric product family design*, respectively. Examples of both approaches follow.

2.2.1   Module-based (configurable) product families

There are numerous examples of module-based product families in the literature; some of the more frequently quoted examples follow.

- *Sony* builds all of its Walkmans® around key modules and platforms and uses modular design and flexible manufacturing to produce a variety of quality products at low cost, allowing them to introduce 250+ models in the U.S. in the 1980s (Sanderson and Uzumeri, 1997).
- *Nippondenso Co. Ltd.* makes an array of automotive components for a variety of automotive manufacturers using a combinatoric strategy that involves several different modules with standardized interfaces; for instance, 288 different types of panel meters can be assembled from 17 standardized subassemblies (Whitney, 1993).
- *Hewlett Packard* successfully developed several of their ink jet and laser jet printers around modular components to gain benefits of postponing the point of differentiation in their manufacturing and assembly processes (Feitzinger and Lee, 1997).

- *Bally Engineering Structures* offers an almost infinite variety of environmentally-controlled structures that are assembled from one basic modular component—the pre-engineered panel—that can be produced in a variety of shapes and sizes and customized with options, attachments, and finishes to fit into any size structure (Pine, 1993b).

These successful examples resulted from careful attention to customer needs and the underlying product architecture in the family. Ulrich (1995, p. 420) defines the product architecture as "(1) the arrangement of *functional elements*; (2) the mapping from *functional elements* to *physical components*; (3) the specification of the *interfaces* among interacting physical components". A product architecture is classified as either *modular*, if there is a one-to-one or many-to-one mapping of functional elements to physical structures, or *integral*, if a complex or coupled mapping of functional elements to physical structures and/or interfaces exists. For example, personal computers (PCs) are highly *modular*, and Baldwin and Clark (2000) trace the development of the IBM's System/360, the first modular computer family. Automotive architectures, on the other hand, are predominantly *integral* (cf., Muffatto, 1999; Siddique, et al., 1998), but modularity has become a major strategic focus for future product development within many automotive companies (Cusumano and Nobeoka, 1998; Kobe, 1997; Shimokawa, et al., 1997). For instance, the rolling chassis module produced by the Dana Corporation (see Figure 1-2) saved DaimlerChrysler nearly $700M when developing their new Dodge Dakota facility (Kimberly, 1999). The rolling chassis module consists of brake, fuel, steering, and exhaust systems, suspension, and drive-line assembled to the frame, and it is the largest, most complex module provided by a supplier, accounting for 25% of the vehicle content. Finally, modularity plays a key role in component reuse (Kimura, et al., 2001) as well as product evolution, upgradeability, and retirement (Ishii, et al., 1995b; Umeda, et al., 1999).

Figure 1-2. Rolling chassis automotive module; adapted from (Kimberly, 1999).

Approaches for developing modular product architectures and module-based product families abound in the engineering design literature. For instance, Mattson and Magleby (2001) discuss concept selection techniques for managing modular product development in the early stages of design. Wood and his co-authors (McAdams, et al., 1999; McAdams and Wood, 2002; Stone, et al., 2000b) present a methodology for representing a functional model of a product in a quantitative manner to assist in developing product architectures and facilitate the identification of a core set of modules for a product family. As part of their work, Stone, et al. (2000a) present a heuristic method to identify modules for these product architectures; this method is later extended by Zamirowksi and Otto (1999) to identify functional and variational modules within a product family. Allen and Carlson-Skalak (1998) develop a methodology for designing modular products that involves identifying and reusing modules from previous generations of products. Martin and Ishii (2002) consider multiple generations of products when presenting their approach for designing modular product platform architectures. Their approach is one of several that uses Quality Function Deployment (QFD) to help identify modules within a product family (Cohen, 1995; Ericsson and Erixon, 1999; Erixon, 1996; Huang and Kusiak, 1998; Sand, et al., 2002).

Modularity is the sole focus in several texts (Baldwin and Clark, 2000; Ericsson and Erixon, 1999; O'Grady, 1999) and is an important topic in many product design textbooks (see, e.g., Otto and Wood, 2001; Pahl and Beitz, 1996; Ulrich and Eppinger, 2004). While several chapters address modularity to a limited extent (e.g., the optimization-based approaches described in Chapter 9), we do not devote much attention to defining product architectures in this book *per se*. The reader is referred to the aforementioned texts as well as the seminal article on modularity by Ulrich (1995) and recent studies by Gershenson and his students (Gershenson, et al., 2003a; Gershenson, et al., 2003b; Guo and Gershenson, 2003; Guo and Gershenson, 2004; Zhang, et al., 2001).

2.2.2   Scale-based (parametric) product families

As stated previously, scale-based product families are developed by scaling one or more variables to "stretch" or "shrink" the platform and create products whose performance varies accordingly to satisfy a variety of market niches. While some consider scale-based product families to be a subset of module-based design (see, e.g., Fujita and Yoshida, 2001), platform scaling is a common strategy employed in many industries. For example:

- *Black & Decker* developed a family of universal electric motors that were scaled along their stack length to produce a range of power output for hundreds of their basic tools and appliances (Lehnerd, 1987).
- *Honda* developed an automobile platform that can be stretched in both width and length to realize a "world car", which was developed after failing to satisfy the Japanese and American markets with a single platform (Naughton, et al., 1997).
- *Rolls Royce* scaled its RTM322 aircraft engine by a factor of 1.8 to realize a family of engines with different shaft horsepower and thrust as shown in Figure 1-3 (Rothwell and Gardiner, 1990).
- *Boeing* developed many of its commercial airplanes, including the 777, by "stretching" the aircraft to accommodate more passengers, carry more cargo, or increase flight range (Sabbagh, 1996).

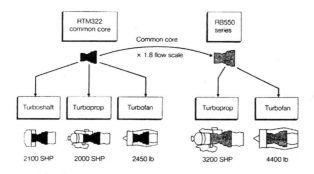

Figure 1-3. Rolls Royce's aircraft engine family; adapted from (Rothwell & Gardiner, 1990).

Scale-based platforms are prominent in the aerospace industry at large as well as small manufacturers. Airbus has recently enjoyed a competitive advantage over Boeing due to improved commonality, particularly in the cockpit. The A330 cockpit is common to all other Airbus types while Boeing's 767-400 cockpit is common only with the 757. This has enabled the A330-200, a less efficient "shrink" of a larger aircraft, to outsell Boeing's 767-400ER, a more efficient "stretch" design of a smaller aircraft (Aboulafia, 2000). Meanwhile, smaller manufacturers such as Embraer seek to exploit scaling and commonality among their aircraft to reduce development and production costs. As discussed on their website (http://www.embraer.com/), the 170 and 175 models have 95% commonality among subsystems as do the 190 and 195 models, and they boast 85% commonality among all four models, including common pilot type rating, avionics systems, fly-by-wire systems, and many high-level components.

Research in scale-based product family design has focused primarily on optimization-based approaches due to the parametric nature of platform scaling (see Chapter 8). For instance, Simpson and his co-authors use optimization-based approaches to design scale-based platforms for families of General Aviation Aircraft (Simpson, et al., 1999), universal electric motors (Simpson, et al., 2001a), and flow control valves (Farrell and Simpson, 2003). Hernandez and his co-authors have also looked at scalable platforms for the universal electric motor family (Hernandez, et al., 2002) as well as for families of absorption chillers (Hernandez, et al., 2001) and pressure vessels (Hernandez, et al., 2003). Fujita and Yoshida (2001) have investigated scale-based optimization methods for sizing families of commercial aircraft. Fellini, et al. (2002a; 2002b) used optimization to help scale automotive platforms for a family of cars. Indices for measuring the *degree of variation* in a scale-based product family have also been proposed (Messac, et al., 2002a; Nayak, et al., 2002; Simpson, et al., 2001b).

3.   ORGANIZATION OF THE BOOK

There has been a flurry of activity that has helped the nascent field of product family design mature in the past decade. In this book, we showcase the efforts of more than thirty experts in academia and industry who are working to bridge the gap between (i) planning and managing families of products and (ii) designing and manufacturing them. Our intent in this book is to share the state-of-the-art in the engineering design community with both academia and industry by providing a collection of the methods and tools that are available to support platform-based product family design.

We have organized the book into four Parts that span the entire spectrum of product realization according to the domain framework (Suh, 2001) as noted in Figure 1-4. Part I focuses primarily on the Customer Domain and its mapping into the Functional Doman. These chapters discuss "front-end" issues related to platform-driven product development, platform planning, platform selection and evaluation, platform leveraging, and product family positioning. In Part II, several optimization-based methods for product family design are presented to address how the Functional Domain impacts the Physical Domain, including methods for module-based and scaled-based product family design as well as methods for requirements flow-down in a product family and platform portfolio planning. The chapters in Part III address "back-end" issues related to the realization of product families and the Process Domain, including techniques for estimating production costs, planning process platforms, and commonalizing shapes to facilitate manufacturing. Finally, Part IV includes four industrial applications that span multiple domains to demonstrate how platform-based product family development can impact product definition, product design, and process design. Detailed discussions of the chapters in each Part follow.

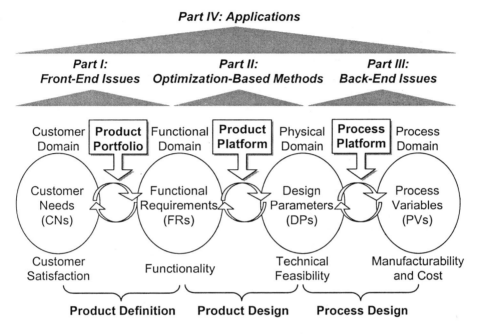

Figure 1-4. Organization of the book.

3.1 Part I: Front-end issues related to platform-based product family development

Of primary importance in product family design and platform development is the interaction with customers and the market. Manufacturers have been seeking for expansion of their product lines and differentiation of their product offerings with the intuitively appealing belief that high product variety may stimulate sales and thus conduce to revenue. At the technical side, designers have always assumed customer satisfaction with the designed product families and platforms is sufficiently high as long as the product meets the prescribed technical specifications. However, what customers appreciate is not the enhancement of the solution capability but the functionality of the product. Therefore, many dimensions of customer satisfaction deserve scrutiny, for example, identifying those product characteristics that cause different degrees of satisfaction among customers; understanding the interrelation between the buying process and product satisfaction; determining the optimal amount of variety and customer integration; explaining the key factors regarding the value perception of product families; and justifying an appropriate number of choices from the customers' perceptive. All these constitute the front-end issues of product platform and product family design.

Part I focuses on such "front-end" issues. Bowman (Chapter 2) discusses the topic of product platform planning from an industry perspective. A product/platform roadmap is introduced as the visual summation for the platform strategy, and for management to guide platform investment or rationalization decisions over the platform's useful life. Halman, Hofer and van Vuuren (Chapter 3) discuss the problems and risks related to implementing and managing product families and their underlying platforms. Using a multiple-case approach, three technology-driven companies are compared in their definitions of platform-based product families, as well as the reasons for and the risks of adopting platform thinking in the development process.

Hölttä-Otto and Otto (Chapter 4) introduce a platform concept evaluation tool that is multi-criteria in nature and scalable to include various alternative criteria as appropriate. This multi-criteria analysis results in a concept phase analysis that helps manage risk by making all aware of the criteria that a development project may need backup plans developed, extra effort applied, and management attention. To help address platform planning, Marion and Simpson (Chapter 5) explore the history of the market segmentation of product platforms. The principles and tools behind market segmentation are introduced, along with several examples, to show how companies have leveraged product platforms successfully into multiple market segments.

Jiao and Zhang (Chapter 6) discuss the issue of product family positioning. An optimization framework is developed by leveraging both customer preferences and engineering costs. Thevenot and Simpson (Chapter 7) discuss several commonality indices found in the literature. Examples are provided on how to use them for product family benchmarking and product family redesign. The study suggests that the combined use of optimization algorithms and commonality indices to support product family redesign provides useful information for the redesign of a product family, both at the product-family level as well as at the component-level.

3.2   Part II: Optimization methods to support platform-based product family development

Although the basic principles of product family design are understood and well documented in literature, quite a few fundamental issues need to be scrutinized. A prevailing principle of product family and platform design is a two-stage process. While product architectures and the range of possible variety are predetermined during a product family architecting stage, a subsequent design and development stage takes place in close interaction between the customers and the manufacturer. Based on what has been learned from the second stage, the product family architecture can be upgraded, which in turn leads to capability enhancement at the manufacturer part. The linchpin is the optimal design of product families and platforms.

Part II is devoted to the methods for optimizing product platforms and families. With emphasis on parameter (detail) design, Simpson (Chapter 8) overviews the fundamental issues and formulations of product platform and product family optimization problems. The design of a family of ten motors is introduced to shed light on the merits and pitfalls of optimization approaches to product platform and product family design. Fellini, Kokkolaras, and Papalambros (Chapter 9) present analytical methods for performing commonality decisions, with an additional design tool derived by combining these techniques. The design methodologies are applied to various automotive examples involving the design of the body and engine. Fujita (Chapter 10) expands the scope of the product family optimization problem and describes several different methods for product family design optimization based on problem classification. A simultaneous optimization method for both module combination and module attributes is introduced. The key in exploring optimal design for product family and platform exists in both development of optimization algorithm and formulation of individual problems. Kokkolaras, Fellini, Kim, and Papalambros (Chapter 11) presents an analytical target cascading (ATC) methodology for translating targets for a family of products to platform specifications for given commonality decisions. The ATC formulation is extended for a single product to a family of products to accommodate the presence of a shared product platform and locally introduced design targets. De Weck (Chapter 12) deals with product family and platform portfolio optimization. He aims to determine an optimum number of product platforms to maximize overall product family profit. A methodology is introduced based on a target market segment analysis, market leader's performance versus price position, and a two-level optimization approach for platform and variant design.

3.3   Part III: Back-end issues related to platform-based product family development

The primary objective in platform-based product family development is providing economical product variety. The underlying idea to achieve this objective by increasing commonality across multiple products through a platform approach. In order to ensure efficient product family, commonality needs to be considered for both product and process issues at component, module, platform, and product family levels. From the manufacturer's point of view, it is essential to design new products with a set of common features, components, and subassemblies that can lead to lowering production cost by eliminating new resource use and sharing existing resources. A firm needs to consider and balance the costs and benefits of all strategic perspectives that a platform-based product development approach generates. A comprehensive product family realization process needs to consider not only customer needs, function requirements and technical solutions, but also incorporate issues related to the backend of the product realization, which includes the production processes.

Perspectives, issues, models, and processes to efficiently consider "back-end" issues are the emphasis in Part III. Fixson (Chapter 13) provides a comprehensive discussion of how individual product architecture characteristics affect specific cost elements over a product's life cycle can serve as a guideline when formulating various tradeoffs. An Activity-Based Costing (ABC) approach is presented by Park and Simpson (Chapter 14) to facilitate use of cost information during product family design and allow designers to investigate possible platforms by examining the effects of differentiated products on activities and resources in production. Siddique (Chapter 15) also uses ABC and extends it to estimate cost and time savings, while considering design and production factors, for implementing a platform-based approach. Generic variety representation, generic structures and generic planning are incorporated by Jiao, Zhang, and Pokharel (Chapter 16) to develop process platforms to configure production processes for new members of product families. Identifying common shapes for components when developing a platform, to facilitate the use of common manufacturing and assembly processes, is discussed by Siddique and Natarajan (Chapter 17). Williams, Allen, Rosen, and Mistree (Chapter 18) discuss the concept and a design methodology for realizing process parameter platforms from which a stream of derivate process parameters can generate a customized product efficiently despite changes in capacity requirement.

3.4   Part IV: Applications of platform-based product family development

Research in platform-based product family development has been driven by the need of industry to compete in the current marketplace and address the problem of providing greater variety, with existing challenges of providing greater quality, competitive pricing, and greater speed to market. Many companies have successfully implemented platform-based product families to satisfy customer needs. These successful implementations provide insight into issues, methods, and benefits related to product family development. Consequently several industry cases are presented in Part IV.

Shooter (Chapter 19) presents a top-down approach to platform-based product development for a family of ice scarpers for a small company. The company started the design with full intent of using platform strategies for developing their product family. Nidamarthi and Harshavardhan (Chapter 20) discuss an approach to architecting successful product platforms within ABB. The platform approach has been applied to allow customers to not only buy products from their catalogues, but also place a turnkey order for a system including design, build, and commissioning. Moreover, the ABB case provides insight into how organizational constraints can be overcome during implementation. Roach and Cox (Chapter 21) describe a web-based tool for a turbine disk product platform, for constructing a web-based product platform customization application to automatically create all of the design artifacts and supporting information necessary for the design of a particular product. Finally, Holmqvist, Lindhe, and Persson (Chapter 22) present a case of creating the platform and the modules for a heat exchanging system by analyzing how the market offer could be achieved by using a smaller assortment of products, modules and components.

PART I: FRONT-END ISSUES RELATED TO PLATFORM-BASED PRODUCT FAMILY DEVELOPMENT

Chapter 2

EFFECTIVE PRODUCT PLATFORM PLANNING IN THE FRONT END

Daniel Bowman
Pittiglio, Rabin, Todd & McGrath (PRTM), 1050 Winter Street, Waltham, MA 02451

1. THE VALUE OF PLATFORM PLANNING IN THE FRONT END

Platform Planning is increasingly being adopted by companies seeking to provide customization while maximizing economies of operation. Platform Planning is defined as the proactive definition of an integrated set of capabilities and associated architectural rules that form the basis for a group of products. When implemented effectively, Platform Planning can provide distinct benefits in cost and market leverage to provide a competitive edge in the marketplace.

Platform Planning is often decoupled from Product Strategy, resulting in platform capabilities that do not meet specific customer needs or are left "dormant" because they do not support a specific product. This misalignment results in dissatisfied customers, stranded investment, and ultimately, missed opportunities in the market.

The benefits of Platform Planning are most effectively realized by implementation in the Front End of product development. The *front end* of product development is where the overall product strategy is defined and the elements of a potential new product or platform are identified. It is here that an alignment between key markets, customer requirements, and underlying platform capabilities can yield the greatest benefits for downstream platform leverage (see Figure 2-1).

Lack of Platform Planning in the Front End can result in a number of pain points. These include:

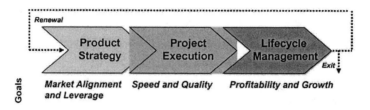

Figure 2-1. Phases of product development.

1. Limited horizons for Platform Planning, resulting in product "one-offs",
2. Technical feasibility is not understood at a sufficiently early date,
3. Concurrent product/platform development decreases product leverage,
4. Inefficient platform leverage and un-integrated architectures,
5. Limited view of platform investments needed for future product growth,
6. Product functionality is compromised as tradeoffs are made and features dropped to maintain schedules, and
7. Unscalable infrastructure that is not able to support growth.

Ultimately, this results in higher capital costs, slower time to market, and lost revenue opportunities. In order to minimize these pain points and maximize value from advanced Platform Planning, PRTM developed a five-step methodology that it uses:

1. Establishing a common language and terminology,
2. Defining a product strategy and value proposition,
3. Tapping the voice of the market,
4. Identifying the vector of differentiation, and
5. Developing product/platform roadmaps.

We explore of each of these steps in the following sections.

2. ESTABLISHING A COMMON LANGUAGE AND TERMINOLOGY

Lack of a common language and set of operating terms for Platform Planning can often derail efforts for engineering, marketing, and product management functions to coordinate their activities effectively. These functions often "talk past" each other, resulting in disagreement and stasis. Alignment on a common set of operating terms is critical before Platform Planning can proceed. Key terms for effective Platform Planning include:

❑   *Market* – A large group of customers who have common set of problems/needs, and who purchase a common group or class of products to solve those problems.

- ❑ *Portfolio* – Groups of projects funded from a common investment pool and managed by a common management team.
- ❑ *Product Platform* – A set of platform elements and architectural rules that enable a group of planned product offerings. Key characteristics of a product platform include: (1) Architectural rules/standards governing how technologies and subsystems ("platform elements") can be integrated; (2) Defines the basic value proposition, competitive differentiation, capabilities, cost structure, and life cycle of a set of product offerings; and (3) Supports multiple product offerings from a single platform, permitting increased leverage and reuse across the product line.
- ❑ *Product* – Products are specific instances of a platform that may have minor or major deviations from the basic platform.
- ❑ *Product Line* – A grouping of products that share similar features, functionality, or lineage to help reach a larger share of the market.
- ❑ *Elements* – Building blocks of a platform that can be varied within certain platform constraints.

Time invested in establishing a working set of terminology and gaining agreement from all the stakeholders involved in a Platform Planning effort will save significant cycles later in the planning process.

3. DEFINING A PRODUCT STRATEGY AND VALUE PROPOSITION

The underpinning of an effective platform plan is a clearly defined Product Strategy (see Figure 2-2). The Product Strategy should be guided by the overall strategy for the company or business unit, including priority markets that should be pursued as well as clear targets for financial returns.

The focus of a platform plan is how to derive value from leverage. Leverage comes in several forms. First is *Cost Leverage,* which is characterized by several qualities. It involves the reuse of product technology across product lines, which includes similar parts, processes, materials, interfaces, and subsystems; the identification of commonality between product lines to enable platform building blocks; and use of platform building blocks to reduce the cost of development, manufacture, and service. *Market* Leverage is the second major form of leverage. It includes reuse of product technology across market or market segment boundaries, a focus on commonalities in customer needs across markets, and development of flexible/modular systems to accelerate time-to-market.

Achieving leverage in Platform Planning is the artful balance between commonality and distinctiveness. Conditions in which platform leverage is difficult to attain are new and undefined markets where specific customer requirements are being satisfied for the first time. In mature markets, platform leverage is more achievable given a known set of customer segments, customer requirements, and track record of product performance. It is in this environment that effective leverage can spawn a whole new category of product without dramatically increasing development cost.

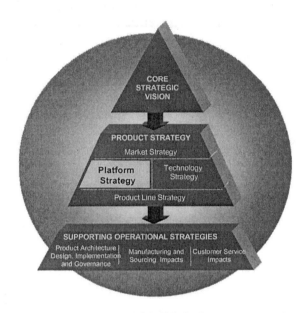

Figure 2-2. Platform strategy in developing a product strategy.

For example, the new VW Beetle reinvigorated a product line that had been dormant for many years. However, the Beetle was based on a platform that supported multiple VW and Audi product lines. As a result, incremental product development costs were kept to a minimum. Incremental investment was isolated to the differentiating elements of the product – styling, interior, and performance. Significant efficiencies were created by sharing the "non visible" elements like the power train, suspension, steering and electrical systems.

The benefits of Platform Planning are measured along several dimensions. These dimensions include cost and complexity reduction, reliability, flexibility, market responsiveness, and simplicity. Cost and complexity reduction are measured by decreases in capital investment required to develop multiple platforms that will in turn only support a limited number of products and product lines. The degree to which a single platform can enable multiple product variants results in lower development cost allocation per variant, and therefore quicker return on investment.

Complexity reduction is a major benefit of effective Platform Planning. Complexity is the "hidden cost" that impacts profit margins as product portfolios proliferate. Through effective Platform Planning, unnecessary and non-valued added complexity can be eliminated from the portfolio. Savings are generated through reduced SG&A (Sales, General, and Administrative) brought about through leaner sales and marketing organizations and reduced product support. Cost of Goods Sold (COGS) are reduced through efficiencies in the supply chain, negotiating better terms with suppliers, and reductions in direct and indirect manufacturing costs. All savings flow directly to the profit line of the business.

Equally as important but perhaps less quantitative are the benefits enabled by simplicity in design and architecture. The same drivers of cost and complexity reduction also allow subsystem design with elegant interface architectures producing rich variety within the sub-function without causing disruptions to any other subsystem. This can result in a customer-pleasing level of features and variability with minimum parts and interfaces.

4.   TAPPING THE VOICE OF THE MARKET

Profit producing and customer pleasing products are lost without tapping the Voice of the Market. Understanding the Voice of the Market ensures that Platform Planning is not overly influenced by internal drives for efficiency. Capturing the Voice of the Market requires a robust process in the front end to use a "customer grounded" ideation process to identify customer requirements and concepts. By customer grounded, we mean a process that allows the company to immerse itself in the customer's environment, and learn about their problems first hand. These customer insights are then translated through "customer voices" into customer requirements. Customer requirements then become the basis for generating concepts that meet the customer requirements. Concepts are then aligned against customer need, company capabilities, and other screening criteria to identify the most promising candidates. The output is customer grounded concepts that are translated into winning solutions by using requirement alignment tools like QFD (Hauser and Clausing, 1988) and S-QFD (Quality, Function, Design).

Concepts can take different "pathways" depending on their level of innovation and scope of impact. For example, a concept might provide a promising new enabling technology that is not ready for commercialization, but holds significant potential. This technology should be "spun off" to a separate yet linked technology development process for nurturing and development.

A concept might be a derivative product that taps into an existing platform capability and allows the company to address a new market segment. This concept then enters the traditional product development process for development.

Finally, a concept might be a "platform", demonstrating the potential to support multiple market segments and as well as meet multiple customer requirements. This type of concept holds the greatest potential for return as well as potentially requiring the greatest level of capital investment. It needs to be carefully vetted in a subsequent more detailed requirements generation process where capital investments are coupled with revenue expectations.

5.   IDENTIFYING THE VECTOR OF DIFFERENTIATION

Effectively capturing the Voice of the Market provides the inspiration for identifying the Vector of Differentiation. The Vector of Differentiation (VoD) is the defining characteristic that the platform will deliver over a period of time that will enable it to meet the target segment's needs while providing a competitive advantage in the market. There is a one-to-one map between the Vector of Differentiation and product platform—if a distinct VoD cannot be defined; there is not a basis for a distinct product platform.

Vectors of Differentiation are usually built along four major competency dimensions (see Figure 2-3). These include:

❑ *Innovation* – unique and fundamental features or capabilities enabled by a robust innovation capability within the company.
❑ *Lower Customer Costs* – lower total cost of ownership which entails both a lower purchase price and lower cost of operation.
❑ *Breadth and Coverage* – products that span a price and performance space, enabling capturing significant market share capture.
❑ *Higher Performance* – increased product performance yielding increased customer productivity and effectiveness.

Vectors of Differentiation are decomposed into Defining, Supporting, and Segmenting Elements within a Platform Planning context. *Supporting Elements* are at the lowest level of the platform architecture, and provide a baseline level of functionality. Without them, the product cannot operate; however, they do not provide a significant competitive edge. *Defining Elements* provide a competitive edge and are the basis for leverage across an entire product line. *Segmenting Elements* address market segment specific customer value propositions and may actually add cost to the platform.

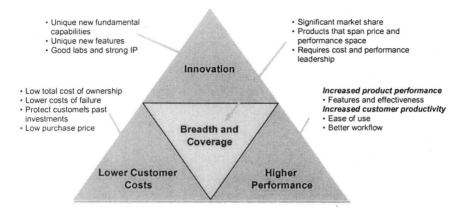

Figure 2-3. Developing the vector of differentiation.

For illustration, think of a financial services product like a credit card. For a credit card product, the Supporting Element is usually the billing or transaction processing system. Without it, the product is inoperable but it does not provide a substantial point of differentiation. The Defining Element could be a unique rewards program capability that when leveraged, enables a series of derivative product line extensions like loyalty cards while being supported by a unique Vector of Differentiation in the market (in this case, Innovation). Finally, the Segmenting Element could be a unique offer to a specific segment like the teen market that adds cost, but provides access to a unique and lucrative audience.

6. EXAMPLE: PLATFORM PLANNING FOR A LARGE AUTOMOTIVE SUPPLIER

A $1B plus OEM automotive supplier need to diversify its customer base in order to ensure its success in the market. However, its product architectures were inflexible, costs were high, and the design bookshelf was not current. In addition, complexity was out of control, driving high manufacturing costs and poor utilization of design engineering resources.

In order to address these issues, a Core Strategic Vision was developed through a comprehensive assessment of the market and internal competencies. In addition, defined product platforms, derived from market requirements and organized by vectors of differentiation within product portfolio, were established. Finally, a platform and product migration plan to balance strategic priorities, short-term commitments, and headcount constraints were developed.

The results were significant. Variable cost savings of 15% were identified while concurrently increasing product flexibility and functionality, by migrating to platform-based strategies. In addition, utilization of product development resources was improved by 18% while unique product architectures were reduced from 17 to 3 globally.

7.   PUTTING IT ALL TOGETHER: PRODUCT/ PLATFORM ROADMAPS

The platform plan comes together in a Product/Platform roadmap. A Product Roadmap (or Release Roadmap) is a planning document that indicates the expected timing of product offerings from product platform(s); examples can be found in Chapter 5. It is usually in the form of a high-level Gantt chart showing the timing of planned future releases and expected duration of major development phases. The roadmap becomes the plan of record for platforms and the products that will be enabled the platforms by showing how product functionality and capabilities are expected to evolve over time for each product.

Most importantly, the product/platform roadmap enables the management team to visualize product/platform timing, cadency, linkages, and synchronization between different levels of the product offering. For example, it can help identify if there is a "one to one" relationship between a platform and a product, indicating little to no leverage. It can identify whether technology development and platform availability are out of cadence, thereby impacting market delivery timing. It can show the products and product families enabled by the platform, their expected life cycle, and when or whether investments will need to be made to the underlying platform. In summary, the product/platform roadmap is the visual summation for the platform strategy and is used by management to help guide platform investment or rationalization decisions over the platform's useful life.

8.   CLOSING REMARKS

In closing, to fully realize the benefits of platform planning, execution needs to begin in the front end of product development. By following the five key steps outlined in this chapter, product development functions can ensure that platforms precede product line plans, resulting in lower overall product development costs and products that better meet customer needs.

Chapter 3

PLATFORM-DRIVEN DEVELOPMENT OF PRODUCT FAMILIES[*]
Linking Theory with Practice

Johannes I. M. Halman[1], Adrian P. Hofer[2], and Wim van Vuuren[3]
[1]*Department of Construction Management & Engineering, University of Twente, The Netherlands;* [2]*Hofer & Partner, Wollerau, Switzerland 8832;* [3]*KPMG Advisory services, Portico Building Marina Street, Pieta' MSD 08, Malta*

1.   SUMMARY

Firms in many industries increasingly are considering platform-based approaches to reduce complexity and better leverage investments in new product development, manufacturing and marketing. However, a clear gap in literature still exists when it comes to discussing the problems and risks related to implementing and managing product families and their underlying platforms. Using a multiple-case approach, we compare three technology-driven companies in their definition of platform-based product families, investigate their reasons for changing to platform-driven development, and analyze how they implemented platform thinking in their development process and which risks they encountered in the process of creating and managing platform-based product families. The field study shows, that the companies involved in the study use a homogeneous concept of platform-based product families, and that they have similar reasons to turn to platform thinking and encounter comparable risks. However, the companies analyzed use mainly product architecture as a basis for their platforms (and ignore many of the platform types advocated in literature), while on the other hand they show divergent applications of the platform concept regarding the

---

[*] This chapter is a modified version of the paper: Halman, J. I. M., Hofer, A. P. and van Vuuren, W., 2003, "Platform-Driven Development of Product Families: Linking Theory with Practice," *Journal of Product Innovation Management,* **20**(2): 149-162. Reprinted with the permission of the PDMA.

combinations of product families and market applications. Through this exploratory study, some important "gaps" in the literature became evident, and in the discussion, these "gaps" are discussed and directions for future platform research are proposed.

2.   INTRODUCTION

In a global, intense, and dynamic competitive environment, the development of new products and processes has become a focal point of attention for many companies. Shrinking product life cycles, rapidly changing technologies, increasing international competition, and customers demanding high variety options are some of the forces that drive new development processes (Wheelwright and Clark, 1992; Pine, 1993a; Ulrich, 1995). In their quest to manage the complexity of offering greater product variety, firms in many industries are considering platform-based product development (Krishnan and Gupta, 1994). Key to this approach is the sharing of components, modules and other assets across the product family as discussed in Chapter 1.

Historical success stories such as the Sony Walkman (Sanderson and Uzumeri, 1995; Uzumeri and Sanderson, 1995), Black & Decker power tools (Meyer and Utterback, 1993), Hewlett Packard's Deskjet printers (Meyer and Lehnerd, 1997), Microsoft's Windows NT (Cusumano and Selby, 1995), and Minolta's "Intelligent lens technology" (Sawhney, 1998) have shown both the benefits and the logic behind the platform concept. Gupta and Souder (1998) even claim that thinking in terms of platforms for families of products rather than individual products is one of the five key drivers behind the success of short-cycle-time companies.

However, a clear gap in literature still exists when it comes to discussing possible limitations of the platform concept and the problems and risks related to implementing and managing product families and their successive platforms. Studies (e.g., Hauser, 2001; Krishnan and Gupta, 2001) have recently started to draw attention upon the significant costs and tradeoffs associated with product platform development. This makes one wonder why and how different types of companies have actually taken up the advocated concepts. Based on the different industrial contexts, one might further expect a variety of applications of platform thinking and product family development, probably far less straightforward as advocated in several of the historical success stories about product platform development.

In this chapter, we analyze and compare how three distinct technology driven companies adopted the concept of platform thinking in their product development process. Before doing this, we will discuss the rationale behind thinking in terms of platforms and product families by reviewing the relevant literature related to these concepts. This discussion is followed by an explanation of our in-depth case study approach. After presenting our case study results, we close with discussing the implications of the main findings of the study and identify some important managerial implications and directions for future research.

3.   PERSPECTIVES FROM LITERATURE

Previous studies (Dertouzos, 1989; Kahn, 1998; Stalk and Hout, 1990; MacDuffie, et al., 1996) have suggested that if companies want to compete more effectively, they have to meet the customer's needs over time better than the competition by offering a high variety of products. More variety will make it more likely that each consumer finds exactly the option he or she desires, and will allow each individual consumer to enjoy a diversity of options over time. In considering the implementation of product variety, companies are challenged to create this desired variety economically. In their quest to manage product variety, firms in most industries increasingly are considering product development approaches that reduce complexity and better leverage investments in product design, manufacturing and marketing (Krishnan and Gupta, 2001). Platform thinking, the process of identifying and exploiting commonalities among a firm's offerings, target markets, and the processes for creating and delivering offerings appears to be a successful strategy to create variety with an efficient use of resources (Wheelwright and Clark, 1992; Meyer, et al., 1997; Meyer and Lehnerd, 1997; Robertson and Ulrich, 1998; Sawhney, 1998).

3.1   Definitions

The terms *product families*, *platforms* and *individual products* are hierarchically different and cannot be used as synonyms. A *product family* is the collection of products which share the same assets (i.e., their platform) (Meyer and Utterback, 1993; Sawhney, 1998); a *platform* is therefore neither the same as an individual product nor is it the same as a product family; it is the common basis of all individual products within a product family (McGrath, 1995; Robertson and Ulrich, 1998). As a consequence, a platform is always linked to a product family, while it can serve multiple product lines in the market. The leading principle behind the platform concept is to balance the commonality potential and differentiation needs within a product family. A basic requirement is therefore the decoupling of elements to achieve the separation of common (platform) elements from differentiating (non-platform) elements.

One possibility to build a *platform* is to define it by means of the product architecture. This *product platform* has been defined by McGrath (1995) as a set of subsystems and interfaces that form a common structure from which a stream of related products can be efficiently developed and produced. Baldwin and Clark (2000) define three aspects of the underlying logic of a product platform: (1) its modular architecture; (2) the interfaces (the scheme by which the modules interact and communicate); and (3) the standards (the design rules that the modules conform to). The main requirements for building a product family based on a product platform are (a) a certain degree of modularity to allow for the decoupling of elements and (b) the standardizing of a part of the product architecture (i.e., subsystems and/or interfaces). A modular product architecture is thus characterized by a high degree of independence between elements (modules) and their interfaces.

The typical inclination is to only think of the product architecture as the basis for a common platform of a product family. In line with recent discussions in literature (Meyer and Lehnerd, 1997; Robertson and Ulrich, 1998; Sawhney, 1998) we argue that a product family should ideally be built not only on elements of the product architecture (components and interfaces) but on a multidimensional core of assets which includes also processes along the whole value chain (e.g., engineering and manufacturing), customer segmentation, brand positioning, and global supply and distribution.

Process platform refers to the specific set up of the production system to easily produce the desired variety of products. A well-developed production system includes flexible equipment, for example programmable automation or robots, computerized scheduling, flexible supply chains, and carefully designed inventory systems (Kahn, 1998). Sanderson and Uzumeri (1995) refer in this respect to Sony's flexible assembly system and an advanced parts orientation system, designed specifically with flexibility, small-lot production and ease of model change in mind. Although the costs of this multi-function machine may be twice as much as a comparable single-function machine, the greater flexibility possible using manufacturing equipment designed with multiple products and rapid changeover in mind offsets its initial cost.

Customer platform is the customer segment that a firm chooses as its first point of entry into a new market. This segment is expected to have the most compelling need for the firm's offerings and can serve as a base for expansion into related segments and application markets (Sawhney, 1998). Established customer relationships and knowledge of customer needs are used as a springboard to expand by providing step-up functions for higher price-performance tiers within the same segment or to add new features to appeal to different segments as discussed in Chapter 5.

Brand platform is the core of a specific brand system. It can either be the corporate brand (e.g., Philips, Toyota, Campbell) or a product brand (e.g., Pampers, Organics, Nivea). From this brand platform sub-brands can be created, reflecting the same image and perceived worth (e.g., Philishave, Hugo Boss perfumes, Organics shampoo). With a small set of brand platforms and a relatively large set of sub-brands, a firm can leverage its brand equity across a diverse set of offerings (Sawhney, 1998).

Global platform is the core standardized offering of a globally rolled out product. As an example, designing software for a global market can be a challenge. The goal is to have the application support different locales without modifying the source code. A global roll out plan details the aspects of the product that can be standardized as well as those aspects that should be adapted to country-specific conditions and customer preferences. Customization can involve physical changes in the product, and adaptation in pricing, service, positioning message or channel (Sawhney, 1998).

3.2   Management of platform-based product families

Cost and time efficiencies, technological leverage and market power can be achieved when companies redirect their thinking and resources from single products to families of products built upon robust platforms. Implementing the platform concept can significantly increase the speed of a new product launch. The platform approach further contributes to the reduction of resources (cost and time) in all stages of new product development. By using standardized and pre-tested components, the accumulated learning and experience in general may also result in higher product performance. Unfortunately this is not a one-time effort. New platform development must be pursued on a regular basis, embracing technological changes as they occur and making each new generation of a product family more exciting and value-rich than its predecessors. Meyer and Lehnerd (1997) propose a general framework for product family development. This framework represents a single product family starting with the initial development of a product platform, followed by successive major enhancements to the core product and process technology of that platform, with derivative product development within each generation. New generations of the product family can be based on either an extension of the product platform or on an entirely new product platform. In case of an extension, the constellation of subsystems and interfaces remains constant, but one or more subsystems undergo major revision in order to achieve cost reduction or to allow new features. An entirely new platform emerges only when its basic architecture changes and aims at value cost leadership and new market applications as discussed in Chapter 2. Systems and interfaces from prior generations may be carried forward into the new design but are joined by entirely new subsystems and interfaces. The more consistent the platform concept is defined and implemented in terms of parts, components, processes, customer segmentation etc., the more effective a company can operate in terms of tailoring products to the needs of different market segments or customers. Since platform planning determines the products that a company introduces into the market during the next five to ten years or beyond, the types and levels of capital investment, and the R&D agenda for the company and its suppliers, top management should play a strong role in this process.

Unlike the benefits of product family development, the risks related to product family development have not been widely and specifically addressed yet in literature. Indirectly some have been mentioned already in the previous sections. Developing the initial platform in most cases requires more investments and development time than developing a single product, delaying the time to market of the first product and affecting the return on investment time. This implies that platform-based development may not be appropriate for all product and market conditions. On top of the fixed investments in developing platforms, platforms may also result in the over-design of low-end variants in a firm's product family to enable subsystem sharing with high-end products (Krishnan and Gupta, 2001). Data collected by Hauser (2001) at one firm over a five-year period further showed the platform-based development approach to be negatively correlated with profitability. Meyer and Lehnerd (1997) address the risk related to the balance between commonality and distinctiveness. A weak common platform will undermine the competitiveness of the entire product family, and therefore a broad array of products will loose competitiveness. Another risk relates to the renewal of product platforms. As pointed out by Meyer and Lehnerd (1997), long-term success and survival require continuing innovation and renewal. A potential negative implication of a modular product architecture approach is the risk of creating barriers to architectural innovation. This problem has been identified by Henderson and Clark (1990) in the photolithography industry and may in fact be a concern in many other industries as well (Ulrich, 1995). The metrics as suggested by Meyer, et al. (1997) can help management to monitor, but they do not explicitly say when to create a new platform and companies can fail to embark in a platform renewal in a timely manner. Robertson and Ulrich (1998) have pointed out organizational risks related to platform development. Platform development requires multifunctional groups. Problems may arise over different time frames, jargon, goals and assumptions. In a lot of cases organizational forces also seem to hinder the ability to balance between commonality and distinctiveness. Engineers, for example, may prepare data showing how expensive it would be to create distinctive products while people from Marketing may argue convincingly that only completely different products will appeal to different markets. One perspective can dominate the debate in the organization.

The concept of building product families based on platforms has been widely accepted as an option to create variety economically. The reasons (or expected benefits) of the concept are mainly greater flexibility in product design, efficiency in product development and realization, and effectiveness in communication and market positioning. The application of the platform principles leads to different platform types according to the kind of assets that can be used as a common basis; however, there are substantial risks and tradeoffs that have to be made in developing and managing platform-based product families.

4.　　RESEARCH

The objective in this chapter is to investigate how and why companies are adopting, developing, implementing and monitoring platform and product family concepts in practice. We used a multiple case study approach. Case study research involves the examination of a phenomenon in its natural setting. The method is especially appropriate for explorative research with a focus on "how" or "why" questions concerning a contemporary set of events (Eisenhardt, 1989). The research design involved multiple cases, generally regarded as a more robust design than a single case study, since the former provides for the observation and analysis of a phenomenon in different settings (Yin, 1994).

4.1　　Sample

We studied three technology-driven companies that have customized platform and product family development to meet their specific product and market needs. These firms represent a variety of product and market contexts and provide examples of a range of platform and product family concepts and implementations. In addition to the technology driven criterion, the following criteria were used for selecting the firms: (1) substantial experience in NPD; (2) developing relatively complex products; (3) experienced in applying the platform and product family concepts; (4) operating in highly competitive markets; and (5) collectively representing a diversity of product and market needs.

We selected three companies that best met our criteria and the additional assumption that these companies would differ in their application of the platform and product family concepts. During the process of data collection, no major deviations were found with regard to these initial assumptions. Before describing our data collection and analysis, we first provide a profile of the companies involved.

4.2   Company profiles

The participant firms were: ASML, a market leader in advanced micro lithography systems; Skil, a power tools division of Bosch; and Stork Digital Imaging (SDI), a worldwide operating company of digital print and pre-print applications for the graphic arts and textile printing markets. All companies have many years of experience with platform-based product family development. Table 3-1 gives an overview of employee numbers and net sales figures for respectively ASML, Skil and SDI.

Table 3-1. Characteristics of the three companies involved in the field study.

	ASML	Skil	SDI
Employees (2001)	7070	445	1595
Net sales (2001)	$1844M USD	$135M USD	$230M USD

ASML's customers, the semiconductor manufacturers, are building increasingly complex ICs. As a result, the critical dimensions of the product design (IC size) are continually reduced. ASML's microlithography is the enabling technology to realize faster and smaller ICs and consequently is under pressure to provide a product which can hold pace with the technological evolution. Skil is oriented to the consumer market for power tools. It is positioned in the low-end segment, where a high pressure on market prices exists, which has to be answered by cost efficient variation of the products offered. SDI produces systems for digital printing technology to its customers. To meet the very high standards in its markets it needs a deep knowledge of different printing processes and has to integrate newest technology in its products.

Although operating in different fields and producing different products, all participating companies are OEMs in a competitive environment that is global, intense and dynamic. As a result, they all share the need to produce a high variety of products at competitive prices, in order to meet customer demands and to face competition.

4.3   Data collection and analysis of our field study

The data collection and analysis was carried out in four phases. The aim of the first phase was to get a general understanding of the companies involved, the products they make and the markets they address.

The second phase consisted of a set of 15 in-depth interviews with employees of a different functional background involved in product family development. The expertise covered Project management, Program management, R&D, Systems engineering, Manufacturing, Marketing, and Customer support. The interview structure consisted of five parts: (1) define the concept of platform-based product families, (2) identify the reasons for changing to platform-based product family development, (3) get insight into how the product family concepts were implemented, (4) identify the perceived risks in the development and management of platform-based product families, and (5) derive the needs for supporting product family development. The average duration of the interviews was two hours.

In the third phase, we performed a content analysis using the procedure recommended by Kassarjian (1977). The aim was to standardize the outcome of the different interviews within and across companies. Three researchers independently performed this analysis and afterwards compared their outcomes and discussed any differences until they reached consensus. After analyzing and comparing the interviews, the results were generalized and served as a basis for identifying gaps in literature and practice.

In the fourth phase, a workshop was organized with participants from all three companies. The aim of the workshop was to confront the company experts with the platform ideas from literature and to present our research findings (i.e., the gaps in practice as well as in theory) concerning the building of product families based on platforms. The results of this phase were the verification of our conclusions, the sharing of platform experience between the company experts, and the identification of further implications for the management of product families and for research.

5.   FIELD STUDY RESULTS

5.1   Definitions of platform and product family concepts used in practice

In the first part of the interview interviewees were asked to give definitions for the concept of platform-based product families used within their company. The terminology and definitions used within each company helped for the proper understanding during the interview. Additionally, it gave insight into how well these concepts are defined and internalized within the company, and which disciplines are most knowledgeable about product family development. The appendix gives an overview of the definitions provided during the interviews.

The majority of the interviewees stated that platform-based product families were a known concept within their company. The analysis shows that although the definitions related to product families differ among the respondents, most refer to a set of related products with different applications based on a common, often physical or technological, "part". Although not specifically defined this way, product families are often seen from a marketing perspective, providing different products to the same or related market segments, using a product platform.

The definitions of product platforms show a similar picture. Although differences in definitions exist both within and between companies, there is an overlap between the definitions with respect to the importance of "basic modules", "a similar concept" or "a core technology". This highlights the technical perspective from which platforms are considered within these three technology-driven companies. Although again semantic differences exist, the same underlying principle applies to most of the definitions used. The appendix does not reveal a clear distinction between disciplines that, over the three companies, appear to be more or less knowledgeable about the discussion on product family or product platform development in literature.

Although many different definitions were encountered in the three companies, the results show that the recursive character of product families and platforms is recognized and accepted in practice. The principles behind building platform-based product families are generally understood and applied. The platform concept is associated with the reuse of elements (modules, components, designs) across multiple products. All the definitions clearly point to a "core" called platform from which specific products are derived. These products constitute a product family. There are differences in the focus and level of detail, but not in the general meaning. The interpretation of the platform concept in practice results in fewer platform ideas than could be expected from literature. Product architecture is the predominant basis for identifying platform potential, while other areas (e.g., processes) remain largely untapped. The answers received lead to the conclusion, that often the platform idea is only applied within a part of the company (e.g., Engineering) and rarely in a cross-functional context.

5.2   Reasons to adopt platform thinking in product family development

All three companies involved have experience with product family development. Table 3-2 gives an overview of their products, market structures, and reasons for changing to a family development approach.

Table 3-2. Specific situations and expectations for product family development.

	ASML	Skil	SDI
Products	Microlithography systems for the semiconductor industry (high-end)	Power tools for the consumer market (low end)	Systems for digital print and pre-print applications (high-end)
Market	1 market segment with different applications (stepper, scanner, twin scan)	Different applications (saws, drills, etc.) each in 2 market segments (opening price point, lower price point)	2 market segments with different applications (graphic arts, textile printing) and a potential new market (photo printing)
Platform potential	High commonality between products within an application (reuse of basic modules)	Very high commonality within applications (across segments), high commonality across all products (components)	Two available technologies for solutions in both market segments (general purpose modules)
Expected benefits	Efficiency (volume and costs, maintenance) Flexibility (time to market, assembly) Effectiveness (training, learning curve)	Efficiency (costs and time; high variety) Flexibility (time to market, styling) Effectiveness (brand identity, understanding the structures)	Efficiency (time and costs for product variation) Flexibility (serving two market segments) Effectiveness (products are easier to explain)

ASML develops platform-based product families for its whole product range. It gave two main reasons for following a family approach. First, a stable platform makes it easier to come up with newer modules and to ramp up volume. Second, from an engineering point of view it is unaffordable to design a new machine from scratch every time a change in a local part of the machine is needed. Besides efficiency in the development process, shorter time to market and ramp-up times, advantages for servicing and maintaining the machines, and improved learning curves during training were mentioned.

Within Skil, efficient use of resources and reducing time to market were seen as the main goal from an engineering and manufacturing perspective. Marketing goals on the other hand were that a product family should be based on commonality in terms of styling, perceived worth and resulting in a strong brand identity. Clearly distinctive product families will help customers to make comparisons and choose tools that fit their needs best.

For SDI the starting point for platform-driven product families was the development of so called "General Purpose Modules" (GPMs). Based on these GPMs, a variety of machines can be easily derived from the same building blocks. The GPM-based platforms leveraged a horizontal expansion for SDI. The GPMs were designed in such a way that it became possible to build machines suitable for both the textile and graphic arts markets. The GPM-based platform approach enabled SDI to maximize its profits while keeping the development budget the same. Besides cost efficiencies in the product development process, and time to market reduction, also a more efficient training program could be developed. According to a SDI Marketing manager, "Once you understand one product, you understand them all".

It can be concluded that, compared to the broad differences regarding the product applications and market structures found in the cases, the reasons and expected benefits from building platform-based product families fall into basically three different categories: (1) enhancing the flexibility in product design, (2) increasing efficiency in product development and realization, and (3) improving effectiveness in communication and market positioning. In a managerial sense, family thinking was found a way of simplification (complexity reduction) for supporting decision making. This argumentation in practice for the development of platform-based product families is very similar to the reasons found in literature

5.3   Implementing platform-based product families in practice

Figure 3-1 shows the evolution of product families within ASML using a product roadmap (see Chapters 2 and 5). A product family is defined on the basis of the available technology and expected market needs. The first developed product type of each family will serve as the platform for subsequent products belonging to the same product family. Follow-on products are enhancements of the initial platform, meaning that one or several modules, mostly related to the optics of the machine, are replaced by an enhanced version without changing the product architecture.

Changes to the product architecture are only made when it is absolutely technically necessary. Since only internal modules are replaced, products belonging to a family look exactly the same from the outside. Approximately 80% of the modules remain similar over the lifetime of a family. This strategy of evolving the product line yields two key benefits. First, customers know that the ASML systems they install today are backward compatible with the manufacturing processes and the installed base of equipment they are already using. Secondly, it enables customers to reduce their manufacturing risks by using the same operator interface, spare parts, and machine-to-machine "mix-and-match" connectivity while adding new imaging capabilities needed to develop more advanced semiconductor devices. In the past years ASML has developed three different platforms (2500/5000 Steppers, 5500 Scanners, and Twinscan), each serving as the basis for subsequent specific product versions. The three product families basically address the same market needs; however, the platforms differ in their performance limits. Products from all three families are still being sold. However, no new models are developed based on the first platform. New products are still being developed from the second platform, while the first products based on the third platform have only just started being produced.

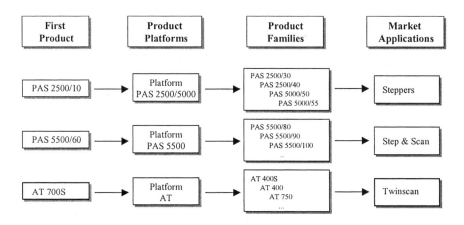

Figure 3-1. Platform-based development of product families within ASML.

Skil develops a range of different power tools for the low end of the do-it-yourself consumer market. The first step in the development process is taken by Marketing. Marketing selects new product ideas and defines the product requirements. At this point the focus is on developing a single product that fits in an existing brand and can be introduced in the market as soon as possible. It is then up to the Engineering department to meet these requirements and to meet cost targets. It is at this (cost reduction) stage that technological commonalities between products (component standardization) are considered.

By reusing existing components instead of developing everything new for every new product, and by (re-) developing components for multiple uses, Engineering intends to achieve cost reduction. As a result, commonality exists (approximately 80%) within the same types of tools (e.g., drills) as well as between different types of tools (approximately 50%) and is found on a component level (e.g., switches, motor, bearings, electronics). Platform thinking in this company therefore relates to the reuse of components, as opposed to the architecture of the complete product in ASML. Family development in this situation is more market driven than technology driven. Multiple product families (consisting of e.g., a set of saw, drill, router and grinder) are developed to address different brand segments, each product family with its own core styling and perceived worth, but all product families utilizing as much as possible the same technical components. Figure 3-2 shows the relationship between components, commonality within the same and between different types of power tools and brand segmentation.

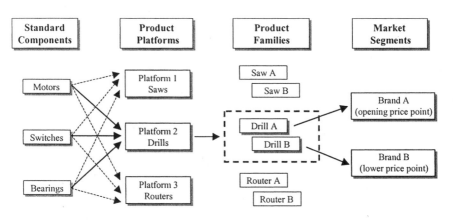

Figure 3-2. Platform-based development of product families within Skil.

Figure 3-3 shows how SDI provides different printing applications for two target groups (i.e., the textile market and the graphic arts market). The company distinguishes between two different platforms underlying their products. Platform thinking in this company is therefore related to the reuse of two different printing technologies: the somewhat older Single Nozzle technology and the new Array technology.

For each of these two technologies General Purpose Modules (GPMs) have been developed, which are used in different products. A modular design is chosen to speed up development and to reduce costs. The single nozzle technology is applied to both the textile and graphic arts market. The Array technology so far has only been used for products for the textile market; however, applications for the graphic arts are on their way. Applications for a totally new market, the photo printing market, are being investigated based on the new possibilities offered by the Array technology. The platform concept emerges during the concept generation phase, where a range of products is defined based on a similar underlying technology. New product ideas are screened on their technological feasibility and their link to customer needs. During design and development, the focus is on keeping as many modules the same among the different applications. The products belonging to the same product family, textile or graphic arts market, are up to 70 to 80% similar, with just small differences in size, color, inking components or frame.

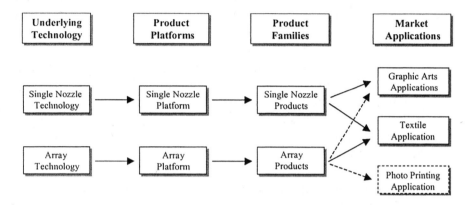

Figure 3-3. Platform-based development of product families within SDI.

Although comparable in terms of their understanding of the platform and product family concepts and the potential benefits resulting from their use, the three companies are quite different in the way they adopted and implemented these concepts in practice so far. Table 3-3 shows the platform types the companies developed, and what level of reusability they achieved *within* product families and *across* them.

Table 3-3. Platform characteristics.

	ASML	Skil	SDI
Platform type	• 3 Product Platforms (basic modules and system architecture) for 3 applications	• Product Platforms per tool type (common components)	• 2 Product Platforms (technology related basic modules and system architecture)
Effects (reuse)	• 80% commonality within products of the same family • low commonality across product families	• 80% commonality within products of the same family (application) • 50% commonality across tool types (components)	• 70-80% commonality within products of the same family

Companies also have specific ways in using their platforms and product families in the market. For example, one market segment can be served by multiple product families (as SDI's textile market), a single platform can serve as a basis for multiple market segments (e.g., drills at Skil), or each platform addresses a single application (ASML). This shows the importance of a company specific definition of platform-based product families.

The high variance in the use of product families and platforms with regard to the market structure, however, contrasts sharply to the few platform *types* used in practice. In all three cases, the predominant framework for finding commonality was the product architecture (i.e., modules and interfaces). This corresponds with the findings from the concept definitions by the companies, and it shows a clear gap compared to the variety of platform types presented in literature.

In two of the three cases, a platform was not planned or defined at the beginning of product development. It emerged later in the product life cycle, when a higher stability in the market requirements made the development of a common platform less risky. This also contrasts with the ideal picture in literature, where product family definition takes place before developing individual products.

5.4 Perceived risks related to platform-based product family development

Developing product families not only provides opportunities for companies; there are also risks involved. As stated earlier, product family development is more strategic and long-term in nature, focusing less on singular opportunities than single product development. Product families require a strong platform on which follow-on products can be built effectively and efficiently and these platforms need to be renewed in time to be able to meet changing customers' demands.

Just like single products, product families have a limited lifetime that needs to be managed. Therefore, decisions have to be made about when to start a new family, which products to launch and in which order, when to move on to an extended or new platform, and consequently a new product family, and where to best allocate scarce resources. Clear metrics or designated methods to take these decisions, as discussed in literature, could not be identified in any of the cases.

For ASML, making revolutionary steps in platform design increases the risk of not getting the platform to work according to specifications or in time (i.e., being too ambitious). It was further brought up that platform development might lead to restrictions on the use of new technologies in a later stage of the product family life cycle (i.e., does not match with the platform), rigidity in design when a lot of choices have to be made in a very early stage, and failure to correctly forecast future user needs. Much of the risk encountered is explained by the need for decisions early in the product life cycle, the analysis of future market requirements, and the high initial investments for platform development.

Skil stressed that product platform development should not be a goal in itself. Developing a product platform should only be considered when there are clear views on reuses in future products. The main risks considered are forecasting future consumer needs, integration of existing elements, and the high impact of mistakes early in development.

For SDI, the challenge of developing product families (in order to make it suitable for a wide range of products) lies in the correct choice of the platform. The major risks encountered are the restrictions for different market segments and the high initial cost and time for platform development.

Table 3-4 gives a summary of the risks and problems facing platform and product family development, as perceived by the interviewees. As can be seen in the table, most companies mention increased development times, costs and complexity of the initial platform as a risk of product family development, reflecting the importance of developing the "right" platform.

Table 3-4. Perceived risks related to platform-based product family development.

	ASML	Skil	SDI
Risks	• Development time and costs of platform • Rigidity in design • Restrictions on the integration of new technologies • Incorrect forecast of future user needs • Change form one platform to another	• High cost and time for integration of existing elements • Platform development becomes easily a goal in itself • Mistakes made in the beginning have a high impact • Failure to forecast customer needs correctly	• Development time and costs to meet specifications of all target markets • Development process becomes more complex • Restrictions for all market segments • Selecting the right platform

There is a general agreement that having the "right" platform-based product family results in substantial competitive advantage. The different tradeoffs in the definition, development and management of platforms, however, have to be considered.

These platform related tradeoffs can be identified within the same categories as proposed in our presentation of reasons to adopt platform thinking (see Section 5.2): (1) flexibility in product (family) design vs. restriction through a platform, (2) efficiency in the development and realization of single products vs. high initial efforts (time and cost) for platform development, and (3) low differentiation through the platform vs. distinct positioning through individualized elements.

Through the balancing of commonality potential and differentiation needs, these tradeoffs can be influenced, and consequently the optimum platform can be determined. As these decisions have to be made early in the product life cycle, they contain a high risk.

It is interesting to see that although the companies analyzed came from clearly different initial situations, they all encountered more or less the same tradeoffs during the definition, development, and management of their product families. These results also fit well into what is found in literature. Table 3-5 summarizes the lessons learned regarding the platform-driven development and management of product families for the three companies.

The effects of developing platform-based product families are dependent on the specific platform definition. All companies agree that making wrong decisions is very expensive. The development of platform-based product families requires a clear concept and a cross-functional understanding of the tradeoffs involved between marketing, sales, engineering, sourcing, manufacturing, etc.

What is generally needed is a better understanding of mechanics and risks involved in platform development and consequently tools for decision-making. Our research indicates that it should be possible to develop a general framework for decision-making, as the underlying principles (mechanic) in building platforms stay the same, and similar tradeoffs and risks were found in all cases.

Table 3-5. Lessons learned from platform-based development of product families.

	ASML	Skil	SDI
Lessons learned	• Definition of a platform requires choosing from alternatives • Development of a platform is a strategic decision • Understanding of market requirements is necessary	• Development of a product family needs a clear concept • A product family makes communication easier • Customer needs have to be identified early	• Having one platform for two market is difficult for the stability of the platform • Market requirements have to be tested before platform development

6.   DISCUSSION AND CONCLUSIONS

The objective in this chapter was to explore the current state of literature concerning the concept for platform-based product family development and management, and to compare this with actual application in practice. When comparing the theoretical and practical perspectives, several observations can be made.

Compared to the literature, the concept of platform-driven product family development was not identically defined either within or across the companies investigated. The underlying ideas and principles, however, are generally agreed upon and lead to the development of platform-based product families in practice. The definition of what constitutes a platform has a much wider meaning in literature than encountered in our case studies. According to the literature (McGrath, 1995; Sanderson and Uzumeri, 1995; Meyer, 1997; Meyer and Lehnerd, 1997; Robertson and Ulrich, 1998; Kahn, 1998; Sawhney, 1998), a platform can be related to product architecture, technology, sourcing, manufacturing and supply processes, customer segmentation, brand positioning, and even people and relationships. In our case studies, respondents predominantly associated this concept with product architecture and technology and in a limited way with customer needs and branding. The interviews did not reveal any structural or planned use of sourcing, manufacturing and supply processes as a base for platform development, which highlights opportunities to further benefit from the principle of platform thinking.

The reasons why companies choose to follow the platform concept when developing product families show a high degree of similarity. Although the companies analyzed differ substantially in the products they offer and in the markets they address, all of them have the same goals in mind when opting for platforms. The resulting commonalities (degrees of reuse) show comparable results in consequence. The platforms they developed, on the other hand, although mainly focused on the means of product architecture (i.e., modules and interfaces), reflect various combinations of product families and market structure. As a result, in one case a single platform can be used in multiple market segments, in another case multiple platforms are applied in a single market segment, and in a third case we found one platform per market application each. Different platform leveraging strategies are discussed in more detail in Chapter 5.

The companies involved in the study acknowledge similar risks as well as opportunities when developing product families. These tradeoffs relate to the increased development efforts for the initial platform and the uncertainty whether the "right" platform is chosen in order to develop enough follow-on products to gain back these extra expenses. Chapter 4 provides some useful metrics for assessing product platform concepts.

The three cases show that gaps still exist between what is written in literature and what is done in practice. Part of this problem originates from the fact that the knowledge transfer from literature to practice has not taken place sufficiently in the companies discussed. None of the respondents expressed any in-depth knowledge about the discussion that takes place in literature.

When seen from applying the concepts in practice, several "gaps" became evident. First, we have seen in our field study a rich and divergent application of platform thinking and product family development, which is far less straightforward as advocated in several of the reported success stories about platform and product development. By understanding and focusing on the different organizational contexts in which platforms and product families are applied, future research may develop categories of options for platform and product family development that are useful in practice given a specific context.

Platform decisions predispose a company's flexibility to react to technological or market changes. Our study showed that, although strongly interested in and convinced about the benefits of product family development, the companies claimed to lack practical guidelines and decision rules to help them in their platform decision-making process. Most platform decisions are not primarily concerned about whether to invest in a platform or not, but about the valuation and strategic selection between platform alternatives. A second important gap in platform literature however is the lack of a sound valuation model, as traditional methods (e.g., NPV) fail to provide the necessary support for valuation and decision-making (Kogut and Kulatilaka, 1994; Baldwin and Clark, 2000; Hofer, 2001). Several chapters in Part III of this book attempt to address these issues.

The companies involved also expressed a great concern about the risks involved in platform and product family development and the lack of knowledge and tools to deal with these risks effectively. Available literature so far has mostly focused on the underlying concepts and benefits of product family development (i.e., effective and efficient product development through reuse) and less on investigating what might be successful strategies to manage the risks and problems related to platform and product family development and implementation. It is suggested therefore to initiate a third stream of research that could provide insight into our understanding of potential successful strategies of how to successfully identify and manage the risks related to platform-driven development of product families.

A fourth direction of future research concerns the variety in the use of platform types in practice. The fact that, compared to literature, only a narrow range of platform types was found in our exploratory study, leads us to the conclusion that continued, expanded research involving companies in different industries should be directed to investigate, whether a substantial platform potential remains undetected and unused in practice.

Filling the aforementioned gaps in the literature would be an important contribution, both from an academic as from a managerial point of view. We are positive that our research has started on the road to answering these pending issues by narrowing the focus for further research and laying a foundation for expanding the investigation to other industries to broaden our knowledge about platform-based product families. This is an essential prerequisite to address industry's needs for continued support in the strategic planning, design, and management of platform-based product families.

APPENDIX DEFINITIONS OF PRODUCT PLATFORMS AND FAMILIES BASED ON RESULTS

Co.	Discipline	Product family and product platform definitions
ASML	Marketing	*Family:* a group of products that have the same outside. *Platform:* the physical shape on the inside (determines the mechanical and electrical lay-out of the inside)
	Customer support	*Family:* related to the body of the product. *Platform:* no definition - technology related term.
	Pilot production	*Family:* no definition. *Platform:* new baseline from which new families can be derived.
	System engineering	*Family:* a family of products with a lot of common modules where you change some of the modules to make a new product. *Platform:* is a family of machines, existing of a lot of different types, with different options and modules.
	Program management	*Family:* a number of modules are basic on which you build a number of products. *Platform:* no definition
Skil	Marketing	*Family:* a group of products based on a similar technical concept with a differentiated look of the product to the end users. *Platform:* no definition
	Project management	*Family:* several children / products going from lower to higher specs (including price) related by appearance (e.g., green housing, similar look) and are not all the same. *Platform:* having as much as possible common parts.
	Manufacturing services	*Family:* different products / models on the highest level (e.g., hammer drills, circular saws). *Platform:* no definition
	Product development	*Family:* no definition. *Platform:* no definition.
SDI	Product development	*Family:* term not used *Platform:* term not used
	Business management	*Family:* kind of basis / technology on which you build different products. *Platform:* software related.
	Business management	*Family:* a group of products that are all based on the same components (building blocks). Look the same, but have small differences to make them suitable for different markets. They have all the same technology inside. *Platform:* more related to underlying technology (less appearance).
	Operations	*Family:* a group of products that is produced for a certain market. *Platform:* term not used.
	Production and process engineering	*Family:* a group of products for the same application field with small changes to the products itself (same underlying principle). *Platform:* underlying core technology.
	Research and development	*Family:* several products based on the same platform. *Platform:* the way separate modules of a system are organized and how the interfaces are arranged.

Chapter 4

PLATFORM CONCEPT EVALUATION
Making the Case for Product Platforms

Katja Hölttä-Otto[1] and Kevin Otto[2]
[1]*MIT Center for Innovation in Product Development, Cambridge, MA 02139 and Helsinki University of Technology, Espoo, Finland 02051;* [2]*Robust Systems and Strategy, LLC, Watertown, MA 02472*

1.　　INTRODUCTION

A platform must support several product variants at any point in time and it must survive several life cycles into the future. The technology composing the platform itself is usually the embodiment of the core value-added capability of the developing company, yet what makes a good platform? This question often arises, for instance, when comparing two alternative platform concepts or deciding whether to update or replace a platform. The decision is more complex than a standard concept comparison exercise, involving forecasts of several applications and alternative technologies. Multiplicity and uncertainty characterize platform concept evaluation.

There has been excellent work in developing product concept evaluation methods, such as Pugh's selection process, concept screening and scoring, or trade-studies (Otto and Wood 2001; Ulrich and Eppinger, 2004); however, these methods evaluate single product concepts. A platform concept has different requirements due to its longer lifetime, and it must enable several derivative products. These added requirements make the single product concept evaluation methods not directly applicable to a platform concept. There is a need for comprehensive platform concept evaluation tools.

The method presented here is designed for evaluating a platform—to determine if a platform is the best possible option for a company in a given situation, which is different from the platform development and optimization methods introduced later in this book. These works are good for initial platform development to, e.g., maximize the commonality while trying to maintain the product performance requirements. However, when making the case for a product platform, the system should be characterized not only by its primary function but also by other "ilities". For example, one may have optimized the performance and cost of the platform, but is it more reliable than another platform? Is it more flexible? Etc.

In this chapter, we introduce a platform concept evaluation tool that is multi-criteria in nature and scalable to include various alternative criteria as appropriate. Several criteria are applied here to demonstrate platform concept evaluation, considering aspects of the product for its entire life cycle. This multi-criteria analysis yields a concept phase analysis that helps manage risk by making all aware of the specific parts of a development project that may require developing backup plans, applying extra effort, and obtaining management attention.

2.   THE PLATFORM ASSESMENT TOOL

We developed a platform assessment tool that makes use of the work of many others in the field of modularity, platforming, and general product development. We created a list of platform metrics from three sources. First, we generated a list from personal experiences of platform development over the last 10 years and over three-dozen platforms. These experiences also included personal mistakes learned from inadequate preparation (e.g., inadequate preliminary assessment). Second, we consulted other executive-level system engineers with an average of 17 years of experience to evaluate and solicit platform metrics based upon examination of their past platforms. Finally, we examined the literature for platform metrics and qualitative criteria used by others. Based upon previous literature and our experience, we have grouped and assembled a total of 19 metrics. We further refined the metrics into six groups, as shown in Table 4-1. Notice that this list may not be appropriate for all industries and should be adjusted when needed. We do not claim comprehensiveness; however, we do claim that these 19 serve many electro-mechanical platforms well.

Interestingly, personal experiences generated several metrics not found in the literature, such as interface adjustments and synergy. Where metrics were available; however, they were easily adaptable to our evaluation tool, which is general enough to any desired metric. We take into account more factors than methods developed to date. This scope is more realistic to industrial practice, where platform decisions must consider a multitude of decision impacts.

Table 4-1. Platform assessment tool summary sheet.

Overall Multi-Criteria Platform Assessment				Score	Grade
				7.5	B - Good

Platform Scorecard	Marginal Corporate Focus	Percentage Corporte Focus	Weighted Contribution	Score	Grade
Portfolio Customer Satisfaction	9	35%	2.5	7.2	B - Good
Product Variety	9	35%	2.6	7.4	B - Good
After Sale Support	3	12%	1.1	9.5	A - Outstanding
Organizational Alignment	3	12%	0.7	5.9	C - Acceptable
Upgrade Flexibility	1	4%	0.4	9.3	A - Outstanding
Development Complexity	1	4%	0.3	7.3	B - Good
		100%	TOTAL	7.5	B - Good

Portfolio Customer Satisfaction Scorecard	Marginal Corporate Focus	Percentage Corporte Focus	Weighted Contribution	Score	Grade
Cost - Worth Distribution	9	50%	4.0	8.0	B - Good
Portfolio Customer Needs	9	50%	3.2	6.3	C - Acceptable
	9	100%	TOTAL	7.2	B - Good

Product Variety Scorecard	Marginal Corporate Focus	Percentage Corporte Focus	Weighted Contribution	Score	Grade
Planned Upgrade Carryover	3	20%	1.9	9.4	A - Outstanding
Common Modules	9	60%	3.6	5.9	C - Acceptable
Specification Variety	3	20%	2.0	10.0	A - Outstanding
	9	100%	TOTAL	7.4	B - Good

After Sale Support Scorecard	Marginal Corporate Focus	Percentage Corporte Focus	Weighted Contribution	Score	Grade
Partitioning for Reliability	3	23%	2.0	8.6	B - Good
Partitioning for Service	9	69%	6.9	10.0	A - Outstanding
Environmental friendliness	1	8%	0.6	8.1	B - Good
	3	100%	TOTAL	9.5	A - Outstanding

Organizational Alignment Scorecard	Marginal Corporate Focus	Percentage Corporte Focus	Weighted Contribution	Score	Grade
Ease of Assembly	3	19%	1.3	6.9	C - Acceptable
Aligned with the Organization	9	56%	3.0	5.4	C - Acceptable
Make-Buy	1	6%	0.6	9.2	A - Outstanding
Testability	3	19%	1.0	5.4	C - Acceptable
	3	100%	TOTAL	5.9	C - Acceptable

Upgrade Flexibility Scorecard	Marginal Corporate Focus	Percentage Corporte Focus	Weighted Contribution	Score	Grade
Unknown Isolation	1	25%	2.5	10.0	A - Outstanding
Change Flexibility	3	75%	6.8	9.1	A - Outstanding
	1	100%	TOTAL	9.3	A - Outstanding

Development Complexity Scorecard	Marginal Corporate Focus	Percentage Corporte Focus	Weighted Contribution	Score	Grade
Function and Form Alignment	1	7%	0.6	9.7	A - Outstanding
Interface Flexibility	3	20%	2.0	10.0	A - Outstanding
Anti-Synergy Avoidance	1	7%	0.0	0.0	F - Unacceptable
1 DOF Adjustments	1	7%	0.7	10.0	A - Outstanding
Limited Extremes	9	60%	3.9	6.6	C - Acceptable
	1	100%	TOTAL	7.3	B - Good

To calculate an overall platform score from the metrics, we project each metric onto a scale from 0 to 10. A metric may sometimes consist of multiple components. We combine these into a single score using a simple weighted sum. We also use a weighted sum to derive the overall platform score from the individual metric scores. We derive the weights based upon percentage of corporate focus as contribution to long-term corporate profit, using corporate metrics methods such as by Hauser (2001). One could also extend this to more refined decision-making methods such as Analytic Hierarchy Process (Saaty, 1980) or Utility Theory (Raiffa and Keeney, 1993). We chose a simpler approach since our aim is not to develop an absolute preference scale, but more simply to determine relative sensitivities – to point out where issues exist in individual alternative platform concepts. These and other research works provide ample means to determine weights for any of the criteria in our framework. Each of the 19 components of the weighted sum are shown in the "Score" column in Table 4-1. Each metric is arranged hierarchically in groups, with a group weight in the top "Platform Score" rows of Table 4-1, and the within-group weights shown in the "Marginal Corporate Focus" column and normalized in the "Percent Corporate Focus" column of Table 4-1. The overall weights of each metric are in the "Weighted Contribution" column, determined by multiplying the group percent weight with the within-group percent weight.

We focus our method on the platform architecture concept phase, where detailed information needed for all metrics is generally not completely available. We suggest using whatever data is available and estimating the rest. Then, one can explore the sensitivity of the analysis to changes in the uncertain data entries. As more detailed data becomes available; this evaluation can be recalculated with the refined data as needed.

3.   EXAMPLE

In order to demonstrate our method and the use of the metrics, we use of a family of five different cordless drills: professional, heavy-duty, value brand, home-use, and a low price model. Here, the platform architecture is the same as the family function structure (see Figure 4-1) since all functions are common between at least two members of the product family. We label each module according to its main component. Note that module "drilling" is a function performed by the user and not by the product, but it is included here to show that not all functions in the architecture have to be realized by the product. We also exclude the battery pack from the analysis, since it is a separate product. The speed changer module is found only in the professional and heavy-duty models. The entire family function structure consists of only two distinct product function structures (merged in Figure 4-1), and the added variety is obtained by altering how the functions are achieved. In addition we show two alternative modularization choices A and B. Alternative A is otherwise same as the current platform except that the motor and transmission modules have been combined into a single module. In alternative B, the clutch is further added to the motor-transmission module and the switch and trigger modules are merged.

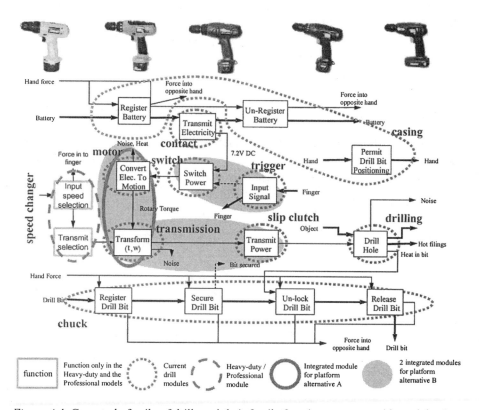

Figure 4-1. Case study family of drills and their family function structure, with modules from (Sudjianto and Otto, 2001) shown.

In this chapter, we evaluate only one platform – the current cordless drill family – in a detailed example. In addition we show results for platform alternatives A and B. We intend that this assessment tool can be used to evaluate multiple alternative platforms. Our framework helps determine and clarify which criteria are strengths for platform and which are weaknesses. This helps manage risk by making all aware what factors may need extra backup plans, effort, or management attention.

4. PLATFORM EVALUATION METRICS

In this section we describe all of the individual metrics in the platform assessment tool.

4.1   Customer satisfaction

Meeting the customer needs is the primary goal of any product. The importance of this criterion is emphasized by a high weight for metrics in this category. This includes checking that each product has a favorable cost-worth distribution and meets the customer needs.

4.1.1   Cost-worth distribution

All modules should add value to the product or they should not be included in a platform. Translating the thought into an actionable concept, each module should have an identifiable function whose customer value is known. To capture this, we apply the cost-worth approach described in (Tanaka, 1989). We calculate the cost and worth of each module, and evaluate based upon the cost-worth difference.

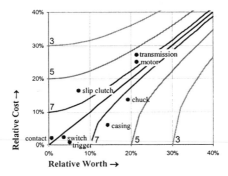

Figure 4-2. Relative cost-worth diagram for a home-use model cordless drill.

To form an interval scored metric from the standard cost-worth graph, we augment it with gradations and then score based on how far a module is from the ideal ratio 1:1, as shown in Figure 4-2. In our example, the transmission module has a score of 7 and the switch a score of 10. The total score for the family architecture is the average of the modules' scores, generating an average measure of the deviation from ideal. For the home use model drill we calculate a value analysis result of 8.1. This reflects what can be seen in the figure as well, that most of the modules are just outside the ideal zone and a little improvement could raise the score closer to 10. For the drill family we calculate 8.0.

4.1.2   Customer needs

The customer need metric measures how well the customer needs are met by the platform. In fact, this is one of the most important metrics; failure here implies the platform's product variants will not sell to the known market segments the company is currently targeting. We compare the product variant's ideal target on each critical requirement to what the platform can actually provide. This should be done in reference to current or competitive benchmarks, such as when using the standard methods of Quality Function Deployment.

A metric scaled to our scale is given by:

$$Y_{CR} = \frac{1}{M} \sum_{\text{variants } i} \frac{1}{K} \sum_{\text{requirements } j} w_{ij} \, R_{ij} \,, \tag{1}$$

where w_{ij} is the revenue weighted importance requirement j for product i; R_{ij} is the score for a customer requirement j for product i on a 0-10 scale, reflecting the gap from its target; K is the number of requirements; and M is the number of variants.

The customer score, R, can be calculated, e.g., by comparing the achieved level of a requirement to the range between the target and the starting level of that requirement. For example, a vehicle platform may support several derivative vehicles, and acoustic noise to the operator may be an important customer need. Several alternative platform concepts can be evaluated through finite element sound transmission codes for percent improvement based on acoustic noise relative to the original platform. This would be a criterion upon which the alternative platforms are assessed.

For the drill platform, the customer need assessment was made drill by drill against the intended market requirements, as assessed by third party commercial consumer groups (consumerreports.com). The drill platform was assessed reasonably at 6.3, with the weakest scores falling under the battery run-time and purchase price, and the strongest scores falling under the weight and charge time.

4.2   Variety

One purpose of a platform is to easily enable product variants. Product variety is a key objective for the example drill platform, and the weights for metrics in this category are high (see Table 4-1). We measure how well a platform achieves this goal using three metrics: carryover, common modules, and specification variety.

4.2.1   Carryover

If a specific function can be incorporated, i.e., carried over, into different products without change and no technology upgrades are expected, then the function should be isolated into a module (Ericsson and Erixon, 1999). A corollary is that when inserted into different products, a module has either 100% carryover of its functions or none at all. To capture this, we define and normalize (see Figure 4-3) a platform metric as follows.

$$Y_{carry} = 10\,(\#\,\text{functions to carryover}/\#\,\text{functions})\cdot 100 \qquad (2)$$

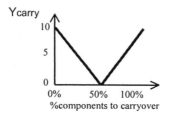

Figure 4-3. Normalization.

The score for the platform is the average of the modules' normalized scores. The metric calculation for each module is shown in Table 4-2. The score for the drill platform is a high 9.4, reflecting the fact that for most drill modules, the entire module will either carry over to the next generation with no change (internal components), or the entire module will be changed (replaced modules or the casing).

Table 4-2. A part of the carryover evaluation score card.

		# fucntions to	% Function		
	# functions	carryover	Carryover	Score	Grade
casing	3	1	33%	3.3	D - Poor
battery contacts	1	0	0%	10.0	A - Outstanding
switch	1	0	0%	10.0	A - Outstanding
motor	1	1	100%	10.0	A - Outstanding
transmission	1	1	100%	10.0	A - Outstanding
slip clutch	1	1	100%	10.0	A - Outstanding
drilling			0%	10.0	A - Outstanding
trigger	1	1	100%	10.0	A - Outstanding
chuck	4	0	0%	10.0	A - Outstanding
speed changer	1	0	0%	10.0	A - Outstanding

(Modules)

4.2.2   Common modules

A common module here is a function that is shared by more than one product in a product family or used more than once in a single product. Kota, et al. (2000) present a method to evaluate a platform based on how well the non-value adding components are shared in a platform. We use a similar approach. We start by identifying the functions that distinguish each product variant. For the remainder (the candidate common modules), we ask how interchangeable they are. For this subset of modules that are to be shared, we ask what differences there are from supported variant to supported variant. For the common core modules, we examine any other module that is attached to the common core, and for these modules, we use the scale:

$$Y_{CM} \begin{cases} 10 & \text{Can be swapped into any variant with no changes} \\ 7 & \text{Can be swapped into at least one other variant with no changes} \\ 5 & \text{Requires different mounting hardware to interchange} \\ 3 & \text{Requires interface design changes} \\ 0 & \text{Requires unique interfaces for each variant} \end{cases}$$

Table 4-3 includes the common module metric scores for the cordless drill family. The overall score of 5.9 for the platform is the average of the modules' scores. The drill platform modules are generally shared with only one other product in the family or are only similar, but not exactly the same, which leads to scores of 7 or under.

Table 4-3. Common modules calculation for the cordless drill family.

Modules	Common Or Unique	Platform	Low price	Home use	Value	Heavy duty	Professional
		Sales: 0	100	100	100	100	100
Casing	U						
Contact	C	5	3	5	7	7	5
Switch	C	7	7	7	7	7	5
Motor	C	5	3	7	7	7	3
Transmission	C	6	5	7	7	5	5
Slip Clutch	C	5	3	7	7	5	5
Drilling	U						
Trigger	C	7	7	7	5	7	7
Chuck	C	7	7	7	7	7	5
Speed Changer	C	5	-	-	-	5	5
Average:		5.9	5.0	6.7	6.7	6.3	5.0

4.2.3 Specification variety

Different specification (Ericsson and Erixon, 1999) means that there is more than one variation of a function or a module. For example, a casing with or without a display has two variations on the display module in the platform architecture. In a well-designed architecture, a function with different specifications is isolated into a module. We define the following metric to illustrate this.

$$Y_{diff} = (\text{\# functions with different specification} / \text{\# functions}) \cdot 100 \qquad (3)$$

The score is normalized as shown in Figure 4-3. The score for the product is the average of the modules' normalized scores. The score for the cordless drill family is a perfect 10, reflecting that different specification functions have been well isolated into modules in the drill family.

4.3   After sale

The responsibility of the developer does not end when a product is launched. We developed three metrics to assess how well the products' after launch life has been taken into account as well as how product disposal by the customer is considered. The metrics are: partitioning for reliability and service, as well as environmental friendliness.

4.3.1   Partitioning for reliability

Platform decisions can have a big impact upon reliability and are often critical. The important decisions for our platform concept evaluation purposes are the module selection and partitioning. A basic heuristic we assert is that one should first apportion functions to modules so that each has an equal reliability. Others assert putting all unreliable functions into one module to increase overall reliability but this has no impact on overall reliability, as a rolled reliability calculation will show.

Lending support to our equal partitioning heuristic, within the software domain under reasonable assumptions, Ferdinand (1993) has mathematically proven that to prevent errors the optimal number and size of any module is the square root of n, for n uniformly generated failure modes. Larger modules have excess internal opportunities for errors, and smaller modules make for excess inter-module opportunities for errors. We extrapolate this principle to more general, non-software design and define an architectural metric based upon how far the number of modules is from this ideal as:

$$Y_{rel\ 1} = 10\min\left(\sqrt{n} \, / \# \, modules, \# \, modules / \sqrt{n}\right). \tag{4}$$

While the previous result is very simple and can be applied early, it assumes equal number of failure modes for each module. On the other hand, at some point failure modes and effects analysis (FMEA) will be done, providing an opportunity to relax the uniform failure mode assumption. One can calculate a scaled metric using risk as determined by the risk priority numbers (RPN) of the FMEA. That is,

$$Y_{rel\ 2} = 10 - \left(RPN_{max} - 1\right)^{\frac{1}{3}}. \tag{5}$$

where RPN_{max} is the highest risk priority number of the modules. As the process proceeds and a bill of material can be formed, past data should be used to evaluate the possible failure modes. For the drill platform, we calculate a score of 8.6, indicating the failure-mode-assessed reliability is reasonable, driven mostly by the chuck and contact modules. Notice these two metrics naturally evolve into a rolled reliability calculation as reliability data becomes available.

4.3.2 Partitioning for service

Service and maintenance on a product is easier when the serviceable functions are isolated into modules (Ericsson and Erixon, 1999). Dahmus and Otto (2001) derive rules for partitioning a product based on service costs, using the service cost model of Gershenson, et al. (1999). Using these rules, it can be shown that to minimize service costs one should isolate two modules as two service modules when

$$C_1 / C_2 \le R_2 (1 - R_1) / R_1 (R_2 - 1), \tag{6}$$

where C_i is the total cost of subset i (isolated and not), including material and any labor, and R_i is the reliability of subset i.

To convert this concept into a platform metric, each module should be analyzed based upon whether it should be combined with its neighboring module and whether it should be split into separate modules. A 0-10 scale can be calculated as the fraction of modules that should be changed. For the drill platform we calculate a score of 10, given that the two modules needing service (chuck and switch) are distinct modules. The metric can be further refined using the actual service cost.

4.3.3 Environmental friendliness

Environmental friendliness has become an increasingly important driver of modularity and, e.g., Smith and Duffy (2001) discuss reuse as a reason to modularize. The environmental friendliness of products in general have been extensively studied. We choose to use the AT&T model developed by Graedel and Allenby (1995) to assess the environmental friendliness of a platform, since it is comprehensive in life cycle scope and has limited data requirements as appropriate for early platform concept evaluation.

The analysis requires giving scores from 0 to 4 (0 being worst) to different matter in different phases of a product's life cycle. The scores for each cell are summed to get an overall 0-100 score that we scale down to 0-10 for our purposes.

Table 4-4. Environmental friendliness of the current cordless drill platform.

Life cycle stage	Environmental concern of module i					Total
	Materials	Energy use	Solid	Liquid	Gaseous	
Resource extraction	2	2	2	3	3	12
Product manufacture	3	2	3	4	4	16
Product packing & transport	4	4	4	4	4	20
Product use	3	2	3	4	4	16
Refurbisment, recycling, disposal	2	3	4	4	4	17
Total	14	13	16	19	19	81

Using available information we calculate the overall environmental friendliness score for the drill platform as 8.1 (see Table 4-4), indicating that the product family does not cause excessive environmental harm.

4.4   Organization

The organization metrics assess how well the platform helps organize the development of the products in question. These metrics include ease of assembly, aligned with the organization, make-buy, and testability.

4.4.1   Ease of assembly

The ease of assembly assessment involves a two-part metric where both parts aim to minimize the assembly time. The first metric can be applied very early in the concept design phase with very little information, whereas the second metric can be applied later, when more geometry and assembly sequence information becomes available. Only one metric should be used, depending on the data available.

The first metric was developed by Ericsson and Erixon (1999) who showed that the assembly time of a product is minimized if the product consists of $K\sqrt{n}$ modules, where n is the number of parts and K is between 1 and 1.5, as fit to a particular company's assembly plant data. Fewer modules prevent parallel assembly, and more modules take too long in the final assembly. This can be recursively applied for sub-assemblies and sub-sub-assemblies. In our scaled metric form, this becomes:

$$Y_{assy\ 1} = 10 \min\left(K\sqrt{n}/\#\,\text{modules},\ \#\,\text{modules}/K\sqrt{n}\right). \qquad (7)$$

With this metric, we calculate a perfect score of 10 for the cordless drill family, indicating that the number of modules has been properly determined.

The second assembly metric is to simply use Boothroyd and Dewhurst (2002) Design For Assembly (DFA) design efficiency metric. Boothroyd claims the assembly time for an ideal module is 3 seconds, when the module's parts are minimally complex to assemble and handle. An assembly metric for any module can then be calculated as:

$$Y_{assy\,2} = \frac{3\,n}{\text{assembly time}} \cdot 10, \tag{8}$$

where n is the number of modules in a product. This design assembly efficiency can then be calculated for each module as used in each product in the platform. One should notice that the ideal assembly time in this metric is difficult to achieve, and benchmarking can help obtain the best estimate of a good score.

As an example, we calculated the assembly score using the latter metric for the cordless drill family and got a value of 6.9. The score shows that the drill is not effectively designed from purely a theoretic assembly point of view. The casing and motor modules drove the low score, since both include multiple screwing operations. Further, improved alignment features could aid in the assembly.

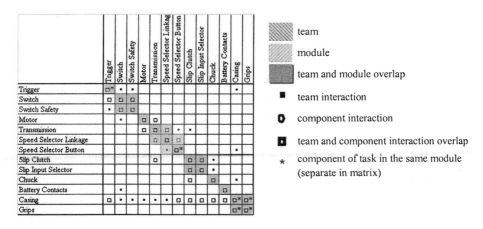

Figure 4-4. Combined component and team interaction DSM.

4.4.2   Alignment with the organization

For short-term platform development to go smoothly, an organizational structure should correspond one-to-one with the product modules. This reduces internal errors during development. More often organizations have existing structures and thereby the organizational and supplier boundaries drive the product partitioning, not vice versa.

In our approach, we extend the Design Structure Matrix (DSM) approach of Sosa, et al. (2003) to define an organization alignment metric. However, we seek a metric for the entire platform and not simply for a single product. We combine a component interaction matrix with a product development team interaction matrix and the degree of overlap of the two then demonstrates the degree of alignment of the architecture with the development organization. With this matrix, we define a metric as follows:

$$Y_{org} = \text{components in teams} \cap \text{components in modules} / \text{components in teams} \cup \text{components in modules} \cdot 10 \qquad (9)$$

In an ideal situation, common modules, i.e., the platform modules, are designed only once and then used in multiple products. This also means that there will be so-called platform teams and tasks. The product specific modules are developed by variant teams. We developed an organizational alignment score for the drill family as 5.4, using the matrix of Figure 4-4. A reorganization of the development teams would improve the score, especially for the transmission and speed changer module development.

4.4.3   Make-buy

Whether to outsource a sub-system or not is an important decision that involves decisions about the ability of the supplier (Fredriksson and Araujo, 2003), level of the technology risk as well as the importance of the sub-system to the company (Fine, et al., 2002). After these decisions are made, the architectural choice is what aspects to include in the outsourced module. In a well-designed architecture, all out-sourced components are separated into modules, distinct from parts made in-house (Ericsson and Erixon, 1999). This allows clear and controlled responsibility for requirements, development, test and quality. We define a scaled metric for any module as:

$$Y^*_{m-b} = (\text{\# outsourced components} / \text{\# components}) \cdot 100. \qquad (10)$$

We calculate this for all modules in the platform, and we take the average of the product scores (see Table 4-5). Since all components in a module should be either out-sourced or made in-house, the ideal values for Y^*_{m-b} are 0 or 100. We normalize the values of Y^*_{m-b} to our scale as shown in Figure 4-3. The overall score is the average of the modules' scores: 9.2. This high score results from most of the drill's components being outsourced and practically no electrical components being made in-house.

Table 4-5. Make-buy calculation for drill.

		# components	# components outsourced	% Outsourced	Score	Grade
	casing	2	0	0%	10.0	A - Outstanding
	battery contacts	1	1	100%	10.0	A - Outstanding
	switch	1	1	100%	10.0	A - Outstanding
	motor	3	1	33%	3.3	D - Poor
Modules	transmission	24	23	96%	9.2	A - Outstanding
	slip clutch drilling	6	6	100%	10.0	A - Outstanding
	trigger	1	0	0%	10.0	A - Outstanding
	chuck	2	2	100%	10.0	A - Outstanding
	speed changer	2	0	0%	10.0	A - Outstanding

4.4.4   Testability

A good module is one that is both easy to test and produces test results that accurately reflect performance in the field. This can be difficult to achieve. For example, the vibration noise of a car engine can be measured on a test stand but that does not correspond to the actual noise the engine produces using the vehicle mounting structure under actual driving conditions. Therefore, it is important to consider testing early in the design process, during architectural concept assessment.

We define a testing metric for any module by considering the flows in an out of the module. Each flow requiring a test measurement is scored according to the following list, and all single module's flow scores are averaged to get a platform score.

$$Y_{test} = \begin{cases} 10 & \text{direct measure of the flow in field conditions} \\ 7 & \text{indirect measure of the flow in field conditions} \\ 5 & \text{on a test stand statistically related to real world} \\ 3 & \text{on a test stand not statistically related to field} \\ 0 & \text{no measurement done} \end{cases}$$

The drill family testing score is 5.4. This low score is a result of many of the components being tested separately.

4.5   Flexibility to change

A product platform will have to adapt to an ever-changing environment. Flexibility is one of the key features of a good platform and the following two metrics help determine whether a platform is flexible or not: unknown isolation, and change flexibility.

4.5.1   Unknown isolation

Unknown and uncertain functionality in a system causes unexpected architectural rigidity, difficulties, and failure. To prepare for these unexpected effects one should isolate all unknowns into modules. To evaluate how well the architecture can accommodate requirement changes, the following steps should be performed for each module interface: (1) Identify the requirements that are subject to change; (2) Identify which flows adjust in range to meet the range of requirement changes; (3) For each interface, identify the design controllable *interface adjustment factor(s)* (IAF) that are free to vary to create the flow range, yet still fit the interface; (4) Determine the domain of the IAF operability; and (5) Ensure the IAF domain provides the product flexibility to cover the range of requirements.

The following example clarifies the concept of an IAF. Suppose we are developing a vehicle with a new hybrid engine versus an internal combustion engine. We know what the new engine should do, and as a result, we design the interfaces of modules adjacent to the internal combustion engine in the old model accordingly to fit both types of engines. However, since the hybrid engine is a new type of engine, we cannot be certain our planned interface will accommodate the hybrid engine. More specifically, the problem could be that the noise level of the new engine together with the rest of the components is unknown. To reduce the uncertainty, we can isolate the unknown module, the engine (whichever type), by, e.g., having scalable damping material at the interfaces to the adjacent modules. We can reduce and add material depending on what is needed as the new engine is installed. This scalability in material is an example of an IAF. Every new uncertain module needs IAFs to permit a seamless introduction of new modules. A metric to address this is:

$$Y_{IU} = \frac{10}{\# \text{ modules}} \sum_{\substack{\text{interfaces} \\ \text{to uncertain} \\ \text{modules}}} \frac{\text{adjustment range}}{\text{required range}}, \qquad (11)$$

which is evaluated only on the modules that have uncertain flows through them. This result can then be scaled by the fraction of modules that are so classified. That is, the drill family has no unknown functions; therefore, the drill platform score is 10.

4.5.2   Change flexibility

A platform must support several product variants at any point in time. Technology evolves, there may be planned upgrades, etc., and the platform must accommodate all of these changes. Rather than requirement-driven upgrades as in the previous metric, this metric considers upgrades to component technology.

Rajan, et al. (2003) developed a metric that is based on possible change scenarios. They identify potential change modes (scenarios) and estimate the readiness of the company to deal with the change as well as flexibility of the product. In addition, they estimate how often or how likely the change is to occur. They combine these four factors into a metric that we use, namely,

$$CPN = \frac{10}{N} \sum_{i=1}^{N} \frac{(R_i + F_i) - O_i + 8}{27} , \tag{12}$$

where:

CPN = Change potential number;

N = Max of (# change modes, #potential effects of change, #causes of change);

R = Readiness (Readiness 1-10, 10 being completely prepared);

F = Flexibility (level of redesign effort 1-10, 10 is no redesign, 1 is new product); and

O = Occurrence (Probability of occurrence, #times in every 10 yrs).

The *CPN* values are calculated for each product, which are then averaged to get the platform score. The drills have potential change modes such as the motor size changing, which causes redesign of the casing. We estimated all change modes, their effects, causes, and the readiness and flexibility of the company, and the occurrence of the changes. We calculate 9.1 as the change flexibility score for the cordless drill family. The high score is mainly due to the low occurrence of changes and low redesign since most parts are available in a range of standard outsourced sizes.

4.6 Complexity

The metrics in the last group aim at reducing the apparent complexity of the architecture and the development. We include five metrics: function and form alignment, interface flexibility, anti-synergy avoidance, 1-DOF adjustment, and limited extremes.

4.6.1 Function and form alignment

A module should have a clear function, and each function should be a clear module. This is supported, for example, by Ulrich and Eppinger (2004) in their definition of modular architecture as a one-to-one mapping from functional elements to the physical components of the product, and Suh's (2001) philosophies of axiomatic design. In a well-designed architecture, a function is not distributed across several modules, since it can require excess design communication and may result in assembly problems. If a function must be shared between modules, it should be driven by large gains on constraints: cost, volume, or mass requirements.

These thoughts have not yet been quantified into a metric that specifically states the degree of function-form independence. The metric we use is two-part. We penalize an architecture that has functions that are performed by more than one module – such as a photocopier with a belt, where the belt is part of both the ink transfer module and the image transfer module; and a module whose parts are in more than one location in the product – such as an elevator, where the control module has a user interface on the cab and a motor controller in the machine room. Both of these architectures require additional coordination between design teams, which may be difficult. The metric for each module is calculated as follows:

$$Y^*{}_{f\&f} = w_1 \cdot \left(\# \text{ parts in more than one module}\right)$$
$$+ w_2 \cdot \left(\# \text{ separate locations with the module's parts}\right), \tag{13}$$

where w_1 is the difficulty weight of having parts in more than one module, and w_2 is the difficultly weight of having a module's parts in separate locations.

As an example, consider the cordless drill family. In general this metric is calculated for each product and the scores are then averaged, using the profit contributions of each product, to get the platform score. We show the result for the cordless drill family in Table 4-6. The scores are normalized according to graph on the right. The score for the cordless drill family architecture is the average of the modules' scores: 9.7, indicating the drill platform is functionally partitioned well. This is intuitive with a teardown of the drill; each function is contained in a module.

Table 4-6. Evaluation of function-form alignment for the drill platform.

	#parts in more than one module	#separate locations with the module's parts	w1	w2	w1*#parts + w2*#locations	Score	Grade	Normalization
casing	0	2	1	1	2	7.0	B - Good	
battery contacts	0	1	1	1	1	10.0	A - Outstanding	
switch	0	1	1	1	1	10.0	A - Outstanding	
motor	0	1	1	1	1	10.0	A - Outstanding	
transmission	0	1	1	1	1	10.0	A - Outstanding	
slip clutch	0	1	1	1	1	10.0	A - Outstanding	
trigger	0	1	1	1	1	10.0	A - Outstanding	
chuck	0	1	1	1	1	10.0	A - Outstanding	
speed changer	0	1	1	1	1	10.0	A - Outstanding	

(Modules)

4.6.2   Interface flexibility

The platform and the variant modules will have to adapt to new, often unexpected, changes. In order for the design changes to be as easy as possible, every module should be isolated so that any changes have only a minimal effect on the other modules. One way of ensuring this is to keep the module interfaces as simple as possible from the redesign point of view. Höltta and Otto (2005) developed a metric for this purpose. They evaluated the redesign effort of different interaction types. Table 4-7 lists the weights of each interaction type and shown a partial example of the drill interfaces.

Table 4-7. Redesign complexity weights and partial drill family structure with weights.

Interaction type	Redesign complexity factor
Material flow	1.1
Acoustic energy	3.8
Electrical energy	1.2
Mech. energy (rot)	1.7
Pneumatic energy	3.2
Thermal energy	2.2
Signal	1.3

The platform redesign complexity score is calculated from the family function structure is defined as follows.

$$Y_{re-d} = 10\min\left(1,2(\#\text{ intermodule interfaces})/\sum_{\text{intermod interf}}\text{redesign complexity weights}\right) \quad (14)$$

We developed this metric with the idea that any interface is sufficiently simplified when it has a complexity factor of 2 or less from Table 4-7. In our example, we calculate a platform score of 10 for the drill family, indicating that the interfaces are all indeed sufficiently simple for design purposes.

4.6.3   Anti-synergy avoidance

As system size grows, stable linear systems often become highly non-linear. This is because the anti-synergies of the multiple flows through each interface combine into an often unexpected response. A way to control these unexpected interactions is to have a design controllable factor at each interface to adjust each flow response going through it. We use, again, the *interface adjustment factor* (IAF). Generally, the problem is that multiple interface flows exists in possibly different spatial directions. Clearly, the optimum is one IAF for each flow, and each IAF ideally influences only one flow and will not impact the remaining interface flows.

Independence is a tall order for an interface design, generally resulting in bulky impractical interfaces. More realistically, when adjusting an IAF to match modules to different product variants or applications, all flows that are affected by the IAF should as a set improve or worsen monotonically as a group. The module is then much simpler to adjust to applications. We find this concept applies equally to software products. Bass, et al. (2003) mention that in order for software to be modifiable the functions should be designed to avoid a ripple effect where a change indirectly affects another module.

We define our metric based on the interactions between the input and output flows of a module and their IAFs. We use a matrix representation (see Figure 4-5), where all the flows in and out of a module are represented with rows and IAFs are represented as columns, and matrix entries are (+)'s and (-)'s to indicate the sign of the change of the flow with an increase in the IAF. As an IAF is changed, the change direction of each impacted flow should be same. That is, for any IAF generally, an ideal design is one where the sign of the sensitivity of every flow is the same.

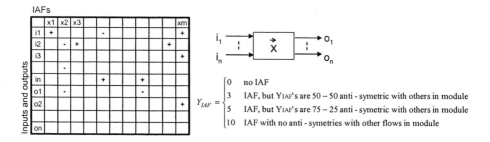

Figure 4-5. A general adjustment matrix.

Therefore, ideally every column has either all (+) or all (-) entries (in addition to the blank entries for the flows an IAF does not impact), as shown in Figure 4-5. The score for the summary platform scorecard is the average of the module scores.

The cordless drill platform does not score high on this metric, as the platform was more focused on cost targets. None of the products in the cordless drill family have IAFs available to permit simplified changes or modifications. All power carrying modules are outsourced and if there is a change, every adjacent component must be changed. Therefore, the anti-synergy score for the cordless drill platform is 0.

4.6.4   One DOF adjustment

This metric is only relevant to architectures of products that require service or swap-outs of modules. This is common in long service life mechanical systems with field-serviced modules. As a part of the service procedures, sometimes the serviced or replaced module must be adjusted to fit well – it must be inserted, measured, and fine-adjusted for a flow or performance metric to be placed on a desired target. Examples include kinematic positioning, current flows with resistor adjustment, fluid flow dampers, acoustic noise or frequency adjustments.

For any serviced module, the ideal number of adjustments is zero, the module can simple be inserted and the insertion automatically aligns the module, followed by an attaching operation such as a snap fit or non-aligning attachment screw. For modules that require precise adjustment, a good interface should have only one adjustable degree of freedom (DOF). More DOFs make the adjustment a difficult and iterative process. In addition, a good interface should always have an indicator directly visible while making the adjustment. For example a current flow gauge should be visible next to a variable resistor, or a bubble indicator should be visible next to a screw positioner.

The score is calculated for each module in the platform that requires periodic service replacements. For any serviced replacement module, the metric for each module is calculated as follows:

$$
Y_{DOF} = \begin{cases} 10 & \text{No adjustments} \\ 7 & \text{1 DOF to align with 1 adjustment with 1 indicator} \\ 1 & \text{2 DOF to align with 2 sequenced adjustments with indicators} \\ 0 & \text{Anything else} \end{cases}
$$

All service replacement modules in the platform are averaged to get the overall platform score. The cordless drill family has only one replaceable part, the battery, related to the battery contact module, which has no adjustments required, and so the platform scores a 10.

4.6.5 Limited extremes

Requirements that are difficult to meet cause problems. An architecture driven by a few extreme requirements forces poor performance on the other requirements. Thus one should limit extreme requirements throughout the development process, and validate them early and continuously.

For this metric we compare the new requirements of the architecture under development to the requirements of the current model in the market. The metric for a product variant in the portfolio is calculated as follows:

$$
Y_{ex} = \min\left(\sum_i \left(10 - \left|\frac{\text{new req}_i - \text{standard req}_i}{\text{standard req}_i}\right| \cdot \text{difficulty}_i\right)\right), \tag{15}
$$

where new req$_i$ is the requirement level i for the product under development; standard req$_i$ is the requirement level for requirement i that is known to be achievable; and difficulty$_i$ is the difficulty of achieving the new requirement *i* (scale 1-10). The value for the platform is the minimum of the supported-product-variant scores: here 6.6, driven by the difficulty in meeting the noise level requirements.

5.   PLATFORM ANALYSIS SUMMARY

Table 4-8 summarizes the platform assessment of all three alternative platforms. The individual metric scores can be summed (weighted sum, see Table 4-1) to obtain first the sub-category scores and then the overall platform scores. The current cordless drill family platform receives a score of 8.0 indicating that the platform is fairly well designed. It is better than the two other alternatives A and B that received total platform scores 7.9 and 7.3 respectively. The overall score, however, is a rough estimate and the difference between 8.0 for the current platform and 7.9 for the alternative A may not be significant. The true value is in the sub-category scores.

The current platform received the highest score in flexibility. This is primarily due to the fact that the drill market is mature and no significant changes are expected. The maturity of the market is accounted for by using a low weight in related metrics. The current platform received the lowest score in category organization alignment. The assembly score is also low, but this is typical.

The alternative A scored similarly to the current platform. The rank order of category scores for platform alternative A is the same as for the current platform but the scores are inferior. The difference in scores, however, is an indicator that our tool can distinguish the contributing factors to two very similar architectures. The current and alternative A architecture differ by only one module. While the bottom line number is of interest, of more importance is the sensitivity contribution to the difference.

The more integral alternative B received different scores. It performed significantly worse in flexibility and variety. This was expected, since the more integral design has larger modules that are more difficult to change if needed. Alternative B received a higher score than the other two platform alternatives in sub-category customer satisfaction. This is because many customer requirements such as weight and performance can be better optimized with an integral design. Alternative B was assessed to have a low score but not substantially lower. This suggests a refined study needs to be subsequently completed, to better quantify the performance difference and the future derivative product option value.

Table 4-8. Platform evaluation for current drill platform and for alternatives A and B.

	CURRENT	A	B
Overall Multi-Criteria Platform Assessment	Score	Score	Score
	8.0	7.9	7.3
Platform Scorecard	Score	Score	Score
Portfolio Customer Satisfaction	7.2	7.2	7.1
Product Variety	8.4	8.4	7.8
After Sale Support	8.9	8.5	7.9
Organizational Alignment	6.7	6.6	5.6
Upgrade Flexibility	9.6	9.5	8.2
Development Complexity	7.2	7.2	7.2
	8.0	7.9	7.3
Portfolio Customer Satisfaction Scorecard	Score	Score	Score
Cost - Worth Distribution	8.0	7.6	7.2
Portfolio Customer Needs	6.3	6.7	7.0
	7.2	7.2	7.1
Product Variety Scorecard	Score	Score	Score
Planned Upgrade Carryover	9.4	9.3	7.9
Common Modules	5.9	5.9	5.4
Specification Variety	10.0	10.0	10.0
	8.4	8.4	7.8
After Sale Support Scorecard	Score	Score	Score
Partitioning for Reliability	8.6	8.5	7.9
Partitioning for Service	10.0	10.0	10.0
Environmental friendliness	8.1	8.1	8.1
	8.9	8.9	8.7
Organizational Alignment Scorecard	Score	Score	Score
Ease of Assembly	6.9	7.4	6.7
Aligned with the Organization	5.4	4.6	1.5
Make-Buy	9.2	8.8	8.6
Testability	5.4	5.5	5.7
	6.7	6.6	5.6
Upgrade Flexibility Scorecard	Score	Score	Score
Unknown Isolation	10.0	10.0	10.0
Change Flexibility	9.1	9.0	6.4
	9.6	9.5	8.2
Development Complexity Scorecard	Score	Score	Score
Function and Form Alignment	9.7	9.6	9.5
Interface Flexibility	10.0	10.0	10.0
Anti-Synergy Avoidance	0.0	0.0	0.0
1 DOF Adjustments	10.0	10.0	10.0
Limited Extremes	6.6	6.6	6.6
	7.2	7.2	7.2

6. CONCLUSIONS

We introduced an assessment tool to technically evaluate a product platform, consisting of a set of metrics. While reasonably comprehensive of what is required for a technical platform evaluation, we find the available metrics non-uniform in their formation and the quality of use. Some metrics are very well researched, validated, and understood. Others have no research available at all. Future research is needed on the importance of each metric for different industries. Further work should also focus on other possible metrics such as intellectual property of a platform.

The multi-criteria nature of the platform tool should enable unbiased evaluation of a platform. As with any tool, a designer might try to modify the metric to give desired results, but we hope the multiple criteria help in preventing the massaging of the results.

Ideally, a simplified analysis would be available at the very early concept phase to evaluate platform alternatives. This screening analysis could then be refined with more detailed analyses in the early development phase, all within this same framework. For example, in our approach the reliability and assemblability analyses have simple metrics to evaluate a platform, based simply on the number of functions and modules. Both also have detailed rollup assessments when more information becomes available on geometry and failure modes. The latter analysis is inappropriate at the early concept phase, while the former is inappropriate for the preliminary design phase. Yet, we have a comprehensive platform evaluation tool that can be used to evaluate different product architecture alternatives during development.

Chapter 5

PLATFORM LEVERAGING STRATEGIES AND MARKET SEGMENTATION

Tucker J. Marion and Timothy W. Simpson
The Harold and Inge Marcus Department of Industrial and Manufacturing Engineering, The Pennsylvania State University, University Park, PA 16802

1. INTRODUCTION

As firms begin to adopt product family and product platform principles in the beginning stages of the product development process, an essential component is to have a cohesive market segmentation strategy for the product family. Managing innovation throughout the product family can be achieved by leveraging three elements within the organization: (1) the market applications for the technology, (2) the company's product platforms, (3) and the common technical and organization building blocks that form the basis of the product platform (Meyer and Lehnerd, 1997). Implementing this strategy can allow the organization to attack different market segments and gain market share while benefiting from the cost advantage of using product families and sharing key common technological modules. This chapter builds upon the product platform planning methods described in Chapter 2 and explores the history of the market segmentation of product platforms. We describe the principles and tools behind market segmentation and include several examples showing how companies have used this process.

While the science of product platforms and product families is relatively new, the benefits of adopting common platform elements to different markets is not. A good historical example of a company specifically designing a product family to service many varied target market segments is General Motors (GM) and their Chevrolet Corvair. Future public relations and opinion aside, the development of the vehicle pertaining to platform techniques and segmentation is valuable.

In the mid 1950's, sales of the rear engine Volkswagon Beetle began to escalate rapidly. Sales increased to the point where America's three largest automakers—GM, Ford, and Chrysler—noticed a burgeoning new market segment, namely, inexpensive compact cars. Up until then, full size cars and trucks were the mainstream vehicle on America's highways. GM decided to enter the market by 1960, with a 'clean sheet' design that would form the basis of an entire line of cars and trucks.

As discussed on the Corvair website (http://www.corvaircorsa.com/), the Corvair project was initiated in 1956, under the direction of GM's Chief Engineer, Edward Cole. The car and platform were unlike any other GM vehicle designed to date. The platform consisted of a rear mounted air-cooled flat six engine, a 108-inch or 95-inch wheelbase with unibody construction, and four-wheel independent suspension with rear trailing arms. The resulting vehicle could be produced in a wide variety of body styles, as shown in Figure 5-1.

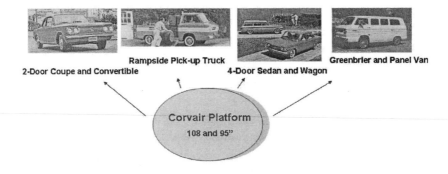

Figure 5-1. Corvair platform derivatives.

It is interesting to note that by 1969, each of the different market segments captured by the Corvair platform were replaced by individual products that were targeted specifically at each segment (i.e., the Camaro replaced the 2-Door Coupe and Convertible, and the Chevrolet Van replaced the Greenbrier). As customers' needs changed, GM thought it could best meet market requirements by diverging from a single platform, which is a challenge that many companies today still face. Strategies for defining the product portfolio (i.e., the number of platform architectures and the number of products derived from each) are discussed in Chapter 12 in this book.

Another more successful historical implementation of product families and product platforms occurred in the 1970's and formed the basis for the segmentation principles outlined in this chapter. According to Meyer and

Lehnder (1997), Black & Decker's product line by the early 1970's was broad and deep, consisting of eighteen power tool groups distributing 122 different models ranging from drills to sanders. By this time, Black & Decker's (B&D) models had evolved into a collection of uncoordinated designs, materials, and technologies. Power tools relied on thirty motors fitting into sixty motor housings. Additionally, B&D relied on 104 different armatures, each requiring their own tooling.

B&D's approach to product development was one of filling holes, i.e., developing single products that meet very specific needs. Little consideration was given to commonality and economies of scale, and this development strategy worked well for a period of time. However, looming government regulations, domestic labor rates, and foreign competition soon forced B&D to reevaluate its entire product line. B&D embarked on an aggressive campaign that sought to reinvent its entire product line with a clear mission: (1) redesign all consumer power tools at the same time, (2) redesign manufacturing simultaneously, and (3) offer products that would meet the new government standards (i.e., double insulation).

B&D, aided by full upper management support, redesigned their product line around a common scalable electric motor. The motor was scalable along the stack length, from 0.8 to 1.75" but was common with its axial diameter, allowing for common housing diameters. These innovations also led to substantial improvements in manufacturing efficiency. The program was completed in three years, for a cost of $17M dollars. Now that a common architecture was developed, new product derivatives could be easily developed from existing components. Cycle times for new products were substantially reduced, with the rate of new product introductions averaging one per week. By 1976, manufacturing savings alone were close to $5M. B&D immediately passed along the cost efficiencies to the consumer, by aggressively cutting costs while still maintaining an internal gross margin of 50%. Market share soared, while competitors fled the market. The case of B&D exemplifies how firms can reinvent themselves through the adoption of a coherent platform strategy. B&D was able to stabilize and maximize its core business, while forming the basis for new market entrants.

2. PLATFORM ROADMAP

Developing cohesive and flexible product architecture is a necessity in successfully implementing a platform strategy. As explained in Chapter 2, the platform should form the basis of an internal product roadmap that outlines future capabilities and functionality. As part of the overall development strategy, the platform should be capable of accommodating new technologies and variations as to develop future derivations with little incremental investment. The ability of a platform to be modified easily and adapted to alternative markets is termed *platform leveraging* (Meyer and Lehnerd, 1997). As shown in Figure 5-2, the basis for the platform is not the platform itself but the foundation of building blocks of R&D and common components and technology. According to Meyer and Lehnerd (1997), "greater power in product development can be achieved if these building blocks are leveraged across the product platforms of different product lines". Building blocks are not only physical components but include core technologies, processes, manufacturing resources, and marketing attributes such as industrial design and branding. The 'building block' foundation that results from a platform and product strategy provides the basis for the core vision of the company. Figure 5-2 shows the product platform as one of the pillars of overall corporate vision.

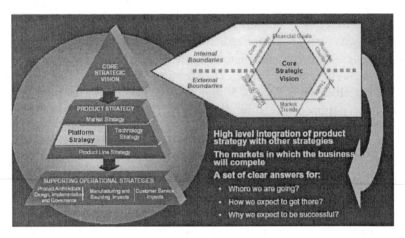

Figure 5-2. Corporate strategic vision (Gordon, 2004).

As noted in Chapter 1, two approaches to platform implementation can be taken, a *top-down approach* or a *bottom-up approach*. In a top-down approach, the company strategically manages and develops a family of products based on a product platform and its modules and/or scale-based derivatives. In a bottom-up approach, the company redesigns and consolidates a group of distinct products by standardizing components to improve economies of scale and reduce inventory. The top-down approach, or *proactive platforming*, is preferred because it develops a cohesive roadmap for commercializing the platform and its offshoots. Bottom-up platforming, or reactive redesign, does optimize current product offerings by streamlining the number of components, but it is not a strategic plan.

Ultimately, the core vision of the company is not to produce one product, but leverage a common platform or key components to other market niches

(Meyer, 1997). A market niche can be described as a particular market area where price and performance requirements are unique. Typically, the price and performance targets are specifications set by the consumer and/or market leaders. For instance, in the MP3 portable player market, a niche can be considered players from $150 - $300 dollars that are small, easy to use, and hold more than a few hundred songs. The current market leader is the Apple iPod, which markets two products in that range, namely, the 20GB iPod for $299.00 (shown in Figure 5-3.) and the Mini iPod for $199.00 to $249.00, depending on memory capacity. Below the 'standard' MP3 player market is a lower cost, lower capacity segment that demands even further simplicity and size constraints. Currently Apple offers the Shuffle line priced between $99.00 and $149.99 (http://www.apple.com/). All told, Apple currently has three product lines (iPod, Mini, and Shuffle) that target three different market segments (low cost, standard, high cost/high feature).

Figure 5-3. iPod.

Overall product strategy is derived from the platform, as the platform should be able to be tailored to meet different market segments and performance targets. The platform draws from supporting elements such as common subsystems and component bases. This allows the platform to be designed for a particular market segment and then be easily modified for different segments and/or higher level tiers within the same segment. If the company is pursuing a redesign of its products in an attempt to achieve a more proactive vision, a product family roadmap should be developed (see Figure 5-4). The roadmap is used to outline the evolution of the product family, platform, and derivatives over time (Wheelright and Sasser, 1989). The roadmap outlines derivative products off of the original platform, and develops an internal guideline for future variations. Ultimately, a new product platform will be developed that reuses the best elements of the original core platform. This continuous renewal allows the company to reach new markets with increasing cost competitiveness and technology performance (Meyer and Lehnerd, 1997). Figure 5-4 shows a product family roadmap over time based on three generations of the product platform.

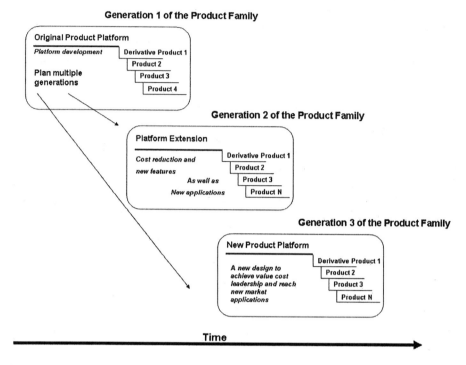

Figure 5-4. Product platform and family roadmap (Meyer and Lehnerd, 1997).

3.   MARKET SEGMENTATION: IMPLEMENTATION AND EXAMPLES

The product platform and product family roadmap should then be transferred into a format that takes into account three major elements of the corporation. These are: (1) the market applications and different segments the platform will target, (2) the company's product platforms, and (3) the common subsystems, components, and development organization (Meyer and Lehnerd, 1997). In terms of market applications, these can be defined as areas and niches in which the product will be sold, to whom, and at what price points. To cover these market areas, the company's product platform(s) are then used to cover the different target segments in the most optimal way possible. Finally, the organization is then aligned to support the product platform with common subsystems, components, and the development organization. Figure 5-5 shows a schematic of the market segmentation grid from Meyer and Lehnerd (1997).

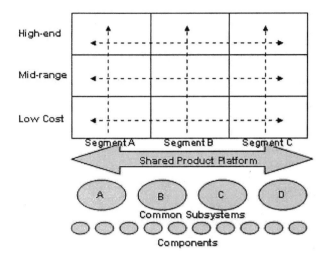

Figure 5-5. Market segmentation grid (Meyer and Lehnerd, 1997).

In drawing from the Apple iPod example in the previous section, there are at least three market tiers in MP3 players (low cost/low feature set, standard feature set, and high-end/high feature set). To cover the high-end market niche, Apple offers increased memory and photo editions of the iPod. As such, if we were to look at the MP3 market, the following tiers could be laid out as shown in Table 5-1.

Table 5-1. MP3 player market chart.

Tier	Features	Cost	Market Leader
High-end	Color Screen, Large amounts of photos and music, portable	$300.00 and up	iPod Photo
Standard	1000's of song storage, easy to use, portable	$199.00 to $299.00	iPod and Mini iPod
Low-end	Simple, inexpensive, 100's of songs	$0.00 to $149.00	iPod Shuffle

For Apple, its market segmentation grid is shown in Figure 5-6. In this example, the vertical axis tiers are graduated by price, and the horizontal axis is comprised of three different segments based on type of memory storage and type of files stored (music and/or photo). The individual products in the iPod family are plotted within the respective market segment in the grid based on the scales used for the vertical and horizontal axes.

If a competitor were to design a platform to compete in the hard-drive-based MP3 segment similar to how Apple introduced the original iPod and iPod Mini, this would be called *vertical scaling* (Meyer, 1997). Drawing from common components and technology, the new market entrant might utilize the same firmware that controls song I/O and playback as the platform base technology. Common components to all three products might be an ASIC chip, scalable PCB boards, and certain housing components. The first to market might be the high-end player, with the other down-market products following shortly thereafter. This is called downward scaling as it involves removal of key features and components. This is the strategy Apple employed with its iPods, first introducing the base iPod with different song capacities then downward scaling the platform into the iPod Mini. Firmware, iTunes software, and control architecture remained the same, while internal components, LCD's and housings were scaled downward in size.

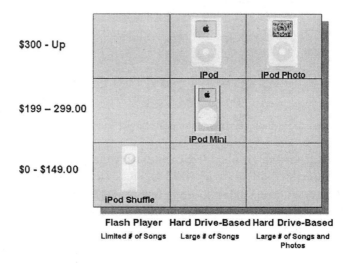

Figure 5-6. Apple iPod market segmentation grid (Photos courtesy of Apple.com)

One risk of this approach is that the lowest tier products might be more costly than required for that particular segment, resulting in lower margins. This is a struggle in the auto industry, as downward scaling automobile platforms can result in an expensive car in lower priced tiers. An example of this is the 2005 Ford Mustang. Ford based the new vehicle on the expensive Lincoln LS/DEW-98 platform, but they had to remove its independent rear suspension to make the car meet its price point (Quiroga, 2005). Higher priced models of the Mustang may eventually be sold with independent suspension, with the suspension being designed as a plug-in differentiation module. According to Meyer (1997), meeting the needs of high-end segments is perceived as the greatest challenge for engineers. As such, it can be difficult to design a low cost platform that can be vertically scaled up to meet the needs of the top tier. A weak base platform can wreak havoc on product derivatives sold at higher-end tiers. Again looking at the automotive industry, GM introduced the compact J-car to the North American market in 1982. Unfortunately, vertical scaling turned into badge engineering, as the low price Chevrolet Cavalier was turned into the higher-priced Cadillac

Cimarron. These vehicles were essentially the same except for ornamental differences (Wilkins, 2005). The car was not a sales success and remains a critical failure of how *not* to vertically scale a platform.

Many companies successfully take a different approach to platform leveraging. For example, in the sub-$30,000 sedan market, there are many brands with a variety of different models and sizes. Toyota uses the same platform to cover both mid-size and large sedans. Their Camry platform not only forms the basis for that model but also supports the full-size Avalon. In doing this, they use common modules, frames, engines, and other components as well as common manufacturing sites. This type of leveraging is called *horizontal leveraging* (Meyer, 1997). According to Meyer, "the benefit of the horizontal leveraging strategy is that a company introduces streams of new products across a series of related customer groups without having to 'reinvent the wheel' for each." A potential downside of horizontal leveraging, which is common to vertical leveraging, is that a low quality or poorly designed platform can negatively impact performance in multiple market segments. An example from the automotive industry occurred in the 1980's with the Chrysler Corporation K-car platform. Models spanned from basic sedans to sports cars, minivans, compact cars, and wagons. While initially successful, the horizontal proliferation (combined with badge engineering) of the models ultimately negatively affected the brand image of Chrysler, Dodge, and Plymouth (Lutz, 1998).

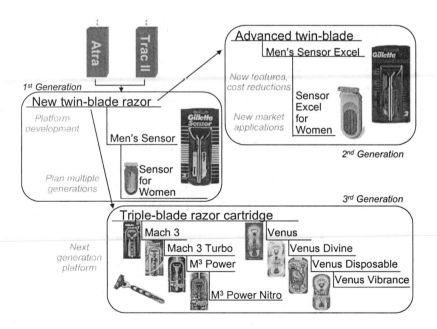

Figure 5-7. Gillette razor catridge platform roadmap.

If done properly, however, horizontal leveraging can be a very powerful concept. A successful example can be found in Gillette's use of the razor cartridge as the platform that they leverage across male and female market segments. Gillette has repeatedly used the razor cartridges (i.e., Atra, Trac II, Sensor, Sensor Excel) as a major leverage point by designing an effective shaving system while lowering manufacturing costs through a highly automated production process (Meyer, 1997). Gillette's newest razor cartridge, the triple-bladed Mach 3, has been developed using a similar leveraging strategy without which they would have had considerable difficulty justifying the $750M in capital investment, the company's largest ever, and billions of dollars used for advertising (Sella, 1998). Gillette is now flooding the market with derivative products based on the Mach 3 as shown in Figure 5-7 using the platform roadmap format from Figure 5-4.

The power of product platforms is magnified when both vertical and horizontal leveraging strategies are combined. This platform leveraging method is called the *beachhead approach* (Meyer, 1997) and is shown in Figure 5-8 along with the vertical and horizontal leveraging strategies. In the beachhead approach, an initial platform is developed, and is horizontally leveraged to other market segments. Then, features and performance are enhanced and the product platform is moved upward into a higher tier. This leveraging strategy is the most effective as well as the most difficult.

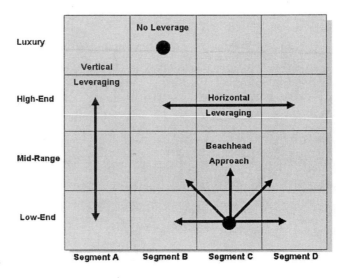

Figure 5-8. Platform leveraging strategies; adapted from (de Weck and Suh, 2003).

There are many good current examples in a variety of industries, but as shown in previous leveraging examples, the use of a beachhead strategy is perhaps most readily apparent in the auto industry. As stated in Chapter 1, competition is severe, product development and life cycles are decreasing, and the market is being divided into smaller and smaller niches. Hence, companies like GM need to develop common platforms and components to remain competitive and regain lost market share. A primary example of a dedicated beachhead approach is GM's new Sigma platform. This is an all-new rear wheel drive/all wheel drive platform developed for Cadillac. The platform was introduced on the CTS in 2002, and has since been leveraged both vertically and horizontally. GM's Sigma Platform market segmentation grid is shown in Figure 5-9.

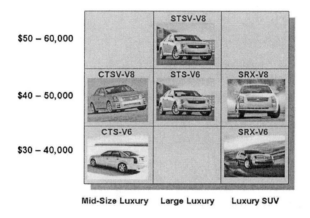

Figure 5-9. GM's Sigma platform for 2005 Cadillac models.

GM has received critical and sales acclaim for its new models, reversing the sales slide it has seen in its Cadillac brand since the early 1980's. It should be noted that the Sigma platform was developed as a high-end platform, even though there are different tiers within the luxury segment. This reduces the risk of stretching the capabilities of the product platform too far, from very low cost into very high cost tiers.

In taking another example from the auto industry, Nissan has approached its high-end vehicles in a similar manner using a high-end product platform. Their FM platform (front amidships) is a front engine, rear and all wheel drive platform that was introduced as the G35 then expanded upwards and across segments. Very similar to Cadillac, the FM platform has been not only a sales success but also a critical success in the marketplace.

There are rewards for a beachhead strategy as well as risks. In the 1990's Volkswagen (VW) developed its popular A-platform, which shared common floor/chassis modules, drive train, and internal cockpit modules among a wide variety of products sold under the Volkswagen, Audi, Seat, and Skoda brands (Wilhelm, 1997). Figure 5-11 shows VW's horizontal leveraging strategy for the A-platform. Additionally, the platform was used to go up market with the Audi TT series. Nineteen models are either in production or planned, and VW estimates development and cost savings of $1.5 Billion/year through the use of product platforms (Bremmer, 1999).

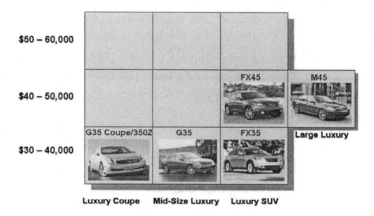

Figure 5-10. Nissan's 2005 FM platform.

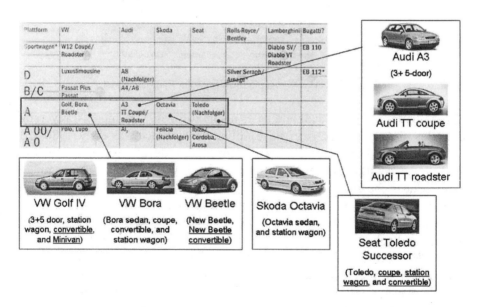

Figure 5-11. Volkswagen A-platform derivatives (Shaukat, 2001).

An issue Volkswagen has seen is an effect on brand worth. Is an Audi worth substantially more in the marketplace than a VW constructed from the same architecture? Or if they are similar in price, is the market driven towards the brand with the greater cache? There has been considerable confusion among the brands due to the high level of commonality, and VW has pledged to start by overhauling its Audi brand in Europe (Miller, 1999; 2002). More recently, the new Phaeton, based on the Audi A8 platform, has been a sales disappointment worldwide (Kiley, 2005). Additionally, there has been a substantial sales slide in sales of all Volkswagen models in the last several years, in part due to products platforms that are similar in price/performance to offerings such as Audi, thus muddling brand distinction. A new product platform must take these marketing and brand issues into account when developing the product roadmap and market segmentation grid.

In the aerospace industry, Lockheed Martin has implemented a successful platform strategy for its family of military transports. With the next series of Block Updates (major aircraft and avionics revisions), software and avionics (major subsystems) will become common for Lockheed's three platforms, the C-130, C-5, and C-27. The common subsystems will form the basis for a new airframe that will eventually replace the C-130. The C-130 replacement airframe platform will be similar to today's aircraft in that is can be easily reconfigured to fill a wide variety of different roles. Today's C-130 fills roles as diverse as Search and Rescue to Gunship applications, all leveraging a common airframe and support systems. Different software and weapons suites can be applied to the airframe to move up the vertical axis (O'Banion, 2004).

Another industry example is Harley-Davidson, which has five vehicle platforms (Sportster, Touring, Dynaglide, Softail, and Custom Vehicles) that compete in different market segments. According to Oosterwal (2004), all five platforms share common subsystems, such as engines. Harley has three main engine architectures, which can be varied by displacement. The five platforms and three engines result in 27 models. This allows Harley to maximize investment and introduce new models with substantially reduced development costs. In addition to the 27 core models, Harley offers consumers near limitless options for customizing their purchases. This mass customization allows consumers to have unique paint, mirrors, gas caps, etc. installed on their bikes. These components are low in cost to develop, but they add high perceived value and increase the selling price of the bike. Customization of the product platform pushes the model into the 'high-end' market tier. In the last fifteen years, Harley Davidson has been able to dramatically increase the number of models, associated sales revenue, and profit due to the implementation of a coherent platform strategy.

Modular platforming techniques are also finding new applications in service industries. According to Schlueter (2004), Cingular Wireless is implementing a platform approach to their Pre-paid and 'Take Charge' cellular plans. These can leverage common technology subsystems such as wireless technology and be applied to different demographic segments.

Additionally, service plans and customer service can be easily reconfigured to meet different market segment needs. These are being used to vertically leverage different customer tiers while still maximizing the Cingular brand.

Market Segmentation and platform leveraging is not only beneficial to large companies as small firms and start-ups can benefit greatly from adopting a platform strategy. As detailed in Chapter 19, a start-up company called Innovation Factory designed a family of ice scrapers around a common blade platform. Based on market analysis, the company segmented the market into hand-held scrapers, mid-size scrapers, and large scrapers. Vertically, the tiers were driven by retail price-points. Figure 5-12 details the Innovation Factory market segmentation grid and the products that were introduced into the respective markets.

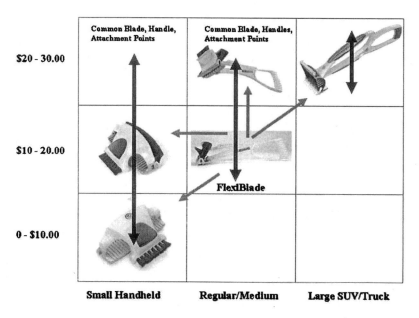

Figure 5-12. Innovation Factory market segmentation grid.

The company was able to introduce two distinct product lines based upon one common blade. A beachhead approach was used, beginning with the IceDozer centered at the $10 - $20.00 price point. Next, the common blade was leveraged horizontally to create the hand-held version. Since the hand-held versions were designed to be modular, this allowed different combinations of parts to be used to vertically leverage the product. This saved the company tens of thousands of dollars in development resources and manufacturing tooling, which it otherwise could not have afforded as a small, start-up company. Additionally, since modular components were designed at the same time, several distinct products could be introduced simultaneously, instantly increasing the number of saleable SKU's, which facilitate discussions with large retailers such as Lowe's. More details on the development of the ice scraper family can be found in Chapter 19.

4. DEVELOPING A MARKET SEGMENTATION GRID

In order to develop successful new product platforms and services, corporations must accurately listen to and identify the needs and expectations of each market segment and tier. This information, combined with the product platform roadmap, can help develop the market segmentation grid. In looking at this competitive landscape, each segment needs to be mapped, and Gordon (2004) suggests asking the following questions for each market segment:

- What is the significance of this segment?
- What are the key products?
- What are their volumes, revenue, and profits?
- What is the outlook for the next 5 years?
- What must the Company do to enter, sustain, and grow in the segment?

In the following sections, a step-by-step method is described that to help a company develop a market segmentation grid.

4.1 Implementing market segmentation

Each market segment needs to be mapped by price and performance tiers, looking at not only current market entrants but also future competitors. Once the axes for the grid are established, the grid can be used by the development team in support of the guiding platform roadmap (see Figure 5-4). A rough sketch of the grid can be developed using employee knowledge and quick Internet searches.

4.2 Detailed key questions and answers

As a subset of market segmentation, the following questions need to be answered: Who are the key players in the industry, and where are they going? What are their margins, and volumes? The team needs to determine the overall growth and potential of each segment. Is it worth the investment? Or, have we discovered a new niche?

4.3   Product platform investigation

The development team must define what the current company product platforms are and define common components and subsystems. According to Meyer, 'defining a product platform for a particular business is not always easy—many companies fall victim to focusing their major efforts in traditional market areas that either plateau or are in decline in terms of new growth.' If this is the case, doing this exercise internally will be enlightening. In looking at the product roadmap (see Figure 5-4), this can be overlaid onto the market segmentation grid to develop a plan for horizontal, vertical, or beachhead leveraging. If no current product platform or product family can be used to cover multiple segments, a totally new product platform should be explored as discussed next.

4.4   New product platform

As with B&D, taking a 'clean sheet' view of the current market and future segments can prove extremely valuable. The product team should work toward defining a product platform that will maximize segment coverage and attack new markets (Meyer, 1997). The product platform specifications and product development process should be developed in conjunction with the product platform roadmap and the market segmentation grid to maximize horizontal and vertical leveraging. Product platform planning during the 'front-end' of the product development process is described in more detail in other chapters in Part I of the book.

4.5   Customer needs for a new product platform

The overriding goal of product line renewal via a new product platform is to bring excitement to the market (Meyer, 1997). The company must develop a '360 degree' view of the customer, and completely understand their needs, requirements, and usage patterns. This 'Voice of the Customer' (VOC) approach has been effective in helping many companies guide the development of product platform specifications and features.

A successful application of using the VOC has been with Case-New Holland (CNH), a world leader in agricultural equipment such as tractors. In developing a new cross segment platform, CNH embarked on an extensive program of interviewing potential customers in each market. In person, one-on-one interviews were held, literally 'in the field' in many cases, to gauge customer feedback on issues ranging from cabin ergonomics to steering mechanisms. Responses were documented, analyzed, and used in the conceptual development process to formulate product solutions. The VOC is an integral part of their product development process, which is termed Customer Driven Product Definition (CDPD) (Kaiser, 2004). In Figure 5-13 is a graphical representation of the CNH's CDPD. Note that the VOC process is implemented during Product Program and Platform Planning.

The result of developing the product planning roadmap and the market segmentation grid should be a well-defined corporate vision for product family and product platform implementation. The strategy should be a multi-year effort that is well supported by upper management, and begins at the building blocks of the organization.

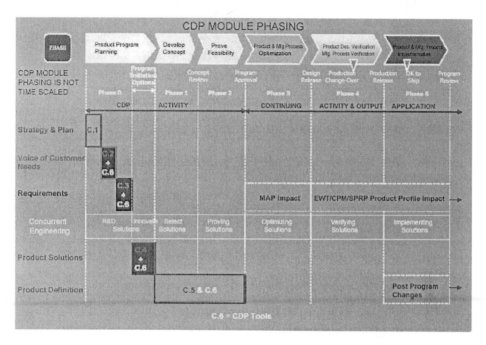

Figure 5-13. CNH Customer Drive Product Definition Process; adapted from (Kaiser, 2005).

5. SUMMARY

Market pressures are forcing companies to change their product development organizations from the front-end through manufacturing. This encompasses the entire development strategy, from market and customer research to supply chain management. Integral to this change among firms in a wide variety of industries is the adoption of platform management and development architectures. Successful traits among industry leaders are the formation of cross-functional development teams, strong management support, and common platform architectures that maximize the sharing of subsystems and components, and the use of market segmentation. Market segmentation is a valuable tool that can be used to identify ways the development team can implement and leverage a product platform. By utilizing the building blocks and core competencies of the corporation, a platform leveraging strategy can be developed to not only exceed the demands of current segments but also attack new and untapped niches. By applying the principles of market segmentation, firms like Apple, Black & Decker, and GM have developed successful, innovative products at lower costs, while increasing revenue and profits. Holistic platform and product development is a key requirement for success in the 21st Century; companies that delay in adopting these principles will be seriously disadvantaged in trying to grow in the global economy. Leveraging your platform through the use of market segmentation is one way to get there.

6.   ACKNOWLEDGMENTS

This work was supported by the National Science Foundation under CAREER Grant No. DMI-0133923 and ITR Grant No IIS-0325402. Any opinions, findings, and conclusions or recommendations presented in this chapter are those of the authors and do not necessarily reflect the views of the National Science Foundation.

Chapter 6

PRODUCT FAMILY POSITIONING

Jianxin (Roger) Jiao and Yiyang Zhang
School of Mechanical and Aerospace Engineering, Nanyang Technological University, Singapore 639798

1. INTRODUCTION

Due to the development of modern technologies and global manufacturing, it becomes harder and harder for companies to distinguish themselves from their competitors. To keep the competitive advantage, the companies intend to provide a variety of products by differentiating their product lines with the belief that product variety may stimulate sales and thus conduce to revenue (Ho and Tang, 1998). A large product variety does improve sales by providing the customers more choices. However, companies with expanding products face with the challenges of controlling costs. The costs exponentially increase with the variety growth. Further, high variety will result in the proliferation of products and processes and in turn inefficiencies in manufacturing (Child, et al., 1991). Mass customization aims at satisfying individual customer needs with the efficiency of mass production (Pine, 1993a). Customization emphasizes the uniqueness of, and the differences among, products (Jiao and Tseng, 2000). To optimize the product variety, a company must assess the level of variety at which customers will still find the company's offerings attractive and the level of complexity that will keep the costs low (Jiao, et al., 1998). Developing product families has been recognized as a natural technique to facilitate increasing complexity and cost-effective product development (Meyer, et al., 1997). In this regard, the manufacturing companies put their effort in organizing, developing, and planning product families to balance the tradeoffs between product diversity and engineering costs.

Based on the reports of marketing analysis, it turns out that some of the product variants may be more preferred as expected, while others, although they may be equally sound in technical terms, may not be favored by the customers. The errors on expectation and achievement mainly result from the diverse customer requirements. Furthermore, it has been reported that not all the existing market segments create the same opportunity for the companies in the same industry due to the discrepancy of their targets, strategies, technologies, cultures, etc. Therefore, it is most important for the manufacturing companies to make the decisions deliberately that which market segment should be targeted and what products should be planned for the target market, namely, product family positioning.

The involved complexity makes product family positioning very difficult. Proper positioning should help leverage the engineering costs and diverse customer preferences. The prediction of the customer preference is difficult because even the customers themselves do not know why they choose specified products. Moreover, the customers always need to make tradeoffs among diverse product features. For example, the customer must make a compromise between "high product quality" and "high price". It is inhibitive to estimate the value that the customers put on every product feature because the values that the customers perceive are based on their perception of the overall products (Green, et al., 1981). Further, cost estimation is deemed to be inhibited. Traditional cost accounting by allocating fixed costs and variable costs across multiple products may produce distorted cost-carrying figures due to possible sunk costs associated with investment into product and process platforms.

Towards this end, this chapter introduces a systematic approach to product family positioning within the context of mass customization. A comprehensive methodology for product family positioning is developed, aiming at leveraging both customer preferences and engineering costs. The remainder proceeds as follows. In the next section, various existing approaches to product family positioning are reviewed. Section 3 presents the formulation of the product family positioning problem. An optimization framework and the according properties of the model are discussed in Section 4. The developed model is represented in Section 5. A case study of notebook computer family positioning is reported in Section 6. The research is concluded with a summary in Section 7.

2. BACKGROUND REVIEW

In general the literature related to product family positioning stems from two broad fields that are closely correlated: product line design and customer preference analysis for optimal product design, as discussed below.

(1) Product Line Design. For product line design, the objectives widely used in selecting products among a large set of potential products include maximization of profit (Monroe, et al., 1976), net present value (Li and Azarm, 2002), a seller's welfare (McBride and Zufryden, 1988), market share (Kohli and Krishnamurti, 1987), and share of choices (Balakrishnan and Jacob, 1996) within a target market. Pullmana, et al. (2002) have combined QFD and conjoint analysis to compare the most preferred features with those profit maximizing features so as to develop designs that optimize product line sales or profits. Kota, et al. (2000) have proposed a product line commonality measure to capture the level of component commonality in a product family. The key issue is to minimize non-value added variations across models within a product family without limiting customer choices.

Another dimension in product line design research is about the price. Robinson (1988) has suggested that the most likely competitive reaction to a new product in the short term is a change in price. Choi and DeSarbo (1994) have applied the game theory to model competing firms' reactions in price and employed a conjoint simulator to evaluate product concepts against competing brands. Dobson and Kalish (1988; 1993) have discussed the tradeoffs involved in price setting and choice of the number of products. Furthermore, product line design basically involves two issues (Li and Azarm, 2002): generation of a set of feasible product alternatives, and subsequent selection of promising products from this reference set to construct a product line. Along this line, existing approaches to product line design can be classified into two categories (Steiner and Hruschka, 2002). One-step approaches aim at constructing product lines directly from part-worth preference and cost/return functions. On the other hand, two-step approaches first reduce the total set of feasible product profiles to a smaller set, and then select promising products from this smaller set to constitute a product line. Following the two-step approach, Green and Krieger (1985; 1989) have introduced several heuristic procedures with the consideration of how to generate a reference set appropriately. On the other hand, Kohli and Sukumar (1990) and Nair, et al. (1995) have adopted the one-step approach, in which product lines are constructed directly from part-worth data rather than by enumerating potential product designs. In general, the one-step approach is more preferable, as the intermediate step of enumerating utilities and profits of a huge number of reference set items can be eliminated

(Steiner and Hruschka, 2002). Only when the reference set contains a small number of product profiles can the two-step approach work well.

(2) Customer Preference Analysis for Optimal Product Design. Measuring customer preferences in terms of expected utilities is the primary concern of optimal product design (Krishnan and Ulrich, 2001) or decision-based design (Hazelrigg, 1998). In typical preference-based product design, conjoint analysis (Green and Krieger, 1985) has proven to be an effective means to estimate individual level part-worth utilities associated with individual product attributes. In order to simulate the potential market shares of proposed product concepts, scaled preference evaluations need to be collected from respondents with regard to a subset of multi-attribute product profiles (stimuli) constructed according to a fractional factorial design. With these preference data, idiosyncratic part-worth preference functions are then estimated for each respondent using regression analysis. Attribute level part-worth utilities can also be computed by respondents' simulated choice data, which is called a choice-based conjoint analysis and hence establishes a direct connection between preference and choice (Kuhfeld, 2004). The conjoint-based searching for optimal product designs always results in combinatorial optimization problems because typically discrete attributes are used in conjoint analysis (Kaul and Rao, 1995; Kohli and Sukumar, 1990; Nair, et al., 1995).

3.   PROBLEM DESCRIPTION

This research addresses the product family positioning problem with the goal of maximizing an expected surplus. A large set of product attributes, $A \equiv \{a_k \mid k = 1, \cdots, K\}$, have been identified, given that the firm has the capabilities (both design and production) to produce all these attributes. Each attribute, $\forall a_k \in A$, possesses a few levels, i.e., $A_k^* \equiv \{a_{kl}^* \mid l = 1, \cdots, L_k\}$. Thus, the product family is embodied in the various combinations of the attribute levels, i.e., $Z \equiv \{\bar{z}_j \mid j = 1, \cdots, J\}$. Each product, $\forall \bar{z}_j \in Z$, is defined as a vector of specific attribute levels, i.e., $\bar{z}_j = \left[a_{kl_j}^* \right]_K$, where any $a_{kl_j}^* = \varnothing$ indicates that product $\bar{z}_j$ does not contain attribute a_k; and any $a_{kl_j}^* \neq \varnothing$ represents an element of the set of attribute levels that can be assumed by product $\bar{z}_j$, i.e., $\left\{ a_{kl_j}^* \right\}_K \in \{A_1^* \times A_2^* \times \cdots \times A_K^*\}$.

The positioned product family, Λ, is a set consisting of a few selected product profiles, i.e., $\Lambda \equiv \{\bar{z}_j \mid j = 1, \cdots, J'\} \subseteq Z$, $\exists J' \in \{1, \cdots, J\}$, denotes the number of products contained in the positioned product family. Every product is associated with certain engineering costs, denoted as $\{C_j\}_j$. The manufacturer must make decisions that what products to offer as well as their respective prices, $\{p_j\}_j$. There are multiple market segments, $S \equiv \{s_i \mid i = 1, \cdots, I\}$, each containing homogeneous customers, with a size, Q_i. Various customer preferences on diverse products are represented by respective utilities, $\{U_{ij}\}_{I \cdot J}$. Product demands or market shares, $\{P_{ij}\}_{I \cdot J}$, are described by the probabilities of customers' choosing products, denoted as customer or segment-product pairs, $\{(s_i, \bar{z}_j)\}_{I \cdot J} \in S \times Z$.

4. FUNDAMENTAL ISSUES

4.1 Objective function

Among those customer preference or seller value-focused approaches, the objective functions widely used for solving the selection problem are formulated by measuring the consumer surplus – the amount that customers benefit by being able to purchase a product for a price that is less than they would be willing to pay. The idea behind is that the expected revenue (utility less price) comes from the gain between customer preferences (utilities indicating the dollar value that they would be willing to pay) and the actual price they would pay, whilst the price implies all related costs. With more focus on engineering concerns, the selection problem is approached by measuring the producer surplus – the amount that producers benefit by selling at a market price that is higher than they would be willing to sell for. The principle is to measure the expected profit (price less cost) based on the margin between the actual price they would receive and the cost (indicating the dollar value they would be willing to sell for), whilst the price implies customer preferences.

Considering both the customer preferences and the engineering costs, the above economic surpluses should be leveraged from both the customer and engineering perspectives. This research proposes to use a shared surplus to leverage both the customer and engineering concerns. Then the objective function can be formulated as the following:

$$Maximize \quad E[V] = \sum_{i=1}^{I} \sum_{j=1}^{J} \frac{U_{ij}}{C_j} P_{ij} Q_i y_j, \tag{1}$$

where $E[\cdot]$ denotes the expected value of the shared surplus, V, which is defined as the utility per cost, is modified by the probabilistic choice model, $\{P_{ij}\}_{I \cdot J}$, and the market size, $\{Q_i\}_i$, C_j indicates the cost of offered product $\bar{z}_j$, and y_j is a binary variable such that $y_j = 1$ if the manufacturer decides to offer product $\bar{z}_j$ and $y_j = 0$ otherwise.

4.2   Customer preference measurement

Volatile market condition, diverse customer preferences, and the competition among similar products make it difficult to measure the customer preference. Conjoint analysis (CA) is perhaps the most widely applied method for modeling consumer preference by marketing researchers. CA is a set of methods originally designed to measure consumer preferences by assessing the buyers' multi-attribute utility functions. The strength of CA is its ability to ask realistic questions that mimic the tradeoffs that respondents make in the real world. In contrast to direct questioning methods that simply ask how important each feature is or the desirability of each level, CA forces respondents to make difficult tradeoffs like the ones they encounter in the real world.

Following the part-worth model that is widely used in CA, the utility of the i-th segment for the j-th product, U_{ij}, is assumed to be a linear function of the part-worth preferences (utilities) of the attribute levels of product $\bar{z}_j$:

$$U_{ij} = \sum_{k=1}^{K} \sum_{l=1}^{L_k} \left(w_{jk} u_{ikl} x_{jkl} + \pi_j \right) + \varepsilon_{ij}, \tag{2}$$

where u_{ikl} is the part-worth utility of segment s_i for the l-th level of attribute a_k (i.e., a_{kl}^*) individually, w_{jk} is the utility weights among attributes, $\{a_k\}_K$, contained in product $\bar{z}_j$, π_j is a constant associated with the derivation of a composite utility from part-worth utilities with respect to product $\bar{z}_j$, ε_{ij} is an error term for each segment-product pair, and x_{jkl} is a binary variable such that $x_{jkl} = 1$ if the l-th level of attribute a_k is contained in product $\bar{z}_j$ and $x_{jkl} = 0$ otherwise.

4.3   Choice model and choice probability

Conjoint analysis yields a preference model, for example a main-effect part-worth model, which defines the functional relationship between attribute levels of a product and a customer's or a segment's overall utility attached to it. Based on this preference model, customers' choices can be modeled by relating preference (utility) to choice. The traditional deterministic first choice rule of preferences assumes that a customer chooses the product from the choice set according to the highest associated utility with certainty. Neglect of uncertain factors in the first choice rule may lead to suboptimal results at the aggregate market level, as market shares of products with higher utilities across customers or segments tend to be overestimated (Kaul and Rao, 1995).

On the other hand, probabilistic choice rules can provide more realistic representations of the customer decision making process (Sudharshan, et al., 1987). Some probabilistic choice rules can offer flexibility in calibrating actual choice behavior such as the option of mimicking the first choice rule (Kaul and Rao, 1995). In general, there are two types of probabilistic choice rules (Ben-Akiva and Lerman, 1985): the generalized (or powered) Bradley-Terry-Luce share-of-utility rule (BTL, also called the α-rule) and the conditional multinomial logit choice rule (MNL).

With the assumption of independently and identically distributed error terms, the logit choice rule suggests itself to be well suited to estimate customer preferences directly from choice data (Green and Krieger 1996).

Under the MNL model, the choice probability, P_{ij}, which indicates how likely a customer or a segment, $\exists s_i \in S$, chooses a product, $\exists z_j \in Z$, among N competing products, is defined as the following:

$$P_{ij} = \frac{e^{\mu U_{ij}}}{\sum_{n=1}^{N} e^{\mu U_{in}}}, \tag{3}$$

where μ is a scaling parameter. As $\mu \to \infty$, the logit behaves like a deterministic model, whereas it becomes a uniform distribution as $\mu \to 0$. Therefore, like with the BTL model, calibration on actual market shares can be carried out subsequently to elaborate preference estimation by post hoc optimization with respect to μ (Train, 2003).

Based on a customer survey, the response rate - how often each product alternative is chosen - can be depicted as a probability density distribution. The demand for a particular product is the summation of the choice frequency of each respondent, $\forall s_i \in S$, adjusted for the ratio of respondent sample size versus the size of the market population (Train, 2003). The accuracy of the demand estimates can be increased by identifying unique customer utility functions per market segment, or class of customers to capture systematic preference variations (Ben-Akiva and Lerman, 1985). Estimates of future demand can also be facilitated using pattern-based or correlation-based forecasting of existing products. Forecasts of economic growth and changes of the socioeconomic and demographic background of the market populations help to refine these estimates.

4.4   Dealing with engineering costs

Traditional cost accounting by allocating fixed costs and variable costs across multiple products may produce distorted cost-carrying figures due to possible sunk costs associated with investment into product and process platforms. It is quite common in mass customization that design and manufacturing admit resources (and thus the related costs) to be shared among multiple products in a reconfigurable fashion, as well as per-product fixed costs (Moore, et al., 1999). In fact, Yano and Dobson (1998) have observed a number of industrial settings, where a wide range of products are produced with very little incremental costs per se, or very high development costs are shared across broad product families, or fixed costs and variable costs change dramatically with product variety. They have pointed out that "the accounting systems, whether traditional or activity-based, do not support the separation of various cost elements".

Furthermore, the cost advantages in mass customization rest with the achievement of mass production efficiency. Rather than the absolute amount of dollar costs, what important to justify optimal product offerings is the magnitudes of deviations from existing product and process platforms due to design changes and process variations in relation to product variety. To circumvent the difficulties inherent in estimating the accurate cost figures, this research adopts a pragmatic costing approach based on standard time estimation developed by Jiao and Tseng (1999). The idea is to allocate costs to those established time standards based on well-practiced work and time studies, thus relieving the tedious tasks for identifying various cost drivers and cost-related activities. The key is to develop mapping relationships between different attribute levels and their expected consumptions of standard times within legacy process capabilities. These part-worth standard time accounting relationships are built into the product and process platforms (Jiao, et al., 2003). Any product configured from available attribute levels is justified based on its expected cycle time. This expected cycle time is accounted by the aggregation of part-worth standard times. The rationale is particularly applicable to family positioning, where "the optimal product profiles are not as sensitive to absolute dollar costs as they are to the relative magnitudes of cost levels" (Choi and DeSarbo, 1994).

Introducing a penalty function, the cost function, C_j, corresponding to product $\bar{z}_j$, can be formulated based on the respective process capability index, PCI_j, that is,

$$C_j = \beta e^{\frac{1}{PCI_j}} = \beta e^{\frac{3\sigma_j^T}{\mu_j^T - LSL^T}}, \tag{4}$$

where β is a constant indicating the average dollar cost per variation of process capabilities, LSL^T denotes the baseline of cycle times for all product variants to be produced within the process platform, μ_j^T and σ_j^T are the mean and the standard deviation of the estimated cycle time for product $\bar{z}_j$.

The meaning of β is consistent with that of the dollar loss per deviation constant widely used in Taguchi's loss functions. It can be determined ex ante based on the analysis of existing product and process platforms. Such a cost function produces a relative measure, instead of actual dollar figures, for evaluating the extent of process variations among multiple products. Modeling the economic latitude of product family positioning through the cycle time performance and the impact on process capabilities can alleviate the difficulties in traditional cost estimation.

5.   MODEL DEVELOPMENT

Treating the price of each offered product as a decision variable will make the problem nonlinear (Yano and Dobson, 1998). To avoid explicitly, nor necessary, modeling of the price, the general practice is to treat price as a separate attribute that can be chosen from a limited number of values for each product (Nair, et al., 1995; Moore, et al., 1999). Adding price as one more attribute, the attribute set becomes $A \equiv \{a_k\}_{K+1}$, where a_{K+1} represents the price possessing a few levels, i.e., $A_{K+1}^{\cdot} \equiv \{a_{(K+1)l}^{\cdot} \,|\, l = 1, \cdots, L_{K+1}\}$. Further, let $\bar{p} = [a_{(K+1)1}^{\cdot}, \cdots, a_{(K+1)L_{K+1}}^{\cdot}]$ be the vector of feasible price levels.

By combining Eqs. (1), (2), (3) and (4), the product family positioning problem can be formulated as a mixed integer program, as below:

$$Maximize \quad E[V] = \sum_{i=1}^{I}\sum_{j=1}^{J} \left(\frac{U_{ij}}{\beta e^{\mu_j^T \frac{1}{L_k L_k^T}} \sqrt[3\sigma_j^T]{}} \right) \left(\frac{e^{\mu U_{ij}}}{\sum_{n=1}^{N} e^{\mu U_{in}}} \right) Q_i y_j , \tag{5a}$$

s.t.
$$U_{ij} = \sum_{k=1}^{K+1}\sum_{l=1}^{L_k} \left(w_{jk} u_{ikl} x_{jkl} + \pi_j \right) + \varepsilon_{ij} , \forall i \in \{1, \cdots, I\}, \forall j \in \{1, \cdots, J\}, \tag{5b}$$

$$\sum_{l=1}^{L_k} x_{jkl} = 1 , \forall j \in \{1, \cdots, J\}, \forall k \in \{1, \cdots, K+1\} , \tag{5c}$$

$$\sum_{k=1}^{K+1}\sum_{l=1}^{L_k} |x_{jkl} - x_{j'kl}| > 0 , \forall j, j' \in \{1, \cdots, J\}, j \neq j' , \tag{5d}$$

$$\sum_{j=1}^{J} y_j \leq J' , \forall J' \in \{1, \cdots, J\}, \tag{5e}$$

$$x_{jkl}, y_j \in \{0,1\} , \forall j \in \{1, \cdots, J\}, \forall k \in \{1, \cdots, K+1\} , \forall l \in \{1, \cdots, L_k\} . \tag{5f}$$

Objective function (5a) is to maximize the expected shared surplus by offering a product family consisting of products, $\{\bar{z}_j\}_j$, to customer segments, $\{s_i\}_i$. Constraint (5b) refers to conjoint analysis – ensures that the composite utility of segment s_i for product $\bar{z}_j$ can be constructed from part-worth utilities of individual attribute levels, $\{A_k^*\}_{K+1}$. Constraint (5c) suggests an exclusiveness condition – enforces that exactly one and only one level of each attribute can be chosen for each product. Constraint (5d) denotes a divergence condition – requires that several products to be offered must pairwise differ in at least one attribute level. Constraint (5e) indicates a capacity condition – limits the maximal number of products that can be chosen by each segment. J' is the upper bound of the number of products that the manufacturer wants to introduce to a product family. Constraint (5f) represents the binary restriction with regard to the decision variables of the optimization problem.

There are two types of decision variables involved in the above mathematical program, i.e., x_{jkl} and y_j, representing the composite attribute levels and the products included in the positioned product family, respectively. Both types of decisions depend on a simultaneous satisfaction of the target segments. The manufacturer's decisions about what (i.e., layer I decision-making) and which (i.e., layer II decision-making) products to offer to the target segments are implied in various instances of $\{x_{jkl} \mid \forall j, k, l\}$ and $\{y_j \mid \forall j\}$, respectively. As a result, the positioned product family, $\Lambda' \equiv \{\bar{z}_j' \mid j = 1, \cdots, J'\}$ is yielded as a combination of selected products corresponding to $\{y_j \mid \forall j\}$, where each selected product , $\bar{z}_j'$, comprises a few selected attributes and the associated levels corresponding to $\{x_{jkl} \mid \forall j, k, l\}$.

6.   CASE STUDY

6.1   Application case

The proposed framework has been applied to the notebook computer family positioning problem. For illustrative simplicity, a set of key attributes and available attribute levels for the notebook computer are listed in Table 6-1. Among them, "price" is treated as one of the attributes to be assumed by a product.

With regard to the class-member relationships, notebook computer family comprises a four-layer AND/OR tree structure, as shown in Figure 6-1. The first layer is the product family, each of which consists of one or more products. Each product consists of a few attributes, thus constituting the second layer. The third layer represents the levels for each attribute, indicating the instantiation of an attribute by one out of many levels.

Table 6-1. List of attributes and their feasible levels for notebook computers.

Attribute		Attribute Levels		
a_k	Description	a'_{kl}	Code	Description
		a'_{11}	A1-1	Pentium 2.4 GHz
		a'_{12}	A1-2	Pentium 2.0 GHz
a_1	Processor	a'_{13}	A1-3	Centrino 1.6 GHz
		a'_{14}	A1-4	Centrino 1.7 GHz
		a'_{21}	A2-1	256 MB DDR SDRAM
a_2	Memory	a'_{22}	A2-2	512 MB DDR SDRAM
		a'_{23}	A2-3	1 GB DDR SDRAM
		a'_{31}	A3-1	60 GB
a_3	Hard Disk	a'_{32}	A3-2	80 GB
		a'_{33}	A3-3	120 GB
		a'_{41}	A4-1	Low (below 2.0 KG with battery)
a_4	Weight	a'_{42}	A4-2	Moderate (2.0 - 2.8 KG with battery)
		a'_{43}	A4-3	High (2.8 KG above with battery)
		a'_{51}	A5-1	Regular (around 6 hours)
a_5	Battery Life	a'_{52}	A5-2	Long (7.5 hours above)
		a'_{61}	A6-1	$800 - $1.3K
		a'_{62}	A6-2	$1.3K - $1.8K
a_6	Price	a'_{63}	A6-3	$1.8K - $2.5K
		a'_{64}	A6-4	$2.5K above

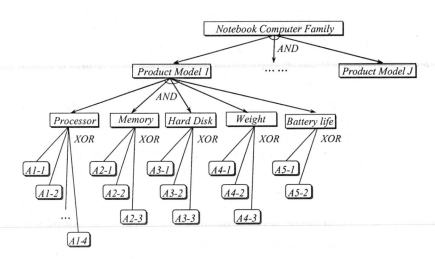

Figure 6-1. Generic structure for notebook computer family.

6.2   Customer preference

It is required to construct testing profiles for conjoint analysis. Given all attributes and their possible levels as shown in Table 6-1, a total number of $4 \times 3 \times 3 \times 3 \times 2 \times 4 = 864$ possible combinations may be constructed. To overcome such a combinatorial explosion, the Taguchi Orthogonal Array Selector provided in SPSS software (http://www.spss.com) is used to generate a total number of 27 orthogonal product profiles. With these profiles, a fractional factorial experiment is designed for exploring customer preferences, as shown in Table 6-2. In Table 6-2, columns 2-7 indicate the specification of offerings that are involved in the profiles and column 8 collects the preferences given by the customers.

Table 6-2. Response surface experiment design.

Conjoint Test							Preference Scale least most
Profile	Processor	Memory	Hard Disk	Weight	Battery Life	Price	1 ⌊⌊⌊⌊⌊⌊⌊⌊⌋ 9
1	P-2.0	256	60	Low	Regular	$800-1.3K	9
2	C-1.7	256	80	Low	Regular	$1.8-2.5K	3
3	P-2.4	512	60	Moderate	Long	$800-1.3K	4
4	C-1.7	512	120	Low	Regular	$1.3-1.8K	7
...	...	...	...	...	...	...	...
25	C-1.7	256	80	Moderate	Regular	$1.8-2.5K	2
26	P-2.0	1	120	Low	Regular	$800-1.3K	8
27	C-1.6	1	80	High	Regular	$1.3-$1.8K	6

A total number of 20 customers are selected to act as the respondents. Each respondent is asked to evaluate all 27 profiles one by one by giving a mark based on a 9-point scale, where "9" means the customer prefers a product most and "1" least. With these data, clustering analysis is run to find customer segments based on the similarity among customer preferences. Three customer segments are formed: s_1, s_2, and s_3, suggesting home users, regular users, and professional/business users, respectively.

For each respondent in a segment, 27 regression equations are obtained by interpreting his original choice data as a binary instance of each part-worth utility. With these 27 equations, the part-worth utilities for this respondent are derived. Averaging the part-worth utility results of all respondents belonging to the same segment, a segment-level utility is obtained for each attribute level. Columns 2-4 in Table 6-3 show the part-worth utilities for three segments with respect to every attribute level.

6.3   Engineering costs

Table 6-3 also shows the part-worth standard times for all attribute levels. The company fulfills customer orders through assembly-to-order production while importing all components and parts via global sourcing. The part-worth standard time of each attribute level is established based on work and time studies of the related assembly and testing operations. With assembly-to-order production, the company established standard routings as the basis for its process platform. Based on empirical studies, costing parameters are known as $LSL^T = 2518$ (second) and $\beta = 0.006$.

Table 6-3. Part-worth utilities and part-worth standard times.

Attribute Level	Part-worth Utility (Customer Segment)			Part-worth Standard Time (Assembly & Testing Operations)	
	s_1	s_2	s_3	μ^t (second)	σ^t (second)
A1-1	0.62	0.71	0.53	485	8.9
A1-2	0.82	0.78	0.81	538	10.3
A1-3	0.80	0.83	1.23	557	11.3
A1-4	0.71	0.73	0.71	521	11.4
A2-1	1.23	1.32	1.25	753	34.2
A2-2	1.26	1.23	1.29	825	36
A2-3	1.32	1.54	1.42	821	35
A3-1	1.17	0.48	0.38	667	23.6
A3-2	1.12	0.88	0.75	703	22.6
A3-3	1.16	1.26	0.89	730	31
A4-1	1.25	0.89	0.72	637	25.5
A4-2	1.44	1.32	0.83	672	27.6
A4-3	1.67	1.21	1.17	715	28.7
A5-1	0.98	0.96	0.85	287	4.32
A5-2	0.82	1.32	0.92	315	5.34
A6-1	0	0	0		
A6-2	-1.55	-0.46	-0.17	N.A.	N.A.
A6-3	-1.38	-0.71	-0.52		
A6-4	-2.21	-2.19	-0.56		

6.4   GA solution

To ensure accurate product family positioning, every possible scenario should be examined. It will result in a combinatorial explosion for the products involved in the product family. Enumeration is inhibitive if a problem is extremely big. Comparing with traditional calculus-based or approximation optimization techniques, genetic algorithms (GAs) have been proven to excel in solving combinatorial optimization problems. The GA procedure is applied to search for a maximum of expected shared surplus among all product alternatives. Assume that each positioned product family may consist of a maximal number of $J' = 4$ products. Then a chromosome string comprises 6 x 4 = 24 genes. Each substring is as long as 6 genes and represents a product that constitutes the product family. During the reproduction process, new product and family alternatives keep being generated through crossover and mutation operations. For every generation, a population size of $M = 20$ is maintained, meaning that only top 20 fit product families are kept for reproduction.

6.5   Results

Adopting the crossover and mutation rate as 0.6 and 0.01, respectively, the results of GA solution are presented in Figure 6-2. As shown in Figure 6-2, the fitness value keeps being improved generation by generation. Certain local optima (e.g., around 100 generations) are successfully overcome. The saturation period (350-492 generations) is quite short, indicating the GA search is efficient. Upon termination at the 492th generation, the GA solver returns the optimal result, which achieves an expected shared surplus of $792K, as shown in Table 6-4.

As shown in Table 6-4, the positioned product family consists of two products, $\vec{z}_1^1$ and $\vec{z}_2^1$. From the specifications of attribute levels, we can see they basically represent the low-end and high-end notebook computers, respectively. With such a two-product family, all home, regular and professional/business users can be served with an optimistic expectation of maximizing the shared surplus.

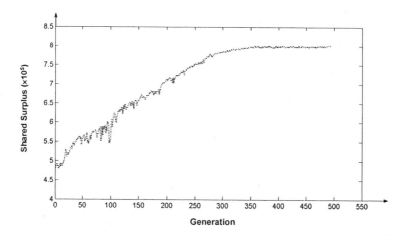

Figure 6-2. Shared surpluses among generations.

Table 6-4. Product family positioning results.

$\Lambda^\dagger$	$\Lambda^\dagger = [1,1,2,3,1,1;4,2,3,1,2,3;0,0,0,0,0,0;0,0,0,0,0,0]$			
$\{\vec{z}_j^t\}_{j'}$	$\vec{z}_1^1 = [1,1,2,3,1,1]$		$\vec{z}_2^1 = [4,2,3,1,2,3]$	
	$a_k^\dagger$	a_{kl}^*	$a_k^\dagger$	a_{kl}^*
$\{a_k^t\}_{(K+l)^t}$	Processor	Pentium 2.4 GHz	Processor	Centrino 1.7 GHz
	Memory	256 MB DDR SDRAM	Memory	512 MB DDR SDRAM
	Hard Disk	80 GB	Hard Disk	120 GB
$\{a_{kl}^*\}_{(K+1)^t}$	Weight	High (2.8 KG above)	Weight	Low (below 2.0 KG)
	Battery Life	Regular (around 6 hours)	Battery Life	Long (7.5 hours above)
	Price	$800 - $1.3K	Price	$1.8K - $2.5K
$E[V^\dagger]$	$792K			

6.6   Performance evaluation

Figure 6-3a compares the results of utility with choice probability, $\sum_{i=1}^{3}\sum_{j=1}^{4}(U_{ij}P_{ij})$, among generations. It is interesting to observe that the distribution of utility with choice probability does not tally with that of the fitness shown in Figure 6-2. The optimal solution (i.e., the last generation) does not produce the best utility performance. On the other hand, a number of high utility achievements do not correspond to high fitness. Likewise, as shown in Figure 6-3b, the distribution of cost performance among generations disorders the pattern of fitness distribution shown in Figure 6-2. This may be explained by the fact that high utility achievement is usually accompanied with high costs to incur. Therefore, the shared surplus is a more reasonable fitness measure to leverage both customer and engineering concerns than either utility or cost alone.

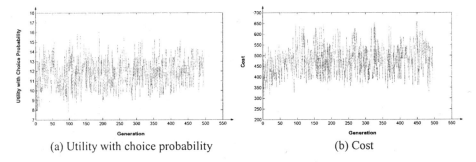

(a) Utility with choice probability   (b) Cost

Figure 6-3. Distribution of performance of GA by generation.

Figure 6-4 compares the achievements, in terms of the normalized shared surplus, cost, and utility with choice probability, of 20 product families in the 492th generation that returns the optimal solution. It is interesting to see that the peak of utility achievement (family #8) does not conclude the best fitness as its cost is estimated to be high. On the other hand, the minimum cost (family #4) does not mean the best achievement of shared surplus as its utility performance is moderate. Also interesting to observe is that the worst fitness (family #20) performs with neither the lowest utility achievement nor the highest cost figure. The best product family (#1) results from a leverage of both utility and cost performances.

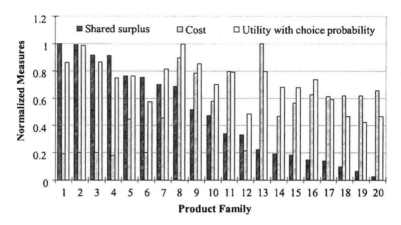

Figure 6-4. Performance comparison of product family population in the 492th generation.

7.   CONCLUDING REMARKS

Differing from the conventional product line design problem, product family positioning optimizes both a mix of products and the configurations of individual products in terms of specific attributes. By proposing a shared-surplus model, this research allows products to be constructed directly from attribute levels. Diverse customer preferences across multiple market segments, customer choice probability, and engineering costs for the composition of a product family are all covered by the shared surplus model. Conjoint analysis is adopted to quantify the customer preference. To circumvent the difficulties due to cost estimation, this research adopts a pragmatic costing approach based on standard time estimation to estimate the engineering costs of the products. To deal with the combinatorial explosion during optimization, the GA is used to position the optimal product family to create the highest shared surplus.

To model customers' choices, we employ the logit choice model and implements it in a segment level. The logit choice model implies a property of independence. In addition, the values of scaling parameter μ applied in the logit model are considered to be equal. The simplistic treatment suggests a possible way to improve the predictive quality of a logit model used in measuring the expected shared surplus.

Chapter 7

COMMONALITY INDICES FOR ASSESSING PRODUCT FAMILIES

Henri J. Thevenot and Timothy W. Simpson
The Harold and Inge Marcus Department of Industrial and Manufacturing Engineering, The Pennsylvania State University, University Park, PA 16802

1.   INTRODUCTION

Manufacturing companies need to satisfy a wide range of customer needs while maintaining manufacturing costs as low as possible, and many are faced with the challenge of providing as much variety as possible for the marketplace with as little variety as possible between products as discussed in Chapter 1. The challenge, then, when designing a family of products is in resolving the tradeoff between *product commonality* and *distinctiveness*: if commonality is too high, products lack distinctiveness, and their individual performance is not optimized; on the other hand, if commonality is too low, manufacturing costs can increase substantially (Simpson, et al., 2001). Commonality has many advantages beyond improving economies of scale: decreased lead-time and risk during product development (Collier, 1980); decreased inventory, handling costs and processing time; reduced product line complexity, set-up and retooling time, and increased productivity (Collier, 1979; Collier, 1981). However, too much commonality within a product family can hinder innovation and creativity and even compromise product performance (Krishnan and Gupta, 2001). Commonality is best obtained by minimizing the non-value added variations across the products within a family without limiting the choices of the customers in each market segment, i.e., make each product within a family distinct in ways customers notice and identical in ways that customers cannot see.

To measure the commonality within a family of products, several commonality indices have been developed. A *commonality index* is a tool to measure the degree of commonality within a product family. It is based on different parameters such as the number of common components, their costs, their manufacturing processes, etc. These indices are often the starting point when designing a new family of products or when analyzing an existing family. They are designed to give valuable information on the degree of commonality achieved within the family and how to improve the design to achieve better commonality in the family and reduce costs. This chapter describes several commonality indices found in the literature and provides examples on how to use commonality indices for (1) product family benchmarking and (2) product family redesign.

2.   COMMONALITY INDICES

This section gives a description of some of the commonality indices found in the literature, how they are computed, their advantages and their limitations. The selected indices are based on a component perspective: they mainly measure the similarities or differences between the components within a product family.   They do not focus on aspects such as their functionality or any parametric variation due to 'scaling'.

2.1   Unique, variant, and common parts

There are three different types of parts: unique, variant, and common. A *unique* part is only used by one product in the family. A *variant* part has the same function between some or all the products of the family, but the design, shape and material differ slightly from one product to the next. Finally, a *common* part is the exact same part shared by some or all of the products in the family. Here, the term *part* denotes any of the smallest decomposable elements within a product, be they components, modules, or subassemblies.

Table 7-1 gives an example of this classification. The columns represent the products within a family, and the rows are the parts. On each row, a number indicates if the part is common between different products. For example, for a given part, if two products share the same number, then they share the same part. If the number is different in each column for a given part, then all the products use different variants of the part. If there is no number, the corresponding product does not contain the corresponding part.

As shown in Table 7-1, all the products share Part 1; hence, this part is common. Part 2 is shared by only two products (Products 1 and 3) whereas Products 2 and 4 do not contain this part. Hence, all the products having this component share the same part, and this part is common. Part 3 differs from one product to another; therefore, it is a variant. Part 4 is also a variant, even though Products 2 and 3 share the same part. Finally, Part 5, only found in Product 2, is unique.

Table 7-1. Example of parts classification.

Part	Type of Commonality	Product 1	Product 2	Product 3	Product 4
Part 1	Common	1	1	1	1
Part 2	Common	1		1	
Part 3	Variant	1	2	3	4
Part 4	Variant	1	2	2	3
Part 5	Unique		1		

2.2 The Degree of Commonality Index

The Degree of Commonality Index (DCI) is the most traditional measure of component part standardization. Presented by Collier (1981), it reflects the average number of common parent items per average distinct part:

$$DCI = (\sum_{j=i+1}^{i+d} \Phi_j)/d, \tag{1}$$

where:

Φ_j = number of immediate parents component j has over a set of end items or product structure level(s);

d = total number of distinct components in the set of end items or product structure level(s);

i = the total number of end items or the total number of highest level parent items for the product structure level(s);

Component item = any inventory item (including a raw material) other than an end item that goes into higher level items;

End item = finished product or major subassembly subject to a customer order or sales forecast; and

Parent item = any inventory item that has component parts.

Equation 2 provides the upper and lower values of the DCI:

$$1 \le DCI \le \beta$$
$$\beta = \sum_{j=i+1}^{i+d} \Phi_j . \tag{2}$$

When DCI = 1, there is no commonality as no item is being used more than once in any of the products. When DCI = β, there is complete commonality.

The example shown in Figure 7-1 illustrates three sets of two end products (numbered 1 and 2). Case I reveals no commonality (DCI=1), as Products 1 and 2 have no common parts. In Case II, DCI is higher, since Products 1 and 2 share Parts 3 and 5. Finally, Case III has the highest DCI since Products 1 and 2 now share most of their components. The DCI can be interpreted as the ratio between the number of common components in a product family and the total number of part in the family. The main advantage of the DCI is its ease of computation, although the moving boundaries make it difficult to estimate the increase in commonality while redesigning a family and to compare different families of products. The Total Constant Commonality Index described next addresses these problems.

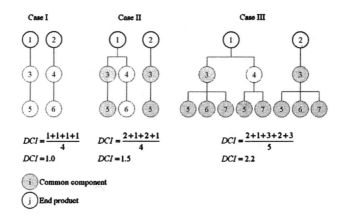

Figure 7-1. Computational examples for the DCI (Collier, 1981).

2.3   The Total Constant Commonality Index

The Total Constant Commonality Index (TCCI) is a modified version of the DCI. Unlike the DCI, which is a cardinal index (and hence an increase in commonality is not possible to measure), the TCCI is a relative index that has absolute boundaries (Wacker, et al., 1986):

$$TCCI = 1 - (d-1)/(\sum_{j=1}^{d} \Phi_j - 1),\qquad(3)$$

where Φ_j is the number of immediate parents component j has over a set of end items or product structure level(s); d denotes the total number of distinct components in the set of end items or product structure level(s); component item refers to any inventory item (including a raw material) other than an end item that goes into higher level items; end item means finished product or major subassembly subject to a customer order or sales forecast; and parent item means any inventory item that has component parts.

Equation 4 shows the lower and upper limits of the TCCI.

$$0 \leq TCCI \leq 1 \tag{4}$$

When TCCI = 0, there is no commonality as no item is being used more than once in any product. When TCCI = 1, there is complete commonality. The example in Figure 7-2 shows different cases, from zero commonality on the left to complete commonality on the right. As the DCI varies from 1 to 10, the TCCI varies within fixed boundaries, from 0 to 1, which is easier to interpret and to use when comparing different families of products.

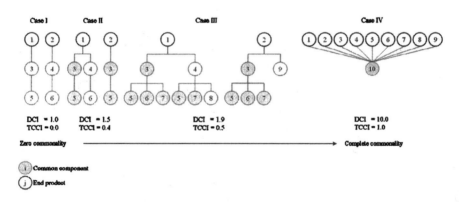

Figure 7-2. Examples for computing the DCI and TCCI (Wacker and Trelevan, 1986).

The TCCI can be interpreted as the ratio between the number of common parts in a product family and the total number of parts in the family. The main advantages of the TCCI are its ease of computation and its fixed boundaries, which gives a better indication of part standardization than the measure provided by the DCI. Its main limitation is the information considered (i.e., only the number of common parts in the family, no associated cost, for example, is included): this index gives a quick but rough estimate of how good the commonality is within a product family.

2.4 The Product Line Commonality Index

Contrary to the indices that simply measure the percentage of parts that are common within a product family (and hence penalizing families with a broader feature mix), the Product Line Commonality Index (PCI) measures and penalizes the differences that should ideally be common, given the product mix (Kota, et al., 2000). The PCI is given by:

$$PCI = 100*(\sum_{i=1}^{P} CCI_i - \sum_{i=1}^{P} MinCCI_i)/(\sum_{i=1}^{P} MaxCCI_i - \sum_{i=1}^{P} MinCCI_i),$$ (5)

where:

$CC_i = n_i * f_{1i} * f_{2i} * f_{3i}$, indicates the Component Commonality Index for component i;

$MaxCCI_i = N$, is the maximum possible Component Commonality Index for component i;

$MinCCI_i = n_i * 1/n_i * 1/n_i * 1/n_i = 1/n_i^2$, means the minimum possible Component Commonality Index for component i;

P = total number of non differentiating components that can potentially be standardized across models;

N = number of products in the product family;

n_i = number of products in the product family that have component i;

f_{1i} = Size and shape factor for component i, indicates the ratio of the greatest number of models that share component i with identical size and shape to the greatest possible number of models that could have shared component i with identical size and shape (n_i);

f_{2i} = Materials and manufacturing processes factor for component i, indicates the ratio of the greatest number of models that share component i with identical materials and manufacturing processes to the greatest possible number of models that could have shared component i with identical materials and manufacturing processes (n_i); and

f_{3i} = Assembly and fastening schemes factor for component i, indicates the ratio of the greatest number of models that share component i with identical assembly and fastening schemes to the greatest possible number of models that could have shared component i with identical assembly and fastening schemes (n_i).

By substituting the values of CCI_i, $MinCCI_i$, and $MaxCCI_i$, the following formula is obtained for the PCI:

$$PCI = 100*(\sum_{i=1}^{P} n_i * f_{1i} * f_{2i} * f_{3i} - \sum_{i=1}^{P} \frac{1}{n_i^2})/(P*N - \sum_{i=1}^{P} \frac{1}{n_i^2}).$$ (6)

Equation 7 gives the lower and upper boundaries of the PCI.

$$0 \leq PCI \leq 100$$ (7)

When PCI = 0, either none of the non-differentiating parts are shared across models, or if they are shared, their size/shapes, materials/manufacturing processes, and assembly processes are all different. When PCI = 100, it indicates that all the non-differentiating parts are shared across models and that they are of identical size and shape, made using the same material and manufacturing process, and the fastening methods are identical. An example of PCI calculation is given in Table 7-2, where each non-unique component is catalogued; the corresponding component commonality index is computed; and finally, the PCI is computed as 85.38 at the bottom of the table.

2.5   The Percent Commonality Index

The Percent Commonality Index (%C) is based on three main viewpoints: (1) component, (2) component-component connections, and (3) assembly. Each of these viewpoints results in a percentage of commonality, which can then be combined to determine an overall measurement of commonality for a platform by using appropriate weights for each item (Siddique, et al., 1998).

The first viewpoint, the percent commonality of components, C_c, measures the percentage of components that are common between the products in the family:

$$C_c = \frac{100 * \text{common components}}{\text{common components} + \text{unique components}} . \tag{8}$$

The greater the value of C_c, the more parts that are being shared.

The component-component connections viewpoint, C_n, measures the percentage of common connections between components:

$$C_n = \frac{100 * \text{common connections}}{\text{common connections} + \text{unique connections}} . \tag{9}$$

Similarly, the assembly viewpoint measures the percentage of common assembly sequences. Two indices are used: C_l, to measure the percentage of common assembly sequences, and C_a, to measure the percentage of common assembly workstations. They are given by the following two equations.

$$C_l = \frac{100 * \text{common assembly component loading}}{\text{common assembly component loading} + \text{unique assembly component loading}} \tag{10}$$

$$C_a = \frac{100 * \text{common assembly workstation}}{\text{common assembly workstation} + \text{unique assembly workstation}} \quad (11)$$

Table 7-2. PCI table for Sony Walkman family (Kota, et al., 2000).

No.	Component	Total No. in Family	MinCCI	Size & Geometry	Material & Manufacturing	Fastening & Assembly	Component Commonality Index
		ui	(1/ni)^2	f1	f2	f3	ui * f1 * f2 * f3
1	Cover spring	4	0.063	1.000	1	1	4
2	Voltage plug contact	4	0.063	1.000	1	1	4
3	Battery lid	4	0.063	0.750	1	1	3
4	Belt clip	4	0.063	0.500	1	1	2
5	Tape clip	4	0.063	1.000	1	1	4
6	Volume knob	4	0.063	0.500	1	1	2
7	Tuning knob	2	0.250	1.000	1	1	2
8	Tuning bridge	3	0.111	0.667	1	1	2
9	Tuning zipper	2	0.250	1.000	1	1	2
10	Zipper roller	2	0.250	1.000	1	1	2
11	Motor	4	0.063	1.000	1	1	4
12	Motor ring	4	0.063	1.000	1	1	4
13	Motor scew	4	0.063	1.000	1	1	4
14	Motor screw	4	0.063	1.000	1	1	4
15	Pulley 1	4	0.063	1.000	1	1	4
16	Pulley 2	4	0.063	1.000	1	1	4
17	Capstan 1	4	0.063	1.000	1	1	4
18	Capstan 2	4	0.063	1.000	1	1	4
19	Capstan 2 gear	4	0.063	1.000	1	1	4
20	Arm 1	4	0.063	1.000	1	1	4
21	Arm 1 cap	4	0.063	1.000	1	1	4
22	Gear 1	4	0.063	1.000	1	1	4
23	Layered gear 1	4	0.063	1.000	1	1	4
24	Gear 1 lever	4	0.063	1.000	1	1	4
25	Spring 1	4	0.063	1.000	1	1	4
26	Spring 2	4	0.063	1.000	1	1	4
27	Gear 2	4	0.063	1.000	1	1	4
28	Gear 2 washer	4	0.063	1.000	1	1	4
29	Layered gear 2	4	0.063	1.000	1	1	4
30	Spring 3	4	0.063	1.000	1	1	4
31	Arm 2	4	0.063	1.000	1	1	4
32	Arm 2 spring	4	0.063	1.000	1	1	4
33	Base board	4	0.063	1.000	1	1	4
34	Plate 1	3	0.111	1.000	1	1	3
35	Plate 2	4	0.063	1.000	1	1	4
36	Lever	4	0.063	1.000	1	1	4
37	Play button	4	0.063	0.500	1	1	2
38	Play plate	3	0.111	1.000	1	1	3
39	Play plate spring	4	0.063	1.000	1	1	4
40	Head	4	0.063	0.750	1	1	3
41	Head screw	4	0.063	1.000	1	1	4
42	Roller fixture	4	0.063	1.000	1	1	4
43	Roller fixture	4	0.063	1.000	1	1	4
44	Roller spring	4	0.063	1.000	1	1	4
45	Rewind button	4	0.063	0.500	1	1	2
46	Rewind plate	4	0.063	1.000	1	1	4
47	Cantilever spring	4	0.063	1.000	1	1	4
48	Fastforward button	4	0.063	0.500	1	1	2
49	Fastforward plate	4	0.063	1.000	1	1	4
50	Stop button	4	0.063	0.500	1	1	2
51	Stop plate	4	0.063	0.500	1	1	2

P = 51
N = 4

Sum (CCI) = 178.000
Sum (MinCCI) = 3.833
PCI = 85.376

These four percent commonality indices can then be combined into an overall platform commonality measure in several different manners. The most popular one is the weighted-sum formulation (Siddique, et al., 1998):

$$\%C = \sum_{i=1}^{4} I_i * C_i = I_c * C_c + I_n * C_n + I_l * C_l + I_a * C_a, \tag{12}$$

where I_i is the importance (weighting factors), $\sum I_i = 1$, and C_i denotes the percent commonality based on each viewpoint as previously described. The resulting %C ranges from 0 to 100.

$$0 \le \%C \le 100 \tag{13}$$

When %C = 0, there is no commonality, and when %C = 100, there is complete commonality. The following example measures the %C between two different automotive platforms from (Siddique, et al., 1998). Table 7-3 gives the details of the computation of C_c and C_n, Figure 7-3 and Table 7-4 explain the calculation of C_a and C_l, and Table 7-5 gives the resulting %C by applying weights to each of the previous described factors.

Table 7-3. Calculation of the C_c and C_n for two different platforms (Siddique, et al., 1998).

Front Structure 3rd Level for Platform A

Component										
Right Apron	1	1	0	1	1	0	1	0	0	0
Right Rail	1	1	1	1	0	0	1	0	0	0
Right Bracket	0	1	1	0	0	1	0	0	0	0
Radiator Support	1	1	0	1	1	0	0	0	1	1
Engine	1	0	0	1	1	1	0	0	0	1
Suspension	0	0	1	0	1	1	1	1	0	0
Dash/Cowl	1	1	0	0	1	1	0	1	1	
Left Bracket	0	0	0	0	0	1	0	1	1	0
Left Rail	0	0	0	1	0	0	1	1	1	1
Left Apron	0	0	0	1	1	0	1	0	1	1

Front Structure 3rd Level for Platform B

Component									
Right Apron	1	1	1	1	0	0	1	0	0
Right Rail	1	1	1	0	1	1	1	0	0
Radiator Support	1	1	1	1	0	0	0	1	1
Engine	1	0	1	1	1	0	0	0	1
Sub-frame	0	1	0	1	1	1	0	1	0
Suspension	0	1	0	0	1	1	1	1	0
Dash/Cowl	1	1	0	0	0	1	1	1	1
Left Rail	0	0	1	0	1	1	1	1	1
Left Apron	0	0	1	1	0	0	1	1	1

	Platform A			Platform B	
A	Total No. of Core Common Components:	6	A	Total No. of Core Common Components:	6
B	Total No. of Unique components including External Components:	4	B	Total No. of Unique components including External Components:	3
C	Percent commonality of components:	60.00%	C	Percent commonality of components:	66.67%
D	Total No. of Common Connections:	14	D	Total No. of Common Connections:	14
E	Total No. of Unique Connections:	5	E	Total No. of Unique Connections:	6
F	Percent commonality of connections:	73.68 %	F	Percent commonality of connections:	70.00 %

☐ Common Connections   1 - connected   0 - not connected

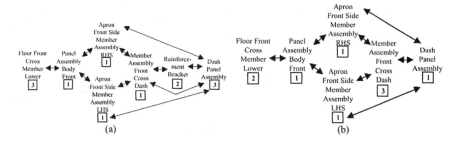

Figure 7-3. Component loading sequence for platforms A and B (Siddique, et al., 1998).

Table 7-4. Assembly attributes between platform A and B (Siddique, et al., 1998).

		Platform A	Platform B
A	Number of components with common loading sequence	3	3
B	Number of components with unique loading sequence	4	3
C	% commonality for loading sequence	42.86%	50.00%
D	Number of common stations	4	4
E	Number of unique stations	6	10
F	% commonality of work stations	40.00%	28.57%

Table 7-5. Percent commonality summary for platform A and B (Siddique, et al., 1998).

	Importance	Platform A	Platform B
% Component Commonality	0.2	60	66.7
% Connection Commonality	0.3	73.7	70
% Assembly Work Station Commonality	0.5	40	28.6
Overall Commonality (weighted sum)		54.2	48.7

The main advantage of this index is its flexibility: it can be adapted to different strategies using the weighting factors. Its disadvantage is that the measure is applied to *each platform and not necessarily to the products in the family*, which requires additional computation.

2.6 The Commonality Index

The Commonality Index (CI) measures the number of unique parts in a product family (Martin, et al., 1996; 1997) and is defined as:

$$CI = 1 - (u - \max p_j) / (\sum_{j=1}^{v_n} p_j - \max p_j), \qquad (14)$$

where u denotes the number of unique parts; p_j is the number of parts in model j; and v_n is the final number of varieties offered. CI ranges from:

$$0 \le CI \le 1. \qquad (15)$$

A higher CI is better since it indicates that the different variants within the product family are being achieved with fewer unique parts. For example, a family of six computer mice was analyzed, each having 20 parts. The denominator of the CI is thus equal to 100 (=6*20-20). If there were no two parts alike, that is to say if there were 120 unique parts, then the CI for the computer mice family would be 0 (= 1-(120-20)/(120-20)), which is the worst commonality possible. On the other hand, if only 70 parts are used to build the family, then the CI would be equal to 0.5 (= 1-(70-20)/(120-20)), indicating a higher degree of commonality. The CI can be interpreted as the ratio between the number of unique parts in the product family and the total number of parts in the family. The main advantage of this index is its ease of computation; however, it only focuses on the number of unique parts, and factors such as the costs of each component are not taken into account.

2.7   The Component Part Commonality Index

The Component Part Commonality Index $CI^{(C)}$ (Jiao, et al., 2000) is an extended version of the DCI. It takes into account product volume, quantity per operation, and the cost of component part and is given by:

$$CI^{(C)} = (\sum_{j=1}^{d}[P_j\sum_{i=1}^{m}\Phi_{ij}\sum_{i=1}^{m}(V_iQ_{ij})]) / (\sum_{j=1}^{d}[P_j\sum_{i=1}^{m}(V_iQ_{ij})]), \qquad (16)$$

where:

d = total number of distinct component parts used in all the product structures of a product family;

j – index of each distinct component part;

P_j = price of each type of purchased parts or the estimated cost of each internally made component part;

m = total number of end products in a product family;

i = index of each member product of a product family.

d_j = part across all the member products in the family;

V_i = volume of end product i in the family;

Q_{ij} = quantity of distinct component part d_j required by the product i;

Φ_{ij} = number of immediate parents for each distinct component part d_j over all the products levels of product i of the family; and

$\sum_{i=1}^{m}\Phi_{ij}$ = total number of applications (repetitions) of a distinct

component $\sum_{j=1}^{d}\sum_{i=1}^{m}\Phi_{ij} = \alpha$.

Equation 17 gives the lower and upper limit of the $CI^{(C)}$:

$$1 \le CI^{(C)} \le \alpha \tag{17}$$

When $CI^{(C)} = 1$, there is no commonality as no item is being used more than once in any of the products. When $CI^{(C)} = \alpha$, there is complete commonality. Figure 7-4 shows three products from the same family. For each product, the parts used are given, as well as the number required. Table 7-6 shows the computation of the $CI^{(C)}$.

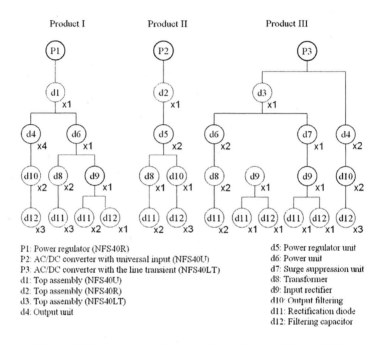

P1: Power regulator (NFS40R)
P2: AC/DC converter with universal input (NFS40U)
P3: AC/DC converter with the line transient (NFS40LT)
d1: Top assembly (NFS40U)
d2: Top assembly (NFS40R)
d3: Top assembly (NFS40LT)
d4: Output unit

d5: Power regulator unit
d6: Power unit
d7: Surge suppression unit
d8: Transformer
d9: Input rectifier
d10: Output filtering
d11: Rectification diode
d12: Filtering capacitor

Figure 7-4. Description of three products (Jiao and Tseng, 2000).

Table 7-6. Calculation of the $CI^{(C)}$ (Jiao and Tseng, 2000).

d_j	P_j	$\sum_{i=1}^{3}\Phi_{ij}$	$V_1 = 100,$ Q_{1j}	$V_2 = 80,$ Q_{2j}	$V_1 = 50,$ Q_{3j}	$\sum_{i=1}^{3}(V_i Q_{ij})$	$P_j \sum_{i=1}^{3}\Phi_{ij} \sum_{i=1}^{3}(V_i Q_{ij})$	$P_j \sum_{i=1}^{3}(V_i Q_{ij})$
d_1	6.4	1	1	0	0	100	640	640
d_2	2.8	1	0	1	0	80	224	224
d_3	7.1	1	0	0	1	50	355	355
d_4	3.1	2	4	0	2	500	3100	1550
d_5	3.9	1	0	2	0	160	624	624
d_6	6.8	2	1	0	2	200	2720	1360
d_7	4.1	1	0	0	1	50	205	205
d_8	3.7	3	2	2	4	560	6216	2072
d_9	1.15	3	1	0	3	250	862.5	287.5
d_{10}	3.5	3	8	2	4	1160	12180	4060
d_{11}	1.2	6	8	4	11	1670	12024	2004
d_{12}	1.4	6	25	6	15	3630	30492	5082
								$CI^{(C)} = 3.79$

If detailed information is available for the family of products, the $CI^{(C)}$ is the most accurate index as it is based not only on the common parts within a family, but also their costs. As such, a very expensive part common throughout a family has much more impact than a part that is very cheap and different from one product to another. The main drawback of this index is the data needed: it requires the costs of each component, which can be hard to obtain or estimate.

2.8   Other commonality indices

Six commonality indices from the literature have been introduced in this section. These indices cover a wide variety of factors, such as the number of common parts in a product family, the number of common connections, the component cost, etc. These indices were chosen due to the fact that they mainly measure component commonality. However, this review is not exhaustive, and other indices can be found in the literature, such as the Generational Variety Index (GVI), a measure for the amount of redesign effort required for future designs of a product, and the coupling index (CI), a measure of the coupling among the product components (Martin, et al., 2002). Maupin and Shauffer (2000) developed other metrics that include direct and indirect costs of production, delayed differentiation, simplicity (i.e., reduction of complexity), and standardization. Although these indices can be more accurate in assessing product families, the data required (such as labor time, assembly times, etc.) can be difficult to collect or estimate.

3.   EXAMPLE OF COMPUTING COMMONALITY INDICES FOR A PRODUCT FAMILY

This section provides an example of computation of the commonality indices previously described in Section 2 for a product family.

3.1   Description of the product family analyzed

The Skil family that we analyzed consists of four cordless screwdrivers shown in Table 7-7. They have two main characteristics: two screwdrivers use a motor with an input of 3.6V while the two others have a 2.4.V motor. Two of them have a flexible head with a lighting system, whereas the two others are simple screwdrivers with a rigid body.

Table 7-7. Characteristics of the Skil screwdrivers.

Product In the family	2.4V Twist Power Driver	3.6V Twist Power Driver	2.4V Twist	3.6V Super Twist
Voltage (V)	2.4	3.6	2.4	3.6
Light	Yes	Yes	No	No
Adjustable collect position	Yes	Yes	No	No

3.2   Product dissection methodology

In order to keep the results homogeneous, the same level of dissection is applied to all of the products in the family. The products are dissected to the lowest level, i.e., to the point where they cannot be divided further and still be manually re-assembled into a functioning product. For example, the electronic printed circuit boards are taken as a single part when analyzed, even if they have several electronics components on them. The analysis does not take any usual fastening methods into consideration, e.g., screws, bolts, etc., and electrical wires are not considered. These parts are easy to share within a family of products, and it would have dramatically (and artificially) increased the values of the commonality indices when computed. After disassembly, each part is photographed. It is used to clearly identify which parts are common within the family. Tables such as Table 7-8 are then created to store this information, including the type of commonality for each part within the family.

Table 7-8. Example of information capture for the Skil screwdrivers.

Module	Module Type	2.4V Twist Power Driver	3.6V Twist Power Driver	2.4V Twist	3.6V Super Twist
		1	2	3	4
Back Panel 1	Variant				
		1	1		
Back Panel 2	Common				
...	...	...	...	...	...
				1	
Front Housing	Unique				

3.3 Computation of the commonality indices

The computation is performed using a spreadsheet program (Microsoft Excel) in order to minimize errors and maximize repeatability. The parts are in rows while the products within the product family are in column (as shown in Table 7-1). This section gives a detailed example of computing one index, namely, the Product Line Commonality Index (PCI), as well as the results of the computations of the other commonality indices. Details of computing the commonality indices described in Section 2 can be found in (Thevenot and Simpson, 2004).

Table 7-9 provides a sample of the data table for the PCI. First, the unique parts are removed from the analysis as they provide unique functionality; only the common and variant parts are kept. The first four columns refer to the parts in each product, and they can be automatically obtained from Table 7-8. The next column, n_i, is the total number of the parts in the family, which indicates how many variants there are for a given part. The next three columns, f_1, f_2 and f_3, are completed by the user. These factors are determined given their definition in Section 2.4. For example, consider the first part, Back Panel 1. All the products in the family have this part, but they all have a different variant with different shape and geometry but with a common material and manufacturing process. Hence, the corresponding f_1 is 0.25 (= 1/4), while f_2 is equal to 1.00 (= 4/4). The last column, '$n_i*f_1*f_2*f_3$', is the 'Component Commonality Index'. Finally, two more values are entered: (1) the number of differentiating parts P, and (2) the number of products in the family N (not shown in the table).

Table 7-9. Computation of the PCI for the family.

Part	2.4V Twist Power Driver	3.6V Twist Power Driver	2.4V Twist Screw Driver	3.6V Super Twist	n_i	$1/n_i^2$	Size & Geom. f_1	Material & Mfg f_2	Fastening & Ass. f_3	$n_i*f_1*f_2*f_3$
Back Panel 1	1	2	3	4	4	0.063	0.250	1.000	1.000	1.000
Back Panel 2	1	1			2	0.250	1.000	1.000	1.000	2.000
Front Housing			1		1					

Using the notation given in Section 2.4, the computation for the PCI is:

$$PCI = \frac{sum\ of\ the\ column\ 'n_i*f_1*f_2*f_3' - \sum \dfrac{1}{n_i^2}}{P*N - \sum \dfrac{1}{n_i^2}} *100. \tag{18}$$

In this example, PCI = 41.20. A similar methodology is conducted for the computation of the other commonality indices; the results can be found in Table 7-10. More details can be found in (Thevenot and Simpson, 2004). These values can be compared between similar product families, and information can be deduced from them, as described in the next section.

Table 7-10. Values of commonality indices for the four product Skil family.

Index:	CI	PCI	TCCI	%C	DCI	CI$^{(C)}$
Value	65.20%	41.20	31.10%	44.50	1.44	1.38

4.   EXAMPLE OF APPLICATIONS OF COMMONALITY INDICES

Commonality indices can provide useful information during (1) product family benchmarking and (2) product family design and redesign by helping the designer choose between different design strategies, and focus on components that are influences the most the commonality. This section presents examples of how commonality indices can be used for product family benchmarking and redesign.

4.1   Product family benchmarking using commonality indices

Commonality indices can first be used to assess the design of a product family and to compare it with either other possible designs or competing families. This section gives an example of comparison of three product families using commonality indices. The product families considered are families of power tools as shown in Table 7-11.

Two types of families are compared. The first one contains products with the same primary function, but specific additional features, such as the Skil screwdrivers, where each product has the same function (screwing) but different features (e.g., adjustable collet position, light, bit holder). This type of family includes the Skil and the DeWalt families. The second type of products analyzed is the Black & Decker Versapak family, which contains different products from the same brand with different functions but shared common components. The values obtained for the commonality indices are summarized in Table 7-12; Figure 7-5 compares these values between the three product families.

Table 7-11. The three product families being compared.

	Product 1	Product 2	Product 3	Product 4	Product 5	Product 6
Skil family	2.4V Twist Power Drive	3.6V Twist Power Driver	2.4V Twist	3.6V Super Twist		
DeWalt family	1/4" Heavy VSR Duty Drill DW217	3/8" Heavy Duty VSR Drill DW226	1/2" Heavy Duty VSR Drill DW231	Heavy Duty Drywall Screwdriver DW25		
Black & Decker Versapak family	Flashlight	Screwdriver	Rotary Tool	Drill	Reciprocating Saw	Circular Saw

Table 7-12. Commonality indices for the power tools.

	CI	PCI	TCCI	%C	DCI	CI^(C)
Skil	65.20%	41.20	31.10%	44.50	1.44	1.38
DeWalt	77.10%	74.70	52.10%	75.90	2.07	2.97
Black & Decker	15.40%	22.70	12.50%	33.90	1.14	1.18

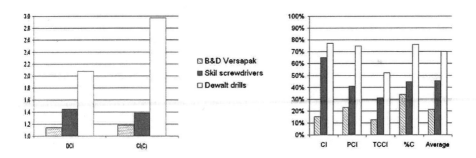

Figure 7-5. Commonality indices for the power tools (the PCI and the %C are drawn on a 0%-100% scale rather than 0-100 scale to be compared to the other indices).

In all cases, the Black & Decker Versapak has the lowest values for the commonality indices, the DeWalt corded drills have the highest values, and the Skil screwdrivers have intermediate values. This trend is consistent between the commonality indices. The poor values for the Black & Decker Versapak can be explained by the design of the family: rather than trying to achieve high commonality between products having the same functions, each product in the Versapak family has specific functions, and they share a limited number of components and connections. Hence, the TCCI and the CI are extremely low compared to the other families, as the Versapak has fewer common components and more unique components than the DeWalt drills and Skil screwdrivers. In this case, the focus should not be on these indices, but more on the PCI and the %C. The PCI, which only considers common and variant parts, has a higher value, which means that an effort was made when designing the non-unique parts. For the %C, the high value is due to the number of common connections between parts.

On the other hand, the DeWalt family has high values for the commonality indices. Most of the parts are indeed identical within the four products. The difference between the PCI and the TCCI can be explained by the fact that the family consists of three drills and one screwdriver. Whereas the design between the three drills is almost the same, the screwdriver has many unique components. While computing the PCI, most of these components are not considered: the resulting PCI is greater than the TCCI (74.70% vs. 52.10% for the TCCI). Although the high values of the commonality indices reveal a good design for the family, a major drawback can also be seen: by standardizing the products, it is hard to differentiate the different products in the family. This exemplifies the tradeoff between commonality and distinctiveness that was mentioned earlier.

Commonality indices provide useful information when comparing product families. By assessing a product family using these indices, the weakness (in terms of component sharing) in a product family can be identified, providing useful information for product family redesign.

4.2   Product family redesign using commonality indices

The indices can be used to obtain relevant information on the level of commonality in a product family. Each index enables designers to identify specific points in the design (such as the number of unique parts, etc.), and comparisons between commonality indices can yield additional information as to the product platform leveraging strategies. For example, as described in (Thevenot and Simpson, 2006), we can identify the different platform leveraging strategies for two different families of single-use cameras, namely, Kodak and Fujifilm. While Fujifilm developed four of their cameras based on two platforms only (with few common components between the two platforms), Kodak created more variety (aesthetically and functionally) by using more platforms (five platforms for seven cameras) but with more common parts shared between the platforms to avoid part proliferation. The methodology shown in Figure 7-6 has been proposed to use the commonality indices described in this chapter for product family redesign. Steps describing its application follow along with an example.

Step 1. Definition of the company's perspective when redesigning a product family: The first step is to define what the perspective of the company is when designing a product family. Depending on the strategy of the company, the focus can be on increasing the number of common components, common connections, reducing costs, etc. From that, the most relevant commonality index or indices can be chosen based on Table 7-13.

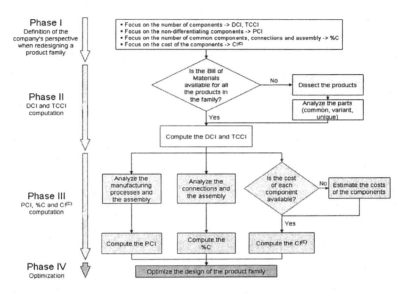

Figure 7-6. Proposed methodology for the computation of the commonality indices.

Table 7-13. Relationship between commonality index and product family design strategy.

Strategy	DCI	TCCI	CI	PCI	%C	CI(C)
Focus on the number of common components	X	X	X			
Focus on the non-differentiating components				X		
Focus on the number of common components, common connections, and assembly					X	
Focus on the cost of the components						X

Step 2. Computation of the TCCI and the DCI: Rather than limiting the computation to one index while designing a product family, the TCCI and the DCI can be computed first, as they give a quick idea of how good the design of the family is. The parts strongly affecting the commonality can be redesign at this stage. A bill of materials of the parts in each product of the family is required to compute the TCCI and the DCI; otherwise, a thorough dissection of the products is needed. Moreover, the fixed boundaries for the TCCI make comparisons between product families easier.

Step 3. Computation of more 'specific' indices depending on the strategy of the company: Based on the perspective of the company (defined in Step 1), the %C, the PCI, and/or the CI(C) should then be computed.

Step 4. Optimization of the design of the family based on these indices: When computed, these indices provide useful information as to how to improve the design of the products in the family. For example, the %C analyzes each product within the family, and it is easy to see which products or components lack commonality and why. Moreover, the commonality indices can be computed and used not only before the optimization, but also during the optimization. Hence, any change during the optimization stage can be directly measured. An example of optimization of a product family using a genetic algorithm (GA) can be found in (Thevenot, et al., 2005). Once the optimization is complete, the GA yields a redesign sequence that can be compared to the original design. This can be viewed as a reduction of the redesign space, where the designer checks the feasibility of the solution *a posteriori* using the list of proposed recommendations, rather than checking the feasibility of a redesign solution *a priori* in a much larger space. Two main types of information can come from using commonality indices. First, *at the product family level*, if there exists more than one design for a particular family, then the GA can assess each design and classifies them. Second, *at the component level*, a list of components to redesign is proposed to achieve the highest commonality with a minimum number of changes.

Recommendations at the product family level: if the designer wishes to assess more than one design for a product family, the GA is used to obtain a graph similar to the one shown in Figure 7-7. This graph aims at evaluating different design strategies of the concerned product family, based on how the factors that are changed influence the selected commonality index.

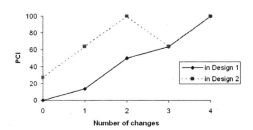

Figure 7-7. PCI versus number of changes in Design 1 and Design 2.

Figure 7-7 is obtained by first categorizing the values obtained for the commonality index based on the number of changes in the parameters. Consider the example shown in Table 7-14, with a product family consisting of three products, each product having two parts. Each part is used in each product. Two different designs need to be assessed. In Design 1, the parts are a variant in each product (i.e., no commonality). This is represented by attributing three different numbers to each part, one for each product (1, 2, and 3). In Design 2, there are two variants for each part, one variant being used by two products (some level of commonality), represented by having the same number for Part 1 – Product 1 and Part 1 – Product 2, and Part 2 – Product 1 and Part 2 – Product 3. The best design (relative to the concerned commonality indices, in this case the PCI) with the minimum number of changes is achieved through Design 3: the parts are common between all the products in the family (complete commonality; in fact, the three products are identical with regard to these two parts). Depending on the commonality index chosen, this may not always be the best design. For example, if the $CI^{(C)}$ is chosen to take the cost of each component into account, and if Variant 2 is cheaper to produce than Variant 1 (provided they both achieve the same function), then the "ideal" design should consist of having only Variant 2 used in all three products.

Table 7-14. Three different designs for two parts in a product family.

		Design 1	Design 2	Design 3
	in Product 1	1	1	1
Part 1	in Product 2	2	1	1
	in Product 3	3	2	1
	in Product 1	1	1	1
Part 2	in Product 2	2	2	1
	in Product 3	3	1	1
	Commonality Level:	None	→	Complete

By running the GA without constraints on Design 1 and Design 2, the optimal value of the commonality index is the one obtained in Design 3 (complete commonality). This value will be identical for both designs, as shown in Figure 7-7; however, the minimum number of changes to achieve this complete commonality is different. In Design 1, a minimum of 4 changes are necessary to achieve Design 3, while only 2 changes are required in Design 2 (see Figure 7-7). For any number of changes, the PCI in Design 1 is higher or equal to the one in Design 2; hence, Design 1 is a "dominated" design relative to the PCI: Design 2 achieves higher PCI (hence higher commonality) than Design 1, for any given number of changes.

Recommendations at the component level: the GA also provides a set of possible changes that could be implemented to maximize the commonality within the product family for a given number of changes. The best combination(s) of parts to redesign is proposed; additionally, the GA provides a ranked list of possible combinations. For a given number of changes between the current and the optimized design, the designer can then choose the feasible combination of parameters that results in the highest PCI value (i.e., highest increase in commonality). If we consider the example shown in Table 7-14 for Design 2, with a maximum number of changes set to 2, the algorithm will return the following information (only the first two sets of recommendations are shown):

1. First set of recommendations: {change Part 1 – Product 3 from Variant 2 to Variant 1, change Part 2 – Product 2 from Variant 2 to Variant 1}. This results in a PCI of 100.
2. Second set of recommendations: {change Part 1 – Product 1 from Variant 1 to Variant 2, change Part 1 – Product 2 from Variant 1 to Variant 2}. This results in a PCI of 63.5.

For a more realistic example, the same methodology has been applied to the Skil family of screwdrivers from Table 7-7. The original designs for the battery and gear case for each product is shown in Table 7-15. By using the PCI to assess the level of commonality in the family within the GA, the following recommendations were generated (only one set of recommendations is given here) for a given number of changes equal to five.

1. Change the battery from the 2.4V Twist: use the same battery as in the 2.4V Twist Power Driver instead.
2. Change the battery from the 3.6V Super Twist: use the same battery as in the 3.6V Twist Power Driver instead.
3. Change the gear case from the 3.6V Twist Power Driver: use the same gear case as in the 2.4V Twist Power Driver instead.
4. Change the gear case from the 2.4V Twist: use the same gear case as in the 2.4V Twist Power Driver instead.
5. Change the gear case from the 3.6V Super Twist: use the same gear case as in the 2.4V Twist Power Driver instead.

Table 7-15. Original design for two parts of the Skil family.

	2.4V Twist Power Driver	3.6V Twist Power Driver	2.4V Twist Screwdriver	3.6V Super Twist
	1	2	3	4
Battery				
	1	2	3	4
Gear Case				

The changes are summarized in Table 7-16. By implementing these five redesign changes, the PCI value increases from 41.20 to 45.50 (an increase of 10.43%), increasing the commonality in the family. Consequently, the use of component-based commonality indices for product family redesign can readily help designers: by giving a list of ranked solutions for possible redesign, the commonality indices-based GA has reduced the redesign space, which helps focus designers on the components that have the most influence on the commonality within the product family.

Table 7-16. Recommended design for two parts of the Skil family.

	2.4V Twist Power Driver	3.6V Twist Power Driver	2.4V Twist Screwdriver	3.6V Super Twist
	1	2	1	2
Battery				
	1	1	1	1
Gear Case				

5. CLOSING REMARKS

Commonality indices to assess product families are important tools when benchmarking and redesigning product families. They provide valuable information on how good the design of a product family is and how to improve it. The combined use of optimization algorithms and commonality indices to support product family redesign provides useful information for redesigning the product family, both at the *product-family level* (assessment of the overall design of a product family) and at the *component-level* (which components to redesign, how to redesign them). The reduction of the redesign space by providing a ranked list of components to modify during product family redesign helps designers focus on critical components that they may not have easily identified without such a systematic approach; however, future work should identify feasibility of the suggested changes. Additional research is also needed to assess the commonality of a product family more accurately based on the components in each product, along with their size, geometry, material, manufacturing process, assembly and costs.

6. ACKNOWLEDGMENTS

This work was supported by the National Science Foundation under Career Grant No. DMI-0133923. Any opinions, findings, and conclusions or recommendations presented in this chapter are those of the authors and do not necessarily reflect the views of the National Science Foundation.

PART II: OPTIMIZATION METHODS TO SUPPORT PLATFORM-BASED PRODUCT FAMILY DEVELOPMENT

Chapter 8

METHODS FOR OPTIMIZING PRODUCT PLATFORMS AND PRODUCT FAMILIES
Overview and Classification

Timothy W. Simpson
Departments of Mechanical & Nuclear Engineering and Industrial & Manufacturing Engineering, The Pennsylvania State University, University Park, PA 16802

1. THE ROLE OF OPTIMIZATION IN PRODUCT FAMILY DESIGN

Optimization has been used for many years during product design to help determine the values of design variables, $\mathbf{x}$, that minimize (or maximize) one or more objectives, $\mathbf{f}(\mathbf{x})$, while satisfying a set of constraints, $\{\mathbf{g}(\mathbf{x}), \mathbf{h}(\mathbf{x})\}$, and the design variable lower and upper bounds, $\mathbf{x}^l$ and $\mathbf{x}^u$, respectively. The typical notation for formulating the optimization problem is as follows:

$$
\begin{aligned}
&\text{Find:} && \mathbf{x} && (1)\\
&\text{Min:} && \mathbf{f}(\mathbf{x}) \\
&\text{Subject to:} && \mathbf{g}(\mathbf{x}) \leq 0 \\
& && \mathbf{h}(\mathbf{x}) = 0 \\
& && \mathbf{x}^l \leq \mathbf{x} \leq \mathbf{x}^u
\end{aligned}
$$

When optimizing a product family, this formulation must expand to include the values of the design variables for each product in the family such that now a *set* of constraints must be satisfied while trying to achieve a *set* of objectives for the family. Thus, the challenge when optimizing a family of products lies in resolving the tradeoff between commonality and individual product performance in the family: companies desire as much commonality as possible within a family without sacrificing the distinctiveness of the individual products in the family as discussed in Chapter 1. In this regard, optimization can be used to help identify the Pareto frontier for this inherent tradeoff. For instance, Simpson, et al. (2001b) examine the tradeoff between different levels of platform commonality within a family of three aircraft, while Nelson, et al. (2001) study the Pareto sets of two derivative products to find a suitable product platform for a family of nail guns. Rai and Allada (2003) present an agent-based optimization framework to capture the Pareto frontier for module-based product families, demonstrating their approach using a family of power screwdrivers and electric knives.

By identifying promising designs along the Pareto frontier, optimization provides useful information to determine the best values for the design variables that define the product platform and the individual products in the family. In some instances, the design variables that define the product platform within the family are known *a priori*, i.e., before performing the optimization, whereas in other instances, determining which variables should be part of the platform and which variables should be unique to each product is a desired output from the optimization. We can thus classify approaches to product family as requiring either *a priori* or *a posteriori* specification of the platform within the family.

Accordingly, we can envision two alternative approaches for optimizing the product platform and corresponding family of products, namely, optimize the platform first and then optimize the individual products or optimize both simultaneously. These two ways of approaching the problem allow us to classify optimization approaches based on the number of stages used. In a *two-stage approach*, for instance, the product platform is designed during the first stage of the optimization, followed by instantiation of the individual products from the product platform during the second stage. In a *single-stage approach*, the product platform and corresponding family of products are optimized simultaneously.

In the next section, an example involving the design of a family of electric motors is introduced to shed light on the merits and pitfalls of both types of approaches and clarify the challenges associated with product platform and product family optimization. Section 3 provides formulations for optimizing the family of motors using two-stage and single-stage approaches and *a priori* and *a posteriori* specification of the platform variables. In Section 4, forty approaches for optimizing product platforms and families of products are classified and reviewed, and closing remarks are offered in Section 5.

2. EXAMPLE: DESIGN OF A FAMILY OF UNIVERSAL ELECTRIC MOTORS

Universal electric motors are so named for their capability to function on both direct current and alternating current. Universal motors deliver more torque for a given current than any single-phase motor (Chapman, 1991). The high performance characteristics and flexibility of universal motors have led to a wide range of applications, especially in household use where they are found in products such as electric drills and saws, blenders, vacuum cleaners, and sewing machines (Veinott and Martin, 1986).

A schematic of a universal motor is shown in Figure 8-1. As shown in the figure, a universal motor is composed of an armature and a field, which are also referred to as the rotor and stator, respectively. The armature consists of a metal shaft and slats (armature poles) around which wire is wrapped longitudinally as many as a thousand times. The field consists of a hollow metal cylinder within which the armature rotates. The field also has wire wrapped longitudinally around interior metal slats (field poles) as many as hundreds of times. For a universal motor, the wire wrapped around the armature and the field is wired in series, which means that the same current is applied to both sets of wire.

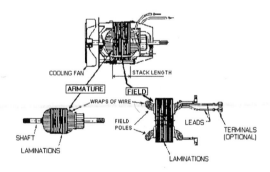

Figure 8-1. Schematic of a universal electric motor (G. S. Electric, 1997).

According to Lehnerd (1987), in the 1970s Black & Decker developed a family of universal motors for its power tools in response to a new safety regulation, namely, double insulation. Prior to that, they used different motors in each of their 122 basic tools with hundreds of variations, from jigsaws and grinders to edgers and hedge trimmers. By redesigning and standardizing the product line, they were able to produce all their power tools using a line of motors that varied only in the stack length and the amount of copper wrapped within the motor. As a result, all of the motors could be produced on a single machine with stack lengths varying from 0.8" to 1.75", and power output ranging from 60 to 650 watts. In addition to significant material and labor savings, new designs were developed using standardized components such as the redesigned motor, allowing products to be introduced, exploited, and retired with minimal extra development cost.

Motivated by Lehnerd's case study, an example problem involving the design of a family of universal electric motors has been created (Simpson, et al., 2001a). The goal in the example is to design a scale-based family of 10 universal electric motors that satisfy a variety of torque requirements based on a single platform. The motor platform consists of the set of common physical dimensions (design variables) that describe the motor while one or more variables are used to 'scale' the motor to satisfy the range of torque requirements. The motor analyses are described next, and specifications for the problem are given in Section 2.2.

2.1   Analyses for the universal electric motor example

The following equations relating the motor design variables to the system responses (i.e., mass, power, torque, and efficiency) are presented in their entirety in (Simpson, et al., 2001a) and are based on analyses from Chapman (1991) and Cogdell (1990). There are eight design variables for each motor:

1. Number of wire turns on the armature, N_c ($100 \leq N_c \leq 1500$)
2. Number of wire turns on each field pole, N_s ($1 \leq N_s \leq 500$)
3. Cross-sectional area of armature wire, A_{wa} ($0.01 \leq A_{wa} \leq 1.0$ mm^2)
4. Cross-sectional area of field wire, A_{wf} ($0.01 \leq A_{wf} \leq 1.0$ mm^2)
5. Radius of the motor, r_o ($0.01 \leq r_o \leq 0.10$ m)
6. Thickness of the stator, t ($0.0005 \leq t \leq 0.10$ m)
7. Current drawn by the motor, I ($0.1 \leq I \leq 6.0$ Amp)
8. Stack length of the motor, L ($0.001 \leq L \leq 0.10$ m)

The *mass* of the motor is the combined weight of the stator (field), the armature, and the windings on both the field and the armature.

$$\text{Mass} = M_{stator} + M_{armature} + M_{windings} \qquad (2)$$

where:

$$M_{stator} = \pi L [r_o^2 - (r_o - t)^2] \rho_{steel}$$

$$M_{armature} = \pi L (r_o - t - l_{gap})^2 \rho_{steel}$$

$$M_{windings} = \rho_{copper} \{ [2L + 4(r_o - t - l_{gap})] N_c A_{wa} + 2[2L + 4(r_o - t)] N_s A_{wf} \}$$

The *power*, P, output for the motor is the power input minus losses in the copper wiring and brushes; mechanical and core losses are assumed to be small and are thus neglected.

$$P = P_{in} - P_{losses} \qquad (3)$$

where:
$$P_{in} = VI$$

$$P_{losses} = P_{copper} + P_{brush}$$

with:
$$P_{copper} = I^2(R_a + R_s)$$

$$P_{brush} = 2I$$

where:
$$R_a = \{\rho[2L + 4(r_o - t - l_{gap})]N_c\}/A_{wa}$$

$$R_s = \{\rho(\#poles)[2L + 4(r_o - t)]N_s\}/A_{wf}$$

The *efficiency,* η, is the ratio of the power output to the power input.

$$\eta = P/P_{in} \qquad (4)$$

Finally, the *torque* generated by the motor is the product of the motor constant, K, the magnetic flux, ϕ, and the current, I.

$$T = K\phi I \qquad (5)$$

where:
$$K = N_c/\pi$$

$$\phi = \Im/\Re$$

$$\Im = N_s I$$

$$\Re = \Re_s + \Re_r + 2\Re_a$$

with $\Re_s = l_c/(2\mu_{steel}\mu_o A_s)$, $\Re_r = l_r/(\mu_{steel}\mu_o A_r)$, and $\Re_a = l_g/(\mu_{steel}\mu_o A_a)$. The μ's are obtained from magnetizing intensity curves in (Chapman, 1991), which requires:

$$H = (N_c I)/(l_c + l_r + 2l_{gap}), \qquad (6)$$

where:
$$l_c = \pi(2r_o + t)/2.$$

2.2   Problem specifications for the motor example

There are two distinct objectives that must be considered when designing the family of universal motors: minimizing the mass (kg) and maximizing the efficiency (%), which is equivalent to minimizing the negative of the efficiency of each motor. There are six constraints for each motor in the family, which are described as follows.

1. Constraint on torque, T_i, for each of the ten motors ($i = 1, ..., 10$):

$$T_i = \{0.05, 0.10, 0.125, 0.15, 0.20, 0.25, 0.30, 0.35, 0.40, 0.50\} \text{ Nm} \quad (7)$$

2. Constraint on power, P, for each motor in the family:

$$P = 300 \text{ W} \quad (8)$$

3. Constraint to ensure a feasible geometry for each motor in the family:

$$r_o/t \geq 1 \quad (9)$$

4. Constraint on the magnetizing intensity, H, in each motor in the family:

$$H \leq 5000 \text{ Amp*turns/m} \quad (10)$$

5. Constraint on the maximum mass of the each motor in the family:

$$\text{Mass} \leq 2 \text{ kg} \quad (11)$$

6. Constraint on the minimum efficiency of each motor in the family:

$$\eta > 15\% \quad (12)$$

Optimizing each motor individually involves 8 design variables, 2 objectives, and 6 constraints, but to optimize the family of 10 motors, the optimization problem, Eq. (1), becomes rather large. It is formally stated as:

Find: $\mathbf{x} = \{N_{c,i}, N_{s,i}, A_{wa,i}, A_{wf,i}, r_{o,i}, t_i, I_i, L_i\}$ $\qquad\qquad$ (13)

Min: $\mathbf{f(x)} = \{\text{Mass}_i, -\eta_i\}$

Subject to: $H_i(\mathbf{x}) \leq 5000$ Amp*turns/m

$\qquad\qquad\quad r_{o,i}/t_i \geq 1$

$\qquad\qquad\quad \text{Mass}_i(\mathbf{x}) \leq 2 \text{ kg}$

$\qquad\qquad\quad \eta_i(\mathbf{x}) \geq 15\%$

$\qquad\qquad\quad P_i(\mathbf{x}) = 300 \text{ W}$

$\qquad\qquad\quad T_i(\mathbf{x}) = \{0.05, 0.1, 0.125, 0.15, 0.2, 0.25, 0.3, 0.35, 0.4, 0.5\} \text{ Nm}$

$\qquad\qquad\quad x_i^l \leq x_i \leq x_i^u$

where $i = 1, ..., 10$ indicates each motor in the family (motor #1 has the lowest torque setting (0.05 Nm) and motor #10 the highest (0.50 Nm)).

All told, there are 80 design variables, 20 objectives, and 60 constraints, which is a challenging problem to solve for many optimization algorithms. Notice, however, that the idea of a platform is nowhere to be found in Eq.

(13); this formulation is simply for the set of 10 motors. Having a platform helps in reducing the size of the optimization problem by splitting the set of design variables, $\mathbf{x}$, into two subsets: one that is common for each product in the family and one that is unique for each product in the family. The set of common variables is usually represented as $\mathbf{x}_c$ where c stands for *common variables* while the set of *unique variables* is usually represented by $\mathbf{x}_{v,i}$, where v stands for each variant ($i = 1, \ldots, \#$ products) based on the platform. Sometimes the notation $\mathbf{x}_p$ is used instead of $\mathbf{x}_c$, where p stands for platform variables (Gonzalez-Zugasti, et al., 2000), but we avoid that notation to avoid confusion as to whether p stands for product or platform.

The designer must now decide how to partition the set $\mathbf{x}$ into these two subsets, $\{\mathbf{x}_c, \mathbf{x}_{v,i}\}$, which can either be specified before (i.e., *a priori*) or be found during (i.e., *a posteriori*) optimization. This gives rise to the two extreme cases of the tradeoff between commonality and distinctiveness: one in which all variable values are common and one in which all variable values are unique. In the first case, every product is the same, which means that none of them are distinct, whereas in the second case every product is unique, and there is no commonality between them. This latter case is referred to as the *null platform*, an important alternative if individual product distinctiveness is critical to market success (Nelson, et al., 2001; Simpson and D'Souza, 2004). While neither case is very practical, they provide the anchor points for the Pareto frontier that is defined by the competing objectives of commonality and individual product performance, and the optimization is used to find the best solution along this frontier for a given product family. The four different formulations follow.

3. PROBLEM FORMULATIONS AND RESULTS

The following four formulations demonstrate how the number of stages used and the specification of the platform variables in the subset, $\mathbf{x}_c$, affect the resulting solution for the family of motors. In Sections 3.1 and 3.2, the platform variables are specified *a priori* to the optimization while the optimization is solved first using a two-stage approach and then a single-stage approach, respectively. In Sections 3.3 and 3.4, more flexible formulations using a two-stage approach and a single-stage approach, respectively, are presented that do not require the specification of the platform variables *a priori*; instead, the optimization determines which variables should be made common and which should be made unique along with the best value for each variable (i.e., *a posteriori* specification of the platform variables). Section 3.5 provides a comparison of all the solutions.

3.1   Two-stage approach with platform variables specified a priori

The first formulation for the motor family followed the description given in the Black & Decker case study (Lehnerd, 1987), which stated that the axial profile of the motor was common and that the stack length was scaled to realize the family of motors. In (Simpson, et al., 2001a), we used this description to partition $\mathbf{x}$ from Eq. (13) into the platform variables, $\mathbf{x}_c = \{N_c,$ $N_s, A_{wa}, A_{wf}, r_o, t\}$, and the variables for each motor, $\mathbf{x}_{v,i} = \{I_i, L_i\}$. Note that I_i, the current in each motor, is best thought of as a state variable that varies for each motor to achieve the desired power. The resulting formulation is:

Find: $\mathbf{x}_c = \{N_c, N_s, A_{wa}, A_{wf}, r_o, t\}$ – *Stage 1*   (14)
             $\mathbf{x}_{v,i} = \{I_i, L_i\}$ – *Stage 2*

Min: $\mathbf{f}(\mathbf{x}) = \{Mass_i, -\eta_i\}$

Subject to: $H_i(\mathbf{x}) \leq 5000$ Amp*turns/m
             $r_{o,i}/t_i \geq 1$
             $Mass_i(\mathbf{x}) \leq 2$ kg
             $\eta_i(\mathbf{x}) \geq 15\%$
             $P_i(\mathbf{x}) = 300$ W
             $T_i(\mathbf{x}) = \{0.05, 0.1, 0.125, 0.15, 0.2, 0.25, 0.3, 0.35, 0.4, 0.5\}$ Nm
             $x_i^l \leq x_i \leq x_i^u$

where $i = 1, \ldots, 10$.

This formulation was solved using a goal programming approach for the two objectives that utilized targets of 0.5 kg and 70% for the mass and efficiency, respectively, and equally weighted deviations from these targets. In essence, once a motor weighed less than 0.5 kg and had an efficiency of 70% or more, it was "good enough" for the family. This approach provides more flexibility when finding solutions since we are not trying to optimize the performance of each individual motor, just reach a suitable target for each. The optimization was completed in two stages using the Generalized Reduced Gradient (GRG) algorithm in OptdesX (Parkinson and Balling, 2002). The first stage involved determining the best settings for the platform variables, $\mathbf{x}_c$, while the unique variables could take on any feasible value. In the second stage, the best values for the platform variables from the first stage, $\mathbf{x}_c^*$, where held constant, and 10 optimization problems were solved to find the best values of the remaining unique variables, $\mathbf{x}_{v,i}^*$, for each motor. The results are summarized in Table 8-1. When compared to a set of individually optimized motors (with no commonality), we found that the motor family based on this platform weigh 9% more, on average, and are 7%

less efficient, on average. Essentially, this compromise in product performance represents the loss of having increased commonality among the family of motors. We refer the reader to (Simpson, et al., 2001a) for more details and the complete formulation for each stage.

Table 8-1. Universal electric motor family based on initial platform formulation.

Motor No.	N_c	N_s	A_{wf} [mm²]	A_{wa} [mm²]	r_o [cm]	t [mm]	I [Amp]	L [cm]	T [Nm]	P [W]	η [%]	M [kg]
			Values of Platform Variables, x_c				Values of $x_{v,i}$		Responses			
1	1062	54	0.376	0.241	2.59	6.66	3.395	0.865	0.05	300	76.8	0.380
2	↓	↓	↓	↓	↓	↓	3.616	1.53	0.10	300	72.2	0.520
3	↓	↓	↓	↓	↓	↓	3.729	1.79	0.125	300	70.0	0.576
4	↓	↓	↓	↓	↓	↓	3.845	2.02	0.15	300	67.9	0.625
5	↓	↓	↓	↓	↓	↓	4.083	2.39	0.20	300	63.9	0.703
6	↓	↓	↓	↓	↓	↓	4.332	2.66	0.25	300	60.2	0.759
7	↓	↓	↓	↓	↓	↓	4.594	2.83	0.30	300	56.8	0.797
8	↓	↓	↓	↓	↓	↓	4.870	2.94	0.35	300	53.6	0.820
9	↓	↓	↓	↓	↓	↓	5.163	2.99	0.40	300	50.5	0.830
10	↓	↓	↓	↓	↓	↓	5.817	2.95	0.50	300	44.8	0.820

To examine this tradeoff in more detail, we examined motors in commercially available drills, and we determined that motor manufacturers vary more than just stack length when they scale their motors to meet a variety of torque and power ratings. In addition to increasing the stack length of the motor, they also allow the number of turns in the field and armature and the cross-sectional area of the wires in the field and armature to vary from one motor to the next. What this means is that the initial set of platform variables, $x_c = \{N_c, N_s, A_{wa}, A_{wf}, r_o, t\}$, may have been too restrictive, hence the loss in mass and efficiency due to the platform. If we reformulate Eq. (14) to reflect this, we get:

Find: $\quad x_c = \{r_o, t\} - Stage\ 1$ $\qquad\qquad\qquad$ (15)
$\qquad\qquad x_{v,i} = \{N_{c,i}, N_{s,i}, A_{wa,i}, A_{wf,i}, I_i, L_i\} - Stage\ 2$

Min: $\quad f(x) = \{Mass_i, -\eta_i\}$

Subject to: $H_i(x) \leq 5000$ Amp*turns/m
$\qquad\qquad r_{o,i}/t_i \geq 1$
$\qquad\qquad Mass_i(x) \leq 2$ kg
$\qquad\qquad \eta_i(x) \geq 15\%$
$\qquad\qquad P_i(x) = 300$ W
$\qquad\qquad T_i(x) = \{0.05, 0.1, 0.125, 0.15, 0.2, 0.25, 0.3, 0.35, 0.4, 0.5\}$ Nm
$\qquad\qquad x_i^l \leq x_i \leq x_i^u$

where $i = 1, ..., 10$.

Using the same two-stage approach and GRG algorithm, we obtain the results shown in Table 8-2. It turns out that these results are essentially equivalent in terms of their mass and efficiency to the set of individually optimized motors, yet the ten motors have the same axial profile (i.e., r_o and t are the same for all 10 motors) and vary in the amount of wire wrapped around each motor and its stack length just like the Black & Decker example (Lehnerd, 1987). Consequently, we have been able to resolve the tradeoff between commonality and individual product performance in a satisfactory manner for this family of motors using optimization.

Table 8-2. Universal electric motor family based on revised platform formulation.

Motor No.	Values of Platform Variables, x_c		Values of $x_{v,i}$						Responses			
	r_o [cm]	t [mm]	N_c	N_s	A_{wf} [mm²]	A_{wa} [mm²]	I [Amp]	L [cm]	T [Nm]	P [W]	η [%]	M [kg]
1	2.59	6.66	970	41	0.306	0.221	3.49	1.18	0.05	300	74.7	0.397
2	↓	↓	981	66	0.306	0.224	3.62	1.37	0.10	300	72.1	0.456
3	↓	↓	986	74	0.306	0.225	3.67	1.44	0.125	300	71.1	0.477
4	↓	↓	990	82	0.306	0.227	3.72	1.51	0.15	300	70.1	0.499
5	↓	↓	999	84	0.307	0.230	3.86	1.81	0.20	300	67.5	0.568
6	↓	↓	1064	80	0.359	0.239	4.03	2.03	0.25	300	64.6	0.646
7	↓	↓	1135	76	0.309	0.257	4.19	2.20	0.30	300	62.2	0.712
8	↓	↓	1166	75	0.282	0.268	4.35	2.42	0.35	300	59.9	0.774
9	↓	↓	1195	72	0.280	0.277	4.51	2.60	0.40	300	57.7	0.833
10	↓	↓	1242	67	0.286	0.293	4.85	2.91	0.50	300	53.8	0.941

3.2 Single-stage approach with platform variables specified a priori

Although the two-stage approach was successful in optimizing the platform and corresponding family of products, we were not certain as to what extent the tradeoff between commonality and individual product performance associated with Eq. (14) was caused by the selection of the platform variables versus the use of the two-stage approach. Consequently, we modified the formulation in Eq. (14) and solved it using a single-stage approach as shown in Eq. (16). This required a different optimization algorithm due to the increased problem size. In particular, Physical Programming (Messac, 1996) was used to formulate and solve the optimization problem in a single stage. The results are summarized in Table 8-3, and details can be found in (Messac, et al., 2002b) along with the complete formulation. When compared to the set of individually optimized motors mentioned earlier, this family of motors weigh 7% more, on average, and are 4.5% less efficient on average. Compared to the two-stage solutions given in Table 8-1, this represents a 2% improvement in mass, on average, and a 2.5% gain in efficiency, on average. While this may not seem like much, it translates into weight reductions in 7 of the 10 motors and increased efficiency in 8 of the 10 motors—results any manufacturer would enjoy.

Find: $\mathbf{x}_c = \{r_o, t\}$, $\mathbf{x}_{v,i} = \{N_{c,i}, N_{s,i}, A_{wa,i}, A_{wf,i}, I_i, L_i\}$ – *Stage 1* (16)

Min: $\quad\mathbf{f}(\mathbf{x}) = \{Mass_i, -\eta_i\}$

Subject to: $\quad H_i(\mathbf{x}) \le 5000$ Amp*turns/m

$\quad\quad\quad\quad r_{o,i}/t_i \ge 1$

$\quad\quad\quad\quad Mass_i(\mathbf{x}) \le 2$ kg

$\quad\quad\quad\quad \eta_i(\mathbf{x}) \ge 15\%$

$\quad\quad\quad\quad P_i(\mathbf{x}) = 300$ W

$\quad\quad\quad\quad T_i(\mathbf{x}) = \{0.05, 0.1, 0.125, 0.15, 0.2, 0.25, 0.3, 0.35, 0.4, 0.5\}$ Nm

$\quad\quad\quad\quad x_i^{1} \le x_i \le x_i^{u}$

where $i = 1, \dots, 10$.

Table 8-3. Motor family using single-stage approach and initial platform formulation.

Motor No.	\multicolumn{6}{c	}{Values of Platform Variables, $\mathbf{x}_c$}	\multicolumn{2}{c	}{Values of $\mathbf{x}_{v,i}$}	\multicolumn{4}{c}{Responses}
	N_c	N_s	A_{wf} [mm²]	A_{wa} [mm²]	r_o [cm]	t [mm]	I [Amp]	L [cm]	T [Nm]	P [W]	η [%]	M [kg]
1	1273	61	0.271	0.271	2.673	7.745	3.432	0.617	0.05	300	76.0	0.395
2	↓	↓	↓	↓	↓	↓	3.618	1.110	0.10	300	72.1	0.513
3	↓	↓	↓	↓	↓	↓	3.713	1.317	0.125	300	70.3	0.562
4	↓	↓	↓	↓	↓	↓	3.810	1.501	0.15	300	68.5	0.606
5	↓	↓	↓	↓	↓	↓	4.010	1.806	0.20	300	65.1	0.678
6	↓	↓	↓	↓	↓	↓	4.219	2.040	0.25	300	61.8	0.734
7	↓	↓	↓	↓	↓	↓	4.438	2.213	0.30	300	58.8	0.775
8	↓	↓	↓	↓	↓	↓	4.668	2.333	0.35	300	55.9	0.803
9	↓	↓	↓	↓	↓	↓	4.912	2.408	0.40	300	53.1	0.821
10	↓	↓	↓	↓	↓	↓	5.451	2.444	0.50	300	47.9	0.830

3.3 Two-stage approach with platform variables determined during optimization

The primary goal when specifying the platform variables *a priori* is to reduce the problem size and resulting computational burden of solving the product family optimization; however, this is when the designer knows the least about which variables have the largest impact on product performance. Selecting the appropriate set of common variables, $\mathbf{x}_c$, for the platform and unique variables, $\mathbf{x}_{v,i}$, for the individual variants within a product family is not an intuitive or trivial task, and we saw the adverse impact that this can have on the overall performance of the product family in Section 3.1. If n is the number of variables that are possible candidates for being made common to a platform (with the remainder being unique among each product variant), then the number of platform alternatives is:

$$\frac{\#\,platform}{alternatives} = \binom{n}{n} + \binom{n}{n-1} + \cdots + \binom{n}{2} + \binom{n}{1} + \binom{n}{0} = 2^n \qquad (17)$$

where $\begin{pmatrix} n \\ c \end{pmatrix}$ is "n choose c", namely, the number of possible combinations of n items (i.e., design variables) taken c at a time (i.e., made common). Note that the alternative $c = 0$ is the *null platform* discussed in Section 2.2.

Ideally, an algorithm for product family design optimization would explore varying levels of commonality to determine the best platform for the family rather than require specifying the common and unique variables *a priori*. Toward that end, we developed a two-stage optimization approach that incorporates a Product Family Penalty Function (PFPF) into the Physical Programming formulation to help determine which variables have the largest impact on performance to drive commonality (Messac, et al., 2002a). The PFPF is used to minimize the variations of the design variables within the family by minimizing the percent variation, $pvar_j$:

$$pvar_j = \frac{var_j}{\overline{x}_j} \qquad (18)$$

where: $\quad var_j = \sqrt{\dfrac{\sum_{i=1}^{p}(x_{ij} - \overline{x}_j)^2}{p-1}} \quad$ and $\quad \overline{x}_j = \dfrac{\sum_{i=1}^{p} x_{ij}}{p} \qquad (19)$

x_{ij} is the value of the jth design variable for the ith product of the p products in the family. The PFPF is an additional objective function that is computed by summing the percent variation of all n design variables within the family:

$$PFPF = \sum_{j=1}^{n} pvar_j \qquad (20)$$

Lower values of PFPF mean more commonality while higher values indicate less. The PFPF is added to the "Min:" statement of Eq. (13) to yield:

Find: $\mathbf{x} = \{N_{c,i}, N_{s,i}, A_{wa,i}, A_{wf,i}, r_{o,i}, t_i, I_i, L_i\}$ (21)

Min: $\mathbf{f(x)} = \{Mass_i, -\eta_i, PFPF\}$

Subject to: $H_i(\mathbf{x}) \leq 5000$ Amp*turns/m

              $r_{o,i}/t_i \geq 1$

              $Mass_i(\mathbf{x}) \leq 2$ kg

              $\eta_i(\mathbf{x}) \geq 15\%$

              $P_i(\mathbf{x}) = 300$ W

              $T_i(\mathbf{x}) = \{0.05, 0.1, 0.125, 0.15, 0.2, 0.25, 0.3, 0.35, 0.4, 0.5\}$ Nm

              $x_i^l \leq x_i \leq x_i^u$

where $i = 1, ..., 10$.

The two-stage optimization approach, which is described in detail in (Messac, et al., 2002a) along with the complete formulation of the optimization problem, uses the PFPF to identify which variables have the largest impact on product performance during the first stage of the optimization, and these variables are selected as the unique variables, x_v, while the remaining variables are taken as platform variables, x_c. The second stage involves finding the best settings for the variables in x_c and x_v using Physical Programming as described in the previous section. In this example, the unique variables were limited to any one variable plus the current, and the results are listed in Table 8-4. Compared to the set of individually optimized motors mentioned earlier, this family of motors weigh only 3% more, on average, and are only 3% less efficient on average, a marked improvement over the results given in Table 8-1, which also scale the platform around a single variable.

Table 8-4. Motor family using two-stage approach and scaling the platform by radius.

Motor No.	Values of Platform Variables, x_c						Values of $x_{v,i}$		Responses			
	N_c	N_s	A_{wf} [mm²]	A_{wa} [mm²]	t [mm]	L [cm]	I [Amp]	r_o [cm]	T [Nm]	P [W]	η [%]	M [kg]
1	1319	68	0.256	0.256	9.22	2.12	3.18	1.46	0.05	300	82.0	0.312
2	↓	↓	↓	↓	↓	↓	3.40	1.83	0.10	300	76.6	0.422
3	↓	↓	↓	↓	↓	↓	3.52	1.98	0.125	300	74.1	0.472
4	↓	↓	↓	↓	↓	↓	3.64	2.11	0.15	300	71.6	0.518
5	↓	↓	↓	↓	↓	↓	3.90	2.33	0.20	300	67.0	0.595
6	↓	↓	↓	↓	↓	↓	4.16	2.48	0.25	300	62.7	0.653
7	↓	↓	↓	↓	↓	↓	4.43	2.59	0.30	300	58.9	0.693
8	↓	↓	↓	↓	↓	↓	4.71	2.65	0.35	300	55.4	0.719
9	↓	↓	↓	↓	↓	↓	4.99	2.68	0.40	300	52.2	0.732
10	↓	↓	↓	↓	↓	↓	5.58	2.69	0.50	300	46.8	0.734

While we expected stack length to have the largest impact on the individual product performance, we were somewhat surprised by these results when we learned that stack length was part of the platform and that the radius was the unique variable used to 'scale' the motors. In talking with practicing motor designers, we confirmed that our finding was true: by varying the torque requirement as we do, it is more effective to scale the radius than the stack length; however, it is much more cost effective to manufacture motors that are scaled along the stack length, which influenced Black & Decker's decision. Furthermore, as we saw in Section 3.1, as long as we also vary the amount of wire wrapped around each motor, we obtain an equivalent set of motors if the axial profile is fixed. Our findings were confirmed in parallel work by Nayak, et al. (2002), who used a commonality goal within their two-stage goal programming formulation. They also found that the platform should be scaled around the motor radius (as well as N_s, A_{wa}, A_{wf} as the selection of the scaling variable is not limited to one variable) not the stack length, to get the best performance within the motor family.

3.4   Single-stage approach with platform variables determined during optimization

We are currently investigating a single-stage approach that uses genetic algorithms (Goldberg, 1989) to examine varying levels of platform commonality during product family optimization (D'Souza and Simpson, 2003; Simpson and D'Souza, 2004). As outlined in (Simpson and D'Souza, 2004), our approach utilizes a set of commonality controlling genes a genetic algorithm (GA) to evaluate varying levels of platform commonality. As shown in Figure 8-2, the chromosome string in the GA concatenates the individual chromosomes strings for each product into one long string and then augments this string with n genes that control the commonality within the individual chromosome strings. The resulting length of the chromosome string is $n + np$, where n is the number of design variables and p is the number of products. Note that if any of these first n genes take the value of 1, then that particular design variable is made common among all of the products in the family; a value of 0 makes that design variable unique within the family. It follows then that if these first n genes are all 1's, there is one hundred percent commonality among the products in the family while a string of all 0's indicates no commonality among the products within the family. As such, varying levels of platform commonality are considered in a single stage process, where the results from the optimization indicate:

1. which variables should be made common (i.e., platform variables),
2. the values that they should take, and
3. the values that the remaining unique variables should take.

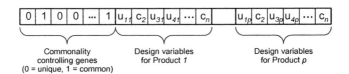

Figure 8-2. GA representation for searching varying levels of platform commonality.

For the universal electric motor family, the resulting chromosome string is 88 genes long, and an example is shown in Figure 8-3. Note that this particular chromosome string represents the motor family listed in Table 8-1. The platform variables, as indicated by the commonality controlling genes, are the first six variables, which equates to $x_c = \{N_c, N_s, A_{wa}, A_{wf}, r_o, t\}$, while the last two variables are unique to each product, which equates to $x_{v,i} = \{I_i, L_i\}$. The values for the corresponding variable for each motor are the same values listed in Table 8-1. This example is only one potential solution that the GA would consider during optimization, as each population in each generation will have different values for the commonality controlling genes as well as the individual variables for each product in the family.

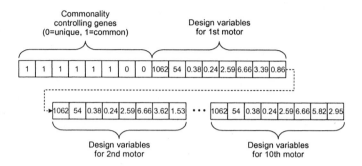

Figure 8-3. Example of GA representation for the universal electric motor family.

Formulation of the product family optimization problem is similar to that of Eq. (21) when using the genetic algorithm. One difference is that the values of the commonality controlling genes are added to the optimization as they dictate how $\mathbf{x}$ is partitioned into $\mathbf{x}_c$ and $\mathbf{x}_{v,i}$. We denote these genes as $\mathbf{x}_{cc}$, where $x_{cc,j}$ indicates the value of the jth commonality controlling gene, which is either 0 or 1, where $j = 1, \ldots, n\ (= 8)$. To solve the problem, we use the NSGA-II, which is available online from the Kanpur Genetic Algorithm Lab in India: http://www.iitk.ac.in/kangal/soft.htm. The NSGA-II is a multi-objective genetic algorithm that can handle multiple fitness functions and constraints (Srinivas and Deb, 1995), and we use three fitness functions (i.e., minimize mass, minimize negative efficiency, and minimize PFPF) to optimize the motor family. The resulting formulation is as follows.

Find: $\quad \mathbf{x} = \{N_{c,i}, N_{s,i}, A_{wa,i}, A_{wf,i}, r_{o,i}, t_i, I_i, L_i\}\ \&\ \mathbf{x}_{cc} = \{x_{cc,j}\}$ (22)

Min: $\quad$ Fitness function 1, 2, & 3 $= \sum_{i=1}^{10} \text{Mass}_i\ ,\ -\sum_{i=1}^{10} \eta_i\ ,\ \&\ \text{PFPF}$

Subject to: $H_i(\mathbf{x}) \leq 5000$ Amp*turns/m

$\qquad\qquad r_{o,i}/t_i \geq 1$

$\qquad\qquad \text{Mass}_i(\mathbf{x}) \leq 2$ kg

$\qquad\qquad \eta_i(\mathbf{x}) \geq 15\%$

$\qquad\qquad P_i(\mathbf{x}) = 300$ W

$\qquad\qquad T_i(\mathbf{x}) = \{0.05, 0.1, 0.125, 0.15, 0.2, 0.25, 0.3, 0.35, 0.4, 0.5\}$ Nm

$\qquad\qquad x_{cc,j} = \{0,1\}$

$\qquad\qquad x_i^l \leq x_i \leq x_i^u$

where $i = 1, \ldots, 10$, and $j = 1, \ldots, 8$.

Two representative sets of results are listed in Table 8-5 and Table 8-6. The motor family listed in Table 8-5 results from the commonality controlling genes, $\mathbf{x}_{cc}$, taking the values $\{1,1,1,1,1,1,1,0,0\}$, which is equivalent to the platform defined in Eq. (14). Consequently, the resultant motors listed in Table 8-5 are very similar to those listed in Table 8-1 in terms of their mass and efficiency. For the family listed in Table 8-6, $\mathbf{x}_{cc} = \{0,0,0,0,1,1,0,0\}$, which equates to the platform defined in Eq. (15) and motor family listed in Table 8-2. Comparing these solutions to those from Table 8-2, we have much less variability in the values of the $\mathbf{x}_{v,i}$ variables as well as slight improvements in both mass and efficiency. This demonstrates the power and flexibility of the GA-based method in that both of these motor families come from the same generation; two separate optimization problems do not have to be solved to find them. Moreover, they are obtained using a single-stage approach that does not require *a priori* specification of the platform.

Table 8-5. Universal motor family from GA equivalent to initial platform.

Motor No.	\multicolumn{6}{c}{Values of Platform Variables, $\mathbf{x}_c$}						\multicolumn{2}{c}{Values of $\mathbf{x}_{v,i}$}		\multicolumn{4}{c}{Responses}			
	N_c	N_s	A_{wf} [mm²]	A_{wa} [mm²]	r_o [cm]	t [mm]	I [Amp]	L [cm]	T [Nm]	P [W]	η [%]	M [kg]
1	1057	55	0.348	0.238	2.54	7.08	3.39	0.878	0.05	300	77.4	0.364
2	↓	↓	↓	↓	↓	↓	3.61	1.542	0.10	300	72.6	0.500
3	↓	↓	↓	↓	↓	↓	3.73	1.806	0.125	300	70.4	0.554
4	↓	↓	↓	↓	↓	↓	3.84	2.043	0.15	300	68.3	0.602
5	↓	↓	↓	↓	↓	↓	4.08	2.412	0.20	300	64.3	0.678
6	↓	↓	↓	↓	↓	↓	4.34	2.689	0.25	300	60.5	0.735
7	↓	↓	↓	↓	↓	↓	4.59	2.860	0.30	300	57.2	0.770
8	↓	↓	↓	↓	↓	↓	4.86	2.975	0.35	300	54.0	0.793
9	↓	↓	↓	↓	↓	↓	5.15	3.028	0.40	300	51.0	0.804
10	↓	↓	↓	↓	↓	↓	5.79	3.005	0.50	300	45.3	0.800

Table 8-6. Universal motor family from GA with radius and thickness as platform.

Motor No.	\multicolumn{2}{c}{Values of Platform Variables, $\mathbf{x}_c$}		\multicolumn{6}{c}{Values of $\mathbf{x}_{v,i}$}						\multicolumn{4}{c}{Responses}			
	r_o [cm]	t [mm]	N_c	N_s	A_{wf} [mm²]	A_{wa} [mm²]	I [Amp]	L [cm]	T [Nm]	P [W]	η [%]	M [kg]
1	2.54	7.03	1057	55	0.366	0.236	3.38	0.88	0.05	300	77.3	0.365
2	↓	↓	1051	55	0.356	0.236	3.61	1.547	0.10	300	72.5	0.499
3	↓	↓	1057	55	0.357	0.238	3.73	1.828	0.125	300	70.3	0.560
4	↓	↓	1057	55	0.367	0.239	3.84	2.044	0.15	300	68.5	0.606
5	↓	↓	1057	55	0.367	0.237	4.08	2.408	0.20	300	64.3	0.679
6	↓	↓	1057	57	0.359	0.238	4.30	2.632	0.25	300	61.1	0.727
7	↓	↓	1057	55	0.368	0.236	4.59	2.855	0.30	300	57.0	0.769
8	↓	↓	1057	56	0.359	0.236	4.84	2.973	0.35	300	53.9	0.793
9	↓	↓	1057	55	0.351	0.238	5.15	3.028	0.40	300	51.0	0.805
10	↓	↓	1057	55	0.353	0.238	5.79	2.995	0.50	300	45.4	0.799

The GA-based method reports motor families for 64 different platforms, which are based on different feasible combinations of $\{0,1\}$ for $\mathbf{x}_{cc}$. Among these are solutions based on the 'null platform', $x_{cc} = \{0,0,0,0,0,0,0,0\}$, and an example is shown in Table 8-7. The performance of the family based on this null platform is very similar to the two families listed in Table 8-5 and Table 8-6 due to the use of the PFPF as a third fitness function in the GA, i.e., all of the solutions are driven to nearly the same region within the design space, and there is not much variation in the values of $x_{v,i}$ as seen in the table. In fact, some might argue that this is not really a 'null' platform since many values are common across several, but not all, of the motors in the family. This gives rise to the question: is platforming an "all or nothing" proposition? The answer is no, which is exactly how variant components come about that are shared between some, but not all, of the products in the family. This is exactly the problem that Hernandez, et al. (2002) tackle, using the electric motor family as an example. Meanwhile, work continues with the GA-based method to improve solution diversity to spread out points in the design space and determine the best settings for the GA parameters to generate more diverse solution sets (Akundi, et al., 2005).

Table 8-7. Universal motor family from GA based on the null platform.

Motor No.			Values of $x_{v,i}$							Responses		
	N_c	N_s	A_{wf} [mm^2]	A_{wa} [mm^2]	r_o [cm]	t [mm]	I [Amp]	L [cm]	T [Nm]	P [W]	η [%]	M [kg]
1	1056	55	0.348	0.234	2.54	6.61	3.38	0.880	0.05	300	76.6	0.365
2	1056	55	0.356	0.236	2.54	6.96	3.61	1.547	0.10	300	72.4	0.501
3	1056	55	0.356	0.236	2.54	6.99	3.73	1.808	0.125	300	70.2	0.554
4	1056	55	0.356	0.235	2.54	6.99	3.84	2.039	0.15	300	68.0	0.600
5	1056	56	0.357	0.236	2.52	6.99	4.08	2.408	0.20	300	64.3	0.670
6	1056	57	0.359	0.237	2.54	6.99	4.29	2.632	0.25	300	61.0	0.726
7	1056	55	0.355	0.236	2.54	6.99	4.59	2.855	0.30	300	56.9	0.768
8	1055	57	0.354	0.236	2.54	6.99	4.83	2.926	0.35	300	54.2	0.784
9	1056	55	0.351	0.236	2.54	6.99	5.15	3.027	0.40	300	50.6	0.803
10	1056	55	0.356	0.235	2.54	6.99	5.79	2.995	0.50	300	44.8	0.795

3.5 Comparison of motor families

Figure 8-4 provides a graphical comparison of the motor families based on how well they achieve their mass and efficiency targets of ≤ 0.5 kg and $\geq 70\%$, respectively, which is labeled the 'Utopia Region'. The results are plotted in the order in which they were presented, progressing from the *a prior* formulations that use two stages ($\bullet$ = Table 8-1 and $\circleddash$ = Table 8-2) and a single stage ($\bigcirc$ = Table 8-3) to the *a posteriori* formulations that use two stages ($\blacksquare$ = Table 8-4) and a single stage ($\square$ = Tables 8-5 to Table 8-7). The results from the single-stage *a posteriori* GA-based method are nearly identical even though the two platforms differ; therefore, only one solution set is plotted in the figure with the $\square$ symbol. The set of individually optimized motors from (Simpson, et al., 2001a) is also included for comparison as indicated by the $\blacklozenge$ symbol.

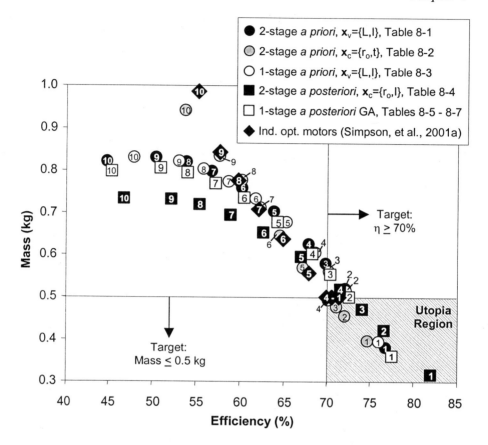

Figure 8-4. Graphical comparison of universal electric motor families.

To facilitate comparison, each symbol is numbered to correspond to a particular motor in the family where a 1 denotes the motor with the lowest required torque setting (0.05 Nm) and a 10 indicates the motor with the highest required torque setting (0.5 Nm). Based on Figure 8-4, it is much easier to visualize the tradeoff between mass and efficiency within the motor family and how the amount of commonality exacerbates this tradeoff. For instance, we can clearly see how the ○ family of motors is nearly equivalent to the set of individually optimized motors: both have four motors in the 'Utopia Region' and the higher torque motors fall very close to one another except for motor #10, which is slightly less efficient but weighs less. Conversely, we can see the extent of the performance loss for the initial platform family (●) when compared to the set of individually optimized motors (◆). If a company wanted to use that platform, we would suggest the family obtained from the GA-based method (□). We can also see the improvement in the results of the single-stage *a priori* formulation (○) to the two-stage formulation (●): most of these motors offer improved efficiency at equivalent mass or offer more efficiency at less mass. The two-stage *a posteriori* motor family (■) yields the best motor family, as these motors tend to have the least mass while having equivalent or higher efficiency.

Several researchers have used the universal electric motor example to benchmark their optimization approaches against published results. For instance, recent work has investigated the use of ant colony optimization (Kumar, et al., 2004) and preference aggregation (Dai and Scott, 2004a) to improve the performance of the motor family. Meanwhile, others have designed the family of motors by solving it as a problem of access in a geometric space (Hernandez, et al., 2002) and by using sensitivity and cluster analysis (Dai and Scott, 2004b) to help identify the platform. A classification and review of many different approaches to product platform and product family optimization is given next.

4.   CLASSIFICATION AND SUMMARY OF OPTIMIZATION APPROACHES

Several optimization approaches have been developed within the engineering design community during the past decade to facilitate product family design and optimization. Table 8-8 classifies 40 approaches from the literature based on the following categories:

1. *Module- or scale-based product family?* – does the problem formulation focus on module- or scale-based families or both? In the universal electric motor example, the emphasis was on scaling the motor around one or more design variables, but the motor could just as easily be taken as a module within a larger problem in designing a family of power tools, for instance. In Table 8-8, 'M' indicates module-based, 'S' scale-based, and 'MS' both.

2. *Single or multiple objectives* – how many objectives are used when formulating the problem? In some cases, only a single objective is used whereas multiple objectives are often considered as evidenced in the universal electric motor example. A 'S' in the table indicates that only a single objective is used while a 'M denotes multiple objectives are considered in the problem formulation.

3. *Model market demand?* – is market demand explicitly modeled and used in the problem formulation? A 'Y' under this heading indicates yes, and a blank indicates that market demand is not being considered. Although not part of the universal electric motor example, this is an important aspect of the problem that should be considered whenever possible.

Table 8-8. Summary of engineering optimization approaches for product family design.

Reference	Module- or scale-based family	Single or multiple objectives	Model market demand	Model manufacturing costs	Consider uncertainty	Specify platform a priori?	Number of stages	Optimization Algorithm(s)	Product Family Example (# of products in the family)
	Details of Formulation								
(Allada and Jiang, 2002)	M	S	Y		Y	Y	2+	DP	Generic modular products (3)
(Akundi, et al., 2005)	S	M					1	GA	Universal electric motors (10)
(Blackenfelt, 2000)	M	S		Y	Y	Y	1	OA	Lift tables (4)
(Cetin and Saitou, 2004)	M	S					1	GA, SA	Welded auto. structures (2)
(Chang and Ward, 1995)	M	S			Y	Y	1	OA	Automotive A/C units (6)
(D'Souza and Simpson, 2003)	S	M				Y	1	GA	General Aviation Aircraft (3)
(Dai and Scott, 2004a)	S	S				Y	1,2	SQP	Universal electric motors (10)
(Dai and Scott, 2004b)	S	S					2	SQP	Universal electric motors (10)
(de Weck, et al., 2003)	M	S	Y	Y			2	SQP	Automotive vehicles (7)
(Farrell and Simpson, 2003)	S	S	Y			Y	2	GRG	Flow control valves (16)
(Fellini, et al., 2000)	M	M				Y	2	NLP	Automotive power train (3)
(Fellini, et al., 2002a)	S	M					2	SQP	Automotive vehicle frames (2)
(Fellini, et al., 2002b)	S	M					2	SQP	Automotive vehicle frames (2)
(Fujita, et al., 1998)	M	S	Y	Y		Y	1	SQP	Commercial aircraft (2)
(Fujita, et al., 1999)	M	S		Y			1	SA	TV receiver circuits (6)
(Fujita and Yoshida, 2001)	B	S	Y	Y			1	SQP, GA, B&B	Commercial aircraft (4)
(Fujita and Yoshioka, 2003)	S	M				Y	1	GA	Auto. lift gate dampers (6)
(Gonzalez-Zugasti, et al., 2000)	M	M				Y	1	NLP	Interplanetary spacecraft (3)
(Gonzalez-Zugasti and Otto, 2000)	M	S		Y			1	GA	Interplanetary spacecraft (3)
(Gonzalez-Zugasti, et al., 2001)	M	S	Y	Y	Y	Y	2	NLP	Interplanetary spacecraft (3)
(Hernandez, et al., 2001)	S	M		Y		Y	2	SA	Absorption chillers (8)
(Hernandez, et al., 2002)	S	S					2+	PaS	Universal electric motors (10)
(Hernandez, et al., 2003)	M	S		Y			2+	ExS	Pressure vessels (16)
(Jiang and Allada, 2001)	M	S	Y	Y	Y		2	SLP	Vacuum cleaners (3)
(Kokkolaras, et al., 2002)	M	M				Y	2	NLP	Auto. vehicle frames (2)
(Kumar, et al., 2004)	S	M				Y	1	Ant	Universal electric motors (10)
(Li and Azarm, 2002)	M	S	Y	Y	Y	Y	2	GA	Cordless screwdrivers (3)
(Messac, et al., 2002a)	S	M					2	NLP	Universal electric motors (10)
(Messac, et al., 2002b)	S	M				Y	1	NLP	Universal electric motors (10)
(Nayak, et al., 2002)	S	M					2	SLP	Universal electric motors (10)
(Nelson, et al., 2001)	M	M				Y	2	NLP	Nail guns (2)
(Ortega, et al., 1999)	S	M		Y		Y	1	SLP	Oil filters (5)
(Rai and Allada, 2003)	M	M	Y	Y			2	NLP	Screwdrivers (3), knives (4)
(Hassan, et al., 2004)	M	M					1	GA	Commercial satellites (3)
(Seepersad, et al., 2000)	S	M	Y	Y	Y	Y	1	SA	Absorption chillers (8)
(Seepersad, et al., 2002)	S	S	Y	Y	Y	Y	2	SA	Absorption chillers (12)
(Simpson, et al., 1999)	S	M				Y	1	SLP	General Aviation Aircraft (3)
(Simpson, et al., 2001a)	S	M				Y	2	GRG	Universal electric motors (10)
(Simpson and D'Souza, 2004)	S	M					1	GA	General Aviation Aircraft (3)
(Willcox and Wakayama, 2003)	S	S				Y	1	SQP	Blended-wing-body aircraft (2)

4. *Model manufacturing cost?* – is manufacturing or production cost explicitly modeled and used in the problem formulation? A 'Y' under this heading in the table indicates yes, and a blank indicates that manufacturing cost is not being considered. As with market demand, it is important to model and include this aspect of the problem whenever possible as it is often an important decision criterion as noted in Section 3.3 for the electric motor example.

5. *Consider uncertainty?* – does the problem formulation take uncertainty into account in either the design, manufacturing, and/or market demand aspects of the problem? A 'Y' in the table under this heading indicates that one or more sources of uncertainty is being considered; a blank indicates that no uncertainty is being incorporated into the problem formulation. While this was not considered in the electric motor example, many researchers have explored the implications of uncertainty as noted in the table.

6. *Specify platform a priori?* – does the designer have to specify the platform variables *a priori* or is the problem formulated so as to identify both the platform and the family during optimization (i.e., *a posteriori*)? A 'Y' under this heading in the table indicates that the platform variables must be specific *a priori* whereas a blank indicates that they do not. Examples of both cases were given for the universal electric motor example in the previous section.

7. *Number of stages* – how many stages are used to solve the optimization problem? A '1' under this heading in the table indicates that a single stage is used, a '2' indicates that two stages are used, and '2+' indicates that more than two stages are used. Examples of two-stage and single-stage approaches were given for the universal electric motor example in Section 3.

8. *Optimization algorithm* – what optimization algorithm is used to solve the problem once it is formulated? One or more of the following acronyms is listed under this heading to indicate the type of algorithm used: B&B = Branch and Bound, DP = Dynamic Programming, ExS = Exhaustive Search, GA = Genetic Algorithm, GRG = Generalized Reduced Gradient, NLP = Non-Linear Programming, OA = Orthogonal Array, PaS = Pattern Search, SA = Simulated Annealing, SLP = Sequential Linear Programming, and SQP = Sequential Quadratic Programming. Many of these algorithms have been applied to the universal electric motor example as noted in the table.

9. *Product family example* – the last column in the table lists the type(s) of product family that is used as an example or test case in the cited work. The number of products in the family is also listed to provide an indication as to the size of the problem being solved.

In looking at the table, the approaches are split evenly between module-based and scale-based product families, while the work by Fujita and Yoshida (2001) specifically addresses both (see also Chapter 10). More than half of the approaches use multi-objective optimization, and three assumptions are often made when using multi-objective optimization:

1. maximizing each product's performance maximizes its demand,
2. maximizing commonality among products minimizes costs, and
3. resolving the tradeoff between (1) and (2) yields the most profitable product family.

Without explicitly modeling *market demand* and associated *manufacturing costs*, however, these assumptions may lead to sub-optimal product families.

The universal electric motor example in the previous sections provides an example of when this can occur. The initial formulation scaled the motors around the stack length of the motor (see Section 3.1), but maximizing commonality in the family using two different approaches revealed that the motor platform should be scaled by the radius to maximize performance. As discussed in Section 3.3, the best choice is stack length, and through discussions with experienced motor designers, we found that production costs, not performance, primarily drove the use of stack length as the scaling variable (Simpson, et al., 2001a). In the table, note that only about half the approaches integrate manufacturing costs directly within the formulation while fewer than one-third incorporate market demand. Also, note that the majority of approaches that include production costs or market demand in their formulation use single objective optimization, rather than multi-objective, where the objective is to either maximize profit or minimize cost.

Although not specifically noted in the table, most of the approaches that incorporate uncertainty in the formulation model it in the market demand and future sales of the products in the family. Uncertainty in customer requirements has also been used to develop robust product platforms. Chang and Ward (1995) were among the first to use robust design techniques to develop a family of products that were insensitive to design changes. Simpson and his co-authors use robust design techniques to develop scale-based platforms for General Aviation Aircraft (Simpson, et al., 1999), electric motors (Simpson, et al., 2001a), and absorption chillers (Hernandez, et al., 2001). Blackenfelt (2000) uses robust design techniques to maximize profit and balance commonality and variety within a family of lift tables.

More than half of the approaches require specifying the platform *a priori* in order to reduce the design space and make the optimization problem more tractable. This is not ideal, however, since most designers use optimization to explore varying levels of platform commonality within the product family as noted in Section 3.3.

Note that single-stage and two-stage approaches are employed almost equally in the literature. While both approaches are effective at determining the best design variable settings for the product platform and product family, single-stage approaches will yield better families of products as discussed in Section 3.2 since the optimization is not partitioned into two or more stages. The dimensionality of single-stage optimization problems, however, is considerably higher than in two-stage approaches, which can lead to computational challenges (Messac, et al., 2002a). A modification to the two-stage approach is introduced by Nelson, et al. (2001) and used by Fellini, et al. (2002a; 2002b; 2000): the first stage involves individually optimizing each product while the second stage involves optimizing the product family with constraints on performance losses due to commonality (see also Chapter 9). Only two multi-stage approaches have been developed. First, Hernandez, et al. (2002; 2003) develop a multi-stage optimization approach by viewing the product platform design problem as a problem of access in a geometric space. Second, Allada and Jiang (2002) introduce a dynamic programming (DP) model for configuring module instances within an evolving family of products. An alternative classification of optimization approaches based on the extent of the optimization (i.e., module attributes, module combinations, or both) is discussed in (Fujita, 2002).

Based on the variety of optimization algorithms listed in the table, there does not appear to be a preferred algorithm for product family design. Both linear and non-linear programming algorithms (e.g., SLP, SQP, NLP, GRG) are employed in many formulations, as are derivative-free methods such as genetic algorithms (GA), simulated annealing (SA), pattern search (PaS), and Branch and Bound (B&B) techniques. When the design space is small, exhaustive search (ExS) techniques (Hernandez, et al., 2003) or orthogonal arrays (Blackenfelt, 2000; Chang and Ward, 1995) can be used to enumerate different combinations of parameter settings and modules. However, very few problems involve so few options that such an approach can be taken, and many researchers advocate the use of GAs for product platform design due to the combinatorial nature of the product family design problems as noted earlier. Finally, algorithm choice is often mandated by the selected framework, e.g., Decision-Based Design (Li and Azarm, 2002), Target Cascading (Kokkolaras, et al., 2002), 0-1 integer programming (Fujita, et al., 1999), Physical Programming (Messac, et al., 2002b), and the Compromise Decision Support Problem (Simpson, et al., 1999).

Finally, these optimization approaches have been tested on a variety of product families as noted in the last column of the table. These product families range from 2-16 products and include *consumer products* such as drills, vacuum cleaners, and automobiles; *industrial products* such as chillers and flow control valves; and *complex systems* such as aircraft and spacecraft.

Detailed analyses for the universal electric motor problem can be found in (Simpson, et al., 2001a); it has been used to benchmark a variety of optimization approaches as noted in the table. The commercial aircraft problem found in (Fujita, et al., 1998; Fujita and Yoshida, 2001) uses aircraft analyses available in the literature in combination with their own models for design and development, facility, and production costs and a profit model for the manufacturer. The nail gun (Nelson, et al., 2001), vacuum cleaner (Jiang and Allada, 2001), and power screwdriver and electric knife (Rai and Allada, 2003) examples are pretty comprehensive as well. The automotive example used in (Fellini, et al., 2002a; Kokkolaras, et al., 2002) is based on a detailed vehicle body structural model that is currently unavailable to the public; simpler models of the automotive vehicle frame can be in (Cetin and Saitou, 2004; Fellini, et al., 2002b). Other analyses are not publicly available.

5.   CLOSING REMARKS

As evidenced by the multitude of approaches listed in Table 8-8, formulations for solving product family optimization problems vary widely. They have been applied successfully to a wide variety of problems as well, but we must bear in mind that optimization primarily supports one aspect of product platform and product family design, namely, parameter (detail) design. New and innovative ways are needed to propagate the use of these techniques into the early stages of design when decision support is critical. Moreover, few, if any, of these approaches have found their way into industrial applications or day-to-day use within industry, and we should strive to educate practicing engineers with the power and potential of these approaches. Finally, we believe that research in this promising area of product platform and product family design will stagnate if test problems and benchmarks are not established and propagated within the community at large. We challenge interested researchers to consider this when devising new and improved approaches for product family optimization.

6.   ACKNOWLEDGMENTS

This work has been supported by the National Science Foundation under CAREER Award No. DMI-0133923. Any opinions, findings, and conclusions or recommendations presented in this chapter are those of the author and do not reflect the views of the National Science Foundation.

Chapter 9

COMMONALITY DECISIONS IN PRODUCT FAMILY DESIGN

Ryan Fellini, Michael Kokkolaras, and Panos Y. Papalambros
Department of Mechanical Engineering, University of Michigan, Ann Arbor, MI 48109

1.   INTRODUCTION

Product variants with similar architecture but different functional requirements may have common parts or elements. We define a product family to be a set of such products, and refer to the set of common elements as the product platform. Product platforms enable efficient derivation of product variants by keeping development costs and time-cycles low. In many cases, however, the individual product requirements are conflicting when designing a product family. The designer must balance the tradeoff between maximizing commonality and minimizing individual product performance deviations. The design challenge is to select the product platform that will generate family designs with minimum deviation from individual optima.

In this section we review the vocabulary and basic model for a product platform. A *component* is defined as a manufactured object that is the smallest (indivisible) element of an assembly and is represented by a set of design variables. A *product* is an artifact that is made up of components. The *product architecture* is the configuration (or topology) of components within the product. A *module* is a component or subassembly that can be interchanged within the product architecture to produce a variety of similar products. A *model* is a mathematical representation of a product that accepts a vector of design variables and returns a vector of responses.

The mathematical notation begins with the set $P = \{p_1, p_2, \ldots\}$ used to distinguish each product $p \in P$. Likewise, the set $C^p = \{c_1^p, c_2^p, \ldots\}$ is defined to represent the components that form a particular product p. Thereby the union of all components, $p = \bigcup_i c_i^p$, constructs the product. Figure 9-1 illustrates the above notation. A product p is associated with a vector of design variables $\mathbf{x}^p$, a vector of responses $\mathbf{R}^p$, and design inequality and equality constraints $\mathbf{g}^p$ and $\mathbf{h}^p$, respectively.

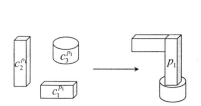

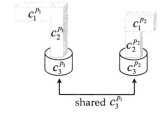

Figure 9-1. Components and assembled product.   *Figure 9-2.* Component sharing within a family of products.

A *product platform* is the set of all components, manufacturing processes, and/or assembly steps that are common in a set of products. We use the following notation to describe a product platform: the set S^{pq} consists of the index pairs of elements that are shared between two products p and q. The set $S = \{ S^{pq} \mid p,q \in P ; p < q \}$ includes all shared elements in the family.

Two types of sharing can be identified when selecting a product platform that is not based on manufacturing processes or assembly steps. In component sharing, one or more components are common across a family of products as shown in Figure 9-2. In addition, it is possible to share "scaled" versions of components. Mathematically this can be described as variable or attribute sharing, where components are based on a platform (of variables) themselves. The example in Figure 9-3 shows the cross-section of two structural beam elements. While the height and width of both parts are same, the thickness is different. The possible manufacturing advantage can be illustrated by this example. By keeping width and height invariant, the same stamping equipment can be used with different gauge steel. In general, manufacturing (and therefore cost) considerations should be taken into account in the design of platforms. We do not address this aspect explicitly, but we attempt to recognize the associated design impact. The methodology presented holds for both component and variable sharing. When defining a platform that includes both component and variable sharing the subscripts c and p are used to denote the individual platform element types, respectively. The information is then held in the set $s = S_c \bigcup S_p$. Finally, a product family is the set of product variants that share a product platform. A family product derived from a platform is also referred to as a product variant.

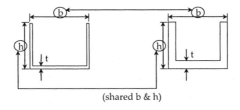

(shared b & h)

Figure 9-3. Variable sharing within a family of products.

Nelson, et al. (2001) proposed a multiobjective optimization formulation for the product family design problem:

$$\max_{\mathbf{x}^{p_1},\mathbf{x}^{p_2},...} \{f^{p}(\mathbf{x}^{p})\} \ \forall p,q \in P, \ (i,j) \in S^{pq}, \ p < q$$
$$\text{subject to } \mathbf{g}^{p}(\mathbf{x}^{p}) \leq 0, \ \mathbf{h}^{p}(\mathbf{x}^{p}) = 0, \ x_{i}^{p} = x_{j}^{q}. \tag{1}$$

The constraint set includes all individual design constraints. Commonality constraints are represented as equality constraints $x_{i}^{p} = x_{j}^{q}$ for each set of shared variables specified in S. An optimization problem is solved for each product separately to determine the null-platform design. The optimal null-platform objective function and design values are denoted by $f^{p,o}$ and $\mathbf{x}^{p,o}$, respectively. Likewise, $f^{p,*}$ and $\mathbf{x}^{p,*}$ are used to represent the optimal objective function and design values for products in the family, respectively.

The null-platform objectives from all family products define the null-platform point, shown in Figure 9-4 as the point with coordinates $(f^{A,o}, f^{B,o})$. The solution of the multiobjective optimization problem in Eq. (1) is a Pareto set. Two different platforms (i.e., different equality constraints) are shown in Figure 9-4. The bounds of the Pareto set determine the utopia point, the best tradeoff design that might be achieved with the platform. In practice, the designer first computes the design penalty from the null-platform point to the utopia point. If that is acceptable, the rest of the Pareto set is computed. The preferred design on the Pareto set can then be selected according to other criteria (e.g., cost or customer preferences).

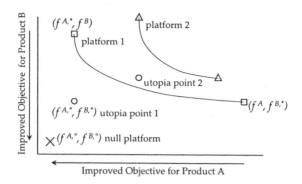

Figure 9-4. Null-platform point and Pareto sets for different platforms.

2.   PLATFORM SELECTION PROBLEM

In many cases functional requirements of product variants are conflicting (Nelson, et al., 2001). Platform-based family designs will then be compromised relative to individually optimized designs due to a tradeoff between maximizing commonality and minimizing individual performance deviations. The design challenge is to determine (select) the platform that will generate family designs with minimum deviation from individual (null-platform) optima.

2.1   Maximizing commonality subject to performance loss constraints

The product family design problem is reformulated to include a) component commonality as an objective term and b) which variables to share in the decision making. This yields a mixed-discrete programming problem due to the presence of the vector of binary (0-1) sharing decision variables η:

$$\max_{\eta, x^{p_1}, x^{p_2},} \quad \{\{f^p(x^p)\}, \sum_{(i,j)_{pq}} \eta_{ij}^{pq}\} \quad \forall p, q \in P, \quad (i,j) \in S^{pq}, \quad p < q$$

$$\text{subject to} \quad \mathbf{g}^p(x^p) \le 0, \quad \mathbf{h}^p(x^p) = 0, \quad \eta_{ij}^{pq}(x_i^p - x_j^q) = 0, \quad \eta_{ij}^{pq} \in \{0,1\}. \tag{2}$$

Sharing design variables η_{ij}^{pq} are equal to 1 if variables x_i^p, x_j^q are shared and 0 otherwise. The introduced objective term sums up all shared variables. Maximizing this term is then equivalent to maximizing commonality.

Moving the terms $\{f^p(\mathbf{x}^p)\}$ to the constraints set using bounds f_l^p, we obtain a single-objective optimization problem:

$$\max_{\eta, \mathbf{x}^{p1}, \mathbf{x}^{p2} \dots} \sum_{(i,j)pq} \eta_{ij}^{pq} \quad \forall p, q \in P, \quad (i,j) \in S^{pq}, \quad p < q$$

subject to $\mathbf{g}^p(\mathbf{x}^p) \le 0, \quad \mathbf{h}^p(\mathbf{x}^p) = 0, \quad \eta_{ij}^{pq}(x_i^p - x_j^q) = 0, \quad \eta_{ij}^{pq} \in \{0,1\}, \quad f^p(\mathbf{x}^p) \ge f_l^p.$

(3)

Simple monotonicity analysis implies that varying bounds systematically will generate the Pareto set, so that the two formulations are equivalent: selecting bounds corresponds to a specific set of objective weights. Further, if we use the function

$$D_o(x_i^p - x_j^q) = \begin{cases} 0 & \text{if } x_i^p = x_j^q \\ 1 & \text{otherwise} \end{cases}, \tag{4}$$

the term $\sum_{(i,j)pq} \eta_{ij}^{pq}$ can be computed based on the values of the design variables:

$$\sum_{(i,j)pg} \eta_{ij}^{pq} = \sum_{pq} \left| S^{pq} \right| - \sum_{(i,j)pq} D_o(x_i^p - x_j^q), \tag{5}$$

where $\left| S^{pq} \right|$ is the number of elements in S^{pq}, which is constant and can be left out. Therefore, maximizing $\sum_{(i,j)pq} \eta_{ij}^{pq}$ is equivalent to minimizing $\sum_{(i,j)pq} D_o(x_i^p - x_j^q)$.

To address the combinatorial nature of the problem, the function D_o is approximated by a function D_α, which should satisfy two requirements: its range should be [0, 1] and it should be continuously differentiable. The function we have selected for D_α is defined as:

$$D_\alpha(x_i^p - x_j^q) = 1 - \frac{1}{\left(\dfrac{x_i^p - x_j^q}{\alpha}\right)^2 + 1}. \tag{6}$$

This function is constructed as a measure of the distance between designs and approaches the function D_o as α goes to zero. Figure 9-5 shows D_α for $\alpha = 0.05$. Since D_α is continuously differentiable, gradient-based algorithms can be used to solve the approximate commonality optimization problem.

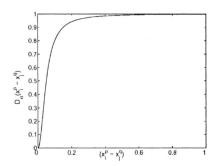

Figure 9-5. The approximation of the function D_o.

To define the performance constraint bounds f_i^p, we define performance loss factors L^p that represent the loss in performance of the family products compared to the null-platform optima $f^{p,o}$. The constraints are then redefined as:

$$f^p(\mathbf{x}^p) \geq (1 - L^p)f^{p,o} = f_i^p. \tag{7}$$

The final formulation of the approximate commonality optimization problem is

$$\max_{\mathbf{x}^{p_1}, \mathbf{x}^{p_2}, \ldots} \sum_{(i,j)pq} D_a(x_i^p - x_j^q) \quad \forall p, q \in P, \ (i, j) \in S^{pq}, \ p < q$$

$$\text{subject to } \mathbf{g}^p(\mathbf{x}^p) \leq \mathbf{0}, \ \mathbf{h}^p(\mathbf{x}^p) = \mathbf{0}, \ f^p(\mathbf{x}^p) \geq (1 - L^p)f^{p,o}. \tag{8}$$

The solution to Eq. (8) may not be unique. Figure 9-6 shows the reduced feasible set resulting from the introduction of performance bounds. Furthermore, multiple combinations of the same number of shared components can exist; these must be differentiated by their relative performance after solving the family problem (Eq. 1). Recall that this step of the methodology aims primarily at selecting the feasible platform set, not at computing final design values $\mathbf{x}^p$.

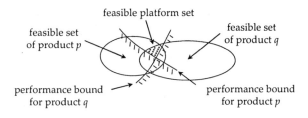

Figure 9-6. Reduced platform feasible set.

The loss factors L^p are considered input parameters specified by the designer. A post-optimal parametric study may be necessary to determine the acceptable tradeoff between performance and commonality, essentially generating the Pareto set of the problem in Eq. (2).

It is possible to include designer's preferences for sharing by modifying the objective function as follows

$$\max_{x^{p1}, x^{p2} \dots (i,j)pq} \sum \omega_{ij}^{pq} D_\alpha (x_i^p - x_j^q) \quad \forall p, q \in P, \quad (i,j) \in S^{pq}, \quad p < q \cdot \tag{9}$$

The scalar weights ω_{ij}^{pq} correspond to the preference on sharing variables x_i^p and x_j^q. Including cost considerations may be straightforward by replacing the ω_{ij}^{pq} with the actual cost of the components: If component "1" costs 200 times more to manufacture than component "2", preference should be placed on sharing the more expensive component. Exploring these multiple layers of multiobjective decisions is beyond the scope of this chapter. Therefore, we will assume no preferences in component sharing, so that all ω_{ij}^{pq} will be equal.

Since the solution of the approximate commonality optimization problem will not yield 0 - 1 values, we must determine which variables will be selected as common. This is done after solving the commonality decision problem of Eq. (8), and in the following manner: the values of the design variables of the candidate components are compared, and assumed to be shared if their relative difference does not exceed a numerical tolerance. The designer must choose an appropriate value that ensures accuracy of the solution for the particular problem being solved. The shared variables are included as commonality constraints when solving the family design problem of Eq. (1). Therefore, one might compare the tolerance on sharing to the tolerance on satisfying constraints. A high tolerance might often lead to the suggestion of more sharing; however, this decreases the chances of satisfying the performance deviation constraints, and thereby achieving a feasible design when the family design problem is solved.

The multiobjective family design problem of Eq. (1) is reformulated to minimize the distance between the null-platform design and the Pareto set corresponding to the selected platform S^{pq*}:

$$\max_{x^{p1}, x^{p2},} \{((f^{p,o} - f^p(x^p)) / f^{p,o})^2\} \quad \forall p, q \in P, \quad (i,j) \in S^{pq*}, \quad p < q$$
$$\text{subject to } \mathbf{g}^p(\mathbf{x}^p) \le 0, \quad \mathbf{h}^p(\mathbf{x}^p) = 0, \quad f^p(\mathbf{x}^p) \ge (1 - L^p) f^{p,o}, \quad x_i^p = x_j^q. \tag{10}$$

The performance bounds are included in case the Pareto point closest to the null-platform point lies outside the area of allowable performance loss.

The proposed family design methodology can be described by the following steps: (1) Determine the optimal null-platform design $f^{p,o}(\mathbf{x}^{p,o})$ for each individual product $p \in P$ by solving the individual optimal design problem; (2) Identify the components that could be shared between products, i.e., define the candidate platform set S^{pq} for any two products p and q in the set P. The candidate platform set for the whole product family is then $S = \{ S^{pq} \mid p,q \in P; \; p < q \}$; (3) Determine the acceptable performance loss factors L^p; (4) Solve the approximate commonality optimization problem to determine components to be shared, i.e., determine $S^* = \{ S^{pq,*} \mid p,q \in P; \; p < q \}$; and (5) Solve the family design problem.

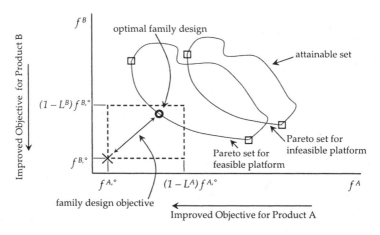

Figure 9-7. Design process of proposed methodology.

Figure 9-7 illustrates the above methodology. The conceptual plot shows a null-platform point $(f^{A,o}, f^{B,o})$. The null-platform objective function values are multiplied by the performance loss tolerances, and so the region where the associated feasible platforms reside is bounded by the points $(f^{A,o}, f^{B,o})$, $((1-L^A)f^{A,o}, (1-L^B)f^{B,o})$, $((1-L^A)f^{A,o}, f^{B,o})$, and $(f^{A,o}, (1-L^B)f^{B,o})$. Solving the commonality decision problem, a feasible platform is found. The performance loss bounds in Eq. (8) will be active unless they are dominated by design constraints, and the obtained designs will correspond to the objective function values $(1-L^A)f^{A,o}$ and $(1-L^B)f^{B,o}$. The family design problem is solved to obtain a Pareto-optimal design, searching for the point closest to the null-platform design.

2.1.1   Example of family with two product variants

We consider a family of side frames for an automotive body with two variants. A variant can be defined by changing the functional requirements and/or the geometry of the model. In this study variants A and B are designed for minimum mass and maximum stiffness (minimum deflection), respectively.

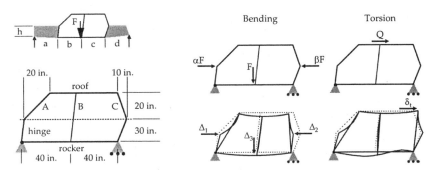

Figure 9-8. Two-dimensional automotive side frame model.

Figure 9-9. Two-dimensional automotive side frame model.

The side frame of the automotive body is modeled simply as an assembly of ten beam elements and seven flexible joints (see Figure 9-8). A finite element solver is used to compute deflections, stresses, and body mass for different values of the rectangular cross-section parameters of the beams (width b, height h, and thickness t). Two loading cases (bending and torsion) are considered, as depicted in Figure 9-9 where $F = 1500$ lbf. and $Q = 1650$ lbf. The vehicle dimensions used here are $h = 30$ in., $a = 20$ in., $b = 40$ in., $c = 40$ in., and $d = 10$ in. To compute α and β, we use the following relations: $\alpha = a\,(c + d)\,/\,h\,(a + b + c + d)$ and $\beta = d\,(a + b)\,/\,h\,(a + b + c + d)$, thus $\alpha = 10/33$ and $\beta = 6/33$. The engine and rear compartments are included in the model as reaction forces applied at the connection of the "A" and hinge pillars and at the center point of the "C" pillar for the bending loading case. Torsion is represented by a horizontal force applied at the joint connecting the "B" pillar and the roof; this force simulates the shear that the structure undergoes under such loading and provides a torsional displacement δ_t. An overall bending displacement is calculated as $\delta_b = \alpha \Delta_1 + \beta \Delta_2 + \Delta_3$. Each component is represented by three design variables. This means that a component can be shared within the family if all three design variables have equal values. However, it my be possible to consider some other form of sharing if only one or two design variables have equal values, for example, from a manufacturing point of view as discussed in Section 1. In this regard, all variables are treated as platform candidates in this study.

Optimal design problems are solved individually for each variant to obtain null-platform optima $f^{p,o}$:

$$\min_{\mathbf{x}^A} \; f^A(\mathbf{x}^A) = m$$

$$\text{subject to } \mathbf{g}^A(\mathbf{x}^A) \le \mathbf{0}, \quad \delta_b \le 0.4 \text{ in.}, \quad \delta_t \le 0.8 \text{ in.} \tag{11}$$

$$\min_{\mathbf{x}^B} \; f^B(\mathbf{x}^B) = \delta_b + \delta_t$$

$$\text{subject to } \mathbf{g}^B(\mathbf{x}^B) \le \mathbf{0}, \quad m \le 250 \text{ lbf.} \tag{12}$$

Maximal stress constraints $\mathbf{g}$ for each beam element are taken into account for all three variants. A modulus of elasticity for steel of 30 Mpsi is used along with a safety factor of 3.0. The computed null-platform optima for the considered variants are 53.4 lbf. for Variant A and 0.587 in. for Variant B.

Having computed the null-platform optima, the commonality decision problem is solved for different values of the loss factors L^p. Experience indicates that 0.025 is a good value for the parameter α in Eq. (6).

In the side frame problem there are 24 design variables representing the cross-sectional variables b (width), h (height), and t (thickness) of the beams. The candidate platform set S is defined by allowing each design variable in Variant A to be shared only with the same variable of the corresponding component in Variant B. Therefore, there are 2^{24} possible sharing combinations (platforms).

The problem in Eq. (8) is solved to determine the "maximal" feasible platform under performance bounds set equal for both variants. A tolerance of 0.5% on the commonality constraints in the objective was used to determine which variables will be shared among the two variants. From the null-platform results, twelve variables were found to be "naturally" shared by inspection, i.e., they had the same optimal values in both designs. Allowing a performance loss of 1%, 5%, 10%, 20%, and 50% resulted in sharing 17, 20, 21, 22, and 24 variables, respectively. The tradeoff between sharing and performance is shown in Figure 9-10. This tradeoff is analogous to the Pareto set that could be generated solving the combinatorial optimal design problem of Eq. (2).

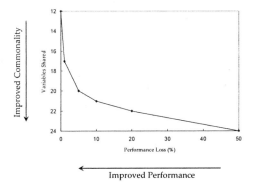

economy sedan (A) / sporty sedan (B)

stretched sedan (C)

Figure 9-10. Tradeoffs between commonality and performance.

Figure 9-11. Side-frames for three different automobiles.

The optimal family design problem is solved next, and results are presented in Table 9-1.

Table 9-1. Optimal product family design results and associated performance losses.

Variant	A	B
Null platform	53.4 lbf.	0.587 in.
Platform of 17 variables	53.9 lbf.	0.593 in.
Performance loss	0.999%	0.972%
Platform of 20 variables	55.6 lbf.	0.616 in.
Performance loss	4.17%	5.01%
Platform of 21 variables	56.0 lbf.	0.646 in.
Performance loss	4.87%	9.96%
Platform of 22 variables	58.3 lbf.	0.687 in.
Performance loss	9.20%	17.00%
Total platform of 24 variables	73.2 lbf.	0.881 in.
Performance loss	37.1%	50.0%

In an attempt to validate the results, a reduced version of the problem in Eq. (2) has been solved. "Naturally shared" variables, as determined by the individual variant optimizations, are shared. Since twelve out of the sixteen width and height variable values of the two variants are always equal, we considered width and height as shared variables, and reduced the size of the combinatorial problem by defining the platform to include these sixteen variables. The problem size was thus reduced to 2^8 platform combinations.

A "top-down" algorithm was implemented by starting with sharing all eight component thicknesses (total platform). If the performance bounds were exceeded, we moved a "level" down by decreasing the number of candidates for sharing from eight to seven; eight different platforms of sharing seven thicknesses were then considered, and so on, until at least one platform was found, for which the performance bounds for a given "level" were not exceeded. Note that it is possible that more than one platforms may satisfy the performance bounds at a given "level".

One platform was obtained for each of the loss factors 1%, 5%, and 50%, by solving the commonality decision problem. For 10% and 20%, two and seven feasible platforms were found, respectively, each once again containing the same number of shared variables as determined by solving the problem of Eq. (8). It is encouraging that for both cases the platform obtained is the one that corresponds to least performance loss where multiple platforms were found. Intuitively this makes sense: since the problem in Eq. (8) is solved over a continuous design space, it is natural that components that have less impact (sensitivity) on the performance will be shared first.

2.1.2   Example of family with more than two variants

When product designers study the implementation of a platform the number of products in the family will likely be more than two. The example of this section demonstrates the use of the methodology for families with multiple products.

The size of the problem increases substantially with the number of variants. The number of possible platforms can be calculated as

$$\text{Number of platform combinations} = \left\{ 1 + \sum_{i=2}^{m} \frac{m!}{i!(m-i)!} \right\}^{n}, \tag{13}$$

where m is the number of variants and n is the number of shareable elements. The assumption made for this calculation is that each variant has the same components to be shared. The unity in the equation allows inclusion of the null platform as a possible combination. Likewise, the total platform is included in the summation for $i = m$.

The number of terms in the objective function of Eq. (8) is equal to the number of sharing possibilities among the variants. Assuming one variant with one shareable component, there exist three sharing possibilities among them for four variants, six sharing possibilities for five variants, ten sharing possibilities for six variants, and so on. The number of sharing possibilities for n shareable components in m products can be calculated by the relation

$$\text{Number of sharing possibilities} = n \sum_{i=1}^{m-1} (m-i). \tag{14}$$

The number of terms will grow significantly with additional products or components, but not as fast as in the original combinatorial problem. The number of possible platforms is reduced by limiting the number of components that the designer considers as shareable.

Consider the design of three automotive body side-frames. Variant A is an economy vehicle, whose weight is minimized with constraints on the bending and torsional rigidity. The design problem is the same as the one given in Eq. (11). Variant B is a sporty vehicle, whose torsional rigidity is maximized with respect to constraints on weight and bending stiffness:

$$\min_{x^B} \ f^B(x^B) = \delta_t$$
$$\text{subject to } \ g^B(x^B) \le 0, \ m \le 250 \ \text{lbf.}, \ \delta_b \le 0.4 \ \text{in.} \tag{15}$$

Variant C is a stretched frame: the lengths of the roof rail and rocker have been extended by 20 in. (symmetrically about the "B" pillar). The objective is to minimize the bending displacement subject to weight and torsional rigidity constraints:

$$\min_{x^C} \ f^C(x^C) = \delta_b$$
$$\text{subject to } \ g^C(x^C) \le 0, \ m \le 250 \ \text{lbf.}, \ \delta_t \le 0.8 \ \text{in.} \tag{16}$$

All three vehicle models use the same α and β values as in the previous example. The optimal objective function values for the three variants are determined to be 53.4 lbf., 0.373 in., and 0.312 in., respectively.

So far we examined what design variables can be shared among products. In fact the designer may need to know explicitly whether an entire component can be shared. The example also demonstrates how this decision can be reached.

The components for each of the body side frames are included in the component set C^P (because all three products have an identical topology, the set C^P is consistent among variants):

$$C^P = \{\text{Rocker, Roof rail, Hing pillar, "A" pillar, "B" pillar}_{(\text{lower})},$$
$$\text{"B" pillar}_{(\text{upper})}, \text{"C" pillar}_{(\text{lower})}, \text{"C" pillar}_{(\text{upper})}\}. \tag{17}$$

The design variables of the product are then mapped into vectors of design variables that correspond to particular product component as follows

$$x^{c_{P_1}} = [x_1^{P_1}, x_2^{P_1}, x_3^{P}]^T, \ x^{c_{P_2}} = [x_4^{P_2}, x_5^{P_2}, x_6^{P_2}]^T, \ x^{c_{P_3}} = [x_7^{P_3}, x_8^{P_3}, x_9^{P_3}]^T, \ x^{c_{P_4}} = [x_{10}^{P_4}, x_{11}^{P_4}, x_{12}^{P_4}]^T,$$
$$x^{c_{P_5}} = [x_{13}^{P_5}, x_{14}^{P_5}, x_{15}^{P_5}]^T, \ x^{c_{P_6}} = [x_{16}^{P_6}, x_{17}^{P_6}, x_{18}^{P_6}]^T, \ x^{c_{P_7}} = [x_{19}^{P_7}, x_{20}^{P_7}, x_{21}^{P_7}]^T, \ x^{c_{P_8}} = [x_{22}^{P_8}, x_{23}^{P_8}, x_{24}^{P_8}]^T,$$

where the three vector entries for each component represent width, height, and thickness. With the design vector defined for each component, the final step is to modify the objective function of the commonality decision problem to deal with decisions based on vectors rather than scalars. This is accomplished by simply using the l_2 norm to compute the difference in design:

$$\min_{x^A, x^B, x^C} \sum_{i=1}^{n=8} \left(D_\alpha \left(\left\| x^{cA_i} - x^{cB_i} \right\|_2 \right) + D_\alpha \left(\left\| x^{cA_i} - x^{cC_i} \right\|_2 \right) + D_\alpha \left(\left\| x^{cB_i} - x^{cC_i} \right\|_2 \right) \right). \tag{18}$$

Since the focus is now on component selection, the roof and rocker of the stretched vehicle is not shareable with the other two vehicles. This is because these are the only two components that do not share a common length with the other two variants. Therefore, when $i = 1, 2$ the last two terms are removed in Eq. (18). The platform selection process is now performed with a loss factor of 5% and 10%. The tolerance of 0.05% is used again for the commonality constraints in the objective. Figure 9-12 represents the shared components between pairs of products for each of the two loss factors. The platform among all three products is the intersection of the shared components for each of the sharing pairs. For a loss factor of 5% there are five components shared, with the upper "B" and "C" pillars being shared among all three variants. For a loss factor of 10% there are seven components shared, now including the "A" pillar in the platform between all three products. Actual performance losses are computed after solving the family design problem and are reported in Table 9-2. They are within the allowable bounds on deviation from the null-platform designs.

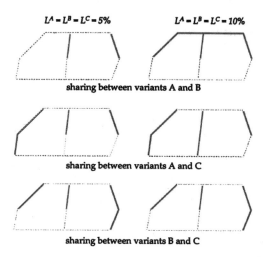

Figure 9-12. Platform results for three variants based on component sharing.

Summarizing, we have applied the methodology to family design problems including multiple products. In addition, the formulation was used for performing platform selection based on component selection (all variables of component shared). These results allow the visualization of modularity as well: components or sets of components which produce variety are defined as modules, and are not part of the platform. A thorough discussion of modularity is beyond the scope of this chapter.

Table 9-2. Optimal product family design results and associated performance losses.

Variant	A	B	C
Null platform	53.4 lbf.	0.373 in.	0.312 in.
Platform with $L^p = 5\%$	55.3 lbf.	0.385 in.	0.325 in.
Performance loss	3.60%	3.18%	4.12%
Platform with $L^p = 10\%$	56.3 lbf.	0.395 in.	0.336 in.
Performance loss	5.49%	5.82%	7.74%

2.2 Reducing the size of the platform selection problem

When the products in the family contain a large number of components that are candidates for sharing, platform selection entails the solution of a large combinatorial problem. With the approach proposed in this section, the problem can be reduced under the assumptions listed below. The derivation presented in this section is based on first order Taylor series approximation. Therefore, in order for the approximation to remain reasonably accurate, the general condition is that the individual optimal designs lie not "too far away" from each other so that a linear approximation is valid in the region between them. For simplicity, the derivation will be presented for a family of two products A and B, but can be generalized readily for more products.

We make the following assumptions: (1) Self-sharing (i.e., component sharing within the same variant) is not possible; (2) Components are either shared by all family members or not at all; (3) Null-platform optimal designs lie "close enough" to each other; (4) The platform design (denoted here by superscript *) lies in the convex hull of the individual solutions (denoted by superscripts ^{A,o} and ^{B,o}). That is, $\exists \lambda_i \in [0,1]$ such that $\forall i \in s,\ x_i^* = \lambda_i x_i^{A,o} + (1 - \lambda_i) x_i^{B,o}$; and (5) Constraint inactivity remains unchanged between individual and family design problems. We will refer to the design solutions that satisfy these assumptions as "mild variants."

As discussed, sharing may cause deviations from individually optimized products design, which is measured by the response representing the functional requirements. In the context of the approach introduced in this chapter, the commonality decision consists in determining which variables have a larger impact on performance. The design variables are arranged in order of increased performance deviation value and the number n of variables to share is determined by a limit on acceptable design deviations. The optimal platform is determined by minimizing the relative deviation, Δ^p, of the design based on any platform with n shared variables with respect to the null-platform design – while remaining in the feasible space for the variants. Formally, this is stated as

$$\min_{s \in S} \Delta$$
$$\text{subject to } |s| = n,$$

(19)

where $\Delta = \Delta^A + \Delta^B$ and

$$\Delta^P = \left| f^P(\mathbf{x}^{P,*}) - f^P(\mathbf{x}^{P,o}) \right| + \sum_{j \in G^P} \max(g_j^P(\mathbf{x}^{P,*}), 0)$$

(20)

for $p \in P = \{A, B\}$. The set G^P includes the constraints that were found to be active during the individual optimization process. Normalization is used to enable the meaningful summation of responses of different nature.

A first-order Taylor approximation of the variation in each response is introduced in agreement with the assumptions listed above:

$$f^P(\mathbf{x}^{P,*}) - f^P(\mathbf{x}^{P,o}) \approx (\nabla f^{P,o})^T \left(\mathbf{x}^{P,*} - \mathbf{x}^{P,o} \right) \quad \text{and}$$
$$g_j^P(\mathbf{x}^{P,*}) \approx (\nabla g_j^{P,o})^T \left(\mathbf{x}^{P,*} - \mathbf{x}^{P,o} \right),$$

where $\nabla f^{P,o}$ and $\nabla g^{P,o}$ are the gradients of f^P and g^P at the null-platform optimal designs.

Furthermore, under assumption 4, the relation between the shared variables, $i \in s$, and the null platform can be rewritten as:

$$\left(x_i^* - x_i^{A,o} \right) = (1 - \lambda_i)\left(x_i^{B,o} - x_i^{A,o} \right).$$

(21)

Consequently, the deviation of the objective f of variant A due to sharing of variables x_i, $i \in s$ is approximated by:

$$f^A(\mathbf{x}^*) - f^A(\mathbf{x}^{A,o}) \approx \sum_{i \in s} \nabla_i f^{A,o} \left(x_i^* - x_i^{A,o} \right) \approx \sum_{i \in s} (1 - \lambda_i) \nabla_i f^{A,o} \left(x_i^{B,o} - x_i^{A,o} \right).$$

Letting $\delta_i = \left| x_i^{B,o} - x_i^{A,o} \right|$, an upper bound and an approximation on the total variation in Δ^A is given by

$$\Delta^A \leq \sum_{i \in s} (1 - \lambda_i) \left(\left| \nabla_i f^{A,o} \right| \delta_i + \sum_{j \in G^A} \max(\nabla_i g_j^{A,o} \delta_i, 0) \right).$$

(22)

A similar upper bound can be obtained for Δ^B. We define the performance deviation vector Π, whose entries correspond to performance deviations due to sharing:

$$\Pi_i = (1-\lambda_i)\left(\left|\nabla_i f^{A,o}\right|\delta_i + \sum_{j\in G^A} \max\ (\nabla_i g_j^{A,o}\delta_i,\ 0)\right) +$$

$$\lambda_i\left(\left|\nabla_i f^{B,o}\right|\delta_i + \sum_{j\in G^B} \max\ (\nabla_i g_j^{B,o}\delta_i, 0)\right).$$

(23)

The l_1 norm of the vector Π provides an upper bound on the actual performance deviation Δ:

$$\Delta \leq \|\Pi\|_1.$$

(24)

The approach for approximating a solution to the original problem is to minimize the upper bound on Δ as given in Eq. (24). In this regard, the choice of the parameters λ_i has to be discussed. These parameters are determined theoretically by the position of the family solution for a given platform S relative to the position of the null-platform solutions for the two variants (Assumption 4). In the framework described here, the exact values of λ_i are not known *a priori* since the solution to the family problem is not available. Therefore, we will assume that $\lambda_i = 0.5\ \forall i$. Hence, there is no bias towards one variant or the other with regard to the family design variables values. The choice of this value does not affect the commonality decisions. However, the validity of the "convex-hull" assumption needs to be checked after solving the family design problem to ensure that commonality considerations are reasonable for the related component or design variable.

The design variables are arranged in order of increasing Π_i. The variables to be shared are the first n variables below some threshold. This minimizes the upper bound on Δ.

2.2.1   Example

A family of automotive body structures is considered. A variant is defined as a structure associated with specific dimensional properties (lengths) and functional requirements. The structures are modeled using finite elements in MSC.Nastran according to modeling approaches described in Fenyes (2000). Modal and static loading cases (torsion on the front and rear shock towers, and bending) are considered, as shown in Figure 9-13. It is assumed that these loading cases give access to the properties that the designer wishes to tailor, and therefore are valid as a basis of the design.

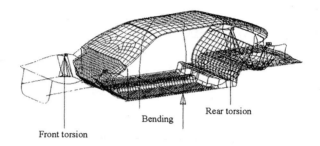

Figure 9-13. Automotive body structure model.

The finite element analysis outputs mass and natural frequencies, as well as displacements and stress responses for static loading cases of front torsion, rear torsion, and bending (denoted d_{ft}, d_{rt}, and d_b, respectively) along with corresponding sensitivity information for all the design variables. These are the cross-sectional dimensions of the beams (width b, height h, and thickness t) and shell thicknesses t. There is a total of 66 design variables.

We used the SCPIP algorithm for solving the optimization problems, which is an implementation of the method of moving asymptotes (MMA), tailored to solve large-scale structural optimization problems efficiently (Zillober, 2001). As mentioned, variants are generated either by implementing dimensional changes or by imposing different design requirements. We will examine both cases.

Let us first consider a family of two variants based on dimensional changes having the same objective functions and constraints. As shown in Figure 9-14, a second variant is generated by stretching the wheelbase and trunk of the baseline vehicle. The engine compartment is shortened, and therefore a smaller engine (and lumped mass representing the engine) is assumed. The models will be referred to as the *short* and *long* wheelbase body models.

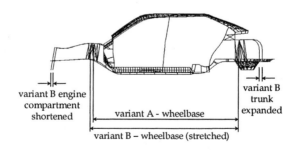

Figure 9-14. Automotive body structure dimensional variants.

The individual optimal design problem for both variants is formulated as

$$\min_{b,h,t} \quad m$$

$$\text{subject to } \omega_1 \geq 21 \text{ Hz}, \ \omega_2 \geq 24 \text{ Hz}, \ d_{fl} \leq 2.9 \text{ mm} \qquad (25)$$

$$d_{rt} \leq 2.9 \text{ mm}, \ d_b \leq 0.2 \text{ mm}, \ \sigma_{max} \leq 25 \text{ MPa}$$

and the null-platform optima are summarized in Table 9-3.

Table 9-3. Null-platform optima (dimensional variants).

	short	long
mass (kg)	715.13	703.36
ω_1 (Hz)	21.00	22.06
ω_2 (Hz)	24.82	27.00
d_{fl} (mm)	2.158	2.170
d_{rt} (mm)	1.905	1.909
d_b (mm)	0.200	0.200

The performance deviation vector $\mathbf{\Pi}$ is computed according to Eq. (23) and the platform is determined using a threshold value of 0.01. 59 variables are selected for sharing and the family design problem of Eq. (1) is solved. The family problem is solved also considering a "total" platform – in which all variables are shared – to assess the usefulness of the approach. Family optima obtained for both platforms are given in Table 9-4.

Table 9-4. Optima for 59-variable and total platforms (dimensional variants).

Variant	59 Var. Platform		Total Platform	
	short	long	short	long
Mass (kg)	715.17	703.54	725.65	703.37
Ω_1 (Hz)	21.00	22.06	21.00	22.24
Ω_2 (Hz)	24.82	27.00	25.83	27.00
D_{fl} (mm)	2.158	2.170	2.082	2.171
D_{rt} (mm)	1.905	1.909	1.837	1.911
D_b (mm)	0.200	0.200	0.191	0.200

Overall, the family based on the 59-variable platform is close to the null platform: the optimal masses of both short and long wheelbase variants are almost identical to the corresponding null-platform designs. The long wheelbases variant using the total platform is still close to the corresponding null-platform variant, compared to a 10.5 kg difference in mass in the short wheelbase variants. The components that are not completely shared among the variants are shown in Figure 9-15.

For each of these components the material thickness is the variable that varies. Overall a large number of variables may be shared with negligible performance deviation (less than 1.5%) relative to the corresponding null-platform designs. This can be traced to the fact that the variants do not have competing design objective functions, and that their geometric configurations are very similar. The combination of these two factors results in relatively close individual optima and family optima.

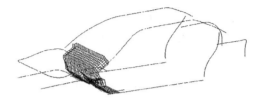

Figure 9-15. Dimensional variants: Non-shared components.

We now look at variants based on the same geometric model (the short wheelbase model) having different design objectives and constraints. Two variants with competing objectives are defined, referred to as "stiff" and "lightweight", respectively. In the former the designer aims at maximizing the stiffness of the structure to improve ride quality, while in the latter the goal is to minimize weight to improve fuel economy.

The flexibility φ is defined as a weighted sum of the displacements d_{ft}, d_{rt}, and d_b. The weights approximate the ratios of the expected displacements (i.e., null-platform optima in Table 9-3) in each loading case; hence flexibility is computed as follows

$$\varphi = d_{ft} + d_{rt} + 10d_b. \tag{26}$$

The optimal design problems for the lightweight variant and the stiff variant are formulated as:

$$\min_{b,h,t} \ m$$
$$\text{subject to } \omega_1 \geq 15 \text{ Hz}, \ \omega_2 \geq 17 \text{ Hz}, \ d_{ft} \leq 2.9 \text{ mm} \tag{27}$$
$$d_{rt} \leq 2.9 \text{ mm}, \ d_b \leq 0.2 \text{ mm}, \ \sigma_{\max} \leq 25 \text{ MPa}$$

and:

$$\min_{b,h,t} \ \varphi$$
$$\text{subject to } \omega_1 \geq 21 \text{ Hz}, \ \omega_2 \geq 24 \text{ Hz}, \ m \leq 822 \text{ kg}, \ \sigma_{\max} \leq 25 \text{ MPa}, \tag{28}$$

respectively. Each variant is optimized individually to obtain a null-platform design. The optimal objective function values for the lightweight and stiff variants are 691.87 kg and 4.4049 mm, respectively. The null-platform optimal designs and sensitivities are used to compute the performance deviation vector $\mathbf{\Pi}$. The design variables are arranged in order of increasing performance deviation. Figure 9-16 depicts a plot of the sorted performance deviation vector.

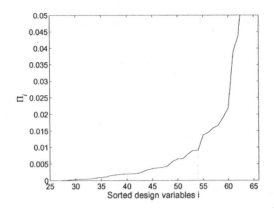

Figure 9-16. Sorted performance deviation vector Π (performance variants).

The graph shows that performance deviation is relatively low for the first 50 variables, and then begins to increase sharply. We chose a 54-variable platform based on the fact that the curve exhibits a sharp increase after 54 variables. The components that are not shared among the variants are shown in the Figure 9-17. As in the previous example the material thickness is most often the dimension that varies. One exception is that the rocker panel differs in width, height, and thickness. We solved the family problem for the 54-variable platform and the total platform by minimizing the distance to the null-platform optimum. The results are summarized in Table 9-5.

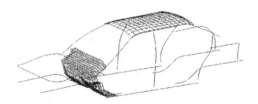

Figure 9-17. Performance variants: Non-shared components.

Table 9-5. Optima for null, 54-variable, and total platforms (performance variants).

Variant	Null Platform		54-Variable platform		Total Platform	
	stiff	*weight*	*stiff*	*weight*	*stiff*	*weight*
Mass (kg)	822.00	691.87	822.00	699.90	822.00	822.00
d_{fl} (mm)	1.581	2.429	1.595	2.270	1.607	1.607
d_{rl} (mm)	1.396	2.148	1.409	1.007	1.419	1.419
d_b (mm)	0.1427	0.2922	0.1429	0.2829	0.1443	0.1443
Flexibility (mm)	4.405	7.499	4.433	7.107	4.468	4.468

Figure 9-18 depicts the obtained Pareto sets for the 54-variable and total platforms. The 54-variable platform shared all but 18% of the variables, with a deviation of 0.6% for the stiff variant and 1.16% for the lightweight variant. In contrast, the total platform has a 1.4% deviation for the stiff variant and an 18.8% deviation for the lightweight variant.

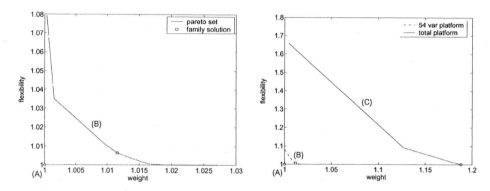

Figure 9-18. Pareto sets for the 54-variable and total platforms, with normalized objectives (performance variants). The plot on the left contains the null platform point (A) and the Pareto set for the 54-variable platform (B). The plot on the right includes also the Pareto set for the total platform (C).

The validity of some of the assumptions described in Section 2.2 can only be checked *a posteriori*, i.e., after solving the individual optimization problems and the family design problem. We checked the assumptions for both case studies. Assumptions 1 and 2 are automatically satisfied by the implementation of the methodology. The distance between the null-platform designs was relatively small for both cases (Assumption 3). Assumption 4 is satisfied; by inspecting the results obtained from solving the family design problem backwards $\exists \lambda_i$ such that $x_i^* = \lambda_i x_i^{stiff,o} + (1 - \lambda_i) x_i^{lightweight,o}$ $\forall_i$ in all cases. This assumption holds for this problem but may not hold for other problems. It is rather strong and further research is needed to relax it.

Assumption 5 is designed to avoid the case where a constraint that is inactive at the individual design becomes active at the family design, a case that is not taken into account in the current derivation. In both cases, no additional constraints became active. In fact, one of the active constraints became inactive in the performance variants case. This is expected, since introducing equality constraints (the commonality constraints) to the family design problem may force some inequality constraints to become inactive at the family design to satisfy the equality constraints.

3. COMBINED METHODOLOGY

A combined methodology that combines the approaches presented in Sections 2.1 and 2.2 is proposed for solving the commonality decision (or platform selection) problem. The approach of Section 2.2 will be used as a "candidate-filtering" step to reduce the problem size to identify components that may be considered as good candidates for sharing. The approach of Section 2.1 is then applied to the candidate platform to complete the commonality decision-making process.

This requires a minor modification so that the two approaches can be integrated efficiently, namely to allow components described as a vector of design variables. This is done by aggregating the performance deviation values of the design variables that define the component into a single value Π^c. Also, previously, the sorted performance deviation vector values were plotted as shown in Figure 9-16. We now look at the increasing cumulative value of the performance deviation as more variables are shared.

The combined methodology can be described in the following steps:

1) Determine the optimal null-platform design $\mathbf{x}^{p,o}$ for each product $p \in P$ by solving the individual optimal design problem.

2) Identify components that can be shared and define an acceptable performance loss factor L^p for each product.

3) Compute the performance deviation vector to determine the candidate platform set.

4) Solve the relaxed platform selection problem (Eq. 8) to determine the optimal platform.

5) Solve the family optimal design problem as formulated in Eq. (1) or Eq. (10).

We will now demonstrate the application of the combined strategy to a family design problem of automotive engines. Engine variants are defined based on different functional requirements.

GT-Power by Gamma Technologies is used as the simulation tool (GTI, 2001). A 24-valve 2.5L V6 engine model, previously validated at various operating points, is used to generate the family. Analysis is performed at a specified operating point of given engine speed and fuel rate, namely at 5000 RPM and wide-open throttle (WOT). The operating point characteristics are:

N_e = 5000 PRM (mean crank speed)

t_p = 90° (throttle angle)

i_{vo} = 331.0° (intake value open w.r.t. CA (crank angle))

i_{vc} = -103.0° (intake valve close w.r.t. CA (crank angle))

e_{vo} = 101.0° (exhaust valve open w.r.t. CA (crank angle))

e_{vc} = 397.0° (exhaust valve close w.r.t. CA (crank angle))

The geometry of components from the intake manifold through the exhaust system is modeled in the simulation. The design variables of particular interest in this study are:

x_1: Bore (b), x_2: Stroke (s), x_3: Connecting rod length (l), x_4: Compression ratio (c_r),
x_5: Intake valve diameter (d_i), x_6: Intake cam-timing angle (i_{cta})
x_7: Intake angle multiplier (i_{am}), x_8: Exhaust valve diameter (d_e)
x_9: Exhaust cam-timing angle (e_{cta}), x_{10}: Exhaust angle multiplier (e_{am})
i_1: Number of cylinders (n_c)

The two exhaust valves are modeled as a single valve by using the equivalent area relation

$$(\text{input diameter})^2 = 2\,(\text{valve diameter})^2. \qquad (29)$$

In addition, the theoretical height of the combustion chamber is computed as a function of the stroke and compression ratio:

$$h_c = s/(c_r - 1). \qquad (30)$$

The simulation responses include

R_1: Brake Power (performance), R_2: Brake Torque (performance),
R_3: Brake-Specific Fuel Consumption (efficiency), R_4: NO$_x$ (emissions),
R_5: dP_{max}/DCA (NVH, knock), R_6: P_{max} (stress/durability).

Brake power and brake torque are measures of the product dynamic performance. Brake-specific fuel consumption (BSFC) is a measure of efficiency, and measured emissions are through NO$_x$ produced. Finally, dP_{max}/DCA, the mean pressure rise with respect to crank angle, and P_{max}, the maximum cylinder pressure, contribute to various measures such as NVH, knock, stress, and engine durability.

The components we focus on are the following: exhaust cams, intake cams, exhaust valves, intake valves, cylinder head, pistons, connecting rod, and engine block (see Figure 9-19).

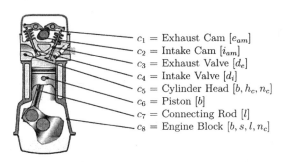

$c_1 = $ Exhaust Cam $[e_{am}]$
$c_2 = $ Intake Cam $[i_{am}]$
$c_3 = $ Exhaust Valve $[d_e]$
$c_4 = $ Intake Valve $[d_i]$
$c_5 = $ Cylinder Head $[b, h_c, n_c]$
$c_6 = $ Piston $[b]$
$c_7 = $ Connecting Rod $[l]$
$c_8 = $ Engine Block $[b, s, l, n_c]$

Figure 9-19. Engine components of interest.

We can map the design variables to components as follows:

$x^{c1} = [x_{10}]$, $x^{c2} = [x_7]$, $x^{c3} = [x_8]$, $x^{c4} = [x_5]$, $x^{c5} = [x_1, h_c, i_1]$, $x^{c6} = [x_1]$, $x^{c7} = [x_c]$, $x^{c8} = [x_1, x_2, x_3, i_1]$.

Note that the number of cylinders, i_1, is fixed at six for all engines. The upper and lower bounds of the design variables were set as follows:

$x_{1,l}, x_{1,u} = 70.9, 95.0$ [mm], $x_{2,l}, x_{2,u} = 70.9, 95.0$ [mm]
$x_{3,l}, x_{3,u} = 105.0, 237.5$ [mm], $x_{4,l}, x_{4,u} = 8.0, 10.0$ [-]
$x_{5,l}, x_{5,u} = 22.0, 35.0$ [mm], $x_{6,l}, x_{6,u} = -10.0, 10.0$ [°]
$x_{7,l}, x_{7,u} = 0.9, 1.1$ [-], $x_{8,l}, x_{8,u} = 30.0, 43.0$ [mm]
$x_{9,l}, x_{9,u} = -10.0, 10.0$ [°], $x_{10,l}, x_{10,u} = 0.9, 1.1$ [-].

The first step in the design process is to define the optimal design model. Various engine design rules of thumb on bore to stroke ratio, connecting rod to stroke ratio, etc. are available in the literature (Heywood, 1998). In additional to geometric constraints, limits are placed on pressure gradients, in-cylinder pressures, and mean piston velocities to maintain the reliability of the engine. The following inequality constraints must be satisfied by all family products:

g_1, g_2: $0.8 \leq b/s \leq 1.2$; g_3, g_4: $350 \leq \pi b^2 s/4 \times 10^{-3} \leq 650$ [cm^3]
g_5, g_6: $1.5 \leq l/s \leq 2.5$; g_7: $d_i \leq 0.37 \, b$ [mm]
g_8: $d_e \leq 0.45 \, b$ [mm]; g_9: $(2 \, s \, N_e)/(60 \cdot 1000) \leq 15.0$ [m/s]
g_{10}: $s/(c_r - 1) \geq 5.0$ [mm]; g_{11}: $1 + s + s/(c_r - 1) + 0.5 \, b \leq 350.0$ [mm]
g_{12}: $dP_{max}/DCA \leq 3.0$ [bar/deg]; g_{13}: $P_{max} \leq 110$ [bar].

We define three variants by means of three functional requirements. The first engine variant is designed to maximize power, the second to minimize fuel consumption, and the third to minimize emissions. The optimal problems are formulated as

$$\max_{x^A} \quad f^A = Power \text{ kW}$$

$$\text{subject to } g_1, g_2, ..., g_{13}$$

$$g_{14}^A : NO_x \leq NO_{x,max} \text{ ppm} \qquad (31)$$

$$g_{15}^A : Power \bullet BSFC \leq 30,000 \text{ g/h}$$

$$\max_{x^B} \quad f^B = Power \bullet BSFC \text{ g/h}$$

$$\text{subject to } g_1, g_2, ..., g_{13}$$

$$g_{14}^B : NO_x \leq NO_{x,max} \text{ ppm} \qquad (32)$$

$$g_{15}^B : Power \geq 80 \text{ kW}$$

$$\min_{x^C} \quad f^C = NO_x \text{ ppm}$$

$$\text{subject to } g_1, g_2, ..., g_{13}$$

$$g_{14}^C : Power \geq 80 \text{ kW} \qquad (33)$$

$$g_{15}^C : Power \bullet BSFC \leq 30,000 \text{ g/h}.$$

The value of $NO_{x,max}$ is based on the baseline value of 25,546 ppm multiplied by 110%, namely, we do not want to produce more than 10% additional emissions with respect to the baseline model.

Following the steps of the combined methodology we first solve the individual optimal design problems to obtain the null-platform optima. The results are shown in Table 9-6.

Table 9-6. Null-platform optima.

Engine	Power (A)	Fuel Usage (B)	Emissions (C)
b	84.43	75.39	95.00
s	79.95	78.40	90.00
l	199.87	196.00	201.02
c_r	10.00	10.00	8.84
d_i	31.24	27.90	27.55
i_{cta}	-10.00	10.00	-10.00
i_{am}	0.99	1.08	1.03
d_e	35.57	30.00	30.00
e_{cta}	-10.00	10.00	-10.00
e_{am}	1.10	1.00	0.90
h_c	8.88	8.71	11.48
disp.	2686	2100	3828
Power	114.20	80.00	80.00
Torque	218.27	152.79	152.79
BSFC	263.50	265.82	331.17
NO_x	26089.02	25717.78	17853.09
dPmx/DCA	1.40	1.28	0.88
P_{max}	52.36	47.97	35.43
Fuel Usage	30000.00	21265.32	26493.90

For the engine designed for maximizing power, three of the inequality constraints are active. These are the upper bounds on the mean piston speed, engine height, and NO$_x$ emissions. For the engine designed for maximizing fuel efficiency four inequality constraints are active (the lower bounds on cylinder displacement and power and the upper bounds on connecting rod to stroke ratio and intake valve diameter with respect to bore). For the engine design for minimizing emissions three inequality constraints are active (the upper bounds on the connecting rod to stroke ratio and intake valve diameter with respect to the bore and the lower bound on power. For the maximum power and fuel efficiency engines the compression ratio is maximized, which in turn increases combustion efficiency. For the low emissions engine the compression ratio is minimized, which corresponds to the strong correlation of NO$_x$ production with the increased heat due to higher compression. There is relatively little "natural" commonality between the three individually optimized engines. The only naturally shared component is the exhaust valve between the engines designed for fuel efficiency and low emissions.

We computed the performance deviation vector for the product family. All functions are normalized and design variables are scaled to be in the range [0, 1]. Figure 9-20 depicts the sorted performance deviation vector in terms of both individual $\prod_i$ and cumulative values. Figure 9-21 illustrates the sorted performance deviation vector with respect to engine components (aggregating deviations that correspond to component variables) in terms of both individual $\prod_i^c$ and cumulative values. The performance deviation vector values sorted with respect to the design variable illustrate which variables could be shared. Our focus is on the deviations sorted with respect to components. From the performance deviation vector values sorted with respect to components, we find that the $\prod_i^c$ values are relatively low for the connecting rod, intake and exhaust valves, and the intake cam. We also observe that after sharing the first four components (components 7, 4, 2, and 3), performance deviation increases significantly. Therefore, these first four components will be shared and the relaxed combinatorial problem will be solved for the remaining components.

We solved the relaxed combinatorial optimization problem for a deviation factor of 5%, 6%, 7%, 8%, 9%, and 10%. The most interesting results were obtained for 6% and 7%, and are presented in Table 9-7, where X denotes a shared component. Note that if a component is shared in all product pairs, then it is shared among all products. The results indicate that we can share the intake and exhaust valves along with the connecting rod and intake cam across all engine variants for both performance deviation bounds. Depending on the allowable deviations from the optimal designs it is also possible to share the exhaust cam across the family. The piston is consistently shareable only between the high power and low emissions engines. With slightly more performance deviation it is also possible to share the entire engine block between these two engines.

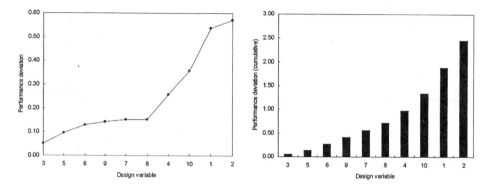

Figure 9-20. Sorted performance deviation vector with the respect to design variables (the plot on the left shows the sorted values of $\prod_i$, while the plot on the right shows cumulative values of the deviation vector).

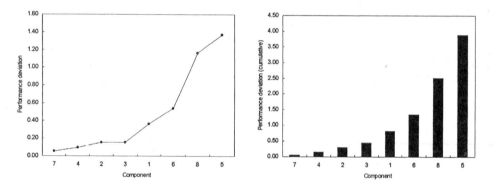

Figure 9-21. Performance deviation vector with respect to engine components (the plot on the top shows the sorted values of $\prod_i^c$, while the plot on the bottom shows cumulative values of the deviation vector.

Table 9-7. Relaxed combinatorial optimization problem results

Shared between:	Loss = 6%			Loss = 7%		
	A & B	A & C	B & C	A & B	A & C	B & C
Exhaust cam			X	X	X	X
Intake cam	X	X	X	X	X	X
Exhaust valve	X	X	X	X	X	X
Intake valve	X	X	X	X	X	X
Cylinder head piston		X			X	
Connecting rod	X	X	X	X	X	X
Engine block					X	

Note that the cylinder head, piston, and engine block are "modules" that happen to be shared in two cases (the piston and engine block between products *A* and *C*). Exchanging these components produces the variety in this engine family.

The final step is to design the product family. We design each of the engines to minimize the performance deviation from the null-platform optima. Performance optima and associated deviations are summarized in Table 9-8. They demonstrate that the bounds on performance deviation due to commonality are not violated. The results have been validated by solving the entire problem using the relaxed-problem formulation. The optimization results are consistent with the combined approach.

Table 9-8. Optimal product family design results and associate performance losses.

Variant	A	B	C
Null platform	114.29 kW	21,265.32 g/h	17,853.09 ppm
Platform with $L^p = 6\%$	109.62 kW	21,704.10 g/h	18,110.20 ppm
Performance loss	40.09%	2.06%	1.44%
Platform with $L^p = 7\%$	106.29 kW	22,685.43 g/h	18,883.23 ppm
Performance loss	7.00%	6.68%	5.77%

4.   SUMMARY

In this chapter we presented two analytical methods for making commonality decisions and demonstrated their use with automotive body structure family design problem examples. The techniques provide a balance on efficiency and accuracy requirements. We then presented a methodology that combines the two approaches and takes advantage of their strengths. The proposed platform selection methodology was then applied successfully to an automotive engine family optimal design example problem.

Chapter 10

PRODUCT VARIETY OPTIMIZATION
Simultaneous Optimization of Module Combination and Module Attributes

Kikuo Fujita
Department of Mechanical Engineering, Graduate School of Engineering, Osaka University, Osaka 565-0871, Japan

1. INTRODUCTION

The design optimization paradigm provides us with a rational synthesis means for the engineering design of products, machines, etc. The essential outcome from computational design optimization is that it can generate the best solution under mathematical representation and procedures, if an original design problem is appropriately translated into a formal style. The outcome is more effective if the original design problem is complicated, since human expertise cannot precisely manipulate such content. This situation is obvious when design concerns shift from component-level to system-level optimality (e.g., Papalambros and Wilde, 2000).

Current trends in manufacturing activities are diverging further from system-level design to multiple-systems-level design. Design and manufacturing activities are restricted by various hidden aspects, such as design and development costs, learning effects and supply-chains in production, production and services inventories, as well as from the direct aspects of single-system-level performance and cost. The viewpoint on multiple products extends the optimality domain to those hidden aspects beyond the traditional ones (Fujita and Ishii, 1997). The engineering challenge of simultaneously designing multiple products has attracted a great deal of attentions in the last decade. This new field is characterized by several terms including product variety, product family, product platform, and modular product.

Among various research activities, Fujita, et al. have been exploring computational optimization methodologies for product variety design under modular architecture. After the task structure of the product variety design was organized as a base for computational design methods (Fujita and Ishii, 1997), they developed an optimization formulation and an optimization procedure for the module attributes of a series of products (Fujita, et al., 1998), and developed another optimization framework for module combination across a series of products through module diversion (Fujita, et al., 1999). They used enumeration and a successive quadratic programming method for the former since each enumerated sub-problem is defined in continuous space, and used a simulated annealing technique for the latter since the problem is combinatorial. Following these, Fujita proposed a classification of product variety optimization problems and indicated the difficulty and necessity of developing simultaneous optimization of both module attributes and combinations (Fujita, 2002). Further, they configured a hybrid optimization method with a genetic algorithm that optimizes module combination patterns, a branch-and-bound technique that optimizes module similarity directions, and a successive quadratic programming method that optimizes module attributes (Fujita and Yoshida, 2004).

This chapter describes inclusively the above series of developments for reviewing the potential roles of design optimization for product families. In the following, the contents and conditions of product variety optimization are surveyed first. A range of product variety design problems are classified into three classes of optimization problems. Second, a general form of the product variety optimality is revealed and mathematical formulation underlying on it is established. Succeedingly, the optimization techniques for three different classes of product variety design problems are shown with associating example cases, respectively. This chapter finally concludes with discussing the limitations and challenges of product variety optimization.

2. PERSPECTIVES ON PRODUCT VARIETY OPTIMIZATION

2.1 Optimization paradigm and its application to product variety design

As mentioned in the introduction, various aspects of product variety design have been intensively studied in the last decade. Since consideration of multiple products is conceptually super-ordinate to every aspect in designing a single product, research topics are widely spread from describing any successful implementation of product variety design in practice, developing tactical methods and tools to facilitate design implementation, to establishing computational supports through design automation. This indicates that the categorization of design research into descriptive, prescriptive and computational (Dixon, 1987) is applicable to product variety design as well, and that the power of computational approaches in engineering design should be transferred from a single product to multiple products. The design optimization paradigm is, needless to say, the most effective and useful among computational design approaches. Its application to practical engineering problems requires the development of mathematical formulation and associated engineering analysis codes.

According to task structure analysis (Fujita and Ishii, 1997), the tasks for product variety design are decomposed into *system structure synthesis level, configuration synthesis level* and *model instantiation level* in sequence. The first two levels explore variety implementation possibilities, while the last level embodies the actual models for providing multiple products. In the sense of design optimization, the first two levels provide a feasible region for optimization, and the last level squeezes the best solution from within it. This indicates that frameworks for establishing optimization formulations are very important since they squeeze a set of latent solutions into an appropriate set of potential solutions where any optimization procedure can be applied. Following elimination of unpromising design solutions, the design optimization paradigm can be effective for product variety design.

The overall circumstance of design optimization and its application for product variety can be summarized as shown in Figure 10-1. That is, since the optimization paradigm consists of the mathematical formulation of an original design problem and the computation with a mathematical programming algorithm for seeking its optimum, the final solution depends on both phases in various aspects such as solution quality and solving efficiency. Optimization formulation eliminates some parts of potential solutions through mathematical modeling, and optimization algorithm further eliminates non-optimal solutions through computation. At least, it is essential that mathematical natures of a formulation and an algorithm are well matched to each other for obtaining the optimal solution. When applying the optimization paradigm to product variety design, since the potential design region is expanded beyond the ordinary design situation for a single product, it is very important to reduce the solution possibility through modeling in the first phase based on the characteristics of a specific situation before optimization computation in the second phase.

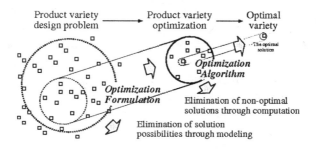

Figure 10-1. The role of optimization in product variety design.

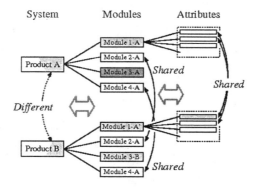

Figure 10-2. Design space modeled with hierarchy of systems, modules and attributes.

2.2 Design space of product variety optimization

Development of general optimization methods for product variety design requires any abstract and general representation of design space. When viewing a product as a system, system-and-subsystems and entity-and-attributes relationships are appropriate means for general representation. Since such means are natively recursive, any proper granularity level of representation must be introduced to assess design possibilities. Under these conditions, similarities and differences between different products are explained as shown in Figure 10-2, that is, different products can share the same modules, and different modules can partially share some attributes. The design optimization determines which part of this representation should be shared or differs and considers the compromise between the merits and demerits in product variety design.

Based on the situation of Figure 10-2, the model instantiation level, which is defined in the task structure (Fujita and Ishii, 1997), is further divided into two sub-levels: *attribute quantification* to develop modules by quantifying attributes and *combinatorial selection* to develop products by selecting practical combinations of modules from potentially feasible ones. These two sub-levels are complementary rather than methodical in their sequence. It means that the product variety optimization includes two sub-problems: *module attribute optimization* and *module combination optimization*. The former problem is defined in continuous space, and the latter problem is defined in combinatorial space.

2.3   Classification of optimization problems

From the above viewpoint on optimization space, Fujita (2002) classified product variety optimization problems into the following three classes:

Class I: Optimize module attributes under fixed module combinations;
Class II: Optimize module combination under predefined module candidates; and
Class III: Simultaneously optimize both module attributes and module combination.

The classification is based on whether the module contents (attributes) are going to be determined or are fixed, and whether module combinations for respective products are going to be determined or are fixed. Among these, the Class III problems belong to the most difficult optimization problems, because the contents of both Class I and Class II problems, i.e., continuous aspect and combinatorial aspect, must be simultaneously optimized. This indicates that when a design problem can be formulated as Class I or Class II, it must be dealt as such a problem rather than as Class III, even though Class III covers all regions. Since all classes are characterized in mathematically different ways, their optimization methods are different each other. Fujita, et al. developed optimization methods for all classes (Fujita, et al., 1998; 1999; Fujita and Yoshida, 2004) as mentioned earlier. They are explained in the following sections of this chapter, respectively.

2.4   Prerequisites of product variety optimization

The issues related to product variety design range over various aspects of engineering problems. Some of them are the concerns of product variety optimization itself and others define rather their prerequisites. Concurrent engineering has tackled the integration of various engineering aspects from 1990s and brought highlights on several methods and tools as its means. Among them, quality function deployment (QFD) focuses on customer's needs in customer's language and provides the quantitative mapping from customer's needs, physical function realization to manufacturing process in engineer's language (e.g., Clausing, 1994). The concept of viewing a product through mapping structure among different aspects is expandable for a series of products (Fujita, et al., 2003a). Figure 10-3 illustrates a circumstance of expanded relationships among those aspects on a product family under the abstracted hierarchical representation shown in Figure 10-2 (Fujita, et al., 1999). In the figure, even though different products are characterized with different primary features, they share some subsidiary features behind. Corresponding to such a relationship, the function structures of different products partially share some common primitive functions each other. Further, it indicates that some common modules or components can be used across the product variety corresponding to those relationships. While the product variety optimization mainly concerns on product realization in the aspects of physical functions or manufacturing modules, the variety of customer's needs are linked with them in the above way. This indicates that interpretation of customer's needs is indispensable as the prerequisites for the optimization. This also concerns on the classification of a specific product variety design problem into either of three classes.

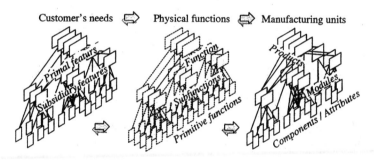

Figure 10-3. Mapping structure among customer's needs, physical function and manufacturing units under product variety.

3. GENERAL FORM OF PRODUCT VARIETY OPTIMALITY

3.1 Objectives of product variety

The optimality domain of product variety is transformed through objectives, while the previous section discussed its region. A major motivation to expand the design target from a single product to multiple products is the benefits expected through expanding volume effect basis. However, there are several disadvantages behind such expectation.

Optimization problems are generally faced on the compromise between different objective items. The tradeoff between performance and direct cost is typical for the optimality of a single product. It is necessary to formalize the corresponding items and their tradeoff patterns in product variety optimality. They include various issues over the production and utilization of all units of different products. Under this reason, total cost or total profit through all designing, manufacturing and utilizing multiple products can be the objective rather than simple performance or direct cost.

3.2   Cost structure

According to traditional definition of cost structure, cost can be decomposed into several ways: direct cost and indirect cost, fixed cost and variable cost, and so forth (e.g., Pahl and Beitz, 1996; Ulrich and Eppinger, 2004). The product variety optimization focus on the effects of the numbers of product kinds and module kinds, the items on which are counted as fixed cost in the cases of a single product. Therefore, the following cost structure is useful for assessment and optimization of multiple products instead of the conventional way:

Cost depended on production volume: This mainly concerns to material cost, fabrication cost, assembly cost and so forth, which can be considered to be basically proportional to the number of production units. The learning effects in fabrication and assembly influence to reduce this category of cost as the number of production units increases. What is more important, the commonalization of modules for different products causes excess cost per unit due to over-specification, which is counted as a disadvantage of product variety design.

Cost depended on the number of product and module kinds: This mainly concerns to design cost, facility cost, etc., which are usually counted as fixed cost for a single product. Since commonalization of modules or else results in reduction of the number of module kinds or else that should be designed and engineered, this category of cost varies by these factors. Further, it is expected that various facilities along production supply chain can be saved through unification of subsystems.

Hidden fixed cost: This corresponds to any other cost that is not mentioned above. This category belongs to sensitive parts of cost estimation and might have influence to product variety cost. However, the optimization viewpoint requires the distinction of controllable and uncontrollable parts through an optimization algorithm. For this reason, the model used in this chapter treats theses costs as an underlying insensitive part of optimization.

The above categorization indicates sufficient tendency of product variety optimality. However, detail cost estimation requires another investigation into the complexity hidden in fixed cost or else (e.g., Martin and Ishii, 1996). For instance, complexity in cost related to R&D, inventory, etc., fluctuates due to commonalization details. Such subsidiary but important issues might be assessed in compliance with a specific situation.

Besides, a firm usually intends to make more profit rather than to reduce cost. The predictive assessment of profit is more complicated than cost. At least, it requires estimating the utility in product use, the selling price, etc., which are much related to business issues beyond engineering issues.

3.3 Tradeoffs in product variety

As the above cost structure already indicates, commonalization of subsystems among multiple products leads both merits and demerits. This is a reason for the necessity of optimization computation as a rational mean for product variety design.

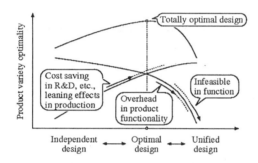

Figure 10-4. Tradeoffs between merits and demerits of commonalization.

Figure 10-4 conceptually illustrates an expected scenario on the tradeoff between the merits and demerits. The horizontal axis means here the degree of commonalization. The left end corresponds to the situation where every product for different segments is independently designed and produced. The right end corresponds to the situation where a single product is introduced to cover all segments. Although the latter is definitely an unlikely case, it is the extreme on the region of product variety design. The vertical axis means here an optimality criterion, where bigger is better. Basically, as a design is shifted from the left to the right, the cost depended on the numbers of product and module kinds decreases (optimality increases). Under the same shift along commonalization, the cost depended on the numbers of units increases (optimality decreases) due to overhead in materials or else, even though learning effects have some influences. Furthermore, as shown in the figure, feasibility in function of each design cannot be satisfied beyond a boundary before the right end. These tendencies correlatively form tradeoffs between the merits and demerits, and deduce the optimal solution within the design region.

4.   FORMULATION FRAMEWORK OF OPTIMIZATION PROBLEMS

4.1   Representation of design space

Under the representation scheme shown in Figure 10-2, the following notation is introduced for generally representing a series of products toward product variety optimization.

(i)   Multiple products, P^1, P^2, ..., P^{n_P}, are simultaneously considered. The total number of products or models is n_P.

(ii)   Each product P^i is composed of a series of modules. The module slots, where modules are installed, are denoted as M_1, M_2, ..., M_{n_M}, respectively. The total number of module slots is n_M.

(iii)   A module that is installed in a module slot M_j of product P^i is denoted as m^i_j (i = 1, 2, ..., n_p; j = 1, 2, ..., n_M).

(iv)   Each module m^i_j is represented by its attribute variables, $x^i_j = \left[x^i_{j,1}, x^i_{j,2}, ..., x^i_{j,n^A_j} \right]^T$. The number of attribute variables representing the module in the j-th slot M_j is n^A_j.

(v)   Under x^i_j, each product P_i is represented by an entire set of attribute variables $z^i = \left[x^{i^T}_1, x^{i^T}_2, ..., x^{i^T}_{n_M} \right]^T$.

All of the variables $\left[z^{1^T}, z^{2^T}, ..., z^{n_P^T} \right]^T$ in the above definitions are native design variables in product variety optimization.

4.2   Module commonality and similarity

Commonalities and similarities among the different products, shown in Figure 10-2, are decomposed into a combination of module commonalities and similarities in respective slots. They are categorized into the following three design schemes (Fujita, et al., 1998):

Independent module design scheme: Totally different modules are used for a unique module slot of different products. That is, sets of attributes are determined independently.

Similar module design scheme: Partially different and partially common modules are used in a module slot of different products. That is, a subset of attributes is constrained to be equal to each other, and the other subset of attributes is determined independently to stretch the original design to another design.

Common module design scheme: A unique module is used in a module slot of different products. That is, all attributes of the module are equal across different products.

Independent module design and common module design are straightly understandable, but similar module design requires some explanation. Since each module is a system as well as a whole product, it consists of subsystems or parts. This indicates that similar module design enables the commonalization of hidden subsystems or parts behind modules. Such commonalization has some effects on the product variety optimality that are similar to common module design, if the direction of variety required in the module slot matches with the key module attribute that differentiates modules. Messac, et al. (2002) referred to 'similar module design' as 'scale-based product family' and 'common module design' as 'module-based product family.'

The similar module design scheme typically corresponds to a '*stretch-based design*,' where module design is scaled up to another module along with a scale variable. When a scale variable is denoted by $x^i_{j,1}$ within $\mathbf{x}^i_j$, it is assumed that the following constraints must be satisfied in a similar module design for module slot M_j from P^{i_1} to P^{i_2}, in which $m^{i_1}_j$ is stretched to $m^{i_2}_j$.

$$x^{i_2}_{j,1} > x^{i_1}_{j,1} \tag{1}$$

$$x^{i_2}_{j,k} = x^{i_1}_{j,k} \qquad (k = 2, 3, \cdots, n^A_j) \tag{2}$$

As for the common module design scheme, when two products P^{i_1} and P^{i_2} share a unique module at a module slot M_j, the following constraints must be satisfied.

$$x^{i_2}_{j,k} = x^{i_1}_{j,k} \qquad (k = 1, 2, \cdots, n^A_j) \tag{3}$$

4.3   System constraints

The design variables for every module and every product are restricted by system constraints. Design conditions for respective products must be different from each other. That is, the following constraint on the design variables z^i of each product P^i must be satisfied:

$$z^i \in Feasible(s^i) \qquad (i = 1, 2, \cdots, n_p) \tag{4}$$

where s^i are the design conditions assigned for product P^i, and *Feasible* $(\cdot)$ means the feasible region of z^i under s^i.

4.4   Cost model

The benefit of product variety design is mainly due to the reduction of what we call hidden costs such as design and development costs, facility costs, inventory reduction, and learning effect enhancement. The discussion on cost structure in Section 3.2 brings a set of quasi-liner monotonic cost equations (Fujita, et al., 1998) for assessing benefits and associated demerits of product family design, which can be used as a part of the objective function in product variety optimization.

4.4.1   Design and development cost

It is assumed that the design and development costs are proportional to the weight of each module, and that cost savings in a similar module design scheme is also proportional to the difference in the corresponding weights. The design and development costs C_D^i for product P^i is represented as:

$$C_D^i(z^i) = \sum_{j=1}^{n_M} C_{Dj}^i(x_j^i) \tag{5}$$

$$C_{Dj}^i(x_j^i) = \begin{cases} \alpha_{Dj} W_j^i & \cdots & \text{in the case of independent module design} \\ \left(\beta_{Dj} \dfrac{W_j^i - W_j^{i_{Base}}}{W_j^{i_{Base}}} + \gamma_{Dj} \right) C_D^{i_{Base}} & \cdots & \text{in the case of stretched module design} \\ 0 & \cdots & \text{in the case of commonalized module design} \end{cases} \tag{6}$$

where α_{Dj}, β_{Dj} and γ_{Dj} are coefficients depending on the module slots. W_j^i is estimated from module attributes, that is $W_j^i = W(x_j^i)$.

In the above equations, a 'stretched module design' means that m^i_j is stretched from $m^{i_{Base}}_j$ under a similar module design scheme, that is, the cost for $m^{i_{Base}}_j$ is measured with j the first equation of Eq. (6) and the cost for m^i_j is measured with the second equation of Eq. (6). 'Commonalized module design' means here that m^i_j and $m^{i_{Base}}_j$ are under the common module design scheme and that the cost of $m^{i_{Base}}_j$ is measured with first equation of Eq. (6) and the cost for m^i_j is measured with the third equation of Eq. (6).

4.4.2 Facility cost

It is assumed that facility cost is similar to the design and development cost. Fundamentally, it is to be proportional to the representative attribute, which is denoted as $x^i_{j,1}$, rather than weight. The facility cost C^i_F for product P^i is represented as follows:

$$C^i_F(z^i) = \sum_{j=1}^{n_M} C^i_{Fj}(x^i_j) \tag{7}$$

$$C^i_{Fj}(x^i_j) = \begin{cases} \alpha_{Fj} x^i_{j,1} & \cdots & \text{in the case of independent module design} \\ \left(\beta_{Fj} \dfrac{x^i_{j,1} - x^{i_{Base}}_{j,1}}{x^{i_{Base}}_{j,1}} + \gamma_{Fj} \right) C^{i_{Base}}_{Fj} & \cdots & \text{in the case of stretched module design} \\ 0 & \cdots & \text{in the case of commonalized module design} \end{cases} \tag{8}$$

where α_{Fj}, β_{Fj} and γ_{Fj} are coefficients depending on module slots. The meanings of 'stretched module design' and 'commonalized module design' are the same as with the design and development costs in Eqs. (5)-(6).

4.4.3 Production cost

It is assumed that the production cost is composed of material costs c_m and processing costs c_p (e.g., labor costs). The learning effect gradually reduced according to the increase in the accumulated number of production units. Therefore, the production cost $c^i_p(\ell^i)$ of the ℓ^i-th unit of product P^i is represented as follows after ℓ' units of the other product P' have been already produced:

$$c_p^i(\ell^i) = c_m^i + c_p^i(\ell^i) \tag{9}$$

The material cost c_m^i is represented by the following equation based on the weight of the respective modules:

$$c_m^i = \sum_{j=1}^{n_M} \vartheta_j W_j^i \tag{10}$$

where ϑ_j is the material cost per weight for module slot M_j.

The processing cost $c_p^i(\ell^i)$ for ℓ^i-th unit of product P^i is represented by the following equation based on the accumulated numbers of production units for respective module slots, ℓ_j^i and ℓ_j^t, rather than ℓ^i and ℓ^t for the overall products:

$$c_p^i(\ell^i) = \sum_{j=1}^{n_M} \kappa_j W_j^i u(\ell_j^i) \tag{11}$$

$$u(\ell_j^i) = \left(\ell_j^i + \sum_{t=1, t\neq i}^{n_p} \varsigma_{j_{t,i}} \ell_j^t \right)^{\frac{\ln r}{\ln 2}} \tag{12}$$

where κ_j is a coefficient depending on the module slot for the unit processing cost. $u(\ell_j^i)$ is a factor representing the learning effect; r is learning curve ratio (Raymer, 1989); and $\varsigma_{j_{t,i}}$ is a coefficient representing the learning effect for m_j^t production from m_j^i production. The value of $\varsigma_{j_{t,i}}$ is 0 in the case of an independent module design scheme, and is 1 in the case of a common module design scheme. The value for a similar module design scheme is between these two values depending on the degree of similarity, as well as Eqs. (6) and (8).

4.5   Profit model

The manufacturer's profit from multiple products is affected by the pricing mechanism for respective products and the capital investment beyond the above cost model. The price C_S^i of product P^i is determined based on the expected utility from the customers' viewpoint. Therefore, C_S^i can be denoted by $C_S^i = C_S(z^i)$, while this function must reflect the direct utility from the product's operation, cost per unit operation, and the capital recovering mechanism of purchase and use.

Under the above cost and price models, all costs that a manufacturer invests before starting production is represented as follows:

$$C_1 = \sum_{i=1}^{n_p} (C_D^i + C_F^i) \tag{13}$$

The balance of profits and expenses after y terms, $C_B(y)$, is represented as:

$$\left. \begin{aligned} C_B(0) &= -C_1 \\ C_B(y) &= C_B(y-1)(1+\rho) - \sum_{i=1}^{n_p} \frac{N^i}{T_p} \int_0^1 c_P^i(\ell^i(t)) dt \\ &+ \sum_{i=1}^{n_p} \frac{N^i}{T_p} C_S(z^i) \qquad (y = 1, \cdots, T_p) \end{aligned} \right\} \tag{14}$$

where ρ is the annual interest rate, N^i ($i = 1, \ldots, n_p$) is the number of total production units of P^i, and T_p is the number of production terms. $\ell^i(t)$ is production unit counter that is given by $\ell^i(t) = \frac{N^i}{T_p}(y-1+t)$. In the above equations, it is assumed that the ratio of production volumes among different products and the production volume per term are kept constant throughout the total production period, and summation operations over production unit numbers are replaced with integral calculus.

The final balance, which corresponds to a manufacturer's final profit, is given by $C_B(T_P)$ from the above recursive equations. Although $C_B(T_P)$ is affected by various uncertain factors such as production volumes, it can be used as a measure for preliminary assessment for product planning purposes.

4.6   Challenges in establishing mathematical formulation

The discussion and assumptions in this section draw some general frameworks for establishing mathematical formulations for product variety optimization. A mathematical formulation of any specific class of design problems should be arranged by considering the following issues:

How to represent and manipulate commonality and similarity of modules among products. While the content of Section 4.1 fully represents the design space, it does not directly model the combinatorial structure underlying in product variety design. When using meta-heuristics based optimization algorithms such as simulated annealing (SA) (e.g., Kirkpatrick, et al., 1983), genetic algorithms (GA) (e.g., Goldberg, 1989), which are well known as powerful methods for complicated combinatorial optimization problems, it is a critical point to embed such structure in perturbation operations, coding systems, etc.

How to construct the measures of constraints and objectives in detail. While their entire shape is already outlined in Sections 4.2-4.5, their details should be reorganized so as to fit with the representation for commonality and similarity, and so forth. Further the model of representing combinatorial structure may bring artificial constraints into the mathematical formulation.

These considerations must be taken in cooperation with formation of any specifical optimization algorithm. It is a nature in developing a practical optimization method for complicated design problems, especially when using meta-heuristics based algorithms.

5.   MODULE ATTRIBUTE OPTIMIZATION ACROSS PRODUCT VARIETY

5.1   Attribute optimization under stretch-based design deployment

Stretch-based design is often introduced to deploy the product variety of large technical systems. It is expected to be effective for cost saving especially in R&D cost. It has been a common design strategy to stretch an airplane for different specifications by expanding or replacing respective modules such as fuselage, main wing, and engine, from early days.

Figure 10-5 illustrates the situation of stretch-based design deployment. In the figure, each product consists of some modules, slots of which are denoted as M_1 to M_m. After the base design is introduced as P^1, a series of products is deployed from P^2 to P^p by replacing or not replacing respective modules in order to effectively cooperate with different requirements and total optimality. Under the meaning of Figure 10-2, the replacement of modules in each module slot is classified into three methods: 'independent module design,' 'similar module design,' and 'common module design,' which are explained in Section 4.2, according to how attributes of modules are different each other.

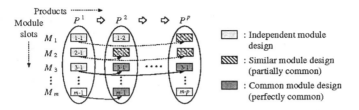

Figure 10-5. Stretch-based design deployment through independent, similar and common module design schemes.

5.2 Mathematical formulation and optimization procedure

When a common module is used across different products or when similar modules are used across different products under the above stretch-based design deployment strategy, merits and demerits must be compromised. The design problem consists of the determination of which module slots are independent, similar or common, that is, commonality and similarity patterns, and the determination of attributes of all modules under respective patterns. This overall situation natively corresponds to a Class III problem. However, when the patterns can be enumerated as a small number, the optimization problem of a product variety can be handled as a set of Class I problems through the enumeration of possible patterns. The formulation of a Class I optimization problem is outlined as follows (Fujita, et al., 1998).

Design variables: All attributes of all modules of all products, which were mentioned in Section 4.1.

Constraints: The constraints among attributes of respective modules that are assigned to be common or similar under the given module commonality and similarity pattern (Equalities in these constraints must be eliminated by the substitution of design variables before optimization computation). They are defined by Eqs. (1)-(3). System constraints for all products, which correspond to the entire set of constraints in the ordinary optimization for a single product. That is, Eq. (4).

Objective: Total profit through the entire production. While system performance or system efficiency is subject to system constraints, selling price of each product must be determined based on its utility. The profit is calculated with the balance between the total sales through the sum of such prices and the total cost for entire production by Eq. (14).

Since this optimization problem can be a constrained real-number nonlinear mathematical programming problem, any conventional optimization technique is applicable. The successive quadratic programming (SQP) method is used in the example shown in the following sections.

5.3   Stretch-based aircraft design

Similarity based design or stretch-based design is often introduced for developing a series of commercial aircraft as aforementioned. For instance, the original design of the DC-9-10 or DC-9-20 was enlarged to the DC-9-30, 40, 50, ..., and to the MD-80. Boeing 777 has a plan to expand a basic model to a different number of seats and a different cruise range by enlarging the fuselage, replacing engines, etc. (Sabbagh, 1996). While this stretch corresponds to what is called the similar module design scheme, the overall situation of aircraft design planning includes the choices of independent, similar and common module design schemes.

The configurations of aircraft are diverse due to their size, purpose, etc. A configuration shown in Figure 10-6(a) is considered here. In this configuration, the aircraft is decomposed into four modules of fuselage, main wing, tail wing, and engine. The main wing is attached to the low part of the fuselage, the horizontal and vertical tail wings are combined into a unique module with a T–type shape, and the two engines are installed in the rear part of the fuselage. Under this configuration, the fuselage can be stretched in its longitudinal direction by increasing l_{cabin} as shown in Figures 10-6(b) and (c) while sharing its sectional structure. This stretch-based enlargement of fuselage is directly effective for the increase of seats with less expense, while it has some side effects on other features through the functional coupling with the other modules. The main wing and tail wing can be stretched in their lateral direction by adding the parts of b' and b'' while sharing their sectional structure as well, as shown in Figures 10-6(d) and (e). This for main wing is effective for increasing lift force with less design changes, while it has some side effects as well. As for the engines, since they are usually purchased from outside, it is assumed that a similar module design scheme is impossible in the optimal design examples. Since these stretches and replacements have mutual side effects beside their primary effects, their compromise on underlying tradeoffs is essential for achieving the optimal design.

An appropriate engineering model for system performance, cost and profit for examining system feasibility and optimality of aircraft design can be formulated (Fujita, et al., 1998) based on some literature (Torenbeek, 1976; Raymer, 1989; Hundal, 1997). The system constraints include cruise range, propulsion economy, stability, takeoff and landing distances. As for the profit model, the utility of an airplane is determined by the gain determined by unit passenger fare, passenger number, cruise length per flight, flight times over its life, the expense determined by fuel consumption ratio that is directly linked to system performance, and other cost items such as crew, maintenance, etc. They are obviously all over the modules of Figure

10-6(a), and define the region of an optimization problem against the necessity of the above compromise. That is, the coupled attributes of different modules must be simultaneously optimized for product variety.

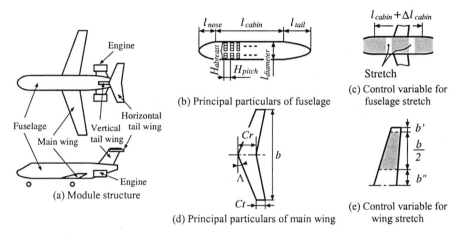

Figure 10-6. Module structure and stretch-based design deployment of aircraft.

5.4   Optimization examples

Figure 10-7 shows the design conditions of example cases, where a pai of airplanes are designed, for demonstrating Class I optimization. The majo design specification of commercial aircraft consists of the number of seat and cruise range. The original design is aimed at 90 seats and 2000 kn cruise range. The stretched design is aimed at the same seat number and the increased different cruise range.

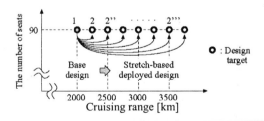

Figure 10-7. Design conditions on aircraft variety in the case studies of Class I.

Figure 10-8 shows the optimal designs for different pairs of cruise range. The horizontal axis is a pair of cruise ranges for two different airplanes, that is, how two design specifications are different each other. The vertical axis is the expected profit of a manufacturer through the production of entire units.

In these examples, since the seat number is the same across airplanes, fuselage should be common. Since stretch of engines does not make any sense, there is a choice of common design or independent design. Thus, 18 module combinations (3 for main wing, 3 for tail wing, 2 for engine) are enumerated. The Class I product variety optimization is executed for every point of the horizontal axis and for all possible module combinations.

The series of optimization results show that a common design is enough in the first third range of requirement differences. A similar design where main and tail wings are stretched is effective for the middle half range, which is named 'Similar design 1' in the figure. Another similar design where different engines are used for different airplanes is necessary for the last fourth range, which is named 'Similar design 2' in the figure. This result indicates that the optimal module combination is varied based on the range of product variety. On the other hand, other commonality and similarity patterns, which are named 'Other similar design' in the figure, are inferior to those patterns. The totally independent design is the worst in most parts expect the last third range.

As the above case study demonstrates, the quantitatively precise situation of product variety optimality is so complicated through various tradeoffs, even though the rough tendency is quite straightforward.

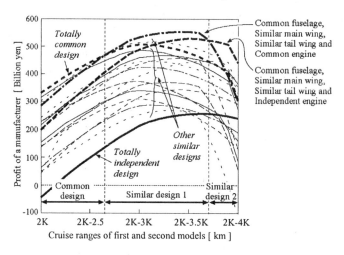

Figure 10-8. Optimality versus variety range in aircraft design in the case studies of Class I.

6. MODULE SELECTION OPTIMIZATION ACROSS PRODUCT VARIETY

6.1 Module combinatorial optimization under module diversion

Product variety design is often demanded from combinatorial explosion of products. For instance, Whitney (1995) revealed that 'assembly-driven manufacturing' as 'combinatoric method' is an efficient manufacturing policy that can tremendously increase flexibility toward a wide variety of products through the on-site investigation of Denso, a Japanese automotive component supplier. Its meaning is that once the modules (parts) are provided under the unified interface among their slots, every combination of them can become products, even though some meaningless combinations are inevitable against customer's needs.

While the above viewpoint is on the combination of modules into products, another viewpoint is the commonalization or diversion of modules among products. As shown in Figure 10-9, when modules for different products are compatible each other in a specific module slot, a unique module can be commonly used across different products, that is, the module of a product can be diverted with one of another product. The compatibility means here that the interface of a module slot against the other slots is compatible across products and that the functional contents of modules in the slot are proportional (not exclusive) between products.

The design problem of module commonalization or diversion, shown in Figure 10-9, is recognized to be typical of the Class II problems under the classification discussed in Section 2.3.

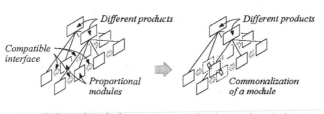

Independent design   Product variety design

Figure 10-9. Possibility of module diversion.

6.2 Mathematical formulation and optimization procedure

The task of Class II problems is to select specific combinations of modules for respective products from potentially possible ones. It is translated to the simultaneous determination of which of module candidates $\tilde{m}_j^k$ ($k=1, 2,..., K_j$) is used for a module slot M_j ($j = 1, 2, ..., J$) of each product P^i ($i = 1, 2, ..., I$) as shown in Figure 10-10. This combinatorial problem is mathematically represented with 0-1 integer variables:

$$\lambda_j^{k^{(i)}} = \begin{cases} 1\cdots & \tilde{m}_j^k \text{ module is implemented in } M_j \text{ slot of } P^i \text{ product.} \\ 0\cdots & \tilde{m}_j^k \text{ module is not implemented in } M_j \text{ slot of } P^i \text{product.} \end{cases} \tag{15}$$

$$(i=1,2,\cdots,I; j=1,2\cdots,J; k=1,2,\cdots,K)$$

and the following constraints (Fujita, et al., 1999):

$$\sum_{k=1}^{K_j} \lambda_j^{k^{(i)}} = 1 \qquad (i=1,2,\cdots,I; j=1,2,\cdots,J) \tag{16}$$

Eq. (16) means that a unique module can be implemented in each slot.

The system constraints on each product are essential in optimization computation as a nature of design problems. They are directly indispensable in Class I problems, since the attributes of modules must be changeable in optimization computation. However, they can be eliminated from optimization computation of Class II problems, since the attributes of modules have been fixed before optimization computation. That is, the feasibility of their combinations can be patterned beforehand, and it is unnecessary to check the system constraints during optimization.

The patterns can be described with three types of constraints on *diversion feasibility*, *diversion simultaneity* and *energy capacity* (Fujita, et al., 1999). Diversion feasibility means whether module diversion, which is shown in Figure 10-9, is possible in each slot or not. Diversion simultaneity means that when a module is diverted in a specific slot, other module(s) in other slot(s) must be simultaneously diverted due to functional coupling between relating module slots. Energy capacity means that energy balance imposes an equation on the sum of input and output energy at respective modules over a product. All of these constraints are represented as simple algebraic equations with the above 0-1 integer variables $\lambda_j^{k^{(i)}}$.

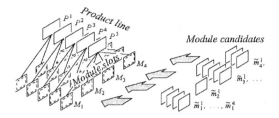

Figure 10-10. Design variables for module selection in Class II problems.

After the objective is defined with the total profit (or total cost), the Class II optimization problem is formulated as a 0-1 integer, constrained, nonlinear programming problem. While this leads combinatorial explosion even under the small number of products, module slots and module candidates, that is, design variables, meta-heuristics based optimization techniques are applicable for this class of optimization problems. Fujita, et al. (1999) have developed an optimization procedure based on the simulated annealing technique (e.g., Kirkpatrick, et al., 1983).

6.3   Television receiver circuit design

Electric or electronic products are often well modularized. Personal computers are typical of such a direction. A virtual design problem of receiver circuits for television sets (Grob, 1975) is used for demonstrating a Class II design problem. Its contents are almost similar to the real circuits, but it is assumed that functional modules are separated through interface connectors for demonstration purpose.

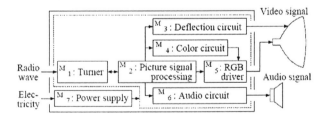

Figure 10-11. Module structure of television receiver circuits.

Figure 10-11 shows the module structure of a television receiver circuit, which consists of seven module slots, turner module, picture signal processing module, deflection circuit module, color circuit module, RGB driver module, audio circuit module, and power supply module. They have relatively clear decoupling and correspondence across three phases of Figure 10-3 due to their functional nature. Customer's needs of television sets range in picture size, picture quality, audio level, power supply voltage (location of use), etc. Against these, for instance, different picture sizes require different deflection circuits due to the magnitude of beam deflection, and different picture quality levels require different picture signal processing circuits and color circuits. Since the picture signal processing circuit and the color circuit are functionally coupled each other, a diversion simultaneity constraint must be considered. Since the power supply circuit provides electricity to the others, an energy capacity constraint is considered on it. Under these relationships, the constraints for diversion, simultaneity, capacity are formulated for a specific set of design specifications.

6.4   Optimization examples

Table 10-1 shows a product variety, which is assumed as an example case for optimization computation. In the table, six models are planned with different features. After production volumes are assumed as 24,000 units respectively, the optimal diversion is searched through the simulated annealing based optimization procedure (Fujita, et al., 1999).

Table 10-2 shows the comparison between the primary module selection based on minimum functional requirements and the optimal module selection gotten by optimization computation. In the table, the first row lists module slots, and the second row lists the module candidates that can be implemented to respective slots. In the middle rows, the symbol • means the primary selection for each product, the symbol ○ means the optimal selection, and the symbol ⊙ means that both are the same. As shown with the symbols •, ○ and ⊙, some modules are diverted, and it results in 2.6 percent reduction of total cost under the assumed situation. While the diversion patterns of module slots M_3, M_4 and M_6 are relatively simple due to less number of candidates, the patterns of module slots M_2 and M_7 are too complicated to predict with insight. The latter patterns indicate that the mathematical optimization for product variety design is necessary and essentially effective.

Table 10-1. Product variety of television sets.

Feature index	Picture size	Picture quality	Audio level	Power supply voltage
P^1	14	Good	Low	100V
P^2	21	Better	Medium	100V
P^3	36	Best	High	100V
P^4	14	Normal	Low	Multi
P^5	21	Good	Medium	Multi
P^6	36	Better	Medium	100V

Table 10-2. Optimization results for the television receiver circuits.

Slot	M_1	M_2				M_3			M_4		M_5	M_6			M_7							
Cand.	$\tilde{m}_1^1$	$\tilde{m}_2^1$	$\tilde{m}_2^2$	$\tilde{m}_2^3$	$\tilde{m}_2^4$	$\tilde{m}_3^1$	$\tilde{m}_3^2$	$\tilde{m}_3^3$	$\tilde{m}_4^1$	$\tilde{m}_4^2$	$\tilde{m}_5^1$	$\tilde{m}_6^1$	$\tilde{m}_6^2$	$\tilde{m}_6^3$	$\tilde{m}_7^1$	$\tilde{m}_7^2$	$\tilde{m}_7^3$	$\tilde{m}_7^4$	$\tilde{m}_7^5$	$\tilde{m}_7^6$	$\tilde{m}_7^7$	$\tilde{m}_7^8$
P^1	◉	◉				•	○		◉		◉	•	○		•					○		
P^2	◉		•	○		◉			◉		◉	◉			•	○						
P^3	◉			◉				◉	◉		◉			◉		◉						
P^4	◉	•	○			•	○		•	○	◉	•	○		•				○			
P^5	◉	◉					◉		◉		◉	◉									◉	
P^6	◉		•	○				◉	◉		◉	◉				◉						
# of kinds	1	4→2				3→2			2→1		1	3→2			5→2							

• : No diversion O : Optimal diversion

Table 10-3. Commonalization effects in the case study of television receiver circuit design.

	Cost Items	No diversion [yen]	→	Optimal diversion [yen]
Cost on production volume	Module material cost	51,762,000	↗	53,088,000
	Module fabrication cost	51,484,037	→	51,226,054
	Product assembly cost	645,703	→	645,703
	Partial Sum	103,891,740	↗	104,959,756
Cost on the number of kinds	Cost dependent on the number of product kinds	8,985,000	↗	9,240,000
	Cost dependent on the number of module kinds	12,375,000	↘	7,560,000
	Partial Sum	21,360,000	↘	16,800,000
	Hidden fixed cost	10,000,000	→	10,000,000
	Total cost	135,251,740	↘	131,759,756

Table 10-3 shows the detail of the above 2.6 percent cost reduction by the contents of cost structure. As aforementioned in the general discussion on cost structure, the cost depended on the number of module kinds is drastically decreased through product variety optimization, because the total number of different modules is decreased from 19 to 11 through module diversion. On the other hand, the cost depended to the number of product kinds and module material costs are slightly increased, because diverted modules have some extra functionality over the minimum requirements. The compromise between the former merit and the latter demerits results in the above 2.6 percent reduction of the total cost.

7.   SIMULTANEOUS OPTIMIZATION OF MODULE COMBINATION AND MODULE ATTRIBUTES

7.1   Simultaneous optimization and hybridization of optimization techniques

The previous two sections demonstrates optimization methods and associated examples of Class I and II problems, respectively. The former is categorized as continuous optimization problems and the latter is categorized as combinatorial optimization problems. The native region of product variety design includes both continuous and combinatorial ones beyond these, as aforementioned. This contrast demands the necessity of discussing Class III problems and developing any optimization method for them.

Development of design optimization has been challenged from two sides; mathematical programming techniques and their application to practical engineering problems. It has been a usual standpoint that the former provides a means for the latter. However, when the possibility of design optimization paradigm is getting to be spread to more practical engineering problems, original contributions from the application side become indispensable for confronting complicatedness of engineering problems. This shift, for instance, has highlighted the field of what we call multidisciplinary design optimization (MDO) in the last decade. While the MDO paradigm mainly focuses on multi-physics design problems, physically large-scale problems, etc., the design problems, the mathematical nature of which are modeled as mixed combinatorial problems, is an essential area under the new trends of design optimization techniques.

Meta-heuristics based optimization techniques, such as SA, GA, are promising for implementing optimization algorithms for complicated mixed combinatorial problems. When applying them to mixed combinatorial problems, it is further necessary to configure representation schemes, optimization operations, etc. according to the relationships underlying on how the optimal solution emerges by combining partial solutions, building blocks, etc. Hybridization of genetic algorithms with any conventional mathematical programming technique would be promising and effective for such configurations. For instance, Fujita, et al. (1993) proposed an optimal nesting method, in which a genetic algorithm optimizes the spatial configuration of pieces and a quasi-Newton method optimizes the precise positions and orientation of respective pieces. Fujita, et al. (1996) developed an optimal planning method for energy plant configuration, in which a genetic algorithm optimizes a plant configuration and a mixed-integer linear programming technique is used to evaluate the optimal operation cost under each fixed plant configuration. Apart from complicated applications, Renders and Plasse (1996) analyzed hybridization of genetic algorithms with traditional hill-climbing methods for global optimization of continuous functions, and reported the tradeoffs between accuracy, reliability and computing time.

7.2   Formulation of simultaneous optimization problem

As a result of the equation development in Section 4, an optimization problem of Class III product variety is formulated as follows (Fujita and Yoshida, 2004).

Design variables: Commonality and similarity patterns meaning which module design schemes are combinatorially used across all module slots and among multiple products, which were mentioned in Section 4.2. Attribute variables of all products, z^i (i = 1, 2, ..., n_p).

Constraints: System constraints for respective products defined by Eq. (4). Constraints among attribute variables defined by Eqs. (1)-(3) under a given commonality and similarity pattern.

Objective function: The manufacturer's expected final profit through all production units $C_B(T_p)$ calculated by Eq. (14).

This optimization problem would be a mixed-integer, nonlinear, constrained-optimization problem, if the commonality and similarity patterns could be represented with a set of integer variables. At least, it includes combinatorial characteristics and nonlinearity within continuous subspaces. It is categorized into difficult optimization problems as mentioned above.

7.3   Hierarchy in optimality

For developing an optimization method for Class III problems, its hierarchical structure in solution space plays an important role. The design space is represented by a combinatorial aspect and a continuous aspect, as aforementioned. The mixture of these two aspects leads to splitting the design space into two layers: combinatorial space and continuous sub-space under each fixed combination. In the former, a similar module design scheme is directional between a pair of products, while independent and common module design schemes are not directional, that is, the direction of similarity, for instance whether $m^{i_1}_j$ is stretched from $m^{i_2}_j$ or $m^{i_2}_j$ is stretched from $m^{i_1}_j$, must be determined, and corresponds to the inequality in the direction of Eq. (1). Thus, the combinatorial space is further split into two layers: the naive pattern of commonality and similarity among products of respective module slots, and similarity directions among similar modules under the former.

As a result, the design space is decomposed into three layers: (i) Commonality and similarity pattern: Which modules are independent, similar or common among products across respective modules; (ii) Similarity directions: Which is the base module for stretching design within similar modules, which are assigned in the pattern of (i) and (iii) Module attributes: Attribute variables of all modules of all products within a continuous subspace under the pattern of (i) and the directions of (ii).

Under the introduction of these layers, the subspace optimization problem of (i) is an optimal grouping problem, that of (ii) is a 0-1 integer programming problem, and that of (iii) is a constrained nonlinear programming problem.

Based on the hierarchical relationship among the above three layers, computational optimization techniques for respective subspaces are combined in the following way. A genetic algorithm (GA) (e.g., Goldberg, 1989) is used for layer (i), since it still includes a huge number of combinations although any continuous optimization nature is excluded. A branch-and-bound technique is used for layer (ii), since its optimization problem is mathematically well-structured, even though it is combinatorial. The successive quadratic programming (SQP) method is used for layer (iii), since its optimization problem is defined in a continuous space. Thus, the subject for developing an optimization method for Class III problems is translated into the development of hybridization of these three techniques.

7.4 Hybridization of optimization techniques

Figure 10-12 shows the overall procedure of the hybridized optimization method developed under the above concept (Fujita and Yoshida, 2004). In the method, the genetic algorithm explores the optimal pattern by manipulating a set of tentative solutions, called a population in the terminology of genetic algorithms, and the optimality of each new solution is evaluated by determining similarity directions and module attributes under its pattern with the branch-and-bound technique. The bounding operation in the branch-and-bound technique uses lower-bound values of the objective function for each node, calculated by solving associated relaxed problems with a SQP method. Through these linkages between the genetic algorithm and the branch-and-bound technique and between the branch-and-bound technique and the SQP method, the converged optimal solution obtained by the genetic algorithm is the optimal solution of not only the genetic algorithm but also a Class III problem of product variety.

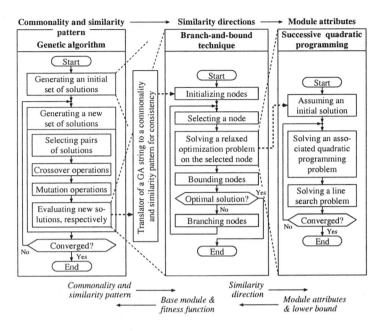

Figure 10-12. Hybridization of optimization techniques in Class III optimization method.

The following sections explain the representation method of a pattern, the contents of a genetic algorithm optimizing it, the formulation of subspace optimization problems on similarity directions and module attributes, its continuous relaxation, and the procedure in the branch-and-bound technique.

7.5   Representation of commonality and similarity pattern

The optimization problem of commonality and similarity patterns is an optimal grouping problem as mentioned in the above. This is decomposed into a set of independent grouping problems in respective module slots. Here a group means a set of products that share a unique module for a common module design scheme or share a base design for a similar module design scheme. Thus, a pattern is represented by the following two steps: (i) Partial relationships between particular pairs of modules are temporarily defined as either of independent, common or similar; and (ii) Consistency over the entire pattern is coordinated by arranging the temporarily defined partial relationships.

The partial relationships of Step (i) are represented by the following variables:

$$\xi_j^{i_1 \leftrightarrow i_2} = \begin{cases} 0 & \cdots & m_j^{i_1} \text{ and } m_j^{i_2} \text{ are independent.} \\ 1 & \cdots & m_j^{i_1} \text{ and } m_j^{i_2} \text{ are similar.} \\ 2 & \cdots & m_j^{i_1} \text{ and } m_j^{i_2} \text{ are common.} \end{cases} \qquad (17)$$

$$(i_1 = 1,2,\cdots,n_p; i_2 = i_1 + 1, i_1 + 2, \cdots, n_p; j = 1,2,\cdots,n_M)$$

where the total number of these variables $\xi_j^{i_1 \leftrightarrow i_2}$ is $n_{M} \cdot {}_{n_p}C_2$.

The pattern defined in the above way is coordinated to be consistent by recursively applying the following rules, until no rule becomes applicable, in Step (ii):

If $\xi_j^{i_1 \leftrightarrow i_2} = 2 \wedge \xi_j^{i_1 \leftrightarrow i_3} = 2 \wedge \xi_j^{i_2 \leftrightarrow i_3} \neq 2$, then $\xi_j^{i_2 \leftrightarrow i_3}$ is set to 2.

If $\xi_j^{i_2 \leftrightarrow i_3} = 2 \wedge \xi_j^{i_1 \leftrightarrow i_2} = 2 \wedge \xi_j^{i_1 \leftrightarrow i_3} \neq 2$, then $\xi_j^{i_1 \leftrightarrow i_3}$ is set to 2.

If $\xi_j^{i_1 \leftrightarrow i_3} = 2 \wedge \xi_j^{i_2 \leftrightarrow i_3} = 2 \wedge \xi_j^{i_1 \leftrightarrow i_2} \neq 2$, then $\xi_j^{i_1 \leftrightarrow i_2}$ is set to 2.

If $\left(\left(\xi_j^{i_1 \leftrightarrow i_2} \neq 0 \wedge \xi_j^{i_1 \leftrightarrow i_3} = 1 \right) \vee \left(\xi_j^{i_1 \leftrightarrow i_2} = 1 \wedge \xi_j^{i_1 \leftrightarrow i_3} \neq 0 \right) \right) \wedge \xi_j^{i_2 \leftrightarrow i_3} = 0$,

then $\xi_j^{i_2 \leftrightarrow i_3}$ is set to 1.

If $\left(\left(\xi_j^{i_2 \leftrightarrow i_3} \neq 0 \wedge \xi_j^{i_1 \leftrightarrow i_2} = 1 \right) \vee \left(\xi_j^{i_2 \leftrightarrow i_3} = 1 \wedge \xi_j^{i_1 \leftrightarrow i_2} \neq 0 \right) \right) \wedge \xi_j^{i_1 \leftrightarrow i_3} = 0$,

then $\xi_j^{i_1 \leftrightarrow i_3}$ is set to 1.

If $\left(\left(\xi_j^{i_1 \leftrightarrow i_3} \neq 0 \wedge \xi_j^{i_2 \leftrightarrow i_3} = 1 \right) \vee \left(\xi_j^{i_1 \leftrightarrow i_3} = 1 \wedge \xi_j^{i_2 \leftrightarrow i_3} \neq 0 \right) \right) \wedge \xi_j^{i_1 \leftrightarrow i_2} = 0$,

then $\xi_j^{i_1 \leftrightarrow i_2}$ is set to 1.

where $i_1 < i_2 < i_3$.

Through these two steps, the following relationships must be concluded as the representation of a commonality and similarity pattern:

If $\xi_j^{i_\alpha \leftrightarrow i_\beta} = 2$ for $\forall i_\alpha, i_\beta \in \{i_i, i_2, \ldots, i_p\}, i_\alpha < i_\beta$, when a set of products, $P^{i_1}, P^{i_2}, \ldots, P^{i_p}$, share a unique module in module slot M_j.

If $\xi_j^{i_\alpha \leftrightarrow i_\beta} = 1$ for $\forall i_\alpha, i_\beta \in \{i_i, i_2, \ldots, i_p\}, i_\alpha < i_\beta$, when a set of products, $P^{i_1}, P^{i_2}, \ldots, P^{i_p}$, share similar modules in module slot M_j.

If $\xi_j^{i_\alpha \leftrightarrow i_\beta} = 0$ and $\xi_j^{i_\gamma \leftrightarrow i_\alpha} = 0$ for $\forall i_\beta; i_\alpha < i_\beta \leq n_p$ and $\forall i_\gamma; 1 \leq i_\gamma < i_\alpha$, when a product P^{i_α} uses a nique module in module slot M_j.

7.6   GA-based pattern optimization

The genetic algorithm optimizes the commonality and similarity pattern that is represented by $\xi_j^{i_1 \leftrightarrow i_2}$, $(i_1 = 1,2,\cdots,n_p; i_2 = i_1 + 1, i_1 + 2, \cdots, n_p; j = 1,2,\cdots,n_M)$. While conventional genetic algorithms use binary strings or a sequence of non-duplicated symbols, each $\xi_j^{i_1 \leftrightarrow i_2}$ is a three-digit number. In order to use binary representation, a two-bit variable $\hat{\xi}_j^{i_1 \leftrightarrow i_2} \in \{0, 1, 2, 3\}$ is introduced for each $\xi_j^{i_1 \leftrightarrow i_2}$. Then the following translation rules are applied from $\hat{\xi}_j^{i_1 \leftrightarrow i_2}$ to $\xi_j^{i_1 \leftrightarrow i_2}$. If $\hat{\xi}_j^{i_1 \leftrightarrow i_2} = 0$, then $\xi_j^{i_1 \leftrightarrow i_2} = 0$. If $\hat{\xi}_j^{i_1 \leftrightarrow i_2} = 1$ or $\hat{\xi}_j^{i_1 \leftrightarrow i_2} = 2$, then $\xi_j^{i_1 \leftrightarrow i_2} = 1$. If $\hat{\xi}_j^{i_1 \leftrightarrow i_2} = 3$, then $\xi_j^{i_1 \leftrightarrow i_2} = 2$. Consequently a design solution at the pattern level is represented by a $2\,n_M\,{}_{n_p}C_2$ -bit binary variable by connecting $\hat{\xi}_j^{i_1 \leftrightarrow i_2}$ $(i_1 = 1,2,\cdots,n_p; i_2 = i_1 + 1, i_1 + 2, \cdots, n_p; j = 1,2,\cdots,n_p)$ into $\vec{\xi}_j$ in consecutive order for respective module slots $M_j (j = 1, 2, \ldots, n_M)$ and further connecting $\vec{\xi}_j$ into a single bit vector $\vec{\Xi}$ in consecutive order.

Furthermore, when a similar module design scheme is eliminated from a possible design space beforehand, $\xi_j^{i_1 \leftrightarrow i_2} \neq 1$ for $\forall\, i_1$, i_2 under the original definition. Since this can be represented by a single-bit variable rather than a two-bit variable, and since a single-bit variable representation shortens the entire string used in the genetic algorithm, the definition of $\xi_j^{i_1 \leftrightarrow i_2}$ is changed in such a case as follows; $\xi_j^{i_1 \leftrightarrow i_2} = 0$ means that $m_j^{i_1}$ and $m_j^{i_2}$ are independent, and $\xi_j^{i_1 \leftrightarrow i_2} = 1$ means that $m_j^{i_1}$ and $m_j^{i_2}$ are common. In this case, any translation rule from $\hat{\xi}_j^{i_1 \leftrightarrow i_2}$ to $\xi_j^{i_1 \leftrightarrow i_2}$ becomes unnecessary.

Under the definition of $\vec{\Xi}$, the so-called Simple-GA (e.g., Goldberg, 1989; Fujita, et al., 1993; Fujita, et al., 1996) is applied to manipulate patterns for obtaining the optimal solution for product variety design. As for genetic operators, uniform crossover is used, and a bit inverse is used as a mutation operator, because the sequence of neighboring bit positions is not significant. Linear scaling, σ-truncation, the expected value plan and the elitist plan are used to enhance its optimization performance.

7.7   Formulation for optimizing similarity directions and module attributes

When the genetic algorithm fixes a commonality and similarity pattern, the formulation of a Class III optimization problem becomes the following.

Design variables: Similarity directions among a set of products that share a similar module design in respective module slots. Attribute variables of all products.

Constraints: System constraints for respective products. Constraints among attribute variables defined with Eqs. (1)-(3) under the given commonality and similarity pattern. The direction of inequality in Eq. (1) is not fixed, because similarity directions are design variables.

Objective function: The manufacturer's expected final profit through all production units. In its calculation, the selection of terms in Eqs. (6) and (8) depends on similarity directions.

The combinatorial part of this mathematical formulation is due to similarity directions. A similarity direction is translated as a determination of the base design for a series of similar modules, that is, it is assumed that products, P^i_1, P^i_2, $...P^{i_n}$, share similar modules in the module slot M_j. Then, when the base design is denoted by $i_{Base} \in \{i_1, i_2, \cdots, i_n\}$, the combinatorial part of the formulation is mathematically described as follows. First, the direction of inequality in Eq. (1) is:

$$x^i_{j,1} > x^{i_{Bse}}_{j,1} \qquad for \; \forall \, i \in \{i_1, i_2, \cdots, i_n\}, i \neq i_{Base} \tag{18}$$

Second, the selection of terms in Eqs. (6) and (8) is:

$$C^i_{Dj}(x^i_j) = \begin{cases} \alpha_{Dj} W^i_j & for \; i = i_{Base} \\[2ex] \left(\beta_{Dj} \dfrac{W^i_j - W^{i_{Base}}_j}{W^{i_{Base}}_j} + \gamma_{Dj} \right) C^{i_{Base}}_{Dj} \\[2ex] \qquad\qquad for \; \forall \, i \in \{i_1, i_2, \cdots, i_n\}, i \neq i_{Base} \end{cases} \tag{19}$$

$$C^i_{Fj}(x^i_j) = \begin{cases} \alpha_{Fj} x^i_{j,1} & for \; i = i_{Base} \\[2ex] \left(\beta_{Fj} \dfrac{x^i_{j,1} - x^{i_{Base}}_{j,1}}{x^{i_{Base}}_{j,1}} + \gamma_{Fj} \right) C^{i_{Base}}_{Fj} \\[2ex] \qquad\qquad for \; \forall \, i \in \{i_1, i_2, \cdots, i_n\}, i \neq i_{Base} \end{cases} \tag{20}$$

In order to exclude the alternative conditions in Eqs. (18)-(20), the following 0-1 integer variables are introduced for the modules, $m^i_j, i \in \{i_1, i_2, \cdots, i_n\}$.

$$\delta^i_j = \begin{cases} 1 & for \ i = i_{Base} \\ 0 & for \ \forall i \in \{i_1, i_2, \cdots, i_n\}, i \neq i_{Base} \end{cases} \tag{21}$$

where the following constraint is applied to the variables,

$$\sum_{i \in \{i_1, i_2, \cdots, i_n\}} \delta^i_j = 1 \tag{22}$$

As a result, Eq. (18) is rearranged into the following form.

$$x^{i_\alpha}_{j,1} + \Delta_j \left(1 - \delta^{i_\beta}_j\right) \geq x^{i_\beta}_{j,1}$$
$$for \ \forall i_\alpha \in \{i_1, i_2, \cdots, i_n\}, \forall i_\beta \in \{i_1, i_2, \cdots, i_n\}, i_\alpha \neq i_\beta \tag{23}$$

where Δ_j is a positive large number. This Δ_j must be large enough to guarantee the satisfaction of Eq. (18) under any values of δ^i_j. However, because too large a value of Δ_j makes the branch-and-bound technique inefficient due to the mathematical nature of a continuous relaxation problem used in it, its magnitude must be moderately small.

Furthermore, Eqs. (19)-(20) are rearranged into the following forms,

$$C^i_{Dj}(x^i_j) = \alpha_{Dj} W^i_j \delta^i_j + \sum_{\substack{k \in \{i_1, i_2, \cdots, i_n\} \\ k \neq i}} \left(\beta_{Dj} \frac{W^i_j - W^k_j}{W^k_j} + \gamma_{Dj} \right) \alpha_{Dj} W^k_j \delta^k_j \tag{24}$$
$$for \ \forall i \in \{i_1, i_2, \cdots, i_n\}$$

$$C^i_{Fj}(x^i_j) = \alpha_{Fj} x^i_{j,1} \delta^i_j + \sum_{\substack{k \in \{i_1, i_2, \cdots, i_n\} \\ k \neq i}} \left(\beta_{Fj} \frac{x^i_{j,1} - x^k_{j,1}}{x^k_{j,1}} + \gamma_{Fj} \right) \alpha_{Fj} x^k_{j,1} \delta^k_j \tag{25}$$
$$for \ \forall i \in \{i_1, i_2, \cdots, i_n\}$$

Finally, the formulation for the subspace optimization problem is obtained as follows.

Design variables: The 0-1 integer variables δ^i_j on similarity directions defined by Eq. (21) for all subsets of products that share a similar module design in respective module slots. Attribute variables of all products.

Constraints: System constraints for respective products. Constraints among attribute variables defined with Eqs. (23), (2) and (3) under the given commonality and similarity pattern. The constraints on 0-1 integer variables $\delta_j^i; \delta_j^i \in \{0, 1\}$ and those defined by Eq. (22).

Objective function: The manufacturer's expected final profit through all production units. In its calculation, Eqs. (24)-(25) are used instead of Eqs. (6) and (8) for similar design modules.

7.8   Branch-and-bound technique for subspace optimization

A branch-and-bound technique requires any associated relaxation problem to examine the possibility of each node for the optimal solution. Under the mathematical formulation developed in the previous section, a continuous relaxation problem is introduced by replacing the constraint on Eq. (21), $\delta_j^i \in \{0,1\}$ with $0 \le \delta_j^i \le 1$. Since this is a nonlinear continuous optimization problem, it can be solved by any conventional method. The method described in this section uses a successive quadratic programming (SQP) method.

The node selection mechanism is a key factor in making a branch-and-bound technique efficient. The node tree structure of the subspace optimization problem is binary because all integer variables are 0-1 variables. Thus the depth-first search strategy is used so as to reach the final node as quickly as possible. When there are two branches on a particular δ_j^i, i.e., $\delta_j^i = 0$ or $\delta_j^i = 1$, priority is given to the node for $\delta_j^i = 1$, because it means $\delta_j^{\tilde{i}} = 0$ for $\forall \tilde{i} \in \{i_1, i_2, \cdots, i_n\}, \tilde{i} \ne i$ under Eq. (22).

7.9   Optimization examples

While Section 5 demonstrated a stretch-based design deployment of aircraft by combining the enumeration of possible patterns and the Class I optimization method, aircraft design planning even with a moderate number of models is categorized into a Class III problem. The following briefly demonstrates such a stretch-based design deployment under the Class III viewpoint by comparing optimal designs among three cases with different patterns of design requirements.

In each case, it is assumed that five airplane models are designed. For each airplane model, the cruise range is assigned to either 1400, 1600, 1800, 2100, 2400, 2600 or 2800 [km] respectively, while how large a difference of cruise range is considered is varied among the cases. In the first case, Case 1, the cruise range is assigned to either 1800, 2100, or 2400 [km]. In the second case, Case 2, the cruise range is assigned to either 1600, 2100, or 2600 [km]. In the third case, Case 3, the cruise range is assigned to either 1400, 2100, or 2800 [km]. The number of seats is assigned to either 90 or 120 [people] respectively. In the optimization computation for each case, the total number of variables $\xi_j^{i_1 \leftrightarrow i_2}$ for representing a design pattern is 40, and the length of an entire string used in the genetic algorithm is 70 bits (this is not 80 bits, because the engines cannot be similar). The production numbers are set to N^i = 200 for different models, and the number of production terms is set to T_p = 15 [years]. As for the coefficients in the profit model, α_{Dj} = 1.6 x 10^7 [yen/kg], β_{Dj} = 0.5, γ_{Dj} = 0.1, α_{Fj} = 1.6 x 10^9 [yen/m] for every module except the engine and α_{Fj} = 1.6 x 10^6 [yen/kgf] for the engine, β_{Fj} = 0.5, γ_F = 0.1, ϑ_j = 4.0 x 10^4 [yen/kg], κ_j =2.8 x 10^5 [yen/kg] for all j, r = 0.8 in the learning effect, $\zeta_{j_{i_1,i_2}}$ = 0.5 ($i_1 \neq i_2$) for all similar module designs, the annual interest rate ρ is 0.03 in the manufacturer total profit. These values are selected for demonstration purposes rather than for precise assessment, since it is difficult to access real cost information for demonstration purposes.

Figure 10-13 shows the design conditions and optimized design patterns for all cases. Each column shows the module commonality and similarity patterns for module slots of a particular case. Each row shows the comparison of such patterns on a particular module slot across different cases. Each compartment is a two-dimensional graph that shows the design pattern on a particular module slot in a particular case over the two axes of cruise range and the number of seats. Each dot indicates a model of products, P^1, P^2, P^3, P^4 and P^5, each hatched set of dots indicates the models that share a common module design, and each arrow indicates the pair of a base module design, which is at the tail, and a stretched module design, which is at the tip, under a similar module design scheme. Each set of dots linked with a fat line means that a unique model is designed for different design conditions because all module slots are commonalized by the design optimization. As for the number of necessary models under the different design conditions, Case 1 requires two different models after design optimization, where P^1, P^2 and P^3 are unified into a single model, and P^4 and P^5 are unified into a single model. That is, the number of models is reduced from five to two by optimization. Case 2 requires four models, and Case 3 requires five models as shown in Figure 10-13. While these tendencies are due to the differences in requirements, the optimal designs are the results of balancing the interaction between commonality and similarity patterns in the respective modules, and they cannot be deduced without any optimization computation. Table 10-4 shows the details of the Case 2 optimal design, which is shown in the second half of the table, in comparison with the totally independent designs of five airlines, which are shown in the first half of the table. While all modules of all airplanes except the engine of P^5 are enlarged through simultaneous optimization, the design and development costs and the facility costs are dramatically reduced, and further the learning effect on the process cost is much enhanced through product variety optimization. As a result, the manufacturer's expected final profit is shifted from minus to plus. Although this result is dependent on the assumed numerical conditions, it reveals that the proposed optimization method can assess the complicated tradeoffs underlying product variety design and that simultaneous consideration of both module combination.

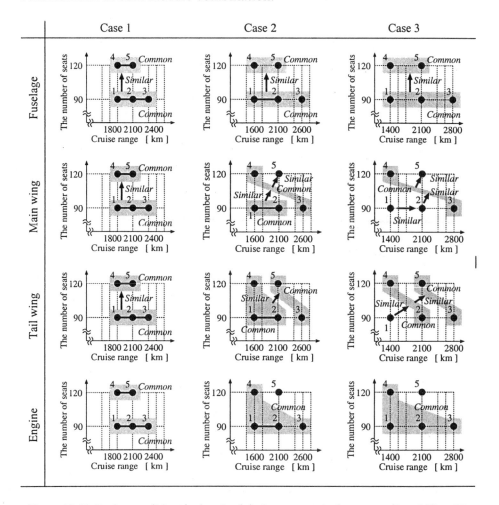

Figure 10-13. Design conditions and optimal design patterns in the case studies of Class III.

Table 10-4. Optimization result of Case 2 in the case studies of Class III

		Totally independent design					Optimal product variety design				
		P^1	P^2	P^3	P^4	P^5	P^1	P^2	P^3	P^4	P^5
Width of main wing	[m]	22.24	23.88	26.09	25.51	27.52	25.58	25.58	27.18	27.18	28.60
Width of horizontal tail wing	[m]	7.76	8.30	9.32	8.97	9.71	9.08	9.08	9.98	9.08	9.98
Height of vertical tail wing	[m]	3.41	3.85	4.84	3.97	4.61	4.22	4.22	5.06	4.22	5.06
Engine power	[kgf]	5,564	6,257	6,934	7,323	8,201	7,269	7,269	7,269	7,269	8,139
Length of fuselage	[m]	27.47	27.47	27.47	30.93	30.93	27.47	27.47	27.47	30.93	30.93
Design and development cost	[×10⁶ yen]	147,897	159,277	172,358	194,297	209,193	169,992		12,250	19,246	39,618
Facility cost	[×10⁶ yen]	106,304	111,608	119,439	122,739	129,557	117,788		10,199	7,164	21,115
Material cost per unit	[×10⁶ yen]	1,600	1,728	1,860	2,106	2,271	1,787	1,823	1,917	2,120	2,276
Process cost for the first unit (without learning effect)	[×10⁶ yen]	2,588	2,787	3,016	3,400	3,661	2,975	2,981	3,160	3,487	3,709
Process cost for the last unit in the 1st year	[×10⁶ yen]	1,423	1,532	1,658	1,869	2,013	1,193	1,195	1,273	1,418	1,609
Process cost for the last unit in the 5th year	[×10⁶ yen]	969	1,043	1,129	1,272	1,370	812	814	867	965	1,096
Process cost for the last unit in the 10th year	[×10⁶ yen]	823	886	959	1,081	1,163	690	691	736	820	930
Process cost for the last unit in the 15th year	[×10⁶ yen]	748	805	871	982	1,058	627	628	669	745	846
Total process cost per model	[×10⁶ yen]	515,306	555,930	599,483	677,720	730,369	521,156	528,685	558,183	618,675	676,027
Total manufacturing cost	[×10⁶ yen]		4,551,478						3,300,099		
Price per unit	[×10⁶ yen]	2,923	3,904	4,811	4,148	5,544	2,860	3,900	4,793	4,195	5,584
Total profit (objective)	[×10⁸ yen]		-865.45						1,037.62		

The computational expense to obtain the optimal solution for a product variety design is another important measure for an optimization method. The genetic algorithm uses the following conditions: the size of the population, i.e., the number of individuals, is 100, the crossover probability is 0.6, the mutation probability per individual is 0.10, and the number of generations is 100. In Case 1 of Figure 10-13, the optimal solution was obtained at the 44th generation, and the entire computation time for 100 generations was ~14.5 hours on a Sun Ultra 10 Workstation (440MHz UltraSPARC-IIi).

8.   LIMITATIONS

The idea expanding the design space from a system to a set of systems leads the challenges in product variety design optimization. Three optimization examples demonstrated in this chapter illuminate the potential of optimal design paradigm in such directions. However, it is necessary to carefully discuss the limitations of product variety optimization. At least, the role of optimization is restrictive within the mathematically formulatable space against the whole design process under the sense of Figure 10-1.

When reviewing three examples shown in this chapter again, each of them does not give a final design plan but a design option or a set of alternatives under a predefined set of conditions. Since the product variety design must be carried out in the early phases of the design process such as planning phase or conceptual design phase, the design conditions cannot be concretely fixed. They are, rather, a part of decision making. For example, the quantity of total production of units or models is affected by final design result, production systems, marketing strategy, competition in the market, etc. Under these reasons, tradeoffs between conditions and optimal solutions should be mathematically revealed and could be directive information for decision makers (e.g., Nelson, et al., 2001).

The time scale of product development is another point concerning the limitations of optimization paradigm. The elements that construct a product are spread from materials, tools, labors, information, knowledge, facilities, capital, etc. under manufacturing systems. This means that some elements are shared among a set of products and some other elements are shared among another different set of products. While the optimization problems and their examples described in this chapter assumes that all parts of products are fully controllable, such a situation might not be realistic in practice. Thus, the optimality under limited controllability on manufacturing systems, that is, the robustness of a controllable part against the other uncontrollable parts, must be concerns toward more effective optimization applications in product development. A design problem of common components for a class of products (Fujita and Yoshioka, 2003b) may be a case under this direction. The framework for designing such common components consists of predefined optimality patterns, which may often take a form of Pareto optimality, design procedure to reach the best option, and computational techniques for revealing potential solutions.

In summary, since product variety optimization tackles on more global optimization than ordinary design of individual systems or components, how to configure an optimization problem itself become a more essential issue in the optimal design paradigm. This means that deeper understanding of a design problem itself becomes essential to accomplish best practices in product family and platform design.

9. SUMMARY

This chapter described optimization methods for product variety design with emphasis on the problem classification and a simultaneous optimization method for both module combination and module attributes. The key in exploring optimal design for product family and platform exists in both development of optimization algorithm and formulation of individual problems. While the methods and formulations described in this chapter are expected to be useful references in implementing optimal design, individual development may be also indispensable for practical applications. It is very important, in this sense, to understand what are the outcomes and limitations of design optimization paradigm in the direction of product variety.

10. ACKNOWLEDGMENTS

The author acknowledges that computer programming and computation of optimization examples were done by Makibi Ishikawa, Tetsuo Yoneda, Hisato Sakaguchi, Hiroko Yoshida, Shin'ichi Yoshioka, and Yusuke Kounoe, who were formerly graduate students of Osaka University, and that discussion with some anonymous engineers in home appliance and automotive companies were directive in this research. The research related to this chapter has been carried out partially at the Strategic Research Base, Handai (Osaka University) Frontier Research Center supported by the Japanese Government's Special Coordination Fund for Promoting Science and Technology.

Chapter 11

ANALYTICAL TARGET CASCADING IN PRODUCT FAMILY DESIGN

Michael Kokkolaras[1], Ryan Fellini[1], Harrison M. Kim[2], and Panos Y. Papalambros[1]

[1]*Department of Mechanical Engineering, University of Michigan, Ann Arbor, MI 48109;*
[2]*Department of General Engineering, University of Illinois Urbana-Champaign, Urbana, IL 61801*

1.   PRODUCT DEVELOPMENT BY HIERARCHICAL DECOMPOSITION

Most products are neither designed nor manufactured as one piece. They are decomposed into parts that are developed individually before they are assembled to form the final product. Typically, this partitioning-based development process matches the hierarchical structure of the product-offering organization. Design tasks are assigned to divisions, departments, and teams according to expertise. An example from the automotive industry is depicted in Figure 11-1. Obviously, this decomposition is not complete and serves only as an illustration of the decomposition paradigm.

Given that the original product design problem has been replaced by a collection of design subproblems, the challenge is to determine appropriate objectives for the latter while taking into account their interactions. The goal is to ensure that concurrent design of parts will not compromise consistency of the final product so that costly iterations at the late stages of the product development process can be avoided.

Analytical target cascading (ATC) is a mathematical methodology developed for translating single product design targets to appropriate part specifications (Kim, 2001). Here, we consider the application of ATC to multiple products, particularly when the latter are based on a common platform. Nevertheless, we will present a brief overview of ATC in single product development to facilitate understanding of the extensions to product family development.

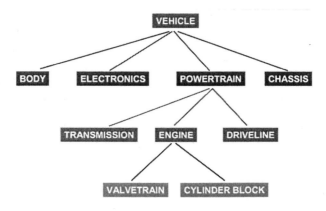

Figure 11-1. Hierarchically partitioned product development in the automotive industry.

2. ANALYTICAL TARGET CASCADING IN SINGLE PRODUCT DEVELOPMENT

The optimal system design problem can be formulated mathematically as a (typically nonlinear) programming problem

$$
\begin{aligned}
&\min_{\mathbf{x}} \quad f(\mathbf{R}(\mathbf{x},\mathbf{p}),\mathbf{T}) \\
&\text{subject to} \quad \mathbf{g}(\mathbf{R}(\mathbf{x},\mathbf{p}),\mathbf{x},\mathbf{p}) \le \mathbf{0} \qquad\qquad (1) \\
&\hphantom{\text{subject to} \quad} \mathbf{h}(\mathbf{R}(\mathbf{x},\mathbf{p}),\mathbf{x},\mathbf{p}) = \mathbf{0},
\end{aligned}
$$

where the function f measures the discrepancy between system responses $\mathbf{R}$ (that depend on design variables $\mathbf{x}$ and parameters $\mathbf{p}$) and design targets $\mathbf{T}$. The vector functions $\mathbf{g}$ and $\mathbf{h}$ are associated with design inequality and equality constraints. A choice for f that is particularly suitable to numerical optimization is the square of the l_2-norm (or some other weighted norm) of the difference vector, i.e., $f(\mathbf{R}(\mathbf{x},\mathbf{p}),\mathbf{T}) = \|\mathbf{R}(\mathbf{x},\mathbf{p}) - \mathbf{T}\|_2^2$.

Let us assume that the system design targets $\mathbf{T}$ are given. Let us further assume that a decomposition of the optimal system design exists, i.e., there is a hierarchical functional dependency between elements in successive levels such that (see Figure 11-2)

$$\mathbf{R}_{ij} = \mathbf{r}_{ij} \left(\mathbf{R}_{(i+1)j_1}, \mathbf{R}_{(i+1)j_2}, ..., \mathbf{R}_{(i+1)j_{n_{ij}}}, \mathbf{x}_{ij}, \mathbf{y}_{ij} \right). \tag{2}$$

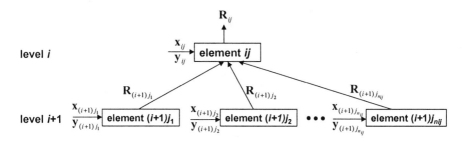

Figure 11-2. Functional dependency between elements in successive levels.

In Eq. (2), subscript index pairs denote level and element, respectively. For example, element j_2 at level $i + 1$ is the second of n_{ij} "children" of element j at level i. The implication of the hierarchical functional dependency is that responses of "parent" elements must depend on at least one response of each of their "children". In addition, they may depend on local design variables $\mathbf{x}$ and/or shared design variables $\mathbf{y}$, i.e., design variables that appear also in the design optimization problems of other elements at the same level. Looking at the example of Figure 11-2, some of the entries of the vector $\mathbf{y}_{(i+1)j_1}$ appear in vector $\mathbf{y}_{(i+1)j_2}$, some in vector $\mathbf{y}_{(i+1)j_3}$, and so on. Finally, we assume that analysis or simulation models $\mathbf{r}$ are available to compute responses $\mathbf{R}$.

Under the above assumptions, ATC operates by formulating and solving a minimum deviation optimization problem for each element in the hierarchy. Since responses of higher-level elements are functions of responses of lower-level elements, it aims at minimizing the gap between what upper-level elements "want" and what lower-level elements "can". Similarly, if design variables are shared among some elements at the same level, their consistency is coordinated at the common parent element.

2.1   Mathematical formulation

Before we proceed with the mathematical formulation of the ATC process, let us introduce some additional notation and definitions.

- We start enumerating levels at the top of the hierarchy with increasing order as we work our way towards the bottom.
- Tolerance optimization variables ε^R and ε^y are used to coordinate responses and shared variables, respectively.
- For response and shared design variables, superscript indices denote the level quantities have been computed at. For example, the vector of

responses of element j at level i computed at level i-1 and cascaded as a target for the optimization problem of element j at level i is represented by $\mathbf{R}_{ij}^{i-1}$. For simplicity, we will omit superscripts when response and shared design variables of an element are computed at the element's level. For example, the vector of responses of element j at level i computed at level i is represented by $\mathbf{R}_{ij}$.

- To coordinate the (possibly interspersed) shared variables of the children of element j at level i, we aggregate all of the former into a single vector represented by $\tilde{\mathbf{y}}_{ij}$. A selection matrix $\mathbf{S}_{(i+1)k}$ is then defined for each child k and used to extract the shared variables vector $\mathbf{y}_{(i+1)k}$ of that child element. A selection matrix is binary (i.e., its entries are either 0 or 1), the number of its rows is equal to the number of the child's shared variables, and the number of its columns is equal to the dimension of the aggregated coordination vector. For example if child k shared only the first, third, and fourth variables of the 5-dimensional coordination vector $\tilde{\mathbf{y}}_{ij}$, then $\mathbf{S}_{(i+1)k} = [1000; 00100; 00010]$, and $\mathbf{y}_{(i+1)k} = \mathbf{S}_{(i+1)k} \tilde{\mathbf{y}}_{ij}$.

The mathematical formulation of the ATC problem corresponding to the j-th element at the i-th level is

$$\text{minimize} \quad \left\| \mathbf{R}_{ij} - \mathbf{R}_{ij}^{i-1} \right\|_2^2 + \left\| \mathbf{y}_{ij} - \mathbf{S}_{ij} \tilde{\mathbf{y}}_{(i-1)l}^{i-1} \right\|_2^2 + \varepsilon_{ij}^R + \varepsilon_{ij}^y$$

$$\text{with respect to} \quad \mathbf{x}_{ij}, \mathbf{y}_{ij}, \tilde{\mathbf{y}}_{ij}, \mathbf{R}_{(i+1)j_1}, \mathbf{R}_{(i+1)j_2}, ..., \mathbf{R}_{(i+1)j_{n_{ij}}}, \varepsilon_{ij}^R, \varepsilon_{ij}^y$$

$$\text{subject to} \quad \sum_{k=1}^{n_{ij}} \left\| \mathbf{R}_{(i+1)j_k} - \mathbf{R}_{(i+1)j_k}^{i+1} \right\|_2^2 \leq \varepsilon_{ij}^R \tag{3}$$

$$\sum_{k=1}^{n_{ij}} \left\| \mathbf{S}_{(i+1)j_k} \tilde{\mathbf{y}}_{ij} - \mathbf{y}_{(i+1)j_k}^{i+1} \right\|_2^2 \leq \varepsilon_{ij}^y$$

$$\mathbf{g}_{ij}(\mathbf{R}_{ij}, \mathbf{x}_{ij}, \mathbf{y}_{ij}) \leq 0$$

$$\mathbf{h}_{ij}(\mathbf{R}_{ij}, \mathbf{x}_{ij}, \mathbf{y}_{ij}) = 0,$$

where l is the parent element of element j at level i.

The formulation of Problem (3) is applicable to any element of the hierarchical decomposition. Nevertheless, top- and bottom-level problems are special cases of this formulation. At the top-level ($i = 0$), there is only one element with n_0 children. Therefore, the element index can be dropped, there are no shared variables, and "cascaded" responses are the given system design targets, i.e., $\mathbf{R}_0^{-1} = \mathbf{T}$. At the bottom level ($i = M$), there are no children responses and shared variables to coordinate, and element responses depend only on local and/or shared design variables. Complete special case formulations are presented in Section 3.

2.2 The ATC process

The ATC process consists of solving all subproblems of the multilevel hierarchy iteratively. Before each optimization subproblem is solved, parameter values (computed at parent and children elements) are updated. After the optimization, the element updates parameter values of its parent and children problems. An ATC iteration is defined as one solution of all the optimization problems. The ATC process converges when the optimization variable values of all subproblems do not change after two successive iterations. The designer must then examine the values of the tolerance optimization variables of all subproblems. If there are unacceptable discrepancies, the designer must either change system design target values or modify the feasible design space to accommodate the conflicting interactions of the system components.

The subproblems must be solved in an appropriate sequence, called a coordination strategy. All subproblems within a level can be solved concurrently, since their parameters do not depend on the solution of same-level subproblems. Therefore, coordination strategies in the ATC process consist of specifying the sequence the levels are visited with. Nevertheless, there exist many alternatives. Since design targets are given at the top of the hierarchy, a top-down direction is typically adopted.

For the three-level hierarchy example shown in Figure 11-3, Figure 11-4 illustrates four different coordination examples. In Scheme I, we start at the top and proceed downwards; after the bottom level is reached, we return to the top, without visiting the intermediate level, and reiterate. In Scheme II, the intermediate level is visited in both directions. In Scheme III, we iterate between the two first levels until convergence, and then proceed to iterations with the third ("nested" coordination).

Experience has shown that the ATC process converges when simple-to-implement strategies I or II are chosen (Kim, et al., 2002, Kokkolaras, et al., 2004). Theoretical convergence properties, however, have been proven under standard convexity and smoothness assumptions only for nested coordination schemes (Michelena, et al., 2003).

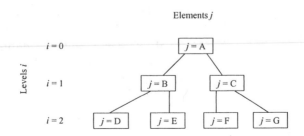

Figure 11-3. Decomposition example with three-level hierarchy.

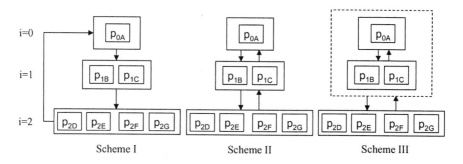

Figure 11-4. Alternative coordination strategies for the three-level hierarchy in Figure 11-3.

3.   ANALYTICAL TARGET CASCADING IN PRODUCT FAMILY DEVELOPMENT

When designing a family of platform-based products, one must quantify the impact of sharing subsystems and/or components on performance. Tradeoffs exist because the products are no longer optimized to satisfy individual design targets only. Product family design is a multiobjective optimization process that combines individual design optimization problems and introduces additional constraints due to sharing.

The ATC process can be used to validate the technical feasibility of implementing commonality decisions. Specifically, it can be employed to derive appropriate platform specifications to satisfy family and individual product design targets while enforcing consistency.

The hierarchical structure of the product family ATC problem is composed by connecting individual product decompositions to a family element that resides at the top level. A simple example, following that of Figure 11-1, is shown in Figure 11-5. Two vehicle variants form a product family; they share the body structure, and their powertrains share transmission and driveline.

Family design targets are defined at the top (family) level, while individual vehicle design targets are defined at the second (product) level. Given a set of family and individual product targets, analysis/simulation models for computing responses of all elements, and a predefined platform, targets are cascaded to elements lower in the hierarchy to determine appropriate subsystem and component specifications.

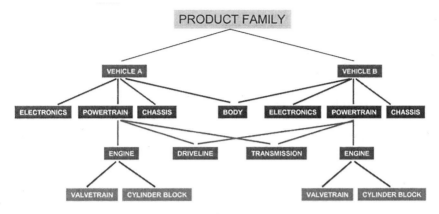

Figure 11-5. Example of a product family hierarchy.

3.1 ATC formulation extensions

The introduction of a family level at the top of the hierarchy has an implication. Design targets **T** are defined both at the top (family) level and at the second (product) level. This concept can be generalized to allow for "local" (as opposed to cascaded) design targets to be introduced at any element of the hierarchy. Therefore, we need to differentiate between responses associated with locally introduced design targets and responses associated with cascaded targets. The former are not inputs of the parent elements, and will be denoted with a tilde, as shown in the example of Figure 11-6.

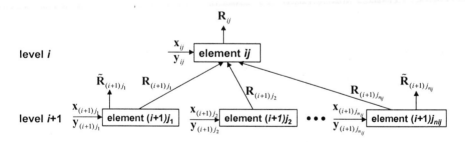

Figure 11-6. Only responses that are not inputs of parent elements can be associated with locally introduced design targets.

The functional dependency relation (2) is rewritten as

$$[\mathbf{R}_{ij}^{\mathrm{T}}, \tilde{\mathbf{R}}_{ij}^{\mathrm{T}}] = \mathbf{r}_{ij}(\mathbf{R}_{(i+1)j_1}, \mathbf{R}_{(i+1)j_2}, \ldots, \mathbf{R}_{(i+1)j_{n_{ij}}}, \mathbf{x}_{ij}, \mathbf{y}_{ij}), \tag{4}$$

where the superscript ^{T} denotes transpose.

Besides introducing an additional family level, the ATC formulation for product family development is extended to enable sharing elements, i.e., to enable children elements to have links with multiple parent elements. To accomplish this, the set P_{ij} is defined for each element j at level i to include its parents. For consistency in the notation, we now also define the set C_{ij} for each element j at level i to include its children. We now present the mathematical formulation of the ATC problems for the four possible cases.

3.1.1   Family level problem formulation

The family ATC problem (at the top level, $i = 0$) is formulated as

$$\text{minimize} \quad \left\| \tilde{\mathbf{R}}_0 - \mathbf{T}_0 \right\|_2^2 + \varepsilon_0^R + \varepsilon_0^y$$

$$\text{with respect to} \quad \mathbf{x}_0, \tilde{\mathbf{y}}_0, \mathbf{R}_{11}, \mathbf{R}_{12}, ..., \mathbf{R}_{1n_0}, \varepsilon_0^R, \varepsilon_0^y$$

$$\text{subject to} \quad \sum_{k \in C_0} \left\| \mathbf{R}_{1k} - \mathbf{R}_{1k}^1 \right\|_2^2 \le \varepsilon_0^R \tag{5}$$

$$\sum_{k \in C_0} \left\| \mathbf{S}_{1k} \tilde{\mathbf{y}}_0 - \mathbf{y}_{1k}^1 \right\|_2^2 \le \varepsilon_0^y$$

$$\mathbf{g}_0(\tilde{\mathbf{R}}_0, \mathbf{x}_0) \le 0$$

$$\mathbf{h}_0(\tilde{\mathbf{R}}_0, \mathbf{x}_0) = 0.$$

3.1.2   Product level problem formulation

At the product level ($i = 1$), all elements share the same parent (the family element). The ATC problem for each product j is formulated as

$$\text{minimize} \quad \left\| \tilde{\mathbf{R}}_{1j} - \mathbf{T}_{1j} \right\|_2^2 + \left\| \mathbf{R}_{1j} - \mathbf{R}_{1j}^0 \right\|_2^2 + \left\| \mathbf{y}_{1j} - \mathbf{S}_{1j} \tilde{\mathbf{y}}_0^0 \right\|_2^2 + \varepsilon_{1j}^R + \varepsilon_{1j}^y$$

$$\text{with respect to} \quad \mathbf{x}_{1j}, \mathbf{y}_{1j}, \tilde{\mathbf{y}}_{1j}, \mathbf{R}_{21}, \mathbf{R}_{22}, ..., \mathbf{R}_{2n_{1j}}, \varepsilon_{1j}^R, \varepsilon_{1j}^y$$

$$\text{subject to} \quad \sum_{k \in C_{1j}} \left\| \mathbf{R}_{2k} - \mathbf{R}_{2k}^2 \right\|_2^2 \le \varepsilon_{1j}^R \tag{6}$$

$$\sum_{k \in C_{1j}} \left\| \mathbf{S}_{2k} \tilde{\mathbf{y}}_{1j} - \mathbf{y}_{2k}^2 \right\|_2^2 \le \varepsilon_{1j}^y$$

$$\mathbf{g}_{1j}(\mathbf{R}_{1j}, \tilde{\mathbf{R}}_{1j}, \mathbf{x}_{1j}, \mathbf{y}_{1j}) \le 0$$

$$\mathbf{h}_{1j}(\mathbf{R}_{1j}, \tilde{\mathbf{R}}_{1j}, \mathbf{x}_{1j}, \mathbf{y}_{1j}) = 0.$$

3.1.3 Intermediate level problem formulation

At the intermediate levels ($i = 2, 3,..., M - 1$), elements can have links to multiple parents. The ATC problem for each element j at an intermediate level is formulated as:

$$\text{minimize} \quad \left\| \tilde{\mathbf{R}}_{ij} - \mathbf{T}_{ij} \right\|_2^2 + \sum_{l \in P_{ij}} \left\| \mathbf{R}_{ij} - \mathbf{R}_{ijl}^{i-1} \right\|_2^2 + \sum_{l \in P_{ij}} \left\| \mathbf{y}_{ij} - \mathbf{S}_{ij} \tilde{\mathbf{y}}_{(i-1)l}^{i-1} \right\|_2^2 + \varepsilon_{ij}^R + \varepsilon_{ij}^y$$

$$\text{with respect to} \quad \mathbf{x}_{ij}, \mathbf{y}_{ij}, \tilde{\mathbf{y}}_{ij}, \mathbf{R}_{(i+1)j_1}, \mathbf{R}_{(i+1)j_2}, ..., \mathbf{R}_{(i+1)j_{n_{ij}}}, \varepsilon_{ij}^R, \varepsilon_{ij}^y$$

$$\text{subject to} \quad \sum_{k \in C_{ij}} \left\| \mathbf{R}_{(i+1)k} - \mathbf{R}_{(i+1)k}^{i+1} \right\|_2^2 \leq \varepsilon_{ij}^R \tag{7}$$

$$\sum_{k \in C_{ij}} \left\| \mathbf{S}_{(i+1)k} \tilde{\mathbf{y}}_{ij} - \mathbf{y}_{(i+1)k}^{i+1} \right\|_2^2 \leq \varepsilon_{ij}^y$$

$$\mathbf{g}_{ij}(\mathbf{R}_{ij}, \tilde{\mathbf{R}}_{ij}, \mathbf{x}_{ij}, \mathbf{y}_{ij}) \leq 0$$

$$\mathbf{h}_{ij}(\mathbf{R}_{ij}, \tilde{\mathbf{R}}_{ij}, \mathbf{x}_{ij}, \mathbf{y}_{ij}) = 0.$$

3.1.4 Bottom level problem formulation

Finally, at the bottom level $i = M$, elements still can have links to multiple parents. The ATC problem for each element j at the bottom level is formulated as:

$$\text{minimize} \quad \left\| \tilde{\mathbf{R}}_{Mj} - \mathbf{T}_{Mj} \right\|_2^2 + \sum_{l \in P_{Mj}} \left\| \mathbf{R}_{Mj} - \mathbf{R}_{Mjl}^{M-1} \right\|_2^2 + \sum_{l \in P_{Mj}} \left\| \mathbf{y}_{Mj} - \mathbf{S}_{Mj} \tilde{\mathbf{y}}_{(M-1)l}^{M-1} \right\|_2^2$$

$$\text{with respect to} \quad \mathbf{x}_{Mj}, \mathbf{y}_{Mj} \tag{8}$$

$$\text{subject to} \quad \mathbf{g}_{Mj}(\mathbf{R}_{Mj}, \tilde{\mathbf{R}}_{Mj}, \mathbf{x}_{Mj}, \mathbf{y}_{Mj}) \leq 0$$

$$\mathbf{h}_{Mj}(\mathbf{R}_{Mj}, \tilde{\mathbf{R}}_{Mj}, \mathbf{x}_{Mj}, \mathbf{y}_{Mj}) = 0.$$

4. EXAMPLE

To illustrate the use of analytical target cascading in designing a family of products, we will demonstrate the application of the methodology on a design problem with two vehicles (Kokkolaras, et al., 2002). The model hierarchy, depicted in Figure 11-7, consists of four levels: the family level, the vehicle level, the system level, and the component level. At the family level, the single family response is defined as the weighted sum of the vehicle responses, in this case the vehicle masses. More emphasis is given on minimizing the mass of vehicle A by assigning a higher weight to that term. Note that in the previous section, the mathematical formulation of the ATC process was presented for the general case using specific symbols to denote responses and local and shared variables, subscripts to identify elements in the hierarchy, and superscripts to specify the level at which quantities were computed. In this section, we will use symbols that are specific to the example. For example, according to the general notation the family response is a function of the vehicle responses: $R_0 = w_1 R_{11} + w_2 R_{12}$; in this section, we use the symbols for vehicle masses instead of the symbols for vehicle responses, i.e., $R_0 = 0.8 m_{v,A} + 0.2 m_{v,B}$. Moreover, we will not use level index superscripts to specify where quantities were computed; instead we will use the superscripts ^{U} and ^{L} to denote quantities that were computed at the upper or lower level, respectively.

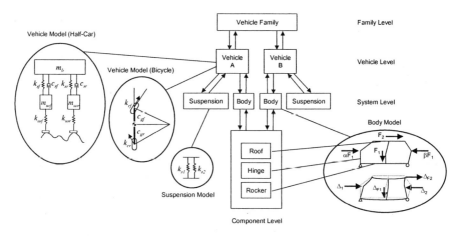

Figure 11-7. Model hierarchy of the two-vehicle example.

Figure 11-8 illustrates the overall analytical target cascading formulation. The individual vehicles are modeled as half-cars and bicycles at the vehicle level. In addition to the mass targets $m_{v,A}^U$ and $m_{v,B}^U$ cascaded from the family level, local targets $\mathbf{T}_A$ and $\mathbf{T}_B$ are set for the ride quality metrics vectors $\mathbf{z}_{v,A}$ and $\mathbf{z}_{v,B}$ of each variant, respectively. Ride quality metrics are defined by the following five responses: front and rear ride frequency, front and rear wheel hop frequency, and under-steer gradient. Vehicle A should have a stiffer ride, whereas vehicle B should have a softer ride. The half-car model computes vehicle mass m_v, ride quality metrics $\mathbf{z}_v$, body-in-white mass m_b, and suspension stiffnesses k_{sf} and k_{sr}. Vehicle responses must meet targets determined at the family and system levels and local targets $\mathbf{T}_A$ and $\mathbf{T}_B$.

Once the responses are computed, they are used as targets at the family and system levels. Note that the subscripts $_A$ and $_B$ included above to denote vehicle variants will be omitted in the remainder of the text for simplicity. Local design variables at the system level include distance of center of gravity (CG) to front end cg_f, distance of center of gravity to rear end cg_r, front tire stiffness k_{usf}, rear tire stiffness k_{usr}, front cornering stiffness k_{cf}, and rear cornering stiffness k_{cr}.

Body and suspensions are modeled at the system level. The local targets $\mathbf{T}_b$ for the body are set to maximize the stiffness of the structure by minimizing the deflection vector δ_b, which is obtained considering the two different loading conditions shown in Figure 11-6, subject to allowable strain energy constraints. The body is represented by a finite element model consisting of ten elements. These ten elements model eight components of a two-dimensional body including the A, B, and C pillars, the hinge pillar, the roof rail and the rocker. Each component i is described by its cross-sectional properties: footprint area A_i, real area A_{Ri}, and moment of inertia I_i, which are functions of sheet metal thickness t_i, height h_i, and width b_i, of the footprint area. The two joint-linking elements are modeled as radial springs. The body model computes body-in-white mass m_b, deflection vector δ_b, and footprint area A_i, real area A_{Ri}, and moment of inertia I_i for each body component i. The suspension model computes sprung stiffnesses k_{sf} and k_{sr} for the front and rear suspensions of the half-car model, respectively, based on the stiffness of two individual springs k_{sf1} and k_{sf2} for the front suspension and two individual springs k_{sr1} and k_{sr2} for the rear suspension. System responses must meet targets determined at the vehicle and component levels and local targets $\mathbf{T}_b$. Once the responses are computed, they are used as targets at the vehicle and component levels.

At the component level, each component i of the body comprising the platform is designed to match the area targets A_i^U and A_{Ri}^U and moment of inertia target I_i^U cascaded from the system level by determining optimal combinations of cross-sectional dimensions (width b_i, height h_i, and thickness t_i). Once the optimal values of these dimensions are determined, analytical expressions are evaluated and optimal values A_i^L, A_{Ri}^L, and I_i^L are passed to the system level for each platform component i. The product platform of the family consists of three body components: the roof, the rocker, and the hinge pillar. To represent the product platform, one common optimal design problem for the three shared components is formulated at the component level. The shared pillars return a common response to the body models of both vehicle variants.

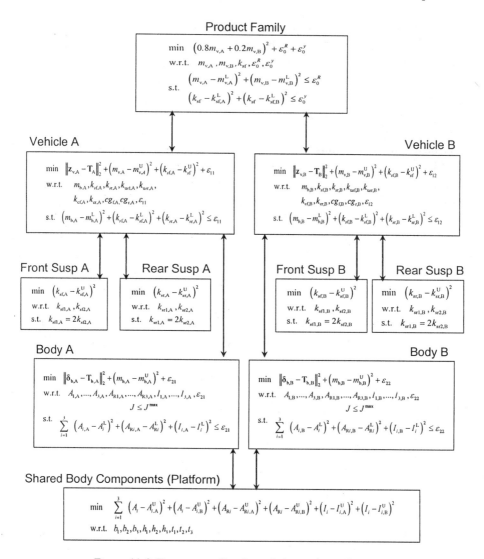

Figure 11-8. Target cascading formulation and coordination.

The front suspension is also shared between the two vehicles to illustrate the concept of using shared variables to represent component sharing in simple cases. This sharing is represented by treating the front suspension stiffness k_{sf} at the vehicle level as a shared variable. This shared variable is coordinated at the family level by computing the suspension stiffness k_{sf} to match the values $k_{sf,A}^{L}$ and $k_{sf,B}^{L}$ determined at the vehicle level for each variant. The computed value is then cascaded to both variants at the vehicle level as k_{sf}^{U}. Note that the front suspension stiffness is also treated as a response at the vehicle level and is cascaded as target to the system level.

4.1   Results

As mentioned above, the family objective was to minimize the weighted sum of the masses of the two vehicles. The target was set to zero in order to achieve the minimum mass possible. Of course, such a target is unattainable, and the obtained optimal value was 2124 kg. The analytical target cascading process converged after ten iterations. The optimal values obtained for the responses for which targets were defined locally at the vehicle level are presented in Table 11-1. It can be seen that all targets are met with a satisfactory accuracy, except for the front ride frequency. Indeed, sharing the front suspension results into the inability to satisfy this target. Therefore, it is necessary to either define another set of target values or reconsider the sharing of the front suspension.

Table 11-1. *Target and optimal values for vehicle level responses.*

Responses z_v	Target value Vehicle A	Optimal value Vehicle A	Target value Vehicle B	Optimal value Vehicle B
Front ride frequency [Hz]	1.273	1.160	0.955	1.120
Rear ride frequency [Hz]	1.592	1.585	1.592	1.592
Front wheel hop frequency [Hz]	10.345	10.348	10.345	10.343
Rear wheel hop frequency [Hz]	10.345	10.347	9.549	9.549
Under-steer gradient [rad/m/s^2]	7.19e-3	7.186e-3	7.19e-3	7.191e-3

Local targets were also set for the bodies at the system level. Bodies were intended to be as stiff as possible by minimizing the deflection vector δ_b. Once again, unattainable zero targets were set and the following values were obtained: The component of the deflection vector due to vertical loading is 0.209 and 0.202 inches for vehicles A and B, respectively; the component of the deflection vector due to horizontal loading is 0.357 and 0.358 inches for vehicles A and B, respectively.

Responses and shared variable values at the family and vehicle levels are compared in Table 11-2. The agreement is satisfactory. Note that the front suspension stiffness is treated as a shared variable. The matching of responses between vehicle and system levels is illustrated in Table 11-3. Once again, deviations are negligible. Note that during the coordination of these two levels the front suspension stiffness is treated as a response. The results in Tables 11-2 and 11-3 demonstrate that the analytical target cascading process yields a consistent design.

Table 11-2. *Vehicle responses and shared variable values at the family and vehicle levels.*

Characteristic	Type	Family level value	Vehicle level value
Mass of vehicle A $m_{v,A}$ [kg]	Response	2139	2139
Mass of vehicle B $m_{v,B}$ [kg]	Response	2162	2163
Front suspension stiffness of vehicle A $k_{sf,A}$ [N/mm]	Shared Variable	35.40	35.49
Front suspension stiffness of vehicle B $k_{sf,B}$ [N/mm]	Shared Variable	35.04	35.50

Table 11-3. *System responses computed at the vehicle and system levels.*

Responses	Vehicle level value	System level value
Front suspension stiffness of vehicle of A $k_{sf,A}$ [N/mm]	35.490	35.499
Front suspension stiffness of vehicle of B $k_{sf,B}$ [N/mm]	35.500	35.499
Rear suspension stiffness of vehicle A $k_{rf,A}$ [N/mm]	39.860	39.790
Rear suspension stiffness of vehicle B $k_{rf,B}$ [N/mm]	36.500	36.617
Body-in-white mass of vehicle A $m_{b,A}$ [kg]	240	239
Body-in-white mass of vehicle B $m_{b,B}$ [kg]	263	263

Table 11-4 presents the results obtained for the product platform, i.e., the three components of the body that are shared between the two vehicles. The agreement between the values obtained at the system level and the values obtained at the component level confirms the ability of the analytical target cascading formulation to account for shared components.

Finally, Table 11-5 presents the optimal values for local design variables for the vehicle, system, and component levels. Although design values are obtained for all optimization problems formulated within the analytical target cascading formulation, it should be emphasized that the main outcome of this process are the design specifications for the elements of the variants at the vehicle, system, and component levels: vehicle masses, body-in-white masses, suspension stiffnesses, and cross-section related properties (areas and moments of inertia) for the platform components of the body. These design specifications correspond to the optimal values of the responses, as presented in the far-right columns of Tables 11-2 to 11-4. For example, the design specifications for the mass of vehicle B is 2163 kg, the design specification for the body-in-white mass of vehicle A is 239 kg, and the design specification for the footprint area of rocker is 8.02 in^2.

Table 11-4. *Platform component values computed at the system and component levels.*

Responses	System level Value (Veh A)	System level Value (Veh B)	Component level
Moment of inertia of rocker I_1 [in^4]	15.387	15.38	15.387
Footprint cross-sectional area of rocker A_1 [in^2]	8.024	8.031	8.020
Real cross-sectional area of rocker A_{R1} [in^2]	5.788	5.788	5.792
Moment of inertia of roof rail I_2 [in^4]	0.162	0.161	0.157
Footprint cross-sectional area of rail A_2 [in^2]	1.548	1.542	1.550
Real cross-sectional area of roof rail A_{R2} [in^2]	1.043	1.044	1.043
Moment of inertia of hinge pillar I_3 [in^4]	13.917	13.859	13.887
Footprint cross-sectional area of hinge pillar A_3 [in^2]	8.152	8.177	8.158
Real cross-sectional area of hinge pillar A_{R3} [in^2]	5.879	5.888	5.883

Table 11-5. *Optimal values of design variables at the vehicle, system, and component levels.*

Design Variable	Level	Optimal value Vehicle A	Optimal value Vehicle B
Distance of CG to front end cg_f [m]	Vehicle	1.39	1.25
Distance of CG to rear end cg_r [m]	Vehicle	2.31	2.45
Front tire stiffness k_{usf} [N/mm]	Vehicle	24.1	24.08
Rear tire stiffness k_{usr} [N/mm]	Vehicle	24.09	20.52
Front cornering stiffness k_{cf} [N/rad/10e $-$ 4]	Vehicle	10.47	11.3
Rear cornering stiffness k_{cr} [N/rad/10e $-$ 4]	Vehicle	12.82	11.84
Front suspension spring stiffness k_{sf1} [N/mm]	System	23.67	23.67
Front suspension spring stiffness k_{sf2} [N/mm]	System	11.83	11.84
Rear suspension spring stiffness k_{sr1} [N/mm]	System	26.53	24.41
Rear suspension spring k_{sr2} [N/mm]	System	13.26	12.21
Width of rocker cross-section b_1 [in]	Component	1.51	shared
Height of rocker cross-section h_1 [in]	Component	5.31	shared
Thickness of rocker cross-section t_1 [in]	Component	0.497	shared
Width of roof rail cross-section b_2 [in]	Component	1.335	shared
Height of roof rail cross-section h_2 [in]	Component	1.161	shared
Thickness of roof rail cross-section t_2 [in]	Component	0.265	shared
Width of hinge pillar cross-section b_3 [in]	Component	1.642	shared
Height of hinge pillar cross-section h_3 [in]	Component	4.969	shared
Thickness of hinge pillar cross-section t_3 [in]	Component	0.530	shared

5. SUMMARY

Most product design tasks require coordinated efforts of multiple teams of expertise according to the organizational hierarchy of the firm. Interactions are more elaborate when the design and development process considers more than one product. Selecting a common product platform is critical to reduce costs when designing product families, upgrading existing products, or deriving new variants. Nevertheless, it is important to make sure that commonality decisions can be implemented and do not introduce inconsistencies.

This chapter presented the analytical target cascading (ATC) methodology for translating targets for a family of products to platform specifications for given commonality decisions. We extended the ATC formulation for a single product to a family of products to accommodate the presence of a shared product platform and locally introduced design targets.

The extended ATC methodology combines given decomposed product hierarchies into a single family hierarchy that takes into consideration the platform, i.e., shared elements. The ATC process is then applied to translate family targets to product targets as well as individual product design targets to subsystem, component, and platform specifications. These design specifications are obtained while taking into account element interactions and sharing tradeoffs to ensure that the overall system design is consistent. In this manner, the ATC methodology is also used to investigate the feasibility of commonality decisions.

Chapter 12

DETERMINING PRODUCT PLATFORM EXTENT

Olivier L. de Weck
Department of Aeronautics & Astronautics, Engineering Systems Division, Massachusetts Institute of Technology, Cambridge, MA 02139

1. INTRODUCTION

Typically, it is assumed that a product family will be derived from a *single* platform once a firm has decided on platforming as an appropriate strategy. The totality of all variants represented in a single-platform family defines the lower and upper performance and value bounds which have to be supported by the platform. This difference between upper and lower bounds of a platform is commonly referred to as *platform extent* (Seepersad, et al., 2000). However, the average number of variants built from a single platform has been steadily increasing in a number of industries (automotive, electronics, aircraft) since the early 1990's. Figure 12-1 shows that the number of models per platform in the automotive industry has been increasing since 2002. This trend had started in the mid 1990's and is likely to continue in the future. The consequence is that each platform has to accommodate a larger number of variants, whereby the extent of the platform is constantly being challenged with each new variant that is assigned to it. There is general agreement that mass customization has led to an increasing fragmentation of the automotive market with the number of individual models for sale in the U.S. rising each year from 33 in 1947, to 198 in 1990 to an estimated 277 in 2009 (Simmons, 2005).

This pervasive phenomenon can be attributed to accelerating mass customization and concomitant market fragmentation. Recently, manufacturers have begun to realize that platforms cannot be "stretched"

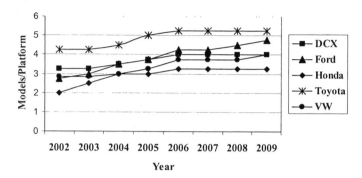

Figure 12-1. Variants per platform trend in the automotive industry (Source: PWC, 2003).

indefinitely, before the competitiveness of some of the associated variants is compromised in a competitive environment. This dilemma has given rise to research in the area of *multi-platform* strategy. The questions posed in this context are not too different from those surfacing in the classic tradeoff between component commonality and product distinctiveness when designing a single product platform. We want to know:

1. Given a set of *m* product variants, what is the optimal number of platforms, *p*, to derive these from?
2. What is the optimal assignment of the *m* product variants to the *p* platforms, given a set of target market segments and competitors?
3. What criteria should be used to decide on platform extent?

The *multi-platform problem* manifests itself at the highest level in the system hierarchy going from individual components to modules and platforms to products to families-of-products. Deciding which components to make common and potentially include in a platform, and which ones to keep unique for each variant is primarily a problem addressed by product designers and manufacturing engineers within the firm (see other chapters). The platform extent problem, on the other hand, is of a more strategic nature since the question cannot be answered without at least a rudimentary understanding of the interplay between product architecture, manufacturing, cost, engineering performance, value, demand and the role of competition. Despite this domain-crossing complexity, we believe that the platform extent problem can be rigorously addressed both quantitatively and qualitatively. This is primarily so because the number of factors to consider are numerous, but finite. The primary challenge in solving the platform extent problem arises from questions of sufficient model fidelity and robustness (depth), rather than questions of comprehensiveness and completeness (breadth).

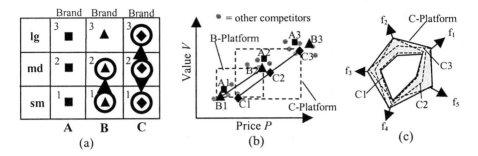

Figure 12-2. Visualization of the platform extent problem: (a) market segmentation grid with connected circles indicating a platform, (b) value-price positioning, (c) notional functional radar plot for C-platform showing platform extent.

Figure 12-2 visualizes the platform extent problem in three domains. First, one may investigate the assignment of variants to platforms in the context of platform leveraging strategies across a market segmentation grid (see Figure 12-2a). Here, we distinguish between three hypothetical manufacturers ("brands"), whereby A offers three variants (small, medium large) without platforming, B combines the first two products in a platform, but not the third, and firm C derives all three products from a common platform. Which of these three strategies is superior cannot be decided from the market segmentation grid alone. Plotting each product in the value-price diagram (Cook, 1997) reveals where each product variant falls with respect to market segment averages and competitors. Platform extent can also be understood as the value-price envelope of all variants associated with a platform (see Figure 12-2b). In this case the C-platform covers the largest value-price region (dashed box). Finally, Figure 12-2c shows the functional performance envelope that contains all variants for the C-platform in a spider chart (radar plot). The gray shaded area can be interpreted as the functional extent of the platform. One can also think of platform extent in the design space (not shown), where the lower and upper bounds of key design variables of the product determine the limits of the platform[1].

This chapter presents a brief review of prior work on the multi-platform (extent) problem and discusses the relevance of the issue to industry (Section 2). Next, in Section 3, we present a modeling framework for single products, which connects the following six domains: (1) product architecture, (2)

---

[1] One has to be careful to distinguish between the *actual platform extent*, which is given by variants actually derived from a platform and the *potential platform extent*, which is defined by the set of variants that could potentially be derived, while still remaining competitive in the market place. The second kind is generally not known a priori.

engineering performance, (3) product value, (4) market demand, (5) manufacturing cost and (6) investment finance. We will argue that such a comprehensive modeling framework is not only desireable, but a necessary – if not sufficient – condition to address the platform extent question. Section 4 shows the extension of the single product model to the multiple-variant, but still single platform case. We will argue that the effect of platforming must be assessed both in terms of its external (market) as well as its internal (manufacturing) consequences. Once the number of variants increases, say beyond $m > 3$, the question of multiple platforms must be addressed (Section 5). We develop the ideas in this chapter based on an actual database of vehicles sold in the U.S. market (2002). Conclusions are drawn in Section 6, whereby the robustness of the answers to various assumptions is discussed. Finally, we conclude with a number of reflections regarding the platform extent problem and recommended areas of future research.

2.   LITERATURE AND REFLECTIONS ON MULTIPLE PLATFORMS

We attribute the first paper on the question of platform extent (multiple platforms) to Seepersad, et al. (2000). In their work they wrestled with the question of finding the optimal platform extent, and therefore the right number of platforms, for a family of industrial absorption chillers. Their method to determine platform extent uses a compromise Decision Support Problem, enhanced by linear physical programming. In this work a manufacturer seeks to offer a family of eight chillers in the capacity range of 600-1,300 tons, in 100 ton increments, based on one or more platforms. Targets are set for manufacturing costs across the family, as well as cycle time for manufacturing of individual variants. A linear physical program attempts to minimize the compound deviation from these targets, Z. The quantity Z represents a weighted sum across the nine targets and four desirability ranges of physical programming. The authors found that a single platform was satisfactory in cases of uniform demand distribution, but that two platforms were advantageous when significant demand gaps created a distance between low and high capacity product variants. Their model addressed the domains of product architecture, engineering performance, manufacturing cost, and capital investment, but did not take into account the effects of customer valuation, market demand and competition.

This last point is important, since it is problematic to assess the potential benefits of a platform strategy in non-monopolistic markets without modeling competition as well. Cook (1997) reports that "Ironically, GM's market share relative to Ford only began to recede in the mid 1980s as GM's brands – Chevrolet, Pontiac, Oldsmobile, Buick, and Cadillac – became less distinctive through the use of common platforms and exterior stampings that reduced product differentiation."

We also found that much of the literature on the multi-platform problem is either engineering-centric or management-centric. It is apparent that engineers rarely have access to and expertise in market related topics. Conversely, management does not have a detailed understanding of the technical issues resulting from various platforming decisions. A practical as well as academic treatment of the platform extent problem needs to eventually bridge the gap between management and engineering.

The ultimate goal of product platforming is long-term profit maximization for the firm, leading to a sustainable increase in shareholder value. We argue that there can only be this one objective, when trying to address the single or multi-platform question. The next section presents a framework that models profit arising from a single product, without platforming. Subsequent sections extend this framework to multiple product variants and platforms.

3. SINGLE PRODUCT MODEL

3.1 Reasoning and modeling framework

Figure 12-3 shows a reasoning and modeling framework that helps clarify the primary domains that exist in the for-profit manufacturing firm. This view is product-centric and does not attempt to capture the myriad of organizational issues present in manufacturing firms. The figure shows the *six primary domain blocks*, as well as the main variables flowing between blocks in addition to important exogenous inputs. Exogenous inputs are those that are not under direct control of the firm.

First, product architecture (1) defines the value-generating functions of the product and maps these to physical components (parts) and modules which are assemblies of parts[2]. Inputs to product architecture are regulations and standards with which the machine must comply. The choice of operating principles of the machine and its decomposition relate the physical components to the vector of independent design variables, $\mathbf{x}$, for which engineers will find the most appropriate values. In order to accomplish this, engineering (2) creates models of functional product performance attributes, $\mathbf{f}$, as a function of the design variables, $\mathbf{x}$.

---

[2] Parts are the atomic units of the product, which cannot be further taken *apart* before they loose their functionality and integrity. Parts can be hardware or software.

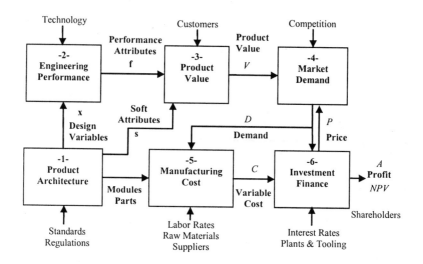

Figure 12-3. Single product reasoning and modeling framework.

The interface between engineering and marketing is primarily concerned with how the vector of performance attributes, **f**, translates to value, V, in the marketplace. The product value model (3) is also impacted by soft attributes, **s**, such as styling, comfort, or dependability, which are only measurable via customer surveys but not directly via (physics-based) performance attributes and engineering models. We subscribe to Cook's (1997) view that value is to be measured in the same monetary unit as price, e.g., [$]. Given a database of existing products grouped into market segments, and given transaction prices, P, demand, D, functional attributes, **f**, and soft attributes, **s**, for these products one can estimate both the predicted value of each product in the marketplace, as well as the relative contribution that each of the functional and soft attributes make to that value. Demand and market share estimation (4) in non-monopolistic markets requires the same information about the competitor's products as about ones own products. This is often not fully available, but specialized service providers (e.g., J.D. Power), government agencies and corporate intelligence are usually in a position to provide ample data in the aforementioned categories. This is particularly true for well established markets (e.g., automobiles, household appliances) but less true for emerging markets (e.g., digital portable devices). It has been our experience that paucity of data is not usually the main problem in enterprises that consider establishing or refining a successful product family strategy. Rather, what is missing is a coherent end-to-end framework, such as the one shown in Figure 12-3, to integrate and interpret the available data. Of all the disciplines involved, demand estimation is the most difficult and uncertain, but it is essential. Much of the engineering literature on platforms bypasses this difficulty by resorting to constructs such as "acceptable performance losses due to platforming", thereby ignoring true market effects. While such simplifying assumptions are understandable in the context of academic research, the for-profit manufacturing firm does not have the luxury of drawing such arbitrary boundaries between domains.

It is interesting to note that when actual sales volumes are available for market segments of interest (actual demand D_a), one has an opportunity to infer the relative weighting factors, γ, between the functional and soft attributes by minimizing the difference between predicted and actual demand. This will be shown later.

Once demand, D, has been estimated in this fashion one can derive the production volume, D_p, which is simply the predicted demand plus scrap and surplus. While discussing platform extent in this chapter, we will assume that demand and production volume are identical. Given labor and material rates, one follows the lower branch in Figure 12-3 to estimate the variable (per unit) manufacturing cost, C, for the products of interest (5). This estimate relies on knowledge of the product architecture, i.e., sets of modules and parts that make up the product as well as a determination of the main fabrication and assembly steps. Estimation of the profit, A, also requires an estimate of the fixed manufacturing cost, F, which – to first order - are assumed to be independent of production quantity, demand D, as well as the investment cost, M. This is done in the investment finance block (6), thus completing the modeling framework. Having obtained estimates for demand, D, price, P, variable cost, C, fixed cost, F, and investment cost, M, one can estimate the profit per unit time, A, as (Cook, 1997):

$$A = D(P-C) - F - M \tag{1}$$

This calculation is typically done on a quarterly, yearly or multi-year net present value basis. Extending the considerations over multiple time periods is done via the discounted cash flow method (*NPV*).

Each of the blocks in Figure 12-3 represents a body of knowledge and a community of practitioners and academics in its own right. Numerous are the arguments that are given (particularly in academia, less in industry) why one cannot or should not consider connecting the six aforementioned domains with simplified models to gain insights into the product realization process. Indeed, one must be careful to apply well grounded methods and appropriately vetted data in such a pursuit. Furthermore, there is no alternative if one wishes to quantitatively answer strategic questions such as the one dealing with platform extent, except to rely exclusively on "instinct" and "experience" of executives. The methods discussed here are primarily meant to augment current practice and should not supplant sound judgment.

We are primarily interested in a *relative comparison* of competing platform strategies such as the ones shown in Figure 12-2, rather, than in absolute profit forecasting. We will carefully stipulate our assumptions and subject the answers provided by such analysis to sensitivity analysis. The next subsections go into more depth on modeling of the six domains. While alternative methods and references exist, we will only illustrate one method in each of the domains. This will be done in the context of the automotive industry, building upon a comprehensive data set (Appendix A). This data is conveniently reused in Section 5 for the multi-platform problem.

3.2   Product architecture

Product Architecture has been practiced since the industrial revolution, but has only recently emerged as a field of serious research (Meyer and Rechtin, 2000; Ulrich and Eppinger, 2004). We have found that thinking about and visualizing simultaneously form (objects) and function (processes) is one of the main challenges in product architecting. Object Process Methodology (OPM) (Dori, 2002) has emerged as an increasingly popular graphical and formal language for representing product architectures. Figure 12-4 shows the basic architecture of an automobile. In this single representation we find the structure of the product, decomposed into assemblies (modules), as well as its main value-delivering function (transporting), along with the operands (driver, passengers, cargo), and internal functions. The relationships between design variables, $\mathbf{x}$, and functional attributes, $\mathbf{f}$, can be formally mapped in this way. While the design variables characterize the parts and assemblies, the functional attributes, $\mathbf{f}$, are associated with the internal functions.

A more thorough discussion of the product architecture shown in Figure 12-4 requires introduction of the object process nomenclature in Figure 12-5 as well as a simplified visualization of the product system itself (in this case the automobile) in Figure 12-6.

The primary value-delivering function of an automobile is transporting certain numbers of passengers and amounts of cargo *comfortably, quickly and economically* from location A to B. This has to be done while meeting all government regulations and standards regarding fuel economy, emissions and safety. We call the passengers and cargo the *operands* of the process, whose attribute state "location" is being transformed from "A" to "B". The driver is both an operand of the process and at the same time the *operator* (agent) executing the process. The driver can be the owner of the automobile, but this is not necessarily so. The automobile is the main instrument of the process called "transporting".

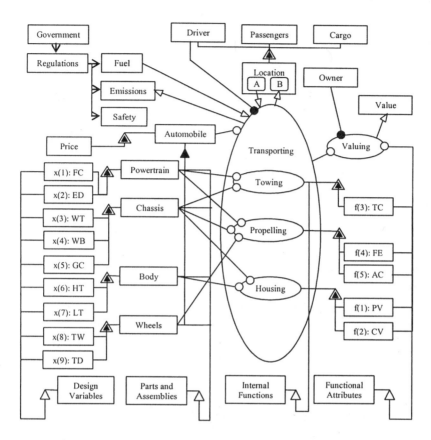

Figure 12-4. Object Process Diagram (OPD) of a generic automobile. See nomenclature at the end of the chapter for details about design variables and functional attributes.

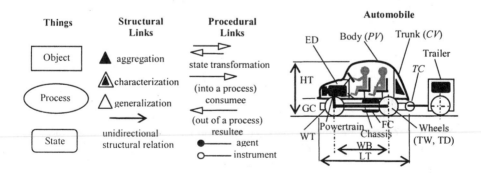

Figure 12-5. OPM nomenclature.   *Figure 12-6.* Automobile schematic.

Moreover, the owner is the individual who made the decision to purchase the automobile at a known price *P*. At the time of the transaction he or she made this decision based on an assessment of the difference between perceived value *V* and price *P*. The value of the product (from a rational utilitarian point of view) is a continuous quantity which depends on *how well* the product fulfills its primary, externally delivered function[3]. The question of "how well" can be quantified via a set of functional attributes, **f**. In Figure 12-4, right hand side, we show five important functional attributes of an automobile: passenger volume (PV), cargo volume (CV), towing capacity (TC), fuel economy (FE) and acceleration (AC). There is no claim that this is a complete list of all relevant functional attributes of an automobile. However, the fact that these attributes are readily available in various consumer publications and that they describe concisely the amounts of operand that can be transported or towed by the automobile as well as the (transient) speed and fuel efficiency is compelling.

In fact, if one "zooms" (Dori, 2002) into the primary value delivering function of the automobile shown as the large oval labeled "Transporting" in Figure 12-4, then lower level internal functions are revealed: Propelling, Housing, Towing. Propelling is the ability of the vehicle to roll on a surface as well as to accelerate and decelerate on command. The primary vehicle modules responsible for this process are the powertrain, the chassis and the wheels. The powertrain comprises, among other parts the fuel tank, engine, transmission, drive shaft and differential. The chassis is made up primarily of the structural underbody (carriage), the braking system as well as the suspension system. The wheels allow the vehicle to roll and transmit the torque generated by the engine to the road. These statements reflect a mapping from internal functions to parts and assemblies. Creating this function-to-form mapping is the primary responsibility of the product architect (Crawley, 2001).

The body of the automobile houses the passengers and cargo, thus shielding them from wind and external elements. It also reduces drag and contributes significantly to the external aesthetic appeal of the vehicle (styling). In a "body-on-frame" (BOF) architecture the chassis and body are clearly separated, whereas in a body-frame-integral (BFI) architecture they are more tightly integrated (Whitney, et al., 2004). Finally, the towing capacity (*TC*) of a vehicle is primarily driven by the power of the engine and the ability of the chassis to transmit the towing load from the hitch through the frame and on to the wheels.

It is apparent that the "design variables" of the product characterize the parts and assemblies of the product system. Thus, when making things

---

[3] We acknowledge that collectors of automobiles potentially have non-utilitarian value functions that are not considered in this work.

"common" in product platforming and module reuse it is not sufficient to make the *values of design variables* common if one intends to actually manufacture common parts and assemblies. In other words, learning curves and economies of scale only materialize by reusing physical components (or software)[4]. We see from Figure 12-4 (lower left) and Figure 12-6 that the powertrain is characterized by the fuel tank capacity (FC) and engine displacement (ED). The chassis is defined by the wheel track (WT[5]), wheel base (WB) and ground clearance (GC). Overall total length (LT) and height (HT) are associated with the body, which can be of type BOF or BFI, while the wheels are characterized by tire width (TW) and diameter (TD).

Obviously, there will be many more design variables than these nine in a comprehensive automobile realization program. For purposes of our illustrations we will remain with this limited set because: (i) they characterize in major part the four modules in Figure 12-4 and, (ii) data on these variables is readily available and (iii) they provide enough complexity to yield interesting results. It is understood that a manufacturing firm will have to design and model the product in much more detail. In fact it has been estimated that the bill of materials (BOM) of a typical automobile contains on the order of 10,000 parts, which leads to a drawing tree on the order of

$$N_{level} = \left\lceil \frac{\log N_{parts}}{\log 7} \right\rceil \tag{2}$$

levels deep. The above equation assumes a 7-tree decomposition-aggregation of the product at each level in accordance with the 7+/-2 rule imposed by human cognition limits (Miller, 1956). Thus, we would expect the drawing tree of an automobile to be about *five levels deep*. The number of design variables for an individual automobile likely exceeds 100,000 (over 10 design variables per part). Note, however, that the design variables called out in Figures 12-4 and 12-6 are associated with modules rather than individual components. For brevity we will treat these modules as the atomic units of the product. If the front-end question of platform strategy and extent has to be answered for a relatively simple product (drill, walkman, etc.) one may be able to decompose and model the product to its lowest level. For complex products above three levels of decomposition (>300 parts) one generally has to work with a more aggregate representation, as is done here.

---

[4] There is a subtler notion of architecture with the use of common "hard points", which could be interpreted as a platform even without sharing physical parts. We will take a more traditional view here and assume that platforming implies physical commonality.

[5] We define WT as the front wheel track in this chapter. The difference between front and rear wheel track is usually very small in passenger cars, but can be more significant in trucks.

It is certainly true that the value of an automobile or any other utilitarian product does not only depend on the functional attributes, **f**, discussed thus far. The marketing literature refers to "soft" attributes (Cook, 1997), i.e., those that cannot be directly predicted by engineering models. Examples of such attributes are: mechanical quality, comfort, styling, mechanical dependability or even the service experience at the dealership. We discuss the importance of these attributes and how they contribute to value in Section 3.4. The soft attributes have been neglected in Figure 12-4 for the sake of clarity.

3.3   Engineering performance

The product architecture is usually an input given to the design engineering team by the product architect(s). This means that the externally-delivered value-generating function as well as the necessary internal functions have been defined and mapped to hardware or software modules and parts to at least two levels of decomposition. It also means that a design vector, **x**, has been defined but specific values for the x_i's are yet unknown.

The main task of design engineers is therefore to determine feasible and perhaps even "optimal" values for the design variables **x**, subject to lower and upper bounds and other technical constraints. This is done by choosing values for the x_i's such that the functional attributes **f** are brought as close to their required targets as possible. These targets are typically defined by (inbound) marketing. Engineers need to also keep track of fixed parameters, **p**, and dependent variables, **y**. Examples of fixed parameters are material properties as well as road and atmospheric conditions. Fixed parameters cannot be directly affected by design engineers. Dependent variables, **y**, are those that depend on design variables or other dependent variables, but which are not directly perceived by the customer and therefore only contribute to product value indirectly. Examples of dependent variables are vehicle curb weight and engine horsepower rating.

In order to establish a mathematical, quantifiable mapping $x : x \mapsto f$ engineers develop models. These models fall into the following three broad categories:

-   Empirical models (analogous prototypes)
-   Physics-based models (virtual prototypes)
-   Actual testable product models (physical prototypes)

Empirical models are essentially "black box" input/output models that approximate the relationships between **x** (inputs) and **f** (outputs). The typical methods used are: linear regressions, Neural Networks, Response Surface Models (RSM) and Kriging models (Srivastava, et al., 2004). What is required in all cases is an extensive an accurate database of existing (or planned) products with sufficient detail to capture the relationships of interest. An example of such a database is provided in Appendix A for the medium sedan/coupe automotive market segment ('MED') in the United States (Autopro, 2002). The database shows the design and dependent variables **x**, and **y**, respectively in Table 12-A1. The functional and soft attributes as well as prices[6] and actual sales volumes (=demand) for the 2002 calendar year are in Table 12-A2.

One may gain initial insights into the relationships between **x** and **f** by producing a set of scatter plots for all pairs of variables $\{x_i, f_j\}$, whereby $i=1,2,..,n$ and $j=1,2,..,m$. In our case we have gathered data for nine design variables, $n=9$, and five functional attributes, $m=5$. The database contains $N=31$ competing models from a number of domestic and foreign brands. Thus, $x_{i,k}$ would be the specific value of the i-th design variable for the k-th model (product). If, for example, we set $i=4$ and $k=7$, we read the total length (LT) of the Chevrolet Impala as $x_{4,7} = 200$ [in] from the database. Appendix B (see Figure 12-B1) shows scatter plots of the design variables x_{1-9} along the rows versus functional attributes f_{1-5} along the columns.

A rather naïve approach to creating an empirical model would be to search for the single design variable, x_i that correlates most strongly with a given functional attribute, f_j:

$$\text{find } i \in \{1, 2, ..., n\}$$

$$\text{such that } \max r^2$$

$$\text{where } r = \frac{1}{N-1} \sum_k \left(\frac{x_{i,k} - \overline{x}_i}{s_{x_i}} \right) \left(\frac{f_{j,k} - \overline{f}_j}{s_{f_j}} \right)_j \qquad (3)$$

where $\overline{x}_i$ and s_{x_i} are the mean and standard deviation of the i-th design variable and $\overline{f}_j$ and s_{f_j} are the mean and standard deviation of the j-th functional attribute, respectively. In this fashion one could create five independent curve fits for each f_j as a function of the most strongly correlated single design variable x_i:

$$\hat{f}_j(x_i) = a \cdot x_i + b \qquad (4)$$

Inspection of Figure 12-B1 reveals that while some statistical relationships exist, there is much scatter in the data. Some pairs exhibit

---

[6] The prices in Table 12-A2 represent manufacturer recommended sales prices (MSRP) and not actual transaction prices. Actual transaction prices are typically lower than MSRP due to dealer rebates, leasing arrangements and other cash incentives, but reliable information on them is difficult to obtain.

relatively strong positive correlation such as total length (LT) versus cargo volume (CV) with r=0.79 and r^2= 0.62, while others exhibit negative correlation such as engine displacement (ED) versus 0-100 km/h acceleration time (AC) with r=-0.72 and r^2=0.52. Yet other combinations appear entirely uncorrelated such as fuel capacity (FC) versus cargo volume (CV) with r^2=0.011 or ground clearance (GC) versus acceleration time (AC) with r^2=0.0006. Table 12-1 shows - for each functional attribute - its strongest regressor, the best linear fit coefficients, a and b, as well as r and r^2 for each fit.

Table 12-1. Parameters for a univariate empirical model to predict f from x.

j	f(j)	i	x(i)	r	r²	a	b	$\varepsilon_{RMS,j}$
1	PV	4	WB	0.75	0.56	-58.7	1.47	3.8
2	CV	7	LT	0.79	0.62	-20.2	0.19	6.3
3	TC	6	HT	0.28	0.08	-2831.5	72.5	23
4	FE	2	ED	-0.68	0.45	31.4	-0.003	6.3
5	AC	2	ED	-0.72	0.52	13.7	-0.002	9.4

The functional attributes of the N database entries can be predicted in this fashion. The goodness of the overall empirical model can be assessed via the root mean square (RMS) error of the residuals for the j-th attribute:

$$\varepsilon_{RMS,j} = 100 \cdot \left[\frac{1}{N} \sum_{k=1}^{N} \left(\frac{\hat{f}_{j,k}\left(x_{i,k}\right) - f_{j,k}}{\overline{f}_j} \right)^2 \right]^{\frac{1}{2}} \tag{5}$$

For example, we see that passenger volume is predicted relatively well by wheel base alone (3.8% error), while towing capacity is predicted poorly (23% error). Across all functional attributes this uncoupled linear model has an average error of ε_{RMS}=9.3%. Figure 12-7 shows the linear regression for all five functional attributes of interest. This certainly is a crude representation of the mapping from **x** to **f**. The largest problem – aside from the magnitude of the prediction error – is that the coupling within the system is neglected entirely. An example would be that one might choose to increase passenger volume by stretching the car, i.e., increasing the wheel base (WB). This will invariably lead to an increase in curb weight, which would lead to an increase in acceleration time, everything else being held constant. This effect is important when making platforming decisions and it is not captured by the linear regression model where acceleration is a function of engine size alone.

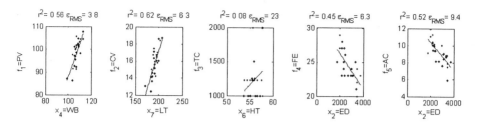

Figure 12-7. Linear univariate regression model for functional attributes (MED segment).

To address this deficiency we will construct a simplified *response surface model* (multivariate polynomial regression) of a vehicle (Srivastava, et. al., 2004). We will choose the regressors based on physical insight and we will introduce two dependent variables for which data is readily available (curb weight and horsepower rating). The model is developed for one functional attribute at a time.

Passenger Volume: The amount of available space in the passenger compartment (see Figure 12-6 center) is expected to scale with wheel track (width), wheel base (distance between axles), height and total length. We stipulate the following response surface:

$$f_1 = a_0 + a_1 x_3 + a_2 x_4 + a_3 x_6 + a_4 x_7 + a_5 x_3 x_4 x_6 \qquad (6)$$

The results using optimal least squares coefficients for passenger volume and the other functional attributes are shown in Figure 12-8.

Cargo Volume: The cargo volume in medium sedans in primarily determined by the trunk space, which in turn depends on the width of the vehicle and the rear overhang (distance from rear axle to rear bumper). For convenience we reuse Eq. (6) for cargo volume, but expect to obtain different least squares coefficients a.

Towing Capacity: In order to provide a large towing capacity a vehicle needs to have a strong engine (high horsepower rating), a sturdy chassis to sustain the axial loads induced by the trailer, a long wheelbase for directional stability, and a reasonably large curb weight relative to the load. Also, more subtly, frontal area (approximated as width times height) and ground clearance (center of gravity location) are relevant factors. We first approximate the horsepower rating [hp] of the vehicle as a function of engine displacement:

$$y_1 = b_0 + b_1 x_2 \qquad (7)$$

Moreover, let curb weight [lbs] be a function of fuel capacity (FC), engine size (ED), width (WT), height (HT) and overall vehicle length (LT):

$$y_2 = b_o + b_1 x_1 + b_2 x_2 + b_3 x_3 + b_4 x_6 + b_5 x_7 \tag{8}$$

Finally, towing capacity is modeled as:

$$f_3 = a_0 + a_1 y_1 + a_2 y_2 + a_3 x_3 + a_4 x_4 + a_5 x_5 + a_6 x_6 \tag{9}$$

Fuel Economy: It is apparent from Figure 12-B1 (4th column) that vehicles with larger engines have inferior fuel economy. Those vehicles also tend to be heavier. Furthermore, frontal area (width times height) contributes to drag increases even though this can be mitigated by efficient aerodynamic styling. Fuel Economy is therefore approximated as:

$$f_4 = a_o + a_1 x_2 + a_2 y_2 + a_3 x_3 x_6 + a_4 x_9 \tag{10}$$

Acceleration: Assuming constant acceleration we can write

$$v = \ddot{x} t + v_0 \quad \text{and} \quad \ddot{x} = F / m \tag{11}$$

with $P = F\dot{x} = Fv$ and $F = P/v$ $\tag{12}$

The following approximation for acceleration time from 0-100 km/h:

$$t = \frac{v^2 m}{P} = f_s = \left(\frac{100000}{3600}\right)^2 y_2 \cdot \frac{0.45}{746 y_1} \tag{13}$$

The correlations between actual functional attributes and those predicted by the RSM model are shown graphically in Figure 12-8 with numerical values provided in Table 12-2.

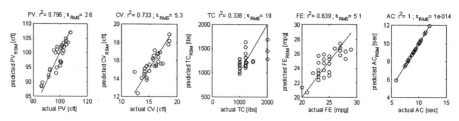

Figure 12-8. Correlations between RSM prediction and actual vehicle attributes (MED).

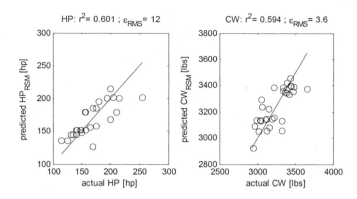

Figure 12-9. Correlations between RSM prediction and dependent variables (MED).

Table 12-2: RSM Prediction with 10% stretched version of a vehicle.

Vehicle	x(4) WB	x(7) LT	f(1) PV	f(2) CV	f(3) TC	f(4) FE	f(5) AC
Actual Honda Accord	105.1	186.8	92.7	13.6	1000	28	9.42
RSM of Honda Accord	105.1	186.8	94.5	14.0	1077	25.72	9.87
10% Stretched Version	115.6	205.5	107.6	16.9	1264	24.67	10.39

We can see from Figure 12-8 that the predictive accuracy of the RSM is 6.3% on average, which is better than the univariate linear regression model. Stretching the vehicle length by 10% increases passenger and cargo volume but worsens fuel economy and acceleration performance slightly which is what we expect (see Table 12-2). This is primarily due to the increased curb weight. Such a model is useful for conducting platform trade studies.

As mentioned previously, one may revert to more detailed physics-based modeling or even to physical prototypes to establish reliable predictions for the functional attributes. For the purposes of this work we will be satisfied with using the response surface models (RSM) developed above. With an average accuracy of 6.3% they represent sufficiently accurate placeholders for the engineering performance domain block shown in Figure 12-3.

3.4 Product value

The primary role of the value model is to translate the functional attributes, **f**, into a scalar quantity called "value", V. We accept Cook's (1997) definition of value as an absolute scalar quantity in monetary units, e.g., [$], which represents the aggregate benefit or worth the consumer derives from a product or service. The perceived net value obtained by the buyer at the time of purchase is $V-P$. The benefit to the seller is $P-C$, where C is the total cost to manufacture, advertise, and distribute the product[7].

---

[7] In some countries the sales price, P, also contains anticipated expenses for end-of-life disposal, but this is not considered here.

The concept of functional value as a continuous function of the performance attributes of the product is given as:

$$V\left(f_1, f_2, \ldots, f_m\right) = V_o v\left(f_1\right) v\left(f_2\right) \ldots v\left(f_m\right) + \Delta V\left(f_1'\right) + \ldots + \Delta V\left(f_k'\right) + \Delta C_{own} \tag{14}$$

where V_o is value of a baseline product (typically defined as the "average" product offered in a market segment), $v(f_j)$'s are the single attribute value curves, $\Delta V(f_k')$ are value additions due to product options and ΔC_{own} is the change in cost of ownership relative to the baseline product. The value curves take the form:

$$v\left(f_j\right) = \left[\frac{\left(f_C - f_I\right)^2 - \left(f_j - f_I\right)^2}{\left(f_C - f_I\right)^2 - \left(f_0 - f_I\right)^2}\right]^\gamma \tag{15}$$

This quadratic equation is derived from Taguchi's loss function (Cook 1997) such that the value of Eq. (15) goes to zero when f_j falls above a critical threshold value, f_C, for smaller-is-better (SIB) attributes, and asymptotes to a maximum value when f_j is below the ideal value, f_I. An example is acceleration time from 0-100 km/h. An ideal value of that attribute might be f_I −2 seconds (similar to Formula 1 race cars), while a critical value might be around f_C =20 seconds above which it would become dangerous to enter a busy highway. If Eq. (15) goes to zero $(f_j=f_C)$ it renders the entire product worthless as shown in Eq. (14), except for the salvage value of the options. If we set $f_j = f_0$, the baseline value, then Eq. (15) becomes unity. The value curve is weighted by an exponent, γ, which expresses the sensitivity of a particular functional attribute relative to others. For our analysis we will ignore the role of product options and cost of ownership and focus on the multiplicative part of Eq. (15). Figure 12-10 shows sample value curves for our functional attributes, f_{1-5}.

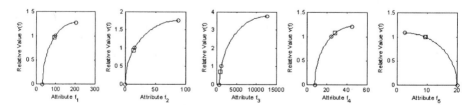

Figure 12-10. Value curves for v(f) for functional attributes f1-f5 (γ=0.5). The first four curves are of the type "LIB" (larger-is-better), while the last is "SIB" (smaller-is-better).

The critical and ideal values for each functional attribute are shown in Table 12-3. These values were found by first determining the minimum and maximum value for each attribute across all 7 market segments considered in this chapter (SML, MED, LRG, SPT, VAN, SUV, TRK) and then going $\pm 20\%$ that extreme value. The baseline values (middle point 'o' in the curves in Figure 12-10) are chosen as the average values for the MED market segment. Finally, the values of the market leader, Honda Accord, were used as the f_i values of interest (square symbols '□' in Figure 12-10).

Table 12-3: Critical, ideal, and baseline value for relative value curves.

Values	f(1) PV	f(2) CV	f(3) TC	f(4) FE	f(5) AC
Type	LIB	LIB	LIB	LIB	SIB
Units	[cft]	[cft]	[lbs]	[mpg]	[sec]
Critical values f_c	32	0	800	8	20
Ideal Values f_i	205	88	12,600	45	2.5
Baseline Values f_0	98.5	15.6	1225	24.4	9.2
Honda Accord f_i	92.7	13.6	1000	28	9.42

Substituting these values into Eq. (15) and then Eq. (14) yields the result that the value of the Honda Accord is 0.66 times the value of the hypothetical baseline (average) product in the market segment. Clearly, this cannot be true, since there must be a reason this particular model leads its segment in terms of sales volume and it doesn't seem to be reflected by our initial value assumptions. There are three potential reasons why this is occurring. First, there are missing product attributes which contribute to value. Second, it is not clear that the weighting factors, $\gamma=0.5$, are correct. Third, there exist factors apart from product attributes such as availability, brand image and intensity of promotion that drive sales. We address the first issue below and the second issue in the next section.

There is ample evidence that "soft attributes" are significant contributors to product value in the automotive industry (Cook, 1997). These attributes cannot be directly predicted by engineering analysis, but can be quantified by customer surveys. We introduce the following "soft" attributes, s_{1-5} (J.D. Power, 2004), all of which are reported on a scale from 1-5:

1. Mechanical Quality (MQ): score that captures owner-reported problems with the engine, transmission, steering, suspension and braking systems in the first 90 days of ownership.
2. Comfort (CO): score that quantifies features that consumers like and dislike about their vehicles with emphasis on comfort, convenience features and seats.
3. Style (ST): this score is based on how consumers rate the interior and exterior styling of their vehicle, uniqueness of styling, and interior and exterior color choices.

4. Mechanical Dependability (MD): captures owner-reported problems after three years of ownership with the engine, transmission, steering, suspension and braking systems.
5. Service Experience (SE): reflects the quality of the repair and maintenance service consumers received at the dealership. This attribute captures enterprise performance more than direct product performance.

The soft attribute scores for the vehicles in the medium car segment are reported in the last five columns of Table 12-A2. The value curves for the soft attributes, assuming exponential weighting factors of $\gamma=0.5$, are shown in Figure 12-11.

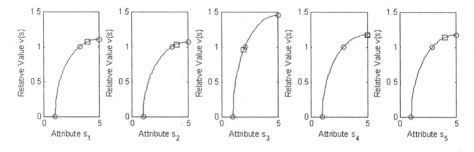

Figure 12-11. Soft attribute value curves: s(1)=MQ, s(2)=CO, s(3)=ST, s(4)=MD, s(5)=SE with critical values at a score=1 and ideal values at score=5; $\gamma=0.5$ for all curves.

Again, the average market segment values of soft attributes are used as the baseline, $s_o= [3.3\ 3.6\ 2.1\ 2.9\ 2.9]^T$. Note that the scores for the Honda Accord are shown as the small squares ('□') in Figure 12-11. This vehicle achieves a relative value of 1.42 on the soft attributes alone and experts in the automotive industry concur that the strengths of this brand indeed lie in the areas of mechanical quality and dependability.

Cook (1997) suggests that an actual revealed market value, V_a, can be found, when both price and demand for a product are known. So, we have two kinds of value which must be brought into equivalence: product value as constructed from aggregation of functional and soft attribute constituents and value as revealed by the marketplace via price and demand. The next section focuses on how demand for a product can be predicted as a function of value and price and how this information can be used to automatically (not arbitrarily) find the appropriate weighting factors, γ.

3.5   Market demand

Cook's (1997) linear demand model predicts for monopolistic (single product) markets that:

$$D = K(V - P) \tag{16}$$

where D is the demand in units sold per unit time, K is the absolute elasticity of demand, V is the product value and P is the sales price.

In a competitive market with $i=1, 2, \ldots, N$ competitors demand increases, as well as price and value differences introduced by one competitor will affect others. According to Cook's S-Model the demand of the i-th product relative to a cartel reference state is captured by the following expression:

$$D_i = K\left[(V_i - P_i) - \frac{1}{N}\sum_{j \neq i}(V_j - P_j)\right] \tag{17}$$

where the summation is over all N competitors in the segment. The coefficient K is the negative of the slope of the demand curve at the reference state. The value of the product in the reference state is equal to the price at which demand for the product, as given by the extrapolation of the linear portion of the demand curve through the reference state, goes to zero. The coefficient K should be set to its value at the reference state, namely:

$$K = F_1\frac{D_R}{P_R} \approx \frac{E_1\bar{D}}{P} \tag{18}$$

where $\bar{P}$ is the average price for the segment and $\bar{D}$ is the average demand for competing products. The price elasticity of demand E_1 is formally defined as:

$$E_1 \equiv \frac{-\delta D_i / D_R}{\delta P_i / P_R} = \frac{-\delta D_i / \bar{D}}{\delta P_i / \bar{P}} = \frac{KP_i}{D_i} \tag{19}$$

when only product i changes price. We assume this elasticity to be the same for other products in the segment. The change in demand for product i when *all* products in the segment are increased in price by δP gives rise to another price elasticity defined by:

$$E_2 \equiv \frac{-\delta D_i / D_i}{\delta P_i / P_i} \tag{20}$$

which is assumed to be independent of the type of product i. Because the elasticity E_2 is independent of the number of competitors, it is more fundamental and is related to E_1 by:

$$E_1 = NE_2 \tag{21}$$

Similar to Cook we will only consider the top seven selling vehicles in each segment, and we select $N=7$. In the medium market segment (see Table 12-A2) the top seven selling vehicles (Honda Accord through Nissan Altima) accounted for only 7/31 (23%) of the models (variants) available for purchase in 2002, but were responsible for 1.87/3.80 million vehicles (49.4%) sold in the MED segment. We therefore restrict our considerations of this (and the other market) segment to only the top seven sellers. A frequent difficulty lies in estimating a reliable value of either E_2 or K without detailed microeconomic data about particular markets. The approach followed here is to set $E_2=1$, which yields the result, after substitution into Eqs. (21), (17) and (18), that

$$\overline{V} = \overline{P}\left(\frac{1+E_2}{E_2}\right) = 2 \tag{22}$$

making product value twice the product price, on average. Using the system of N linear Eq. (17) for demand we can then solve for the "revealed" value in terms of actual demands and prices occurring within a market segment:

$$V_i = P_i + \frac{N\overline{P}[m_i + 1]}{[N+1]E_2} \tag{23}$$

where:

$$m_i = \frac{D_i}{\displaystyle\sum_{j=1}^{N} D_j} = \frac{D_i}{D_T} \tag{24}$$

is the market share of the i-th product, i.e., the fraction of demand for product i relative to the total demand for the N competing products in the market segment. Equation (23) and the data in the columns labeled "D" and "P" in Table 12-A2 give us the opportunity to estimate the market revealed value of the N products in this market segment, see Table 12-4.

Table 12-4. Market revealed value estimation for medium car segment (MED).

i	Brand – Model	D_i Table 12-A2	m_i Eq. (24)	P_i Table 12-A2	V_i Eq. (23)	V_i/P_i
1	Honda Accord	414,718	0.22	18,890	38,547	2.041
2	Toyota Camry	390,449	0.21	18,970	38,419	2.025
3	Ford Taurus	353,560	0.19	19,035	38,167	2.005
4	Chevrolet Impala	208,395	0.11	20,325	38,210	1.880
5	Pontiac Grand Am	182,046	0.10	17,135	34,794	2.031
6	Chevrolet Malibu	176,583	0.09	17,760	35,372	1.992
7	Nissan Altima	148,345	0.08	16,649	34,019	2.043
	Average	267,728	0.14	18,395	36,790	2.0

At this point we discussed two different ways to predict product value. Eq. (23) predicts the value of a product as it has been revealed in the marketplace by consumer purchasing patterns. In Eq. (14), we stipulated a multiplicative model of product value as an aggregate of functional (and soft) attributes. Both values have to be equivalent. This leads us to solve for the optimal least-squares weighting coefficients, γ, that will minimize the difference between the demand predicted from the aggregate value (see Eq. 14) and the demand actually revealed by the market (see Eq. 23):

$$\text{find } \gamma_j^*, j = 1, 2, ..., m$$

$$\text{s.t. } \min \varepsilon_{RSS} = \sum_{i=1}^{N} \left(D(V_i)_{market} - D(V_o v(f_1)...v(f_m))_{predicted} \right)^2 \qquad (25)$$

This procedure was followed for the MED market segment and the resulting optimal weighting coefficients are shown in Table 12-5.

Table 12-5. Optimal weighting coefficients γ^* to predict product value and demand[8].

f_1=PV	f_2=CV	f_3=TC	f_4=FE	f_5=AC	s_1=MQ	s_2=CO	s_3=ST	s_4=MD	s_5=SE
0.560	0.030	0.033	0.030	0.03	0.261	0.061	0.134	0.030	0.641

Figure 12-12 shows the resulting value curves for the MED car segment, where the optimal γ^*'s from Table 12-5 and the critical, ideal and baseline values from Table 12-3 have been used. These are indicative of consumer preferences in this particular segment. The magnitude of the exponents alone

---

[8] The lower bound for γ was set to 0.03, the upper bound to 1.0. These bounds correspond to lower and upper values of the exponent γ found in Cook (1997, Table 5.3).

is somewhat misleading in understanding the 'most important' attributes, since the shapes of the value curves are also influenced by the positions of the critical, ideal and baseline values for each attribute.

From Table 12-5 and Figure 12-12 we learn that this segment appears to be very sensitive to passenger volume (f_1), initial mechanical quality (f_6) and service experience (f_{10}) and somewhat sensitive to styling and comfort. The demand predicted by the value model and the actual demand match quite closely as is seen in Figure 12-13. Because some exponents are at their lower bounds, a small correction factor for $K=K*K_{corr}$, with $K_{corr}=1.046$ has to be introduced to maintain a constant size of the predicted market segment.

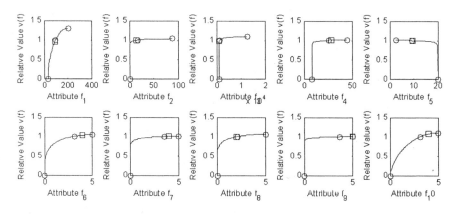

Figure 12-12. Market-revealed value curves for all m=10 attributes in the 'MED' segment. Attributes s1-5 are treated as f6-10 during analysis.

The benefit of the value and demand models is that demand can be predicted as a function of perturbations to the functional and soft attributes. Such changes are introduced indirectly via manipulation of the design variables associated with various product design and platform strategies. These physical changes in the product can now be propagated to estimate changes in demand across multiple market segments. This obviates the need for introducing artificial concepts such as "acceptable performance losses" when comparing platform design alternatives.

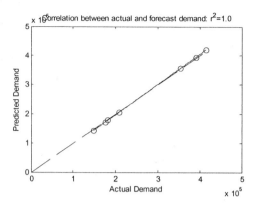

Figure 12-13. Correlation between forecasted and actual demand for top 7 sellers in MED market segment. Demand prediction made with Eq. (17).

3.6 Manufacturing cost

The main function of the manufacturing cost model is to estimate the total cost of manufacturing the product. The total cost of manufacturing D_i instances of the i-th product in a unit time period is given as (Cook, 1997):

$$CT_i = D_i C_i + F_i \tag{26}$$

where D_i is the demand (sales volume) discussed in the previous section, C_i is the variable (per unit) cost, referred to as "marginal costs", and F_i is the fixed cost. The relationship between operating revenues, $P_i D_i$, total cost CT_i and operating earnings (pre-tax profit), A_i, is shown in Figure 12-14.

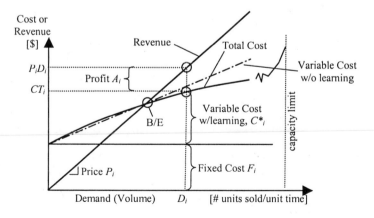

Figure 12-14. B/E Diagram: Cost and Revenue as a function of demand (volume).

We can rewrite the operating profit of the i-th product from Eq. (1) as:

$$A_i = \alpha_i P_i D_i \tag{27}$$

where α is the operating (profit) margin. Thus, total costs are estimated as:

$$CT_i = [1 - \alpha_i] P_i D_i \tag{28}$$

The next important piece of information is the operating leverage, ϕ, i.e., the relative split between fixed and variable costs, see Eq. (26). We can approximate this as:

$$F_i = \phi_i CT_i = \phi_i [1 - \alpha_i] P_i D_i \tag{29}$$

Since building cars and trucks is a capital intensive business we expect the average operating leverage to be at least $\phi_i = 0.3$.[9] This yields an estimate of the total variable cost at the nominal production operating point, D_i of:

$$C_i = (1 - \phi_i)(1 - \alpha_i) P_i D_i \tag{30}$$

Similarly, the per-unit variable cost is:

$$c_i = (1 - \phi_i)(1 - \alpha_i) P_i \tag{31}$$

Modeling of the variable costs of a product is typically done by one of the following three methods:

1. Bottom-up process oriented: model individual fabrication and assembly steps
2. Cost-estimation-relationships (CER's): fit regression curves to historical cost data of precursor products/systems.
3. Costing-by-analogy: take a known product and its cost as a reference baseline and calculate differential costs with respect to the baseline by adjusting for changes in design variables or product options.

We will pursue the third approach, assuming knowledge of α and ϕ. Moreover, an important reason for platforming is the promise of being able to achieve a significant learning curve. This learning curve manifests itself as progressively decreasing variable per unit cost, shown by the flattening

---

[9] Later we discuss a sensitivity analysis with respect to profit margin α, operating leverage ϕ, and the cost breakdown coefficients, β_{ik}.

total cost curve in Figure 12-14. Thus, we rewrite the variable cost per unit as a function of production volume as:

$$c_i(D_i) = c_{tfu,i} \cdot D_i^B = \sum_{k=1}^{n_{mod}} c_{tfu,i,k} \cdot D_i^B \qquad (32)$$

$$B = \log S / \log 2 \qquad (33)$$

where S is the learning curve factor. The factor S represents the average relative variable cost of a unit, each time the number of production units is doubled. Typically, S is on the order of 0.9 for (parts) fabrication, 0.75 for assembly and 0.98 for material expenditures. Overall one often assumes $0.8 < S < 1.0$. The variable $c_{tfu,i,k}$ is the theoretical first unit cost of the k-th module of the i-th product, where n_{mod} is the number of modules in the product. One can first solve for the theoretical first unit cost of the entire product by using Eqs. (31)-(32) and rearranging

$$c_{tfu,i} = c_i(D_i) \cdot D_i^{-B} = (1 - \phi_i)(1 - \alpha_i) P_i D_i^{-B} \qquad (34)$$

The variable cost breakdown for the main modules of a vehicle (see Table 12-6) is assumed as follows:

$$c_{tfu,i,k} = \beta_{ik} c_{tfu,i}, \quad \text{where } k = 1, 2, ..., n_{mod} \qquad (35)$$

Table 12-6. Assumed cost breakdown coefficients β_{ik}.

i	Market Segment	β_{i1} Powertrain	β_{i2} Chassis	β_{i3} Body	β_{i4} Wheels	Total
1	SMP	0.35	0.30	0.25	0.10	1.0
2	MED	0.30	0.35	0.30	0.05	1.0
3	LGP	0.25	0.35	0.35	0.05	1.0
4	SPT	0.40	0.30	0.22	0.08	1.0
5	SUV	0.30	0.35	0.30	0.05	1.0
6	VAN	0.25	0.30	0.40	0.05	1.0
7	TRK	0.35	0.4	0.20	0.05	1.0

An estimate of the average, theoretical first unit costs of each module, in each product, in each market segment can therefore be obtained—using market segment averages—as:

$$c_{tfu,i,k} = \bar{\beta}_{ik} (1 - \bar{\phi})(1 - \bar{\alpha}) \bar{P} D_i^{-B}, \quad \text{where } k = 1, 2, ..., n_{mod} \qquad (36)$$

For the medium passenger car segment (see Appendix A) with $D_i=$ 267,728 (market segment average sales volume among top 7 selling vehicles) we arrive at:

Table 12-7. Theoretical first unit cost calculations for 'MED' market segment (i=2).

S	$\bar{\phi}$	$\bar{\alpha}$	$\bar{P}$	$c_{tfu,2,1}$ Powertrain	$c_{tfu,2,2}$ Chassis	$c_{tfu,2,3}$ Body	$c_{tfu,2,4}$ Wheels
0.95	0.3	0.1	$18,395	$8,766	$10,227	$8,766	$1,461

This results in a theoretical first unit cost per vehicle of $c_{tfu,2}=$ \$29,221, according to Eq. (34). The cost breakdown shown in Table 12-7 is only valid for an average product whose design variables all have the average settings of the first $N=7$ competitors. Because variable cost is not only affected by production quantity, but also by other factors such as the amount of material used, we scale the theoretical first unit cost of each module with respect to the average design variable settings as follows:

$$c'_{tfu,i,1} = c_{tfu,i,1} \cdot \frac{x_{i,2}}{\overline{x}_2}$$

$$c'_{tfu,i,2} = c_{tfu,i,2} \cdot \frac{x_{i,3}x_{i,4}}{\overline{x}_3\overline{x}_4}$$

$$c'_{tfu,i,3} = c_{tfu,i,3} \cdot \frac{x_{i,3}x_{i,6}x_{i,7}}{\overline{x}_3\overline{x}_6\overline{x}_7} \tag{37}$$

$$c'_{tfu,i,4} = c_{tfu,i,4} \cdot \frac{x_{i,8}x_{i,9}}{\overline{x}_8\overline{x}_9}$$

The variable powertrain cost scales linearly with engine displacement[10], the variable chassis cost scales linearly with footprint area (WB times WT), the variable cost of manufacturing the body scales with enveloping volume and the variable wheel costs scale with wheel diameter and tire width.

Given this information, one may now estimate the total variable cost of a vehicle product (C'_i), given its market segment (*i*), production volume (D_i) and design variable settings ($\mathbf{x}_i$) as:

$$C'_i(D_i,\mathbf{x}_i) = c'_{tfu,i} \cdot D_i^B = \sum_{k=1}^{n_{mod}} c'_{tfu,i,k}(\mathbf{x}_i) \cdot D_i^B \tag{38}$$

---

[10] In a more detailed cost model one would want to capture whether or not the engine is naturally aspirated or boosted (turbo charged).

The total cost of production is then:

$$CT_i = D_iC'_i + F_i \qquad (39)$$

The per unit variable costs, C', fixed costs, F, operating profit, A, and margin, α, were simulated for the $N=7$ competitors in the MED market segment in this fashion (see Table 12-8). The same assumptions were made as in Table 12-7. In particular we assume that a hypothetical manufacturer who sells an average vehicle and achieves an average sales volume in this segment achieves a profit margin of $\alpha=10\%$, which includes the 2% average margin of the dealers in the sales network (Harris, 2002)[11].

The results predict that 4 out of 7 models will generate a profit, under the assumption that the fixed cost, F, is the same to produce each model[12]. Also, we see that a large production volume (assuming the product is actually sold at price P) is beneficial in two ways: (i) it generates significant revenue, and (ii) the per-unit cost C' is lowered due to the learning curve, Eq. (32). Figure 12-15 shows where the 7 competitors fall on the break-even diagram corresponding to the hypothetical average manufacturer. It is interesting to note that the actual average profit margin, α, among the seven models is only 4.2%, which is lower than the 10% stipulated for an average manufacturer. This difference is explained by the non-linearity introduced through the learning curve and the uneven distribution of sales volumes in the market segment.

Table 12-8. Manufacturing cost simulation for 'MED' market segment.

j	Model	P [$]	D	F [$10^9$$]	C' [$]	A [$10^9$$]	α
1	Honda Accord	18,890	414,718	1.3297	10,407	2.1883	0.28
2	Toyota Camry	18,970	390,449	1.3297	10,861	1.8363	0.25
3	Ford Taurus	19,035	353,560	1.3297	11,984	1.1632	0.17
4	Chevrolet Impala	20,325	208,395	1.3297	13,323	0.1485	0.04
5	Pontiac Grand Am	17,135	182,046	1.3297	10,834	-0.1826	-0.06
6	Chevrolet Malibu	17,760	176,583	1.3297	12,276	-0.3614	-0.12
7	Nissan Altima	16,649	148,345	1.3297	12,086	-0.6528	-0.26

3.7 Investment finance

In order to compare different platforming strategies with each other one must take into account capital investments, costs (both fixed and variable) as well revenues from sales. The typical cash flow profile of a generic product life cycle is shown in Figure 12-16.

---

[11] Recent profit margins in the automotive industry have fallen significantly below 10%.
[12] The fixed cost is mainly due to recurring, but volume independent charges related to manufacturing plants, non-hourly labor as well as equipment, tooling and facilities.

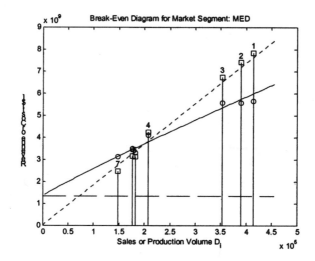

Figure 12-15. B/E diagram for medium car segment: '□'= revenue, 'o'= cost. Model numbers (1-7) correspond to the top seven selling vehicles shown in Table 12-A2.

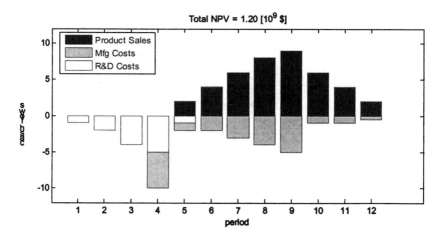

Figure 12-16. Typical product cash flow profile over a 12-year life cycle, units [$B]. Product launch is assumed to occur in year 5, with tooling being procured in year 4.

A number of periods, N_{RD}, is spent developing the product, which includes component and module design, integration, prototype manufacturing as well as laboratory and field testing. It used to take at least 48 months (N_{RD}=4) to develop a completely new vehicle in the automotive industry, but this number has been dropping to below 36 months and is approaching 24 months thanks to more efficient CAD/CAE/CAM processes, increasing design reuse and platforming. Typical R&D budgets of automotive companies, M_{RD}, range between 2-6% of yearly sales revenues.

After product launch, the sales revenues from a product, D_iP_i, usually ramp up at some rate, peak after a number of periods and then start to decline as the product gradually loses ground to newly introduced competitors. Manufacturing costs are incurred for tooling and plant equipment, fixed expenditures as well as variable costs such as labor and material. The variable costs can be affected by the learning curves (see Eq. 32). It is usually advantageous to separate out the investment required in upkeep and modernization of plants and equipment, M_P, as a percentage of annual sales. As we will see later, the development of one or multiple product platforms can be seen as such an investment, whose primary target is to lower variable costs.

The annual forecasted profit for the j-th year and the i-th product can then be estimated as:

$$A_{i,j} = D_{i,j}\left[P_{i,j} - C'_{i,j}\right] - F_{i,j} - M_{RD,i,j} - M_{P,i,j} \tag{40}$$

Demand $D_{i,j}$ is predicted from Eq.(17), price $P_{i,j}$ is chosen as a free variable based on pricing strategy, the variable costs $C'_{i,j}$ are computed using Eq. (38), fixed costs, $F_{i,j}$, are estimated from Eq. (29) and R&D costs, $M_{RD,i,j}$, and plant investment costs, $M_{P,i,j}$, are percentages of sales (e.g., 5% for R&D). We assume N_{RD} periods for development and N_P periods of production. The merits of a particular product realization program can be assessed by its net present value (NPV), also referred to as net present worth (Cook 1997):

$$\Pi_i = \sum_{j=1}^{N_{RD}+N_P} \frac{A_{i,j}}{(1+r)^j} \tag{41}$$

where r is the discount rate which captures the time value of money, at a minimum the risk-free interest rate.[13] The NPV analysis is the final step in the product framework shown in Figure 12-3. The predictions for yearly profit, $A_{i,j}$, operating margin α_j and the expected NPV, Π_i, of new investments (i.e., new product development, new platforms, etc.) are the ultimate decision metric of the for-profit firm and its shareholders.

An NPV analysis was conducted for the market leader in the medium ('MED') passenger car segment and the resulting (non-discounted) cash flow profile is shown in Figure 12-17. From a comparison of Figure 12-16 and Figure 12-17 one can see that a number of simplifying assumptions were

---

[13] An example of a "risk free interest rate" is the interest yielded by U.S. government bonds for the period equivalent to the total duration of the project or product life cycle.

made. The majority of these relate to ignoring (unknown) future market dynamics:

- The R&D expenditure profile is flat and the total R&D budget is equivalent to $M_{RD} = m_{RD} N_P P_i D_i$, spread out over N_{RD} = 3 years.
- The sales volume D_i predicted by Eq. (17) remains valid for the N_P =8 years of production, which implies time invariance of prices (P), customer preferences (γ), and technologies (*f(x)*).
- The fixed costs, F_i, and variable costs, C'_i, are also assumed to be time invariant.

We accept these assumptions for now and will discuss the potential impact of market dynamics later. The simulated cash flow profile for the market leader of the ('MED') segment is shown in Figure 12-17, predicting a NPV of $5.62 billion over 11 years.

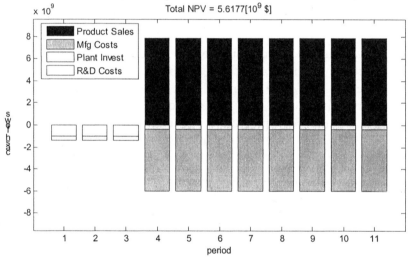

Figure 12-17. Hypothetical NPV analysis for market leader (see Appendix A) with N_{RD}=3, N_P=8, r=0.06, m_{RD}=0.049 (PSA 2004), m_P=0.068 (PSA 2004) and sales and manufacturing cost estimates shown in Table 12-8.

3.8   Modeling accuracy

With the six sub-models of our modeling framework (see Figure 12-3) completed, we can assess its accuracy by benchmarking against the *N*=7 best selling products of the MED passenger car market. For each product we input the actual design vector, **x**, and predict the functional attributes, **f**, the product value, V_i, demand, D_i, profit, A_i, and operating margin, α_i. Prediction errors according to the root-mean square (RMS) metric defined in Eq. (5) can be used as a measure of accuracy (see Table 12-9).

Table 12-9: Model predictions and accuracy for 'MED' segment.

j	Model	P [$]	f: ε_{RMS} (%)	D actua	D predict	m_D: (%)	A [$10^9$$]	α
1	Honda Accord	18,890	3.41	414,718	517,397	0.35	2.1883	0.28
2	Toyota Camry	18,970	13.95	390,449	298,486	-7.87	1.8363	0.25
3	Ford Taurus	19,035	2.28	353,560	301,825	-5.75	1.1632	0.17
4	Chevrolet Impala	20,325	12.23	208,395	315,252	2.58	0.1485	0.04
5	Pontiac Grand Am	17,135	0.13	182,046	217,998	-0.24	-0.1826	-0.06
6	Chevrolet Malibu	17,760	19.70	176,583	374,956	6.87	-0.3614	-0.12
7	Nissan Altima	16,649	23.06	148,345	275,882	4.07	-0.6528	-0.26
	Average	18,395	10.68	267728	328828	3.96	0.591	0.043

We see that, on average, the model predicts functional attributes within 10.68% for the first $N=7$ competitors in this market segment, given only knowledge of the design vector, **x**, for each vehicle. This confirms that the engineering performance model developed in Section 3.3, using response surface models (RSM), is accurate within 6-11%. The largest error is incurred in predicting towing capacity (TC), but Figure 12-12 shows that value is relatively insensitive to this functional attribute, at least in the MED market segment. The demand, *D*, in terms of actual numbers and predictions is shown in columns 5 and 6 of Table 12-9, respectively. Note that total market size is not preserved when we add up the predictions in column 6, because the vehicles were evaluated using Eq. (17) one-at-a-time and not concurrently. The corresponding prediction errors in % market share, m_i, see Eq. (24), are shown in column 7. While errors in predicting the absolute number of units sold per unit time can be substantial, we find that market share is predicted with an average accuracy of 3.96% for this market segment. It is difficult to compare predictions of profit on a model-by-model basis because most firms only present aggregate data in their annual reports and do not disclose their profit for individual vehicles, either for accounting or tactical reasons. We therefore use this model for further analysis.

3.9 Single product optimization

The question one might care to answer at this point is the following: What is the design of the "optimal" vehicle in the MED car market segment that will theoretically maximize NPV? The designers and managers of a single vehicle (without platforming) can choose the following quantities, assuming that the vehicle architecture (see Figure 12-4) remains fixed:

- Design variables: $x_{i,1} - x_{i,9}$
- Soft attributes: $s_{i,1} - s_{i,5}$
- Price: P_i

The single vehicle design optimization problem for *NPV* maximization then becomes:

$$\max \Pi_i\left(\mathbf{x}, \mathbf{s}, P_i\right)$$

$$s.t. \quad x_{i,j,L} \le x_{i,j} \le x_{i,j,U} \quad j = 1, 2, \dots, 9$$

$$1 \le s_{i,j} \le 5 \quad j = 1, 2, \dots, 5 \tag{42}$$

$$P_{\min,i} \le P_i \le P_{\max,i}$$

As upper and lower bounds of the design variables we choose the minimum and maximum values encountered in the (*N*=7) market segment (see Table 12-A1). The price is also bounded below and above by the maximum and minimum price encountered in the market segment. These restrictions are necessary because, as stated above, the value and demand models are only valid in the vicinity of the cartel reference state (Cook, 1997). Furthermore, it is assumed that the product we are designing is introduced as a direct competitor to the other *N*=7 products, which remain fixed at their current settings.

Equation (42) is solved using a gradient-based sequential quadratic programming (SQP) algorithm[14]. The algorithm is provided with the settings of the current market leader as the starting point (see Table 12-10).

Table 12-10. Single vehicle design (initial guess – top), optimal design (bottom).

Initial guess				
Po=$18,890	Xo(1)=17.1 g	Xo(2)=2254 ccm	Xo(3)=61,2"	Xo(4)=105.1"
Xo(5)=3.9"	Xo(6)=54.9"	Xo(7)=186.8"	Xo(8)=195 mm	Xo(9)=15"
So(1)=4	So(2)=4	So(3)=2	So(4)=5	So(5)=4
Optimal design				
P*=**$20,325**	X*(1)=**20.0 g**	X*(2)=**2189 ccm**	X*(3)=**59.0"**	X*(4)=105.1"
X*(5)=3.9"	X*(6)=**57.9"**	X*(7)=**200"**	X*(8)=195 mm	X*(9)=15"
S*(1)=**5**	S*(2)=**5**	S*(3)=**5**	S*(4)=5	S*(5)=**5**

The differences between the "optimal" design[15] (bottom) and the current market leader (top) are highlighted in bold in Table 12-10. First, as expected, all the soft attributes are set to their highest level since there is nothing in the model that would trade against this change. The optimized design has a somewhat larger fuel capacity (FC), smaller engine (ED), smaller wheel track (WT), but is higher (HT) and longer (LT) than the current leader. These changes can be explained by the strong sensitivity of the medium segment to passenger volume (see Table 12-5) relative to the other functional attributes. Without bounds on the design variables the optimizer

---

[14] MATLAB V7.0.1 (R14) is used as the modeling environment, with the function *fmincon* serving as the SQP constrained optimization algorithm.

[15] Due to non-convexity of the problem, we cannot claim global optimality.

would attempt to further increase the gap between value and price in Eq. (17) by increasing the size of the vehicle further. This is prevented by the bounds, as otherwise the vehicle specifications would start to place it in the next larger car segment. This hypothetical vehicle is predicted to achieve an NPV of Π^*=60.8 \$B over 11 years compared to an estimated NPV of Π_o=9.2 \$B for the initial guess over the same period. How can this large difference be explained?

Table 12-11 shows a comparison of the initial vehicle and the optimized vehicle in terms of functional attributes, **f**, and value constituents, **v**. The optimized vehicle achieves a value of 1.334, relative to the market segment average, resulting in a value of V^*=\$49,078 and a predicted sales volume of D^*=1.628 million vehicles. The current market leader on the other hand achieves a relative value of 1.047 with an absolute value of V_o=\$38,520 and a predicted sales volume of D_o=0.517 million units. Moreover, we notice that both the soft attributes (1.218 vs. 1.103) as well as the functional attributes (1.095 vs. 0.949) contribute to the relative value increase of the optimized vehicle relative to the initial guess. While it might not be realistic to expect a 5-point J.D. Power rating in all the soft attributes in practice, we will allow for this possibility during subsequent platform optimization.

Table 12-11. Comparison of initial vehicle design (x$_o$) and optimized design (x*).

	f_1=PV	f_2=CV	f_3=TC	f_4=FE	f_5=AC	s_1=MQ	s_2=CO	s_3=ST	s_4=MD	s_5=SE
γ_i	0.560	0.030	0.033	0.030	0.03	0.261	0.061	0.134	0.030	0.641
f_o	94.4	14.0	1077.1	25.7	9.9	4	4	2	5	4
v_o	0.966	0.996	0.987	1.000	0.999	1.029	1.003	0.997	1.002	1.071
f^*	110.1	19.1	1882.9	25.0	10.7	5	5	5	5	5
v^*	1.061	1.004	1.032	0.999	0.997	1.040	1.005	1.058	1.002	1.100

4. SINGLE PLATFORM OPTIMIZATION

In the previous section we discussed the single product framework shown in Figure 12-3, developed the underlying mathematics for each of the six sub-models, and exercised the framework with help of an automotive database (see Appendix A) for the medium passenger car market segment. So far, however, we have taken only a single product view. A manufacturing firm wanting to satisfy a larger number of customers whose preferences (both in terms of functional and soft attributes) do not cluster tightly will attempt to satisfy demand with a family of products. Therefore, we first develop an understanding of clustering of design variables, functional attributes, and value-vs-price for all seven market segments.

Next, we ask the question how this variety of preferences might be satisfied by a new entrant who wishes *to compete with exactly one model in each market segment*. There are two fundamental strategies in this respect.

First, one may want to treat each product independently of the others such that its characteristics and manufacturing processes may be uniquely tailored. Second, one may build them from one (or more) platforms by reusing common modules or scaling the product in one or more physical or functional dimensions. We first establish a baseline by designing a new product for each market segment, without a platform. Then, we select one of the modules as the platform and optimize the product family based on it.

4.1   Multiple market segment view

A database of six additional market segments each with its top selling, N=7, competitors has been developed. Table 12-C1 in Appendix C shows an overview of average, minimum and maximum values for the design variables and dependent variables in each market segment. Table 12-12 shows demand elasticity, total market segment size, average price and average value for the top seven vehicles in the given market segment.

Table 12-12. Comparison of seven market segments (only top 7 products).

j	Market Segment	K elasticity	D_T actual	$\overline{P}$ [$]	$\overline{V}$ [$]
1	SML	113.99	1,459,000	12,769	25,539
2	MED	106.60	1,874,096	18,395	36,790
3	LRG	19.09	632,894	29,928	59,856
4	SPT	10.46	382,473	21,330	42,661
5	VAN	39.97	918,193	23,197	46,394
6	SUV	60.05	1,488,058	26,492	52,983
7	TRK	133.06	2,778,964	16,587	33,173

The value-vs.-price position of the top seven selling vehicles in each segment is shown in Figure 12-18. It is interesting to note that the small sedans (SML) are all clustered together in the lower left corner, followed by the medium sedans (MED) along the diagonal to the upper right. The large sedans are typically in the $22,000-$30,000 price range with the exception of two outliers (Cadillac Deville, Lincoln Towncar) in the upper right of Figure 12-18. These two vehicles would probably be broken out into a luxury sedan segment if the market segmentation were done at a finer level of granularity. The outlier in the sports segment is the Chevrolet Corvette. The seven top-selling SUVs do not cluster nicely but span a relatively large price-value range. The (pickup) trucks fill the gap between the small and medium sedans in the price-value space.

How many platforms would be needed to span the space shown in Figure 12-18 if the manufacturer decided to compete with one model in each market segment? The first step is to optimize a vehicle for each segment (as done in Section 3.9) without any platforming commonality constraints.

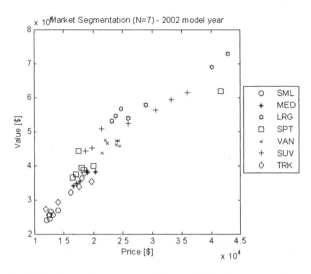

Figure 12-18. Position of seven top-selling vehicles in seven market segments.

4.2   Baseline Case 1: no platforming

This subsection establishes a baseline by computing the "optimal" vehicle design that maximizes NPV (see Eq. 41) in each market segment, independently. This optimization is done by assuming that all design variables can be chosen independently of other market segments, constrained only by the bounds given within each segment (see Table 12-C1). This therefore reflects the "null platform" case. Or said otherwise, each product is its own platform ($p=m$) and there is no explicit reuse between segments.

Table 12-13 shows the initial product family (current market leaders) on top and optimized product family on the bottom. Simulated Annealing (SA) was used as the optimization algorithm[16]. The bounds on the design variables **x** were set by the minimum and maximum occurrences in the product database for each market segment (see Table 12-C1).

The optimization suggests that a new market entrant could hypothetically develop a new product family that exceeds the sales volume (D) and net present value (NPV) of the product family comprised of the current market leaders. The optimized product family (without platforming) is predicted to achieve a sales volume of 13.2 million vehicles versus 8.8 million vehicles for the current family, an increase of 50.3%. The NPV over an 11 year program life (see Figure 12-17) increases from an estimated current 242.00 $B to 764.8 $B (+241.4%).

---

[16] The optimal MED vehicle settings in Table 12-10 and Table 12-14 (6th row from bottom) are slightly different because SQP was used in the former case, while SA was the optimizer used in the latter case.

Table 12-13. Initial (o) versus optimized product variants without platforming.

	Segment	P [$]	V [$]	D [units]	NPV [$B]
Initial Product Family	SML_o	12,810	26,389	354,650	1.92
	MED_o	18,890	42,531	1,006,100	26.98
	LRG_o	24,660	63,162	391,120	7.13
	SPT_o	17,475	72,194	440,810	6.21
	VAN_o	21,980	48,743	304,490	5.03
	SUV_o	21,355	71,430	2,307,000	61.04
	TRK_o	18,540	58,564	3,993,200	133.69
	Total			8,797,370	242.0
Optimized Product Family (no platform)	SML*	12,609	27,781	562,217	6.04
	MED*	20,270	49,047	1,631,753	57.68
	LRG*	38,043	80,795	483,815	40.83
	SPT*	32,768	72,974	267,309	22.24
	VAN*	22,907	51,707	397,535	9.79
	SUV*	35,658	103,215	3,506,711	325.21
	TRK*	19,569	75,294	6,380,642	303.01
	Total			13,229,982	764.8

How could this be achieved? There are three quantities that the firm can affect directly in tailoring their product offerings: (1) the design variables for each vehicle, x_i, the values of the soft attributes, s_i, and the price, P_i. From Table 12-13 we see that the optimizer chose to lower the price for one vehicle (SML), while increasing the price moderately for some vehicles (MED, VAN, TRK) and significantly for others. In most cases these changes are justified by also increasing the value of the vehicle. All soft attributes, s_i, have been set to their maximum value of 5. Recall that the optimizer attempts to maximize the NPV for each vehicle, independently of all other vehicles, based on Eq. (41). This equation takes into account changes throughout the product framework (see Figure 12-3) that affect the NPV.

Table 12-14 shows how the optimizer chose to perturb the design of each vehicle from the initial guess to the optimized design. In some cases a vehicle was downsized (e.g., SML car) in terms of its wheelbase (WB) and wheel track (WT) in other cases the vehicle was significantly stretched (e.g., SUV). These changes can only be understood by considering the values of the γ-weighting factors for each vehicle in each market segment. A detailed discussion of these weighting factors is possible, but beyond the scope of this chapter.

As we can see from Table 12-14, each vehicle has its own, freely chosen design variables and is not constrained by platforming in any way. It is interesting to note that the engine was downsized for most vehicles (except SML and MED), presumably due to the benefits of better fuel economy. We will take the values of the design variables in the bottom half of Table 12-14 as the baseline product family, without platforming. The financial performance of this product family is estimated at an aggregate NPV of $B 764.8 over 11 years.

Table 12-14. Original vehicle designs (top), optimized designs (bottom).

Segment		x(1) FC [g]	x(2) ED [ccm]	x(3) WT" [in]	x(4) WB [in]	x(5) GC [in]	x(6) HT [in]	x(7) LT [in]	x(8) TW [mm]	x(9) TD [in]
Initial Product	SML_0	13.2	1668	57.9	103.1	5.27	55.1	174.7	185	14
Family	MED_0	17.1	2254	61.2	105.1	3.90	54.9	186.8	195	15
	LRG_0	18.5	3790	62.3	112.2	5.40	57.0	200.0	215	15
	SPT_0	15.7	3802	60.2	101.3	4.30	53.1	183.2	225	16
	VAN_0	20.0	3301	63.0	119.3	5.60	68.9	200.5	215	15
	SUV_0	16.8	4011	58.5	101.8	6.70	68.4	180.4	235	16
	TRK_0	25.0	4196	65.4	119.9	7.30	72.7	206.9	255	16
Optimized	SML*	14.1	1720	58.8	97.0	4.9	53.0	168.1	186	14
Product Family	MED*	20.3	2421	59.0	106.1	4.9	57.9	200.3	212	15
(no platform)	LRG*	19.2	3283	61.1	108.1	5.2	56.7	210.5	214	16
	SPT*	14.5	3106	60.6	104.8	6.3	49.7	191.5	200	15
	VAN*	20.0	3076	66.5	118.0	7.3	69.1	196.9	212	15
	SUV*	15.0	2679	62.0	129.5	7.7	68.2	173.0	236	15
	TRK*	24.8	3504	62.4	104.3	6.7	67.0	207.4	236	16

4.3 Case 2: Single platform optimization

The question posed here is whether the use of a product platform could potentially improve the financial performance of the product family. If this could be achieved it would primarily be via lowering of the variable costs associated with the module which is chosen as the platform, see Eq.(38). The improvement in variable costs would be captured via the manufacturing learning curve, Eq.(32). Thus, with the designation of one module as the product platform, which is shared by at least two products, the estimation of variable costs for the product family has to be revised accordingly:

$$ C_{fam} = \sum_{i=1}^{m} D_i \cdot c_{tfu,platform} \cdot \left[\sum_{i=1}^{m} D_i \right]^{B} + \sum_{i=1}^{m} D_i \sum_{k=1}^{n_{mod}-1} c_{tfu,i,k} \cdot D_i^{B} \qquad (38b) $$

In Eq. (38b) the first term is the variable cost contributed by the platform, which is used by all variants. The learning curve benefits occur from summation of the production volumes over all m variants that reuse the platform. This is one of the main mechanisms by which the use of platforms can lower the cost of the entire product family. The second term in Eq. (38b) is contributed by all the other unique (non-common) modules which are not reused between variants.

The second mechanism by which product platforms lower manufacturing cost is by allowing multiple variants to be produced from the same platform on the same assembly line. Figure 12-19 shows how three different models follow each other on the same assembly line.

Figure 12-19. Three models on one assembly line (Daimler-Chrysler, courtesy: Magna).

This effect is captured in our model, by adjusting the fixed costs from their original formulation for the i-th product:

$$F_i = \phi_i CT_i = \phi_i [1 - \alpha_i] P_i D_i \qquad (29)$$

to a formulation where the fixed cost is estimated for the entire family.

$$F_{fam} = \left\lceil \sum_{i=1}^{m} D_i \Big/ D_{plant} \right\rceil \cdot \phi_{fam} [1 - \bar{\alpha}] \bar{P} D_{plant} \qquad (43)$$

Here the first term (in the ceiling function) estimates the number of manufacturing plants that are needed to produce the entire product family by diving the total production volume over all m variants by the plant capacity, D_{plant}, and rounding up to the next integer. Here we assume that a vehicle manufacturing plant has a yearly capacity of 250,000 vehicles. The fixed cost per plant are estimated via the average revenue generated by the plant at full capacity, $\bar{P} D_{plant}$, the average profit margin, $\bar{\alpha}$, and the operating leverage for the product family, ϕ_{fam}.

Which module should be chosen as the platform? Other chapters in this volume address the question of how a platform should be optimally chosen in depth. We will not dwell on this point, but rather adopt the traditional approach in automotive engineering, whereby the chassis (see Figure 12-6) is used as the platform. There are two distinct steps in the single platform problem: (i) choosing the module which is to be used as the platform, and (ii) choosing "optimal" values for the design variables associated with the platform. Of the three design variables associated with the chassis, we let ground clearance (GC) be adjusted freely via the suspension and wheel size, while wheelbase (WB) and track (WT) are constrained as platform variables.

Case 2: We will first assume that we have only a single platform (p=1) and that we will derive all 7 variants from it. This is implemented by forcing all variants to use the same settings for WT and WB, while allowing unique settings for all other design variables. Table 12-15 shows the settings for an initial guess of the platform, whereby the WB and WT were chosen as the (demand) weighted average of the WB and WT over all optimized variants shown in Table 12-14 (bottom):

$$WT_\alpha = x(3)_\alpha = \frac{1}{D_T} \sum_{i=1}^{m} D_i x_{3,i}^*$$

$$WB_\alpha = x(4)_\alpha = \frac{1}{D_T} \sum_{i=1}^{m} D_i x_{4,i}^*$$

(44)

All variants built from this single α-platform are assessed in terms of functional performance, value and demand as described in the single product model described in Section 3. It is assumed that the only effect that the platform has on the upper branch of the model framework (see Figure 12-3) is to constrain the value of the platform variables. The lower branch is modified in terms of variable costs and platform costs according to Eq. (38b) and Eq. (43). A tradeoff therefore exists between value (and concomitant demand) losses of individual variants in the upper branch of the framework and variable and fixed cost savings in the lower branch of the framework.

Table 12-15 shows the design variable settings for the single platform case (α), where the platform variables have been chosen – somewhat naively – according to Eq. (44). The platform is highlighted in gray.

How well would such a platform-based product family perform? We can evaluate the performance of the product family by computing the expected sales and revenue of each vehicle as before, except that we replace the "optimal" settings of variables WT and WB from Table 12-14 with the settings for the platform (gray shaded) shown in Table 12-15. We therefore expect each variant to be somewhat suboptimal compared to the optimized variants, which is expected to result in a reduction in value and sales of each variant since the price has been kept constant. On the other hand there will be a benefit through lowering the variable cost of the platform module (chassis) across the product family (see Eq. 38b) and concomitant lower fixed costs (see Eq. 43). Table 12-16 compares the performance of the optimized product family without platforming (left) to a family where all variants use the initial α-platform design (right) according to Eq. (44).

Table 12-15. Initial guess at a single vehicle platform (α).

	Segment	x(1) FC [g]	x(2) ED [ccm]	α-platform WT [in]	WB [in]	x(5) GC [in]	x(6) HT [in]	x(7) LT [in]	x(8) TW [mm]	x(9) TD [in]
α-	SML	14.1	1720	61.8	111.5	4.9	53.0	168.1	186	14
Platform-	MED	20.3	2421	61.8	111.5	4.9	57.9	200.3	212	15
based	LRG	19.2	3283	61.8	111.5	5.2	56.7	210.5	214	16
product	SPT	14.5	3106	61.8	111.5	6.3	49.7	191.5	200	15
family	VAN	20.0	3076	61.8	111.5	7.3	69.1	196.9	212	15
	SUV	15.0	2679	61.8	111.5	7.7	68.2	173.0	236	15
	TRK	24.8	3504	61.8	111.5	6.7	67.0	207.4	236	16

The result is that the use of the single platform shown in Table 12-15 leads to a 20.5% drop in NPV and a 22.8% drop in sales volume, while the estimated cost benefit of the platform is only estimated to be about 3.6%.

Table 12-16. Comparison of no-platforming with single vehicle platform (α) strategy.

	No Platforming	Single α-Platform
Total NPV (11 y)	$B 764.8	$B (577.3) 598.1
D_{fam}	13,229,982	10,208,757

Table 12-17. Demand and value comparison.

	No Platforming		Single α-Platform (Table 12-15)	
	Demand [units]	Value [$]	Demand [units]	Value [$]
SML	562,217	27,781	0	0
MED	1,631,753	49,047	1,355,984	46,784
LRG	483,815	80,795	389,732	76,482
SPT	267,309	72,974	278,151	73,880
VAN	397,535	51,707	189,251	47,148
SUV	3,506,711	103,215	2,383,818	86,853
TRK	6,380,642	75,294	5,611,820	70,238
Total	13,229,982		10,208,757	

How can this be explained? Table 12-17 shows a demand and value comparison of the product families without (left) and with platforming (right). The use of the ill-conceived initial platform has caused a loss of value for most vehicles, with the exception of the SPT car. In the case of the small sedan (SML) the model predicts that there would be no demand at all for such a car in its particular market segment.

Why is this so? Plugging in the values of the design vector for the platform-based SML vehicle in the engineering performance model (Section 3.3) leads to the functional attributes shown in Table 12-18. The critical, baseline and ideal values for this particular market segment are also shown. We see that the platform is oversized for this small vehicle and causes one of the functional attributes, $f(3)$, to fall outside the critical range which renders the vehicle value-less according to the model. One may argue whether or not a small sedan with a towing capacity of only ~475 lbs, but otherwise satisfying attributes, is indeed without value.

Table 12-18. Functional attributes and critical, ideal, and baseline values for α-platform based SML vehicle.

Values	$f(1)$ PV	$f(2)$ CV	$f(3)$ TC	$f(4)$ FE	$f(5)$ AC
Type	LIB	LIB	LIB	LIB	SIB
Units	[cft]	[cft]	[lbs]	[mpg]	[sec]
Critical values f_C	32	0	**800**	8	20
Ideal Values f_I	205	88	12,600	45	2.5
Baseline Values f_0	88.9	13.2	1,143	31.4	9.6
SML (α-based) f_i	85.9	24.9	**474.8**	28.0	10.75

Should we therefore conclude that platforming is not appropriate for this particular product family? Not necessarily. First, the α-platform has not (yet) been optimized for the product family, since we have only chosen an initial guess, Eq. (44), for the platform design. Second, one would have the opportunity to adjust the non-platform design variables in response to the platform settings.

We first consider the performance of the product family over a wide range of platform designs, by varying WB and WT over the entire range of occurrences in the vehicle database (see Table 12-C1). The narrowest vehicle has a wheel track of 54.4" (TRK), the widest one has a WT of 68.4" (SUV). The shortest wheelbase is 89.2" (SPT), the longest one is 137.10" (SUV). We accept these as the lower and upper bounds on the dimensions that any platform can take.

Figure 12-20 shows the result of a comprehensive search over platform designs between the lower and upper bounds for the two platform variables (WB and WT). The plot is shown as a contour chart. The result suggest that the best possible product family NPV that can be achieved with a single platform is $B 649.4 at platform settings of WTα*=57.9" and WBα*=104.2". The figure suggests that this "optimum" single platform design is relatively flat, but that it does not achieve as good a performance as the family of point-optimized vehicles without platforming ($B 764.8).

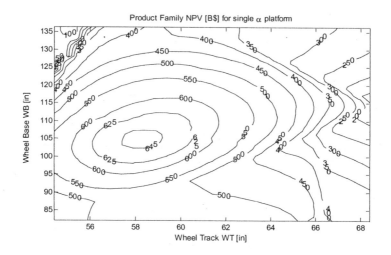

Figure 12-20. Product family NPV [$B] as a function of α-platform design (WT, WB).

Again, we might conclude that platforming is not an appropriate strategy at this point. Note however, that the individual variants have not been re-optimized for the use of this optimized platform. At this point one may enter an optimization loop where the platform and variants are optimized in turn until convergence is achieved. Several single-level and bi-level approaches have been proposed to solve this classical problem (Simpson 2003).

Instead, we focus on the features of Figure 12-20, which suggest that the decision space is not smooth and that the use of *multiple platforms* might be more appropriate. This is motivated by the fact that the "extent" of the single α-platform suggested so far might be too large. In fact, industrial practice confirms that no single automotive manufacturer offers a product family across the seven market segments (see Figure 12-18) that is built from a single platform, primarily because the requirements of the different variants would be too diverse.

5.   MULTI-PLATFORM STRATEGY

5.1   Case 3: Platform extent optimization

In this section we consider the results when multiple platforms are allowed. In this case there could be α, β, γ, ... platforms. The questions that arise are situated one level above the single platform problem. In addition to choosing the "optimal" design variable settings for each platform, we ask:

- What is the best number of platforms, p, to implement?
- What is the optimal assignment A_p of the m variants to the p platforms?

We recognize that any multi-platform strategy occupies the intermediate space between two extremes: no-platforming ($p=m$) and single-platforming ($p=1$). A number of manufacturing firms with $m>3$ product variants are using multiple platforms to support their product families. However, oftentimes the use of multiple platforms arises historically, e.g., via acquisition of other firms rather than based on systematic considerations.

Here, we formulate the multi-platform (platform extent) problem as a weighted least squares optimization problem. This is an approximation to what could be solved as a tri-level optimization problem. In a tri-level formulation the individual variants would be optimized at the lowest level, the platforms at the intermediate level and the variant-platform assignment would be solved at the highest level. Such a multi-level scheme is likely to be intractable, or at a minimum it will converge poorly. Instead, the weighted least-squares problem is first solved for the single platform case ($p=1$) and subsequently for each case of an additional platform ($p=p+1$) until we reach the case where $p=m$ (= no platforming):

find $\mathbf{x}_{\text{platform}}$, A_p such that

$$\min \sum_{i=1}^{p} \sum_{j=1}^{m} A_{p,i,j} \cdot D_j P_j \left[\sum_{k=1}^{n_{\text{xplatform}}} \left(\mathbf{x}_{\text{platform},i,k} - \mathbf{x}^*_{j,k} \right)^2 \right]^{1/2} \tag{45}$$

$$\forall p = \{1, 2, ..., m\}$$

The cost function contains the Euclidian distance between the design variable settings of the j-th optimized variant (without platform) $\mathbf{x}^*_j$ and the platform settings $\mathbf{x}_{\text{platform},i}$ of the i-th platform to which the j-th product variant has been assigned. The assignment is done via a binary assignment matrix A_p, which has p rows and m columns. For example, if variant $j=2$ is assigned to platform $i=3$, then the entry $A_p(3,2)=1$. Assuming that a product variant can only be built from a single platform, all other entries in the 2nd column of A_p will be zero. The Euclidian distance between "ideal" variant settings and the platform settings are weighted by the expected revenue (demand times price) of the variant.[17]

The result of this optimization is, for each number of allowed platforms, an "optimal" design of each platform as well as the corresponding platform-

---

[17] Note that the demand and revenue of the j-th variant in Eq. (45) is estimated based on the baseline (Case 1) without platforming.

to-variant assignment, A_p. We demonstrate this procedure for the automotive platforming problem developed in Section 4.

Figure 12-21 shows the result of optimizing a single product platform (α) with the weighted least squares objective (see Eq. 45). The hexagons labeled 1-7 in Figure 12-21 indicate the position of the ideal settings for the variants (SML through TRK) from Table 12-14 (bottom). Three different potential single platforms are shown:

-   x_{p1}^*: optimal single platform from NPV exhaustive search, Figure 12-20 (NPV= $B 649.4)

-   x_{p2}^*: optimal single platform with Simulated Annealing, Eq. (45) (NPV= $B 605.1)

-   x_{p3}^*: optimal single platform with gradient search, Eq. (45) (NPV= $B 603.2)

In the figure the variant-to-platform assignment is shown by connecting the variants to their respective platform ($\mathbf{x}_{p3}$ in this case). We note that using the approximate least squares objective metric, Eq. (45), causes a 7.1% penalty compared to using the more expensive NPV metric directly. We will accept this penalty in order to solve the multi-platform problem.

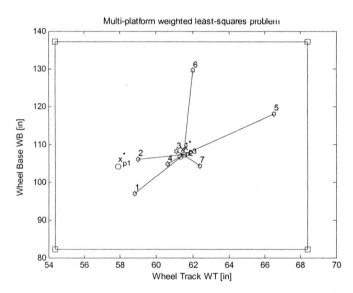

Figure 12-21. Weighted least-squares optimization of product platform (p=1).

Figure 12-22 shows a block diagram of the multi-platform (platform extent) solution method. First, it is recommended to solve the two bounding cases ($p=1$, $p=m$). This gives an initial guess for the product platform designs for the intermediate cases where $(1<p<m)$. For a given number of platforms, the platform designs are placed in the design space (see Figure 12-21) and are perturbed, e.g., with Simulated Annealing (SA) or another search algorithm. Next, the optimal assignment A_p from variants-to-platforms is found that will minimize the weighted objective function (see Eq. 45). This loop is repeated until convergence is achieved and the optimal platform designs $\mathbf{x}_{p*}$ and assignment A_p is obtained for $p= \{2, 3... m-1\}$.

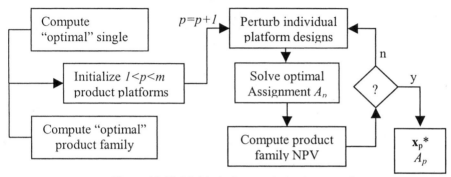

Figure 12-22. Multi-platform optimization procedure.

Figure 12-23 shows the result of solving the multi-platform problem with the above method. The predicted product family NPV is shown as a function of the number of allowed product platforms. On the left side $(p=1)$ we find the case of a single platform from Section 4.3, on the right we find the no-platforming case from Section 4.2. The latter case is identical with the multi-platform case for $m=p=7$, as each platform, automatically converges to the optimal settings for each variant. In that case each "platform" only supports a single variant.

The bar chart (see Figure 12-23) shows the average results for five runs of the simulated annealing algorithm. We can clearly see that the single-platform case shown in Figure 12-21 is the worst strategy. The results are relatively flat for $2<p<6$, primarily because the WB and WT requirements for variants 3, 4 and 7 are relatively similar, as can be seen in Figure 12-21. The best strategy appears to be case where 6 platforms are used. In that case only the MED and LRG sedans share a common platform. The strategy with $p=5$ platforms is nearly equivalent.

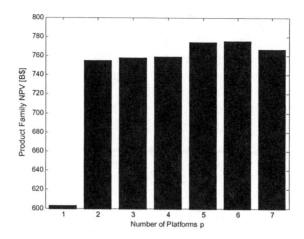

Figure 12-23. NPV as a function of the number of platforms.

Table 12-19. Optimal variant-to-platform assignment.

A_p	1-SML	2-MED	3-LRG	4-SPT	5-VAN	6-SUV	7-TRK
$p=1$	1	1	1	1	1	1	1
$p=2$	2	2	2	2	1	1	2
$p=3$	2	3	3	2	1	1	2
$p=4$	3	4	4	3	1	2	3
$p=5$	5	3	3	1	4	2	1
$p=6$	2	6	6	3	4	1	5
$p=7$	5	7	1	3	6	4	2

Perhaps more interesting than the raw NPV predictions for each platform strategy (which are subject to model uncertainty), is the suggested variant-to-platform assignment. For the cases shown in Figure 12-23 we can find the best assignment of variants to platforms (see Table 12-19):

In the single platform case ($p=1$) all variants are forced to use the same platform (1) and there is no choice, except for the choice of design variables of that single platform. When a second platform is allowed ($p=2$), the VAN and SUV are assigned a new platform that is optimized for a larger wheel track and wheel base. This is intuitive as these two particular vehicles (5-VAN, 6-SUV) can also be seen as the outliers in Figure 12-21. With $p=3$, the MED and LRG sedans share a platform, while the SML, SPT and TRK products share another platform. With $p=7$, every variant is customized and has its own "platform". A deeper understanding of variant-to-platform assignment can be gained by plotting the position of the p platforms and assignment of variants in the same graph (see Figure 12-24).

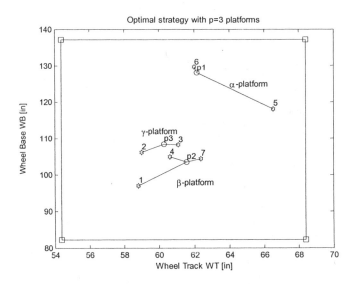

Figure 12-24. Optimal strategy with p=3 platforms.

Figure 12-24 shows that in the case of *p*=3, the α-platform would support the VAN (5) and SUV (6), but would be heavily weighted towards the SUV based on its more significant generated revenue, P_6D_6. The β-platform would support the SML (1), SPT (4) and TRK (7) vehicles and be weighted towards the TRK, since this represents one of the largest market segments (in the North American database). Finally, the γ-platform would support the MED (2) and LRG (3) sedans with a longer wheelbase, but slightly narrower wheel track than the β-platform.

5.2   Sensitivity analysis

From a decision theory standpoint one would like to know how robust the answers in Figure 12-23 are to various assumptions made throughout the modeling framework outlined in Section 3 and shown in Figure 12-3. We first list what are believed to be the key assumptions that are likely to affect the optimal multi-platform strategy (see Table 12-19). We then modify one of those assumptions to gauge its effect on the NPV results.

Table 12-20. Assumed parameters with potential impact on optimal platform strategy.

Parameter	Eq.	Potential Effect on Platforming
a_0-a_6	(6)-(13)	RSM for engineering performance can be of limited validity if a variant is built on a platform that is far from the average of the data that was used to construct the original RSM.
f_C, f_I	(15)	Critical and ideal values for the product value model can impact the valuation of a platform-derived product variant by more or less penalizing deviation from the optimal design of non-platformed variants.
N	(17)	The number of competitors considered in the demand prediction model will affect the degree to which platforming can lead to a market share penalty or benefit.
E_2	(20)	Price elasticity affects demand via $K(V\text{-}P)$
α_i	(28),(43)	The amount of operating profit (before platforming) will influence how helpful platforming is for improving NPV. Platforming is much more interesting when α_i is low, because manufacturing cost savings through platforming can make a more substantial contribution to profit.
ϕ_i	(29),(43)	The operating leverage affects the value of platforming.
S, B	(32),(33)	A strong manufacturing learning curve (S small) is expected to strongly favor platforming, because the benefits of reuse are captured in the variable manufacturing costs.
β_{ik}	(35)	The % content that the platform constitutes relative to the whole product will impact the estimated benefits of platforming. We assumed $0.3 < \beta_{i2} < 0.4$ for the chassis, but if this is substantially smaller, then the choice of the chassis as the platform would be less beneficial.
M_{RD}, M_P	(40)	The R&D budget and budget for facility and tooling upkeep and maintenance will impact platform benefits. Platforming can also impact what those budgets need to be.
r	(41)	The discount rate will affect the benefit of platforming in the future. A platform is an investment in future production capability. So if r is very large (>10%), future benefits of platforming will be washed out of the NPV calculations.
D_{plant}	(43)	Platforming is beneficial in industries where plant capacities have to be large (e.g., D_{plant}>100,000) due to economies of scale and where multiple variants can be built in the same plant. A larger D_{plant} will benefit platforming.

We investigate the effect that the learning curve factor, S, has on the optimal platform strategy. So far, we have assumed that $S=0.95$ across all fabrication and assembly processes, see Eq. (33). This is a relatively conservative value, as it assumes that doubling the production quantity will allow to lower the per-unit cost to 95% of its previous value. We set $S=0.80$ and repeat the multi-platform optimization process shown in Figure 12-22.

The results with $S=0.80$ are shown in the bar chart of Figure 12-25. We can see that the magnitude of the predicted NPV's is significantly higher than in Figure 12-23. This is not due to increased sales volumes, but rather to substantial savings in variable manufacturing costs across the family, C_{fam}, Eq. (38b). The second interesting effect is that the no-platforming strategy ($p=7$) is now clearly inferior, and one should focus on strategies where $3<p<6$. The actual choice of strategy in practice will also depend on factors that are not represented in the model. This includes dealing with legacy platforms and variants that cannot be replaced quickly, mainly due to large capital investments (e.g., tooling) that have to be amortized.

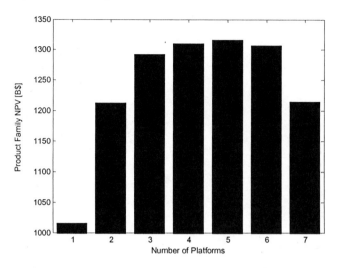

Figure 12-25. NPV as a function of the number of platforms, S=0.8.

At the beginning of the chapter we had suggested that there are four ways of viewing and interpreting the concept of "platform extent":

1. **design space view**: range of design variables **x** covered by platform, see Figure 12-24;
2. **market segmentation view**: see Figure 12-26a for β-platform from Figure 12-24;
3. **value-price view**: *V-P* space spanned by platform, see Figure 12-26b for β-platform; and
4. **functional view**: radar plot of variant functional attributes, see Figure 12-26c.

Figure 12-26 shows the extent of the β-platform from Figure 12-24. This platform was suggested by the optimized multi-platform strategy where *p*=3 platforms were utilized. Rather than drawing notional pictures (as in Figure 12-2), platform extent can now be crisply defined, computed and visualized.

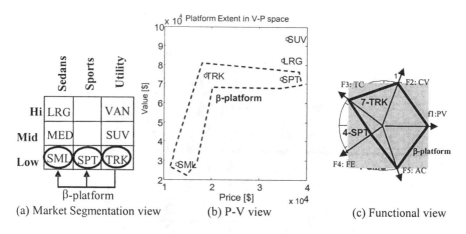

Figure 12-26. Three different views of product platform extent.

6.   DISCUSSION AND SUMMARY

6.1   Discussion

While the product framework (see Section 3), single platform case (see Section 4) and multi-platform method (see Section 5) presented an end-to-end view, it must be acknowledged that a significant number of simplifying assumptions were made. In some cases such assumptions are necessary and without significant consequences on the final answer. In other circumstances one must be more cautious when interpreting the results of such analysis. What follows is a list of aspects that need to be considered beyond what was discussed in this chapter:

-    **Pricing Strategy**: Demand (and revenue) are significantly impacted by transaction prices for individual product variants. Instead of leaving prices at their "optimal" values as was done here, they are often adjusted based on seasonal factors, competitive action or inventory levels.
-    **Iteration Loops:** Setting platform variables and variant variables is often done in an iterative fashion. So, the optimal variants shown in Table 12-14 could be re-optimized after finding the optimal single or multi-platform strategy. This variant re-optimization could in turn trigger a revision of the platform strategy.
-    **Electronics & Software:** The software and electronics content of many products (including automobiles) is constantly increasing. The product modeling framework (see Figure 12-3), however, is mainly geared

towards electro-mechanical systems. Multi-platforming for products with significant software and electronics content needs more research.

- **R&D-centric businesses**: The benefits of platforming in this chapter were mainly captured via reduction of variable (see Eq. 38b) and fixed manufacturing costs (see Eq. 43). In a number of businesses, however, the main reason for platforming is related to saving time and cost in research and development. It is still unclear how multi-platform strategies could be evaluated in that context.

- **BOF vs. BFI Architecture**: In Section 3.2, we alluded to the fact that cars and trucks are typically built from either a body-frame-integral or body-on-frame architecture. It is difficult, if not impossible, to merge vehicles of such different architectures onto the same platform, even if their "raw" design variable settings might suggest to do so.

- **Product detail**: As mentioned previously it should be possible to solve the platform extent problem directly (i.e., down to individual parts and details) for simple products with up to $7^3 \sim 300$ components (e.g., coffee makers, simple cameras, etc.). For systems that are more complex (e.g., cars, airplanes, complex electronics) it is likely that one has to resort to an abstracted, higher level representation of the product.

- **Technical model fidelity**: It was demonstrated that the RSM engineering performance model used her was accurate within 6-11%, on average. For some industries the use of such simplified models might be acceptable, for others it might be misleading.

- **Soft attributes**: The model suggested that a firm should strive to set all "soft" attributes to the maximum value, e.g., a J.D. Power rating of 5 in all categories. In reality this will be difficult to achieve and improvement of soft attributes will require resources. This tradeoff has not been modeled, but it has been suggested that one of the true benefits of platforming (and multi-platforming) is that a firm can dedicate more attention and resources to the soft attributes (e.g., styling, interiors) because the main functionality of the product variants has already been "built-in" to the one or more platforms.

- **Functional attributes:** One needs to be careful to interpret and monitor the evolution of the functional value weighting factors, γ. First, these can (and do) evolve over time and new functional attributes can emerge that were not considered when the platform(s) were originally conceived.

- **Flexible product platforms**: All platforms considered in this chapter were considered as fixed, i.e., once WT and WB were chosen they could not be changed. An important research topic is the embedding of flexibility in product platforms. For example, we see in Figure 12-24 that the β and γ-platforms are not too different in terms of ΔWT ($\sim$2") and ΔWB ($\sim$8"). One could conceivably agree on a compromise width (WT) and design the platform flexibly such that it could be stretched lengthwise (WB), say from 100-110" without significant switch costs. Such flexibility might require upfront investment, but might allow reducing the "optimal" number of platforms suggested under the assumption that each platform is fixed.

6.2   Summary

In this chapter we have discussed the relevance and challenges of the multi-platform problem that all manufacturing firms face who offer many product variants ($m>3$). The problem is challenging and multi-faceted and requires the adoption of an end-to-end modeling framework (see Figure 12-3) that connects six different domains: product architecture, performance engineering, value modeling, market demand modeling, manufacturing costing and investment finance. Various methods and techniques exist to populate these sub-domains, but we believe that the interface quantities (design variables $\mathbf{x}$, functional attributes $\mathbf{f}$, value V, cost C, demand D and NPV) are invariant.

The first step in answering the platform extent problem (how many platforms?, what is the optimal variant-to-platform assignment?) is to solve the two extreme cases: $p=m$ (no platforming) and $p=1$ (single platform). These cases can be solved with a number of single and bi-level optimization schemes that have been proposed in the literature. To solve the multi-platform problem ($1<p<m$) we suggest a weighted least squares formulation (see Eq. 45) rather than setting up an intractable tri-level formulation. In the least squares problem, we allow p platforms and attempt to position these in the design space such that the m variants can be optimally assigned to minimize the objective function. The objective function is a compound Euclidian distance metric, whereby the distance of each variant to its assigned platform is weighted by its projected revenue per unit time.

The framework has been demonstrated for a (hypothetical) product family of 7 automotive model variants. The results indicate that a 5-platform strategy appears most suitable. The model suggests that forcing these diverse products onto a single platform is counterproductive because such a platform extent would be too large, leading to significant value losses of the associated variants in their particular segments. Industrial practice mirrors this result as automotive platforms are typically leveraged only within larger market segments (e.g., mid-size sedans) and rarely between segments.

The framework presented here is believed to be generally applicable, subject to the caveats mentioned in the previous section. Figure 12-27 shows that the "Platform Family Plan" occupies a central place in the corporate strategy of the for-profit manufacturing firm.

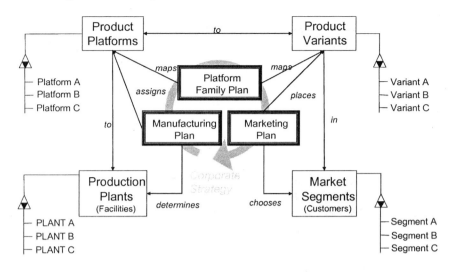

Figure 12-27. Platform Family Plan and corporate strategy.

The Marketing Plan assigns product variants to market segments and decides which of the firm's products will compete against which of the competitor's products. The platform extent problem therefore requires that both the market segments and the types and number (*m*) of desired product variants are known. The Platform Family Plan assigns variants to platforms and embodies key engineering decisions about architecture and design of the platforms and variants. Finally, the manufacturing plan assigns platforms to manufacturing plants. This is important as plants are typically equipped with machines and tooling that are geared towards a limited set (mostly one) platform. One of the most important complications in the manufacturing plan is that plants should be made to run at or near capacity. One of the main reasons for success (or failure) in achieving profitability in the automotive industry relates to whether or not plant capacity can be utilized at a high rate (say >80%). This depends not only on the market success of individual models, but also on the flexibility to balance loads of a multi-platform product family across multiple manufacturing facilities.

NOMENCLATURE

AC	acceleration time from 0-100 km/h, [sec]
CO	comfort, [1-5]
CV	cargo volume, [cft]
CW	curb weight, [lbs]
ED	engine displacement, [ccm]
FC	fuel tank capacity, [gallons]
FE	combined fuel economy, [mpg]
GC	ground clearance, [in]
HP	horsepower rating, [hp]
HT	total height, [in]
LT	total length, [in]
MD	mechanical dependability, [1-5]
MQ	mechanical quality, [1-5]
PR	price, [2002 U.S. $]
PV	passenger volume, [cft]
SE	service experience, [1-5]
ST	styling, [1-5]
SV	sales volume, [# units/year]
TC	towing capacity, [lbs]
TD	tire diameter, [in]
TW	tire width, [mm]
WB	wheel base, [in]
WT	front wheel track (width), [in]

A	profit, $
B	learning curve exponent
$c_{i,tfu}$	theoretical first unit costs (variable) for the i-th product
$c'_{i,tfu}$	scaled theoretical first unit costs (variable) for the i-th product
$c_{i,tfu,k}$	theoretical first unit costs (variable) for the k-th module of i-th product
C_i	variable cost of the i-th product, $
C'_i	variable cost of the i-th product with scaling and learning curve, $
D_a	actual sales volume (=demand), # units per unit time
D_i	predicted demand for the i-th product, # units per year
f_i	i-th functional performance attribute
$\mathbf{f}$	vector of functional performance attributes
F_i	fixed cost for i-th product, $

m	number of variants, number of functional attributes
m_P	Plant and tooling investment costs a percent of sales, %
m_{RD}	Research and development costs as a percent of sales, %
$M_{P,i,j}$	Plant investment costs for the i-th product in the j-th year, $
$M_{RD,i,j}$	Research and development costs for the i-th product in the j-th year, $
N	number of competitors in a market segment
N_P	number of time periods of production
N_{RD}	number of time periods for research and development
n	number of design variables
n_{mod}	number of modules in product
NPV	net present value, $
p	number of platforms
$\mathbf{p}$	vector of fixed design parameters
P	product price, $
r	discount rate, typically 0-20%
s_i	i-th soft attribute, 1-5 (J.D. Power Rating)
$\mathbf{s}$	vector of soft attributes
S	learning curve factor, typically 0.8-1.0
x_i	i-th design variable
$\mathbf{x}$	vector of design variables
$\mathbf{y}$	vector of dependant variables
V	product value, $
α	operating margin, 0-1
$\beta_{i,k}$	variable cost breakdown coefficient for i-th product and k-th module
ϕ	operating leverage (ratio of fixed to variable costs)
γ_i	weighting factor for i-th product attribute
γ	vector of attribute weighting factors
Π_i	net present value of i-th product, $

APPENDIX A

Automotive database (Autopro, 2002) for the United States of America. Market Segment: medium cars (MED), sorted in order of decreasing sales.

Table 12-A1. Design variables x(1)-x(9) and dependent variables y(1)-y(2).

Brand	Model	C FC [g]	x(2) ED [ccm]	x(3) WT" [in]	x(4) WB [in]	x(5) GC [in]	x(6) HT [in]	x(7) LT [in]	x(8) TW [mm]	x(9) TD [in]	y(1) HP [hp]	y(2) CW^ [lbs]
Honda	Accord	17.1	2254	61.2	105.1	3.9	54.9	186.8	195	15	150	3036
Toyota	Camry	18.5	2362	60.8	107.1	5.4	57.9	189.2	205	15	157	3142
Ford	Taurus	18	2982	61.6	108.5	5.4	56.1	197.6	215	16	155	3336
Chevrolet	Impala	17	3350	62	110.5	NL	57.3	200	225	16	180	3389
Pontiac	Grand AM	14.3	2189	59	107	NL	55.1	186.3	215	15	140	3118
Chevrolet	Malibu	14.3	3136	59	107	5.5	56.4	190.4	215	15	170	3053
Nissan	Altima	20	2500	61	110.2	4.1	57.9	191.5	205	16	175	3048
VW	Jetta	14.5	1984	59.6	98.9	4.1	56.7	172.3	195	15	115	2945
Buick	Century	17.5	3130	62	109	5.7	56.6	194.6	205	15	175	3368
Pontiac	Grand Prix	17.5	3136	61.5	110.5	5.5	54.7	196.5	205	15	175	3384
Chrysler	Sebring	16.3	2351	59.4	103.7	6.2	53.7	190.2	205	16	142	3099
Dodge	Stratus	16.3	2351	59.4	103.7	6.2	53.7	190.9	205	16	147	3115
Oldsmobile	Alero	14.1	2196	59.1	107	5	54.5	186.7	215	15	140	3010
Dodge	Interpid	17	2736	61.9	113	5.1	55.9	203.7	225	16	200	3469
Mercury	Sable	18	2982	61.6	108.5	NL	55.5	199.8	215	16	157	3379
Nissan	Maxima	18.5	3500	60.2	108.3	5.5	56.3	191.5	215	16	255	3218
Saturn	L-100	15.7	2198	59.8	106.5	6.3	56.4	190.4	195	15	135	2989
Subaru	Legacy	16.9	2457	57.5	104.3	6.1	55.7	184.4	205	15	165	3320
VW	Passat	16.4	1781	59.6	106.4	4.9	57.6	185.2	195	15	170	3322
Mitsubishi	Galant	16.3	2350	59.4	103.7	6.2	55.7	187.8	195	15	140	3031
Toyota	Avalon	18.5	2995	61	107.1	5.1	57.7	191.9	205	15	210	3417
Chevrolet	Monte-Carlo	17	3350	62	110.5	5.9	55.2	197.9	225	16	180	3340
Hyundai	Sonata	17.2	2351	60.6	106.3	NL	56	186.9	205	15	149	3217
Mazda	626	16.9	1991	59.1	105.1	5.2	55.1	187.4	205	15	125	2961
Buick	Regal	17.5	3790	62	109	5.7	56.6	196.2	215	15	200	3438
Oldsmobile	Intrigue	17	3472	62.1	109	5.8	56.6	195.9	225	16	215	3434
Chrysler	Concorde	17	2736	61.9	113	5.1	55.9	207.7	225	16	200	3479
Kia	Optima	17.2	2351	60.6	106.3	6.1	55.5	186.2	205	15	149	3190
Daewoo	Leganza	15.8	2198	59.6	105.1	NL	56.6	183.9	205	15	131	3157
Hyundai	XG350	18.5	3467	60.6	108.3	6.3	55.9	191.5	205	16	194	3651
Mitsubishi	Diamante	19	3497	60.8	107.1	4.6	53.9	194.1	215	16	205	3439
Mean		17.0	2714	60.5	107.3	5.42	55.9	191.5	209	15	168	3242
Min		14.1	1781	57.5	98.9	3.9	53.7	172.3	195.0	15.0	115	2945
Max		20.0	3790	62.1	113.0	6.3	57.9	207.7	225.0	16.0	255	3651

Market Segment: Medium Sedans (MED)

Table 12-A2. Functional attributes f, demand D, price P, soft attributes, s.

Brand	Model	f(1) PV [cft]	f(2) CV [cft]	f(3) TC [lb]	f(4) FE* [mpg]	f(5) AC [sec]	D SV [#]	P PR [$]	s(1) MQ 1-5	s(2) CO 1-5	s(3) ST 1-5	s(4) MD 1-5	s(5) SE 1-5
Honda	Accord	92.7	13.6	1000	28	9.42	414718	$18,890	4	4	2	5	4
Toyota	Camry	101.7	16.7	2000	27	9.31	390449	$18,970	3	4	2	3	3
Ford	Taurus	104.7	17	1250	23	10.02	353560	$19,035	3	3	3	3	3
Chevrolet	Impala	104.5	18.6	1000	25	8.76	208395	$20,325	4	4	2	3	3
Pontiac	Grd AM	92.2	14.6	1225	28	10.37	182046	$17,135	3	4	2	3	3
Chevrolet	Malibu	98.6	17.1	1000	23	8.36	176583	$17,760	3	3	2	4	3
Nissan	Altima	103.2	15.6	1000	26	8.11	148345	$16,649	2	3	2	5	3
VW	Jetta	86.9	13	1225	27	11.92	145221	$16,850	3	5	3	2	2
Buick	Century	101	16.7	1000	23	8.96	142157	$20,535	4	3	2	4	5
Pontiac	G Prix	98	16	1000	23	9.00	128935	$21,230	3	3	2	3	3
Chrysler	Sebring	86.3	16.3	1000	24	10.16	118459	$20,390	3	4	2	2	3
Dodge	Stratus	86.3	16.3	1000	25	9.86	111125	$18,690	3	4	2	2	3
Oldsmobile	Alero	91.2	14.6	1225	27	10.01	109302	$17,805	3	3	2	2	3
Dodge	Interpid	104.5	18.4	1500	23	8.07	109098	$20,810	4	3	2	4	3
Mercury	Sable	102.5	16	1250	23	10.02	102646	$20,020	3	3	2	3	4
Nissan	Maxima	102.5	15.1	1000	22	5.87	102535	$24,699	3	5	2	3	3
Saturn	L-100	96.9	17.5	1000	27	10.31	98227	$16,370	4	4	2	2	4
Subaru	Legacy	91.4	12.4	2000	23	9.37	95291	$19,295	3	3	2	2	2
VW	Passat	95.4	15	1225	25	9.10	95028	$21,750	4	5	2	2	2
Mitsubishi	Galant	97.6	14.6	1225	24	10.08	93878	$17,707	3	3	2	2	2
Toyota	Avalon	105.6	15.9	2000	24	7.57	83005	$25,845	5	5	2	5	3
Chevrolet	M Carlo	98.2	15.8	1000	25	8.64	72596	$20,425	3	4	3	2	3
Hyundai	Sonata	100	14.1	1225	25	10.05	62385	$15,499	2	3	2	2	2
Mazda	626	97.1	14.2	1225	29	11.03	50997	$18,785	3	4	2	3	3
Buick	Regal	101.8	16.7	1000	23	8.00	49992	$23,485	5	3	2	4	5
Oldsmobile	Intrigue	101	16.4	1000	24	7.43	39395	$23,160	4	3	2	3	3
Chrysler	Concorde	107.6	18.7	1500	23	8.10	32331	$22,790	4	3	3	5	3
Kia	Optima	100	13.6	1225	24	9.96	25910	$14,899	2	3	2	2	2
Daewoo	Leganza	101	14.1	1225	23	11.22	18347	$14,599	3	3	2	2	2
Hyundai	XG350	102	14.5	1225	21	8.76	17884	$23,999	2	5	3	2	2
Mitsubishi	Diamante	100.9	14.2	1225	20	7.81	17227	$25,687	3	3	2	2	2
Mean/Tot.		98.5	15.6	1225.0	24.4	9.2	3796067	$19,809	3.3	3.6	2.1	2.9	2.9
Min		86.3	12.4	1000.0	20.0	5.9	17227	$14,599	2.0	3.0	2.0	2.0	2.0
Max		107.6	18.7	2000.0	29.0	11.9	414718	$25,845	5.0	5.0	3.0	5.0	5.0

APPENDIX B

Scatter plots for vehicles in automotive product database (Appendix A).

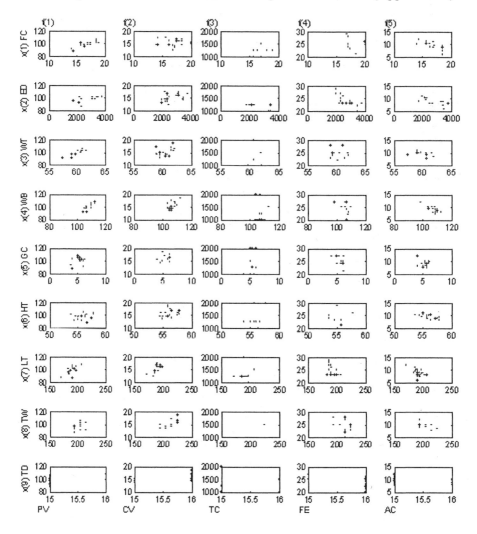

Figure 12- B1. Scatter plots of x versus f for medium sedan (MED) market segment.

APPENDIX C

Design variable bounds given by vehicle product database.

Table 12-C1. Overview of design and dependent variable for seven market segments.

Segment	x(1) FC [g]	x(2) ED [ccm]	x(3) WT [in]	x(4) WB [in]	x(5) GC [in]	x(6) HT [in]	x(7) LT [in]	x(8) TW [mm]	x(9) TD [in]	y(1) HP [hp]	y(2) CW [lbs]
SML – avg	**13.37**	**1860**	**57.73**	**99.9**	**5.27**	**55.50**	**172.18**	**187.08**	**14.38**	**122**	**2535**
SML – min	11.9	1468	55.30	93.3	3.90	53.00	161.10	175.00	13.00	92	2035
SML - max	15.9	2457	59.60	105.0	7.70	59.40	182.00	205.00	16.00	180	2965
MED – avg	**17.0**	**2714**	**60.5**	**107.3**	**5.42**	**55.9**	**191.5**	**209**	**15**	**168**	**3242**
MED – min	14.1	1781	57.5	98.9	3.9	53.7	172.3	195.0	15.0	115	2945
MED - max	20.0	3790	62.1	113.0	6.3	57.9	207.7	225.0	16.0	255	3651
LRG – avg	**19.40**	**3737**	**61.57**	**112.1**	**5.40**	**56.57**	**197.99**	**221.00**	**16.13**	**243**	**3754**
LRG – min	16.9	1985	57.2	102.6	3.9	53.6	182.3	205.00	15.00	185	3220
LRG - max	23.7	4605	63.4	121.5	6.7	58.9	215.3	245.00	18.00	340	4376
SPT – avg	**15.05**	**2743**	**58.54**	**98.2**	**4.68**	**51.36**	**170.25**	**211**	**15.83**	**200**	**2879**
SPT – min	12.70	1781	55.60	89.2	4.00	47.70	153.00	185	14.00	115	2195
SPT - max	18.50	5665	61.90	106.4	6.00	56.00	193.50	245	17.00	350	3323
VAN – avg	**21.70**	**3397**	**63.17**	**114.8**	**6.17**	**68.89**	**194.24**	**214.33**	**15.13**	**189**	**4041**
VAN – min	19.80	2429	60.60	111.2	4.30	64.20	186.90	205.00	15.00	150	3699
VAN - max	27.00	4300	66.10	120.7	8.70	75.00	201.50	225.00	16.00	240	4709
SUV – avg	**21.06**	**3491**	**61.61**	**108.7**	**8.14**	**70.68**	**185.39**	**237.16**	**15.95**	**202**	**4075**
SUV – min	14.70	1983	55.00	92.9	6.30	65.00	155.40	195.00	15.00	120	2777
SUV - max	44.00	5408	68.40	137.1	10.20	77.20	226.70	275.00	17.00	300	6650
TRK – avg	**22.8**	**3457**	**61.7**	**119.1**	**8.2**	**68.2**	**203.3**	**235.0**	**15.6**	**180**	**3774**
TRK – min	15.8	2189	54.4	103.3	6.7	62.0	184.4	205.0	15.0	120	2750
TRK - max	31.0	5326	68.0	131.1	10.4	74.4	221.7	265.0	17.0	285	5437

PART III: BACK-END ISSUES RELATED TO PLATFORM-BASED PRODUCT FAMILY DEVELOPMENT

Chapter 13

A ROADMAP FOR PRODUCT ARCHITECTURE COSTING

Sebastian K. Fixson
Department of Industrial & Operations Engineering, University of Michigan, Ann Arbor, MI 48109

1. INTRODUCTION

In recent years many markets have exhibited increasing demand heterogeneity; they are fragmenting into more and smaller market niches. This development threatens the large-scale assumption of many mass production processes. As a result, firms face the dilemma of how to provide a wide variety of goods for prices that can compete with mass produced products. To respond to these challenges, many firms have begun searching for ways to combine the efficiency of mass production with the variety of customer-oriented product offerings. A major focus of these efforts has been the fundamental structure of the product: the product architecture. Examples for this development are Sony's personal music players (Walkman) that use common drives across different models (Sanderson and Uzumeri, 1995), different power tools that use similar motors (Meyer and Lehnerd, 1997), PDAs (personal digital assistant) that can be turned into an MP3 player, a camera, or a telephone with different attachments (Biersdorfer, 2001), and automobiles with common components across models (Carney, 2004).

Researchers of disciplines ranging from engineering to management have focused their attention on these phenomena, and have developed tools to guide the difficult process of providing variety to the customer while maintaining near-mass production efficiency, i.e., to 'mass customize' (Pine, 1993a). The approaches vary in their perspective and level of analysis. Some focus more on ways to increase external product variety while maintaining low costs, while others target their efforts on internal variety reduction without losing the variety appeal for the customer. The underlying idea of most of these approaches is to increase commonality across multiple products. The level in the product hierarchy at which commonality is pursued varies: it can be focused on common components (Eynan and Rosenblatt, 1996; Fisher, et al., 1999), on modules (Chakravarty and Balakrishnan, 2001; Dahmus, et al., 2001; Sudjianto and Otto, 2001), on product platforms and product families (Gonzalez-Zugasti, et al., 2000; Jiao and Tseng, 2000; Simpson, et al., 2001) or on production processes (Wilhelm, 1997; Siddique, et al., 1998), although the lines between these levels are sometimes blurred.

From an overall strategic perspective a firm needs to balance all benefits it can achieve by increasing commonality across products with all the costs this approach creates. For example, it needs to weigh the revenue decreasing effects through cannibalization that product commonality can cause against the cost savings that commonality can achieve (Robertson and Ulrich, 1998). Ideally, this multi-objective problem requires the balancing of cost, revenue, and performance effects when selecting a product architecture from a set of candidates. Although cost is only one of these variables, there are at least two major reasons that make it worthwhile to explore this problem with a focus on the cost portion alone. The first reason is that in many cases cost is a major, if not the most important, decision variable. More specifically, most products contain two types of components: those with a strong influence on product quality and those with only a weak influence on product quality (Fisher, et al., 1999). For components of the latter type cost becomes the only decision variable, provided that the components' performance level is sufficient (Thonemann and Brandeau, 2000). The second reason for building a roadmap focusing on cost is that it can—once established—serve as a building block for the development of more sophisticated design support tools such as product architecture design guidelines or optimization models. These tools often build on existing cost estimation models which in turn incorporate known or assumed relationships between product architecture and costs as well as cost allocation rules, and to interpret the results of the models requires a thorough understanding of how the problem has been framed. In other words, what are the multiple and complex relationships between various product architecture characteristics and various costs along the product life cycle? The existing research is somewhat inconclusive. For commonality decisions, one aspect of product architecture, effects on individual costs have been demonstrated (Park and Simpson, 2003), whereas for modularity, another aspect of product architecture, no general relationship with cost has been found (Zhang and Gershenson, 2003). In other words, the complex relationship between product architecture and costs is still insufficiently understood (Simpson, 2004). Similarly, what is the impact of the applied allocation rules on the cost models, and consequently, their results? Finally, to develop the deep understanding of the relationships between product architecture and costs in turn requires a good understanding of the input data, i.e., how is the product architecture described and what types of costs are considered? Figure 13-1 illustrates this chain of requirements for building design support tools with respect to costs. The remainder of the chapter develops a roadmap that helps covering all requirements from input data to the cost estimation models.

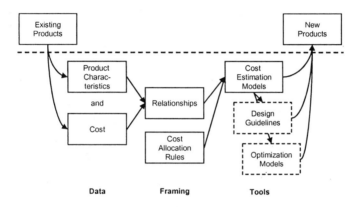

Figure 13-1. Requirements chain for developing product design support tools for cost.

2. DEVELOPING A ROADMAP FOR PRODUCT ARCHITECTURE COSTING

The roadmap comprises four steps (see Figure 13-2). The first step is an assessment of the differences in product architecture between potential candidates. This step is crucial because in order to make the analysis of cost consequences of different product architectures possible requires the ability to distinguish different product architectures in the first place. The product architecture costing roadmap builds on a multidimensional product architecture description methodology. In the second step of the roadmap the relevant life cycle phase, or phases, with respect to costs have to be identified. The question of relevance hinges on a variety of factors such as product lifetime, production volume, total value, and cost ownership. The third step requires determining the cost allocation rules to be used for the costing procedure. The choice of certain accounting decisions can have a profound effect on how the product architecture-cost relationship is modeled. In its fourth step, the roadmap calls for the selection of suitable cost models. Existing models differ in their requirements for data accuracy and sample size, as well as their ability to predict cost differentials of product architectures differences. Each step of the proposed roadmap is discussed in more detail in the following sections.

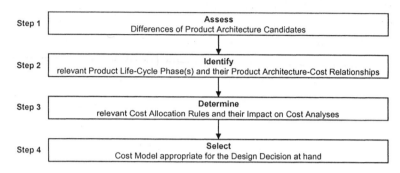

Figure 13-2. A roadmap for product architecture costing.

2.1 Step 1: Assess differences of product architecture candidates

2.1.1 The special role of product architecture as a design variable

Product designers make numerous decisions throughout the design process. Each of these decisions has consequences for some costs along the product life cycle. Two characteristics label the links between these decisions and their cost consequences. The first characteristic describes how difficult it is to construct the link; the second how valuable it is to know it (see Figure 13-3).

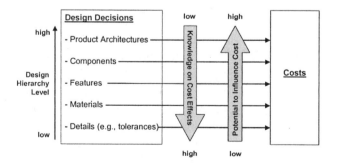

Figure 13-3. Product architecture decisions in the design hierarchy.

The level of difficulty to establish a link between design decisions and their cost effects depends on the hierarchy level at which the decisions are made. On a very detailed level, it is fairly straightforward to construct a link between the design decision and its cost implication for two reasons. First, on the detailed level it is often clear on what costs to focus on, and second, for well known links historical data often exist. For example, there is ample data on how a more stringent surface smoothness requirement affects the manufacturing cost to create that surface. Design textbooks typically provide cost tables or functions to guide these type of design decisions (Michaels and Woods, 1989; Pahl and Beitz, 1996). On the next higher level of abstraction, design decisions affect the choice of materials, production processes, or part features. Materials have been used as a cost determining decision variable for a long time since in many mass production environments material costs represent a significant fraction of total production costs (Ostwald and McLaren, 2004). For this reason, rules-of-thumb have been developed to allow approximate but quick cost estimates. For example, to assess the cost impact of selecting manufacturing processes, Esawi and Ashby (2003) have developed a simple model that requires the input of only a few parameters. The primary aim of that method, however, is the relative ranking of multiple processes with respect to cost, not to predict exact costs. Product features have also been used as decision variables for which cost models have been developed. Often the models combine cost estimations on the feature level with cost estimations on component and assembly levels (Weustink, et al., 2000) and the product family level (Park and Simpson, 2003). Yet another level up in abstraction is populated by design guidelines such as Design for Manufacturing (DFM) or Design for Assembly (DFA). They represent codified knowledge of links between design decisions and production costs. However, they are not cost prediction tools but present the knowledge in a condensed form such that they direct the designer's attention to cost creating design issues, and lead him towards (relatively) lower cost solutions (Boothroyd, et al., 2002). Finally, on the level of product architecture there are numerous examples for relationships between *individual* aspects of the product architecture and *individual* costs, but no approach exists that provides a generic yet comprehensive description of this multidimensional relationship.

The second characteristic that describes the link between design decisions and cost elements along the product life cycle is the leverage to influence the costs if the link is actually known. It is generally assumed that earlier design decisions have greater potential to influence costs than later design decisions.[18] This creates a dilemma for the designer: it is the early phase of design decisions where the potential to influence total life cycle cost is the greatest, and yet early in the design process is where the fewest data for detailed cost estimations exist. How, then, can this link be constructed? This roadmap builds on a methodology that can distinguish between different product architectures along multiple dimensions on a relatively abstract level.

2.1.2 A multidimensional method to assess product architecture differences

The product architecture is the fundamental structure and layout of a product and is defined during concept development (Ulrich and Eppinger, 2004). Building on Ulrich's description of product architectures (Ulrich, 1995), a multi-dimensional product architecture description method has been developed (Fixson, 2005). The method relaxes three fundamental assumptions of earlier work. First, it allows for independent assessments of the two main product architecture dimensions: *function-component allocation* and *interfaces*. Second, it acknowledges that these two dimensions are themselves multidimensional constructs. Third, it assesses the product architecture for each function separately—in contrast to most product architecture descriptions in the literature that essentially provide *average* assessments of a product's architecture.

The first of the two dimensions, *function-component allocation (FCA),* is concerned with the extent to which a product's functions are isolated on physical components. It measures for each function (on the selected architecture level) the degree of function-component allocation. More specifically, each function is assigned two indices that determine its position relative to the extremes of 1-to-1 and many-to-many relationships between functions and components. A 1-to-1 measurement indicates a situation in which the function under consideration is provided exclusively by one component, and this component provides exclusively this function only. This style of FCA is called modular-like. In contrast, a few-to-many measurement indicates a situation in which a function is provided by many components (an integral-fragmented style). A many-to-few measurement

---

[18] Various authors present the idea that somewhere between 60% and 90% of the total life cycle cost are committed during product design. Interestingly, although these numbers are used by a variety of authors from diverse fields ranging from accounting to engineering to management, e.g., (Smith and Reinertsen, 1991:100; Anderson and Sedatole, 1998:231; Blanchard and Fabrycky, 1998:561; Clancy, 1998:25; Knight 1998:21; Sands, et al., 1998:118; Buede 2000:7; Weustink, et al., 2000:1; Bhimani and Muelder, 2001:28), nowhere is real data presented as evidence. One exception exists that models costs in more detail, however, it also does not specify a particular fraction of the total life cycle cost that is committed during design, but rather assesses the cost influence potential of the design phase versus the one of the production phase (Ulrich and Pearson, 1998).

denotes an integral-consolidated style where one component provides multiple functions. Finally, a many-to-many measurement represents an integral-complex FCA style. It is important to take the FCA measure for each product function individually because the reuse of a component across a product family depends to a large degree on the role a component can play in different products. The second dimension of the product architecture description method, *interfaces,* is itself multidimensional and is concerned with three characteristics of the interfaces that connect the components. The first characteristic, interface intensity, describes in detail the role each interface plays for the product function. Interfaces can be spatial, or they can transmit material, energy, or signals or any combination of the above (Pimmler and Eppinger, 1994). The second characteristic, interface reversibility, describes the effort it requires to disconnect the interface. This effort depends on two factors: the difficulty to physically disconnect the interface, and the interface's position in the overall product architecture. Finally, the third characteristic, interface standardization, depends both on product features as well as the population of alternatives. While some researchers have used different types of interfaces to categorize types of modularity like swapping, sharing, bus, and sectional (Ulrich and Tung, 1991; Pine, 1993a), the method presented here views the extent to which an interface allows different kinds of interchangeability as a matter of perspective. In other words, the level of standardization can be different for any component that is involved in the interface. Standardization is a function of the number of alternatives that exist on either side of the interface.

As example, compare the two trailers in Figure 13-4 (top). They both provide the same functionality. However, they exhibit very different product architectures. Figure 13-4 (bottom) shows two different patterns of how each component provides one or more functions. Figure 13-5 illustrates the same information with the help of product architecture maps. Each function is assessed separately and along multiple dimensions. The location on the x-y plane identifies where each function is positioned in between the extreme points of 1-to-1 and many-to-many function-component allocations. To put it differently, the position describes each function's FCA style. The three interface sub-dimensions (intensity, reversibility, and standardization) are independently scaled on the vertical axis. The value of this assessment independence can be seen by comparing individual functions for the two trailers. For example, the function *transfer loads to road* exhibits identical product architecture characteristics for both trailers whereas all other functions show significant differences along the multiple dimensions.

The following sections refer to these dimensions of product architecture when discussing the remaining three steps of the roadmap for product architecture costing: identifying the relevant life cycle phases and costs and their product architecture-cost relationships, determining the relevant allocation rules, and selecting appropriate cost models.

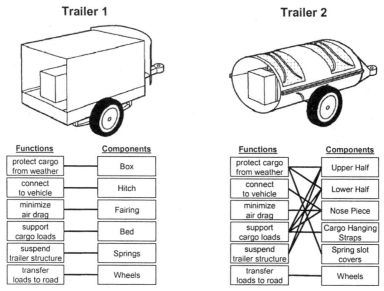

Source: Ulrich (1995) "The role of product architecture in the manufacturing firm"

Figure 13-4. Two trailers with different product architecture.

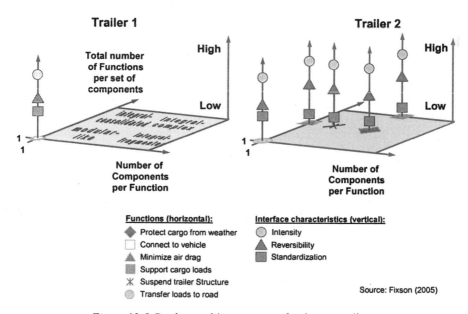

Source: Fixson (2005)

Figure 13-5. Product architecture maps for the two trailers.

2.2 Step 2: Identify relevant product life cycle phase(s) and their product architecture-cost relationships

Before one can begin to investigate the cost implications of differences in product architecture one has to decide on which costs to incorporate in the analysis. This problem has two components. The first is concerned with the decision of which life cycle phase(s) are relevant with respect to cost for the decision at hand, and the second strives to identify the relevant product architecture-cost *relationships* within the selected life cycle phase or phases. The factors used to identify relevant product life cycle phases are discussed next, followed by a detailed account of known effects that individual product architecture characteristics have on costs for each life cycle phase.

2.2.1 Which life cycle phase matters?

Every product and system, regardless of size, value and lifetime, progresses through different phases during its life: design, development, production, use, and retirement. In each of these phases, different processes and activities are performed with and on the product (see Figure 13-6). Each of these processes and activities creates a cost that occur at different points in time, at different locations, and can be borne by different constituents.

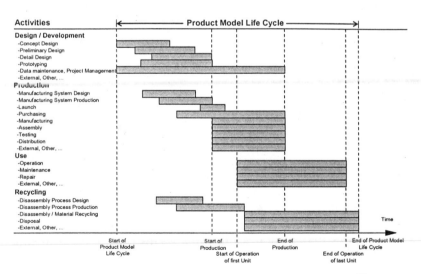

Figure 13-6. Activities throughout the product model life cycle[19].

---

[19] Note that the diagram depicts the product life cycle of all units produced during a model's life. In case of only one unit produced (e.g., expensive or special equipment), the diagram collapses into the individual product's life cycle. In this chapter, the term life cycle refers to the life of a single product.

Given that various costs occur in different phases of a product's life, one of the first decisions of a cost evaluation is to determine those costs that are relevant for the design decision at hand. The relevance of individual costs depends on the life cycle cost profile and the ownership of the costs. A product's life cycle cost profile is determined by both absolute values and relative distributions of the costs over the life cycle, the durations of the individual phases, and the production volume. To separate products according to their absolute values of total lifetime and total life cycle costs, it has been suggested to cluster the universe of different products into three major categories: large-scale, medium-scale and small-scale systems (Asiedu and Gu, 1998). Large-scale systems can have total lifetimes of several decades and total life cycle costs of billions of dollars. Lifetimes of medium-sized products are typically measured in years, with total life cycle costs ranging from thousands to millions of dollars. Small-scale products can have lifetimes as short as a few months and life cycle costs as low as a few hundred dollars (see Figure 13-7).

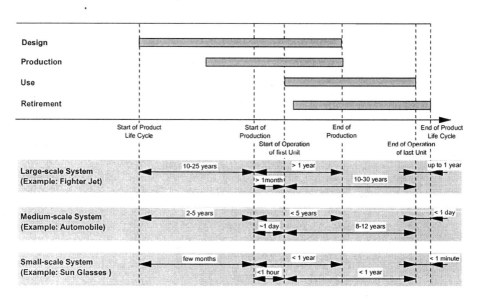

Figure 13-7. Lifetimes of different product categories.

In addition to the absolute values, the relative distribution of time and cost over the different life cycle phases also plays an important role in determining on which costs to focus. These differences in relative distribution can be caused by differences in scale and technical complexity. For example, a small product, say a radio clock, will require very few maintenance and support activities, which translate into low costs during its use, whereas for long living and large scale products as, for instance, a navy ship, these costs can represent almost 2/3 of total lifetime costs (Sands, et al., 1998). Another factor that influences the relative size of the costs of the individual life cycle phases is the production volume per model. A small production volume results in relatively higher development cost per unit compared to the situation in which the total development costs can be spread over a large production volume. The consequence of the differences in total life cycle cost, total life time, life cycle cost distributions, and production volumes are different life cycle cost incurrence curves (see Figure 13-8).

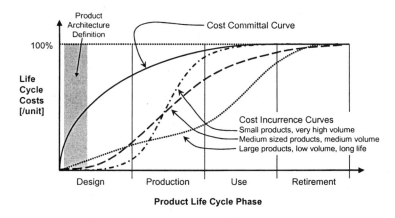

Figure 13-8. Cost committal and incurrence curves.

Finally, the life cycle phase in which certain costs occur does not necessarily determine who bears these costs. For example, warranty policies can transfer costs between producer and user (Blischke and Murthy, 1994), and most of so-called external costs are often borne by the society at large while the product user pays only a fraction of it directly. More generally, depending on a variety of additional factors such as market dynamics, level of competition, or institutional environment, a number of different cost distribution schemes are conceivable, enforced by different contractual agreements. Since most of these factors are not decision variables for the designer, the following discussion of each life cycle phase individually looks at costs independently from the ultimate ownership. Also note that while the primary focus of this chapter is on cost effects triggered by product architecture design decisions, other performance measures—such as time, and to some extent revenue—that are impacted by these decision, are discussed where relevant.

2.2.2   Product architecture effects on costs of the product development phase

The first phase of a product's life encompasses activities such as conceptual and preliminary design, detail design and prototyping, testing, as well as supporting functions such as data maintenance and project management.   For engineered products, the costs for these processes represent primarily engineering resources, i.e., personnel.   To address the question of how differences in product architecture affect the resource consumption during the design phase some researchers have linked the task structure of the design process to the product architecture (von Hippel, 1990; Eppinger, et al., 1994; Gulati and Eppinger, 1996).   Over time, a firm's organizational structure often mirrors the product structure of the products the firm produces (Henderson and Clark, 1990).   Thus, the design decision on the number and size of 'chunks' (subsystems, modules, parts, etc.), i.e., the *function-component allocation scheme*, translates into the number and size of teams working to develop the product (Baldwin and Clark, 2000). The number and size of the teams determines their internal complexity as well as their external communication requirements.   Both factors in turn determine the teams' efficiency.   Either extreme, i.e., one very large team or many, very small teams, appears to be a relatively inefficient organizational form, the former requiring many internal iterations, the latter producing a long sequence of information transfers.   Therefore, creating product architectures that balance the design complexity that incurs between the chunks (integration effort) on one hand, with the sum of the design complexity within the chunks on the other, by designing chunks of medium complexity, seems to be a resource efficient approach.   This effect has been found empirically for complex software development projects (MacCormack, et al., 2004).   For the second product development performance measure next to cost, total development time, a similar effect has been demonstrated: for the development of a turbopump of a rocket engine it has been shown that there is a number of blocks of the product architecture (modules, chunks, etc.) that translates into a medium number of teams that minimizes the duration of the project development project (Ahmadi, et al., 2001).   Apparently, both costs and time functions exhibit a minimum if the product is decomposed into a medium number of subunits; and increases when fewer but larger subunits are chosen, and increases when more but smaller subunits are selected.

The relative value of time compared to cost depends on a number of market parameters as well as the ratio between revenue and costs.   For example, companies operating in fast pace market environments will especially value a product architecture's potential to reduce the time-to-market. Product architectures that allow conducting much of the design process in clusters in parallel to arrive at the shortest possible total design time are of particular value to them. In a specific case about a Polaroid camera housing, for example, it has been found that the foregone sales in case of a longer development time far outweigh any achievable cost savings in manufacturing (Ulrich and Pearson, 1993). In a case like this, a product architecture that helps to reduce development time is much more valuable than one that focuses on cost savings in the production phase.

Also, strictly speaking, the design phase is only one component of the time-to-market. If 'market' is understood as sale (or start of operation) of the first unit, then production preparations become part of the time-to-market, in particular tool design and manufacturing. Hu and Poli (1997) have compared assemblies made from stampings with injection molded parts regarding their effects on time-to-market. They find that parts consolidation, i.e., the reduction of the number of chunks the product consists of can be disadvantageous with regards to time-to-market when the time to produce the tool for larger, more complex parts extends the total time-to-market.

In addition to the particular product function-component allocation scheme, the characteristics of the *interfaces* between the chunks are likely to affect the efficiency of the design process, and thus its costs. The weaker the interface connections are, i.e., the lower their intensity, the more the different design teams can be working independently on different subsets of the product. This can reduce the number of iterations between the teams, and thus increase overall design process efficiency. In a case study of the development of an automotive climate control system, strong coupling between components has been identified as one reason for development cost increases (Terwiesch and Loch, 1999). Weaker interface dependencies may also improve the second performance indicator, total development time (time-to-market), because it allows the design tasks to proceed in parallel (Baldwin and Clark, 2000). For example, analyzing the product development of integrated circuits it has been found that higher levels of interface independence increase the design flexibility and reduce the risk of having to repeat experiments (Thomke, 1997).

Finally, both characteristics of a product's architecture, i.e., its function-component allocation and its interfaces, affect development costs in a particular way as a consequence of the nature of design work. Design costs are one-time costs in the product model's life, i.e., their relative contribution to the unit costs is highly sensitive to changes in the production volume. If only one product is ever produced, say, a racing boat, then this single unit has to bear all development costs which makes the cost for design and development a substantial portion of the unit's life cycle costs. In contrast, for mass-produced products like vacuum cleaners the design costs are shared by potentially millions of identical products, which makes the design and development costs per unit relatively small. For the assessment of cost implications of architectural decisions this issue is also relevant when product architectures allow sharing of portions (platforms, modules, components) of a product across product families, and, therefore, allow the sharing of their development costs. The savings through the reuse of designs affect both development cost and time (Nobeoka and Cusumano, 1995; Reinertsen, 1997; Siddique, 2001; Siddique and Repphun, 2001). Figure 13-9 shows the mechanisms by which individual (and combinations of) product architecture characteristics express their relationship to performance measures such as product development costs and time.

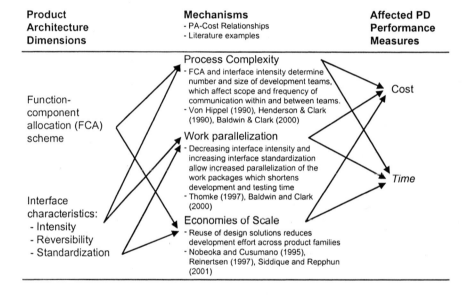

Figure 13-9. Effects of product architecture characteristics on development costs and time.

2.2.3   Product architecture effects on costs of the production phase

With respect to the impact of product architecture decisions on costs that occur during the production phase two sub-sets of processes require separate discussions: (1) manufacturing and assembly, and (2) logistics.

To understand how the first dimension of product architecture, i.e., the size and number of components (function-component allocation scheme) affects manufacturing and assembly costs it is helpful to review the basic idea behind design-for-manufacturing (DFM) and design-for-assembly (DFA) guidelines. Both guidelines help the designer to focus on product characteristics that consume avoidable resources during manufacturing and assembly, respectively, but each with a different rationale. DFM aims at simplifying manufacturing processes, which results—in addition to lower investment—in reduction of process variability and ultimately in faster process rates and higher yields, and thus lower cost. In contrast, DFA generally emphasizes part count reduction, the use of only one assembly direction and the preference of symmetrical parts (Boothroyd, et al., 2002). Empirical evidence exists that supports both claims individually. In case of automobile rear lamp production, for instance, it has been found that complex products requiring complex manufacturing processes result in higher costs compared to simpler parts producible with simpler processes (Banker, et al., 1990). On the other hand, in an analysis of the costs of electromechanical assemblies it has been found that the assembly cost savings through part count reductions can be significant (Boer and Logendran, 1999). Part count reduction is generally seen as a cost reduction tool (Schonberger, 1986; Galsworth, 1994). These findings result in cost curves that increase in opposite directions with respect to the optimal number, and thus complexity, of modules into which a product should be decomposed. The minimum of the sum of the two curves depends on their specific shapes (see Figure 13-10).

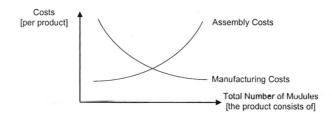

Figure 13-10. Manufacturing and assembly cost behavior with respect to number of modules.

The argument for products requiring simpler manufacturing processes rests essentially on the idea that these processes perform faster and more reliably than their more complex counterparts. Assuming that simpler products require simpler manufacturing processes, this means the product feature complexity affects the efficiency of the process, which in turn directly affects the costs via process speeds and yields. In other words, a design that allows processes to be robust is more likely to consume fewer resources. With respect to product architecture, this observation means that the designer should strive to keep the size of modules or chunks below a complexity level that makes them difficult to manufacture. On the other hand, the argument for products requiring fewer parts (and, as a consequence, fewer manufacturing processes and assembly steps) to achieve lower costs is immediately obvious, as long as the reduction of the number of processes is not paid for with lower yields in the remaining ones. A shift from one manufacturing process to another to reduce part count can have a dramatic impact on assembly time and cost. For example, the instrument panel for the cockpit of the commercial aircraft Boeing 767-4ER used to be manufactured from 296 sheet metal parts and assembled with 600 rivets. A move to precision casting has reduced the part count to 11 and the assembly time from previously 180 hours to 20 hours (Vollrath, 2001). In sum, the product's function-component allocation, i.e., its number and size (complexity) of components, affects both manufacturing and assembly costs, typically in opposing directions, and designers need to develop an understanding of the relative importance of these cost elements in their particular environment.

From a unit cost perspective there is one other effect of product architecture on production costs: this is the use of common components across product families. If the fixed cost portion of manufacturing and assembly can be distributed across a larger number of units, the unit production costs decrease. However, the magnitude of these savings needs to be compared with the potential cost penalties for over-designing a sub-unit or module. For example, products whose costs are dominated by materials costs, i.e., variable costs, such as automotive wire harnesses, may not gain much through the use of commonality (Thonemann and Brandeau, 2000). More generally, the resource use-rate typically decreases with component commonality, but the cost-rate (per cost driver) often increases (Labro, 2004); the final outcome depends on the specific circumstances.

In addition to the product architecture characteristic number and complexity of chunks, the characteristics of the interfaces between the chunks influence the production costs. Interfaces preferred from the low cost production perspective are such that they minimize complexity and uncertainty within the production process. This means, the better the process is known and the more likely it can be performed successfully, and the lower the total number of different processes in the production system is, the lower the expected production costs. The nature and intensity of the interfaces can also be relevant to the production. For example, electronic interfaces consisting of only a plug and a socket may be easier to assemble error-free than a complex mechanical rod connection.

The second subset of production costs is concerned with the aspects of logistics. For the purpose of this chapter, logistics costs encompass costs for storage, transportation, inventory, and work-in-process (WIP). Storage and transportation need to be considered between suppliers and plant, inside the plant, and between plant and customers. Product architecture decisions—the specification of the product's function-component allocation and its interfaces—are most likely to affect these costs to the extent to which they determine packing space and product protection requirements.

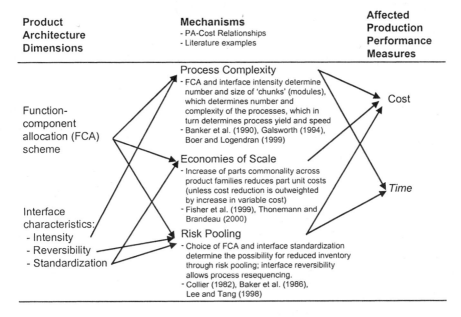

Figure 13-11. Effects of product architecture characteristics on production costs and time.

Product architecture differences also impact the costs for inventory and WIP. The more a product architecture allows late customization or postponement strategies, the more it can contribute to savings in storage and WIP costs through pooling effects. Parts commonality has been identified as a way to reduce the safety stock level for a given service level (Collier, 1982). Others have shown, however, that while the stock for a common part can be lower compared to the unique parts it replaces, the safety stock of the remaining unique components increases if a certain service level is to be maintained (Baker, et al., 1986). These findings have been confirmed for an arbitrary number of products and joint distribution as long as the costs for the product-specific components (that are replaced by a common one) are the same (Gerchak, et al., 1988). For the two-product case, cost ratios have been derived that bound the advantage of the use of common components (Eynan and Rosenblatt, 1996). Another strategy to reduce inventory is to move the common inventory as much upstream in the supply chain as possible to wait with the product customization as much as possible. This strategy might require a re-sequencing of the operations (Lee and Tang, 1998). The key product architecture characteristic for this strategy is the interface reversibility. If it is low, an operation reversal may not be possible because the technical nature of the operations prohibits a reversal (e.g., in the case of steel components welding has be completed before painting). In sum, the use of common parts can reduce inventory, but it needs to be investigated with the specific demand pattern, the relative costs of the components, and other product architecture constraints in mind.

The product architecture's effect on time can have an additional impact on costs via the detour of increasing demand volatility. Because demand volatility increases upstream ('bullwhip-effect'), product architectures can reduce this effect if they allow for parallelization of production to achieve short lead times. Long lead times, together with high levels of demand uncertainty, can amplify the bullwhip-effect and create significant additional costs in the supply chain (Levy, 1994). Overall, a complete assessment of the impact of architectural characteristics on production costs should incorporate manufacturing, assembly, and logistics costs, and evaluate how to balance these different effects. Figure 13-11 summarizes the effects of individual product architecture characteristics on production cost and time.

2.2.4   Product architecture effects on costs of the use phase

In general, three types of costs occur during product use: (1) the costs for operation, (2) the costs for maintenance, and (3) all external costs incurred by the operation of the product.

Most products require some input to operate them. The costs for these inputs can be for fuel or utilities like energy, water, or pressurized air, or costs incurred by the product's characteristic, for instance labor requirements for a machine operation. While it is very difficult to make a general statement about the relationship between product architecture characteristics and operation costs, some issues can be pointed out. Operating costs typically contain two types: costs for standard operation and costs for preparation activities, for example training. The training of personnel is analogous to the setup of a machine: a process necessary to begin operation. Similar to the production arena, if the set-up time, i.e., training time, can be reduced, then the system's productivity increases. A product architecture can contribute to this reduction in 'set-up time' by utilizing common components across members of a product family (which requires proper function-component alignment). For example, aircraft producers are trying to install similar, if not identical, cockpits into airplanes of different sizes to reduce the airlines' need to retrain their crews (Anonymous, 2005). Similarly, if it is not the operator that changes (as in the airplane case) but the task that the product has to accomplish (e.g., a machine tool that is planned to produce a variety of components) then a product architecture that supports to reconfigure the product quickly is advantageous (Landers, et al., 2001). Proper function-component alignment and high degrees of interface reversibility are key in this situation to improve the productivity of the product by reducing its set-up costs and, thus, its operating costs as measured by units produced per time unit.

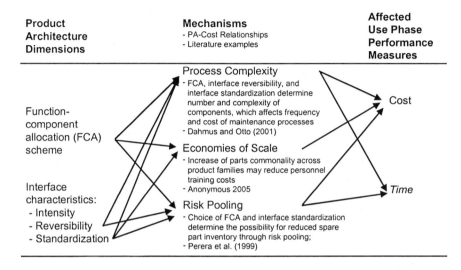

Figure 13-12. Effects of product architecture characteristics on use phase costs and time.

With respect to maintenance costs that occur during a product's use phase two major questions are relevant. First, what is the likelihood that maintenance (and its costs) will occur during the product's use phase and, second, what will be the anticipated costs for this maintenance procedure? Grouping parts with similar expected lifetimes together is likely to reduce the repair and replacement costs by minimizing the required parts replacement processes (Dahmus and Otto, 2001). A proper module definition (function-component allocation) can help achieving this goal. In addition, a product architecture that allows easy and fast access for maintenance and repair requires less time to execute the actual maintenance procedure and, consequently, leads to lower maintenance costs. The product architecture characteristic interface reversibility is the important design variable in this case. Also, in case that a product has multiple identical parts (function component allocation) fewer parts need to be stocked in inventory (compared to unique parts) for providing the same level of availability (Perera, et al., 1999). Like risk pooling across products in production, this strategy translates into lower spare part inventory costs as part of the maintenance costs. Note that the different elements of maintenance costs described above may react differently to the same product architecture design decision.

Finally, the operation of any product may also cause so-called external costs, e.g., damages to public health or the environment through emissions. A link between product architecture design decisions and external costs is very difficult, if at all, to establish and goes beyond the scope of this work. Figure 13-12 recapitulates the effects of individual product architecture characteristics on costs and time in the product use phase.

2.2.5   Product architecture effects on costs of the retirement phase

In the last phase of a product's life cycle, costs are created through activities like disassembly or disposal. In addition to these direct costs, external costs, like degradation of the environment or air quality, can occur.

To estimate disassembly costs as a function of the product architecture is very difficult, particularly since it is often unclear which disassembly sequences is the most economically viable one. The reverse of the assembly process may, or may not, be the most cost effective way to disassemble the product. Researchers have suggested a number of scoring processes to compare disassembly efforts for different designs. Some suggest comparing disassembly costs for different designs on a relatively high level of aggregation. Emblemsvag and Bras (1994), for instance, propose to list all activities the disassembly of various products would require, compute the costs for each activity per time unit, determine the time each design requires each activity, and compare the results. This type of analysis, however, does not reveal specifically which architectural features make one design more costly to disassemble than another. To answer this type of question more detailed analyses are required. Das, et al. (2000), for example, propose to compute a disassembly effort index based on seven factors, such as time, tools, fixtures, access, instruct, hazard, and force requirements. The fact that both the score for each of these factors as well as the weights among them are based on qualitative assessments demonstrates the difficult nature of the task to estimate disassembly costs unambiguously. Others have extended this work to include bulk recycling in addition to disassembly activities (Sodhi and Knight, 1998). However, while the product architecture affects disassembly costs (via the dimensions function-component allocation scheme and interface reversibility), its impact on bulk recycling is only relevant together with the specific values of the materials involved. Finally, while determining the costs to landfill a product (or parts of it) is relatively straightforward, the results, however, are unlikely to depend on architectural characteristics of the product (leaving material consideration aside). Figure 13-13 summarizes the product architecture's effects on costs in the product retirement phase.

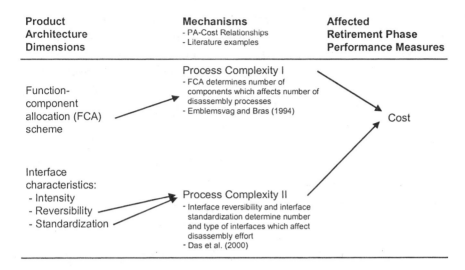

Figure 13-13. Effects of product architecture characteristics on retirement phase costs.

This section has demonstrated that the designer responsible for defining the product's architecture faces a difficult task. Since the analysis for product architecture costing requires a decision on which life cycle phase to include, the designer must develop an understanding of the longitudinal tradeoffs that product architecture design decisions face between life cycle phase and within individual life cycle phases. The second step of the roadmap presented in this section provides a guideline to develop this understanding.

2.3 Step 3: Determine relevant cost allocation rules and their impact on cost analyses

Once the various cost types that can occur over a product's life and the relationships between product architecture design decisions and these costs are identified, the third step of the roadmap requires to determine the rules for the cost allocation procedures. Particularly relevant for the results of any cost analysis are the—often only implicit—assumptions on the analysis boundaries, on the overhead allocation mechanisms, and on the dynamics of the process under investigation.

2.3.1 Unit of analysis

Typically, product unit costs are chosen for cost comparisons of assembled products. There are, however, other units of analysis that could be selected alternatively: product families, product programs, departments, factories, companies, or entire economies. The order of this list of potential levels of analysis indicates an increasing distance from the physical object itself. While a cost analysis focusing on a product makes it easy to assess costs that are directly related to the product (e.g., material consumption), it makes the allocation of more 'distant' costs (e.g., factory guards) very difficult. On the other hand, for cost analyses on a company level, almost all costs are somewhat 'direct' (see Figure 13-14).

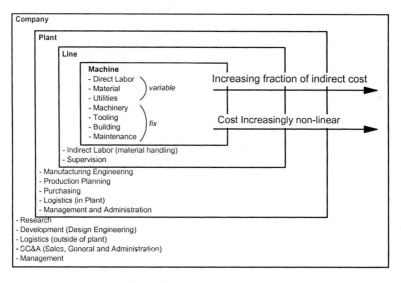

Figure 13-14. Different levels of cost analysis.

The direct-indirect cost classification depends on the choice of the cost object. "A useful rule of thumb is that the broader the definition of the cost object, the higher the proportion of its total costs are its direct costs—and the more confidence management has in the accuracy of the resulting cost amounts. The narrower the definition of the cost object, the lower the proportion of its total costs are its direct costs—and the less confidence management has in the accuracy of resulting cost amounts." (Horngren and Foster, 1991:28) Since product architecture costing is concerned with the cost effects that product architecture choices trigger, it is logical to focus the cost analysis on a level where product architectures can be distinguished, i.e., on the product or product family level. This in turn creates the above mentioned allocation problem of how to allocate the significant indirect cost portion, often called 'overhead.' Overhead usually encompasses costs with various levels of 'indirectness.' For the interpretation of cost consequences of product architecture design decisions it is very important to understand the mechanisms by which these overhead costs are allocated.

2.3.2   Allocation of overhead costs

The accounting literature employs two distinctions for costs: direct versus indirect costs and fixed versus variable costs (Horngren and Foster, 1991). While the first uses the cost traceability to separate direct from indirect costs, the second uses the dependency with regards to changes in production volume as a measure to classify fixed and variable costs.

In the production arena, costs that are typically considered variable are costs for direct labor, materials, and utilities. In contrast, machinery, tooling, and building costs are usually considered fixed costs. These distinctions, however, are not clear-cut, but depend on the chosen time horizon, the chosen manufacturing technology, and the chosen accounting principle. A change in the chosen time horizon can turn the same costs from fixed into variable costs. Labor costs are typically viewed in short time frames as fixed costs whereas in the long run they are typically treated as variable in nature. The choice of a manufacturing technology may determine whether a specific or a generic tool is deployed. A shear as a cutting tool that can be used to produce other products as well exhibits variable cost behavior whereas a specific cutting die that does the same job, but can only be used for this specific product becomes fixed costs. Finally, certain accounting principles can shift costs from the fixed costs category into the variable cost category, and vice versa. The assumption, for example, that free machine capacity can be employed for other jobs turns the allocated machine costs effectively into variable costs whereas the assumption that the machine is dedicated to a specific product results in fixed cost behavior.

In sum, what is typically called overhead is a broad category with often fuzzy boundaries. It is, however, a category that becomes increasingly important due to increasing product and process complexity, shrinking direct labor content, shorter product life cycles, and increasingly heterogeneous markets (Miller and Vollmann, 1985; Doran and Dowd, 1999; Cokins, 2000). Table 13-1 gives an overview of the magnitude of some overhead costs found in recent studies.

One characteristic feature of overhead costs is their lack of direct dependency on production volume. Activities that support in various ways the actual production processes do not necessarily vary in direct proportionality with the production volume. It has been argued that the costs for these activities vary with the intensity or frequency of these activities. For example, the time and manpower to write a purchasing order does not vary with the number of equal parts ordered, but each order incurs an average cost for the transaction 'write purchasing order.' This insight triggered the development of activity-based costing (ABC) (Kaplan, 1991; Kaplan and Cooper, 1998). ABC promotes a cost allocation process in proportion to the activities consumed by the products produced. The basic idea of ABC is to calculate the costs of activities (cost drivers) and 'charge' products with the time with which they consume an activity times the use rate per time unit. The cost drivers can be on various levels in the firm: "While some activity cost drivers are unit-related (such as machine and labor hours), as conventionally assumed, many activity cost drivers are batch-related, product-sustaining, and customer-sustaining" (Cooper and Kaplan, 1992:4).[20]

Table 13-1. Overhead costs found in recent studies.

Author(s)	Total Costs (=100%)			Activities considered	Industry
	Direct Material	Direct Labor	Mfg. Overhead (MO)		
(Banker, et al., 1995)	65.4%	8.9%	25.7%	Plant level study	Electronics, Machinery, Automobile components (mean values of 32 facilities)
(Foster and Gupta, 1990)	54.3%	6.6%	39.1%	Procurement, Production, Support	Electronics (mean values of 37 facilities)
(Galsworth, 1994: 85)	40%-65%		35%-60%	Total Costs: Function cost: 40% Variety cost:25% Control cost: 35%	Manufacturing
(Hundal, 1997)	45%-65%	8%-20%	22%-40%	Not specified	Aerospace, Computers, Electronics, General Equip., Automobiles
(Miller and Vollmann, 1985)	20%-40%		60%-80%	Overhead Costs: G & A 20% Indirect Labor 12% Engineering 15% Equipment 20% Materials OH 33%	Electronics

While ABC represents an invaluable step towards a better understanding of how to allocate what used to be called 'overhead,' it is still helpful to review some of the assumptions that underlie even ABC with respect to product variety. More specifically, these assumptions are concerned with linearity of activity-cost relationships, with different types of variety, and sequence-dependent variety costs.

---

[20] Some have criticized ABC as leading to poor short-term decisions, and suggested the Theory-of-Constraints (TOC) as a better tool for short-run cost allocations. TOC assumes all costs other than direct material as fixed (Goldratt and Cox, 1984). Then, to maximize profitability, TOC seeks to maximize throughput. TOC promotes finding the bottleneck in an existing system and adjusting all other production to it to eliminate inventory. In the debate about whether ABC or TOC is the superior way of interpreting costs, various authors argue to understand both methods as opposing ends of a continuum with respect to planning time horizon: ABC for long-range planning, TOC for short-term decisions (Fritzsch, 1997; Cooper and Kaplan, 1998; Kee, 1998). Since the choice of the product architecture is a rather long-term decision, ABC is the more relevant method for our purposes here.

ABC argues that many overhead costs are related to activity type and activity frequency rather than production volume. Standard ABC typically assumes a linear relationship between activity and cost. The limits of this assumption, however, become apparent in case of product variety. Product variety often causes additional work in activities such as planning, control, monitoring, and coordination (Lingnau, 1999). Not only does this cost propagation effect make it more difficult to trace individual costs, it often creates also an additional allocation problem: if product variety creates costs above the sum of the costs of the individual products, how are these variety-related extra costs allocated to the individual products?

To make matters more complicated, product variety can also take on different forms each of which has a different effect on costs. For example, Ittner and MacDuffie (1994) defined three levels of product variety in their study of overhead costs in automotive assembly plants: core or fundamental variety (model mix complexity), intermediate variety (parts complexity), and peripheral variety (option complexity). They find empirical support only for the latter two affecting overhead costs, ".. reflecting the considerable logistical, coordination, and supervisory challenges that accompany an increase number of parts and more complex manufacturing tasks." (Ittner and MacDuffie, 1994:29) Another approach to specify product variety has been followed by Anderson (1995) who measures the impact of product mix heterogeneity on manufacturing overhead costs by identifying seven independent product attributes, using engineering specifications. By measuring on the attribute-level, Anderson finds that increased overhead cost "is associated with increases in the number and severity of setups and increased heterogeneity in process specifications (expected downtime) and quality standards (defect tolerance heterogeneity) of a plant's product mix" (Anderson, 1995:383).

Finally, how product variety is distributed over time can affect the effort to balance and sequence a production line. For example, taking a production perspective a study of product variety finds that "[o]ption variability has significantly greater negative impact on productivity than option content in automobile assembly" (Fisher and Ittner, 1999:785). In this case, variety's impact on indirect and overhead labor is much greater than it is on direct labor. The authors explain this with the built-in slack in automotive assembly lines that allows handling option variation in the first place. They point out that because these costs are born through the variability complexity it is difficult to allocate these excess costs to any specific product.

With respect to the question of how the link between product architecture characteristic and cost is influenced by the cost allocation procedure some general observations can be made. A product architecture that allows operations conducted closer on a per-unit basis allows more precise cost allocation. For example, a process that produces only one part at a time allows easy allocation of all non-direct costs (setup, purchasing, etc.). In contrast, product architectures that cause complex logistical, balancing, sequencing, or quality processes may make the cost allocation more difficult. Within limits, these arguments call for products with architectures consisting of fewer, more modular-like components (dimension function-component allocation) and with high levels of interface standardization.

2.3.3   Process dynamics

The third issue of the roadmap's third step: 'determination of cost allocation rules,' is concerned with the extent to which the processes under consideration are considered static or dynamic. There are two cases of non-static situations: (i) a one-time change followed by a static period, and (ii) a change over longer periods of time. In the first case, the relevant issue is the ratio of 'ramp-up period' to 'normal production period.' If, for example, an entire production run will extend over several years and the ramp-up takes only a few days, the cost analysis focus can be put on the system costs assuming it in its static condition. In contrast, if the production run is relatively short and the ramp-up takes up a significant portion of it, the systems costs are not well represented by the production run alone. In some production environments the ramp-up time can represent a significant fraction of total production time, e.g., it can take up to six months to bring an automotive assembly plant up to full production load (Almgren, 2000).

Cost changes over longer periods of time can occur in two ways: the change itself can either be constant or variable. The case in which the change is (for the most part) constant is often caused by what has come to be known as the learning curve effect. The argument is that with accumulating production volume workers and engineers are getting better in what they are doing. They improve the processes and their work environment in a manner that continuously improves their overall productivity. Often times the learning effect is measured as a constant fraction of cost reduction, e.g., 20%, with every cumulative doubling of the production volume. Empirical evidence has been presented that this effect indeed exists (Anderson, 1995). Activity-based costing systems can help to detect these learning effects (Andrade, et al., 1999).

In the second case of changing unit costs the change itself is dynamic, i.e., unit costs do not change by a constant rate but follow dynamic patterns. An example of this phenomenon is the case of non-constant unit costs as a result of different ways of sequencing different products through jointly used production processes. Flexible manufacturing systems (FMS), for example, can manufacture different products on the same machine. The set-up time, however, may depend on what product has been produced prior to the one under consideration. Will the same tool be used? If not, is the tool change time dependent on what tool was used for the previous product? This problem has been addressed through the use of ABC systems in conjunction with production planning models (Koltai, et al., 2000).

With respect to the effects of product architecture choice on unit cost, the phenomenon described in this section cannot be determined with product architecture data alone, but requires data (or assumptions) on the production environment including scheduling and the production program information.

2.4 Step 4: Select cost model appropriate for the design decision at hand

The fourth step of the roadmap for product architecture costing requires the selection of one or several cost models that are appropriate for the design decision at hand. A number of cost models have been developed to help designers to assess the economic consequences of design decisions. The existing models can be grouped into three categories: parametric, analogous, and analytical. Parametric models aim at establishing scaling factors of cost drivers found through analysis of historical data. Regression analysis is a typical method to extract such scaling factors. Due to the simplicity in use, parametric techniques are used in many industries (Bielefeld and Rucklos, 1992; Uppal 1996). Non-parametric methods such as neural networks have also been applied to find design variable-cost relationships (Bode, 2000).

The underlying idea for analogous models is to search for similarities between the design at hand and a large number of historical cases stored in databases. To be able to compare products on multiple levels (product, subassembly, part, etc.) hierarchically structured approaches have been developed (Liebers and Kals, 1997; Rehman and Guenov, 1998; Ten Brinke, et al., 2000). Other approaches focus more on abstract elements like features (Brimson, 1998; Leibl, et al., 1999).

Finally, cost models in the third category, analytical cost models, come in two very different flavors. One category is represented by abstract mathematical models, often used to generate insights into general questions. Their emphasis is mostly on structural tradeoff modeling, while the functions of relationships between individual design decisions and costs are typically assumed to be known in their shape (Roemer, et al., 2000; Thonemann and Brandeau, 2000; Krishnan and Gupta, 2001). The other flavor of analytical models is represented by detailed technical cost models of (mostly) manufacturing processes to estimate the associated costs (Clark, et al., 1997; Locascio 1999; Locascio, 2000; Kirchain, 2001). Technical cost models model manufacturing processes based on the process physics and establish links between a few critical design parameters and the process dynamics, which in turn determine the costs. Existing cost modeling techniques are not discussed here in detail; the interested reader is referred to recent reviews in (Asiedu and Gu, 1998; Layer, et al., 2002).

Instead of presenting the different cost models in detail, this section presents four criteria to help thinking about making the appropriate cost model selection when assessing cost implications of product architecture design decisions. First, does the cost model or technique require a substantial data set of similar cases? Regression analyses or neural networks, for example, usually require sufficient cases to be able to produce relevant cost predictions. Second, how large is the number of acceptable cost drivers? Most cost modeling techniques allow only a limited number of cost drivers. To some extent this question is related to the previous one in that the number of available cases restricts the number of acceptable cost drivers. Third, how large are the acceptable differences between the product architecture candidates under investigation? This criterion is particularly relevant if substantially different product architectures are to be analyzed. Modeling techniques that build on a set of known cases are usually limited when applied to entirely new cases. Finally, what certainty level is required for the input data? As indicated earlier, cost analyses in early design stages typically lack detailed and accurate product design date. The assessment of the cost models along this fourth criterion reveals the underlying modeling philosophy. Some models use search procedures to find relevant data among existing cases (e.g., analogous models) whereas others build the cost analysis for every case anew (e.g., process-based cost models). Depending on the goal of the product architecture analysis and the available data, different methods are advantageous. Table 13-2 summarizes the various cost modeling approaches with respect to product architecture costing.

Table 13-2. Assessment of various cost estimation models along four application criteria.

Cost Estimation Models	Application Criteria	Data Set Requirement (min case base)	Acceptable Number of Cost Drivers	Acceptable Difference in Architecture Decomposition	Required Certainty of Data Input
Parametric	Regression Analysis	Large	Low	Small	Medium
	Complexity-Theory Based	Medium	Low	Small	High
	Neural Networks	Large	Low	Small	Medium
Analogous	Feature-Based	Medium	Low	Small	High
	Expert Systems	Large	Medium	Medium	High
Analytical	Abstract Modeling	Small	Small	Small	None
	Process-Based Cost Models	Small	Medium	Large	Medium

3. CONCLUDING REMARKS

This chapter has introduced a roadmap for product architecture costing. Each step of the roadmap prior to the actual modeling of a specific situation, i.e., (1) to assess differences of product architecture candidates, (2) to identify relevant product life cycle phase(s) and their product architecture-cost relationships, (3) to determine relevant cost allocation rules, and (4) to select cost models appropriate to the situation at hand have been discussed in detail. This comprehensive discussion of how individual product architecture characteristics affect specific cost elements over a product's life cycle can serve as a guideline when formulating various tradeoffs. For example, a manufacturer of long-living products, e.g., a ship builder, might want to tradeoff costs for building the ship with the costs for operating it. In contrast, a manufacturer of mass-produced consumer goods might be more interested in the cost tradeoff between the costs for parts fabrication and the costs for assembly. For any given firm, the determination of the relevant tradeoffs is impacted by such factors as the firm's business model, its warranty policies, and its competitive and legal environment. The roadmap also provides an overview of how cost allocation rules can affect cost analyses results, and thus the cost advantage of one product architecture over another. Finally, the roadmap includes a categorization of existing cost models, and illustrates which one to select depending on the size of the available data set, the given data set's level of variation and accuracy, and the number of acceptable cost drivers.

This roadmap for product architecture attempts to provide a comprehensive consideration of the relevant questions when conducting an analysis of the cost consequences of product architecture differences. The relationships identified and cost models presented can now serve as stepping stones for the development of user-friendly design guidelines as well as of more complex optimization models (see Figure 13-1). Some thoughts on how these next steps could proceed follow.

The development of product architecture design guidelines that lead the designer towards 'better' product architectures, given the requirements that the product faces, can be envisioned similar to the development of the well-known DFM/DFA guidelines. The DFM/DFA guidelines represent the condensed experience across many cases of design changes with respect to manufacturing and assembly. Similarly, a database containing the results of many specific product architecture-cost analyses could be used to search for more general patterns of cost effects that are due to differences in product architecture. As a step in this direction a firm might build a repository of their own cost data and associate the data with the corresponding product architecture characteristics. This way the firm might populate the product architecture-cost space with more of its own data points. Over time, this would offer the chance to introduce internal learning into the product architecture design process (Anderson and Sedatole, 1998) and would foster the construction of product architecture design guidelines.

The product architecture-cost relationships identified by the roadmap can also inform the process of building more complex models that can support the product architecture development process more dynamically. While knowing the cost effects, allocation rules, and cost models discussed in this chapter allows evaluating cost consequences of differences along individual product architecture characteristics, this knowledge does not automatically feed back into the product architecture design process. If it were possible to turn product architecture characteristics into variables that exist across the entire solution space—which they currently often do not—they could be used to find optimal architectures, optimal with respect to the cost determined as relevant. With respect to the product architecture development process, this would replace the process of selecting among product architecture candidates with one that helps designers to develop more cost effective product architectures by giving immediate feedback to product architecture design suggestions. One particularly promising extension of this research direction on product architecture costing is the treatment of uncertainty. While uncertainty is inherent in any estimation of future data, the way in which it is modeled might provide additional insights for the product architecture selection and development decisions. While deterministic cost models can be augmented with sensitivity analyses, more sophisticated measures of risk and uncertainty could advance the cost modeling tools, and by extension, the product architecture creation.

4.   ACKNOWLEDGMENTS

For helpful comments and discussions on cost modeling the author would like to thank Joel Clark, Frank Field, Randolph Kirchain, and Rich Roth. In addition, the chapter benefited from helpful comments from three anonymous reviewers. Research funding from the International Motor Vehicle Program (IMVP) at MIT is gratefully acknowledged.

Chapter 14

AN ACTIVITY-BASED COSTING METHOD TO SUPPORT PRODUCT FAMILY DESIGN

Jaeil Park and Timothy W. Simpson
The Harold and Inge Marcus Department of Industrial and Manufacturing Engineering, The Pennsylvania State University, University Park, PA 16802

1.   INTRODUCTION

As companies are being challenged to produce a wider variety of products to satisfy customers that have different needs while maintaining competitive prices, platform-based product family development has become a cost-effective method for reducing production costs (Roberson and Ulrich, 1998). In general, production costs are generated by production activities ranging from purchasing raw materials to distributing finished products, and those activities consume direct and indirect resources (Horngren, et al., 2000). These costs are identified and collected through management accounting systems that companies have developed for accounting purposes and used to estimate the production costs of existing products. However, many management accounting systems are incapable of providing the necessary information to support platform-based product development because many companies have developed their own accounting systems to help them remain profitable and eliminate unnecessary costs in production. In many cases, the primary objective of management accounting systems is to support management to control overall equipment efficiency (OEE)[21] and keep it as high as possible.

---

[21] Overall Equipment Effectiveness (OEE) helps managers focus on improving the performance of machinery and plant equipment they own. OEE = Availability x Performance Rate x Quality Rate. This simple formula provides an excellent benchmarking tool to see how companies are doing in terms of overall equipment utilization, production speed, and quality. For more details, we refer the reader to: http://www.bin95.com/Overall_Equipment_Effectiveness_OEE.htm.

As designers become more aware of how much it costs to develop a specific product platform or product family, informed product decisions can be made. A cost system should be constructed in such a way to facilitate decision-making during product family design. Traditional accounting systems adopted by many companies focus more on direct costs such as material and labor costs, which are traceable to products, and less on indirect costs, which are shared by more than one product. It turns out that indirect costs are frequently lumped together and allocated to products arbitrarily by averaging them across products. Traditional accounting systems are appropriate when direct costs becomes dominant in production, or indirect costs change proportionally to production volumes. In situations where indirect costs become a large portion of production cost due to a proliferation of products (Anderson, 1997) and where they do not always vary proportionally to production volumes, traditional accounting systems are no longer appropriate for cost estimation. If designers nonetheless count on traditional accounting systems, indirect costs that are allocated to individual products based on volume-based cost drivers might distort cost information for a family of products, which can lead to making inappropriate decisions (Cooper and Kaplan, 1988).

During product family design, customers' needs in various market segments are translated into the individual products in the family, and common components or subsystems of the products are combined as a product platform that supports the individual products (Simpson, 2004). Increasing the number of individual products in the family increases indirect costs relating to inventory, setup, inspection, maintenance, material handling and storage, and rework while the product platform serves as a source of cost savings by reduction of indirect costs as well as direct costs associated with it. In this context, estimating cost savings from the product platform becomes an emerging challenge for designing a cost-effective product platform and corresponding family of products.

Activity-based costing (ABC) is a useful costing method to help measure indirect costs more accurately by classifying activities, assigning indirect costs in traditional accounting systems to the activities (i.e., activity costs), and then allocating the activity costs to products by measuring the cost drivers of the activities. For illustrative purposes only, traditional and activity-based costing are shown in Figure 14-1 (Bruns, 1989). The overhead costs account for 53% of total expenses (i.e., a summation of labor, material, and overhead costs), which are incurred by producing three different products such as valves, pumps and flow controllers, and the standard unit costs decrease or increase depending on ways in which the overhead costs are allocated.

	Valves	Pumps	Flow Controllers
Monthly Production	7,500 units	12,500 units	4,000 units
Material	$16.00	$20 00	$22 00
Direct labor	$4 00	$8 00	$6.40
Overhead @ 439% of direct labor	$17.56	$35 12	$28.10
Standard unit cost (Tradtional costing)	$37.56	$63.12	$56.50

Overhead

Machine depreciation	$270,000	
Set-up labor	$2,688	
Receiving	$20,000	
Material hanlding	$200,000	
Engineering	$100,000	
Packing and shipping	$60,000	
Maintenance	$30,000	
Total overhead costs	$682,688	

Total labor costs = 9,725 hours x $16 = $155,600
Total Material costs = 7500 x $16 + 12500 x $20 + 4000 x $22 = $458,000
Overhead rate = $682,688 / $155,600 = 439%

	Valves	Pumps	Flow Controllers
Monthly Production	7,500 units	12,500 units	4,000 units
Material	$16	$20	$22
Direct labor	$4	$8	$6.40

Overhead

	Valves	Pumps	Flow Controllers
Machine Depreciation	$12 50	$12.50	$5.00
Set-up labor	$0.02	$0.07	$0.44
Receiving	$0.08	$0 31	$3.88
Material hanlding	$0 83	$3.10	$38.76
Engineering	$2.67	$2.40	$12.50
Packing and shipping	$0.27	$1.12	$11.00
Maintenance	$1.39	$1.39	$0.56
Overhead total	$17.75	$20 89	$72.13
Standard unit cost (ABC)	$37.75	$48.21	$100.53

Figure 14-1. Traditional (left) and ABC (right) systems.

Let us take a closer look at the problems of traditional costing systems. Suppose that a company produces Products A and B at the costs shown in Table 14-1. All the production expenses are aggregated and allocated to products based on their direct labor content. The unit costs of Products A and B are estimated at $3.90 and $4.50, respectively, with an indirect rate of 300%, which is computed by dividing the total indirect cost ($162,000) by the total direct labor cost ($54,000). The company attempts to develop a new product whose direct labor cost is $8,000, which is estimated from previous engineering experience. The question is: how to estimate the indirect cost of the new product in a circumstance where the indirect cost is a significant portion of total cost? If the same indirect rate is applied, the production cost of the new product is estimated at $6.20.

Table 14-1. Traditional product costing.

	Product A	Product B	Total	New Product
Sales Volume	$50,000	$35,000	$90,000	$10,000
Material Costs	$75,000	$60,000	$135,000	$30,000
Direct Labor	$30,000	$24.000	$54,000	$8,000
Overhead @ 300%	$90,000	$72,000	$162,000	$24,000
Unit Cost:	$3.90	$4.50		$6.20

Now suppose that a designer collects the activity costs for Products A and B. The indirect costs (overhead costs) of Products A and B are assigned to each activity, and each activity cost is allocated to Products A and B via its cost driver as shown in Table 14-2. For instance, $80,000 is allocated to the activity cost of handling, and the total quantity of the cost driver is 90 for Products A and B. Its cost driver rate is given by $889 ($80,000/90) per production run. For the New Product, which requires 15 production runs, its handling cost can be estimated at $13,333 (15 x $889). The estimated costs of the new product are shown in Table 14-3. In this example, the production cost of the new product is estimated to be even higher ($7.20) because those activities associated with the new product are more involved compared to the production volumes of other products.

As such, ABC is a more appropriate costing method for product family design because it helps measure indirect costs more accurately, which are not necessarily proportional to the volume of products produced and cost savings by platforms. Let us suppose that Products A and B are designed as a family based on a platform and that this platform reduces activity use such that the quantities of the cost drivers shown in Table 14-4 are halved. The activity costs and unit costs of the family are estimated in the same fashion as before and shown in Table 14-5. The costs are significantly reduced from $8.30 ($4.00 + $4.30) to $6.80 ($3.20 + $3.60) by the platform. The costs do not include direct material cost savings from the platform. With this cost information, product designers can make better decisions about the product platform for a family of products. The next section discusses a way to implement ABC in more detail when considering multiple products.

Table 14-2. Activity costs and their allocation to products.

Activities	Cost	Cost Drivers	Product A	Product B	Total	Cost Driver Rates	New Product
Handling	$80,000	# production runs	50	40	90	889	15
Setup	$30,000	Setup time (hours)	200	50	250	120	60
Support	$20,000	# products	1	1	2	10,000	1
Machining	$32,000	# machine hours	5,000	5,000	10,000	3.2	1,000

Table 14-3. Activity-based costing for the same three products.

Activities	Product A	Product B	New Product
Material Costs	$75,000	$60,000	$30,000
Direct Labor	$30,000	$24,000	$8,000
Handling	$44,444	$35,555	$13,333
Setup	$24,000	$6,000	$7,200
Support	$10,000	$10,000	$10,000
Machining	$16,000	$16,000	$3,200
Unit Cost:	$4.00	$4.30	$7.20

Table 14-4. Cost drivers for the product family.

Activities	Cost Driver Rates	Cost Drivers	Product A	Product B
Handling	889	# production runs	25	20
Setup	120	Setup time (hours)	100	25
Support	10,000	# products	0.5	0.5
Machining	3.2	# machine hours	5,000	5,000

Table 14-5. Cost drivers for the product family.

Activities	Product Family	
	Product A	Product B
Material Costs	$75,000	$60,000
Direct Labor	$30,000	$24,000
Handling	$22,225	$17,780
Setup	$12,000	$3,000
Support	$5,000	$5,000
Machining	$16,000	$16,000
Unit Cost:	$3.20	$3.60

2.   ACTIVITY-BASED COSTING FOR MULTIPLE PRODUCTS

In a cost structure where indirect cost is a large portion of total cost due to diverse products with large volume variations, the main objective of using ABC is to provide designers with relevant cost information regarding the contributions that each product makes to the overall production cost. ABC is represented by linkages between activities and the resources consumed by these activities. The general principle is to use separate activity costs if the cost or productivity of resources is different and if the pattern of demand varies across resources (Atkinson, et al., 2004). The following steps are used to implement ABC for multiple products.

- *Step 1 - Describe the production system based on its activities:* The first step is to describe the production processes that designers are interested in or concerned about using activities
- *Step 2 - Classify major activities and the resources used by them:* The second step is to classify major activities and the resources consumed by the activities to produce each product.
- *Step 3 - Collect costs (expenses) for each activity (i.e., activity costs):* Collecting the costs generated by each activity is crucial in ABC. Since most accounting systems are designed to collect the costs by functional areas in the production department, the cost information necessary to determine activity costs may not be readily available. Additional efforts on collecting costs on activities should be made to investigate relative activity percentages of indirect resources.
- *Step 4 - Select activity cost drivers:* An activity cost driver is a measurable unit of performing an activity. Cost drivers should relate to ways in which activity costs are consumed.
- *Step 5 - Assign the activity costs to each product:* The activity cost driver rates of the activities are calculated by dividing each activity cost by the total quantity of each activity cost driver. Activity cost driver

rates are then multiplied by the quantity of each activity cost driver used by each of products.

To demonstrate the aforementioned steps, a simple example follows.

- *Step 1* – A production system can be described in a graphic expression of activities using a flow diagram, see Figure 14-2 (Park and Simpson, 2004). This example system is currently producing three products.

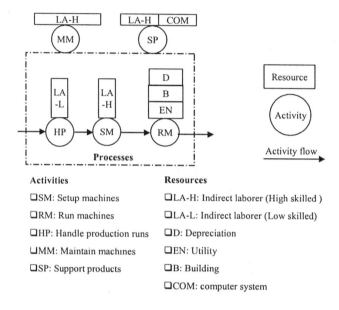

Activities

- ☐SM: Setup machines
- ☐RM: Run machines
- ☐HP: Handle production runs
- ☐MM: Maintain machines
- ☐SP: Support products

Resources

- ☐LA-H: Indirect laborer (High skilled)
- ☐LA-L: Indirect laborer (Low skilled)
- ☐D: Depreciation
- ☐EN: Utility
- ☐B: Building
- ☐COM: computer system

Figure 14-2. Example of an activity flow.

- *Step 2* - An extensive study should be undertaken to identify the indirect costs (i.e., supporting costs) that designers believe are driven by certain activities. These activities are product-specific and depend on the level of difficulty to establish a link between activities and their costs, but these activities can usually be reduced to 20% of the activities consuming 80% of the cost[22]. From Figure 14-2, the following activities are identified: setup machines, run machines, handle production runs, maintain machines, and support products.
- *Step 3* - Costs are collected through the accounting system. For example, indirect labor of $1000 is needed to setup machines from one product to another on a monthly basis. The activity costs and resources consumed by the activities in Figure 14-2 are shown in Table 14-6.

---

[22] In ABC, it is common to find that 20% of the activities consume 80% of the cost. Best practice organizations focus ABC cost object allocation methods for the top 20 most expensive activities.

Table 14-6. Activity costs and resources consumed by the activities.

Activities	Costs	Resources Consumed by Activities
Setup	$1,000	Indirect laborers
Machining	$30,000	Depreciation, utilities, building
Handling	$10,000	Indirect laborers
Maintaining	$1,000	Indirect laborers
Support	$3,000	Indirect laborers, computer system

- *Step 4* - Cost drivers are selected to represent the quantity of activities used to product individual products. The cost drivers for this example are listed in Table 14-7.

Table 14-7. Activity cost drivers.

Activities	Cost Drivers
Setup	Set up hours
Machining	Machining hours
Handling	# production runs
Maintaining	Machining hours
Support	# products

- *Step 5* - Once the activity cost drivers are determined, designers obtain quantitative information on the quantity of each activity cost driver and the quantity of cost drivers used by each product. This information is summarized in Table 14-8. Each activity cost is assigned to the products by activity cost driver rates that are calculated by dividing the activity costs by the total quantity of the activity cost drivers. The unit indirect costs of Products A, B, and C are estimated at $24.40, $36.80, and $122.00, respectively including tooling costs. The total indirect unit cost of all three products is estimated at $183.20.

Table 14-8. Cost drivers and cost allocation for each product.

Activities	Costs	Cost Drivers of Product A	Product Costs of A	Cost Drivers of Product B	Product Costs of B	Cost Drivers of Product C	Product Costs of C
Sales Volume		100		500		100	
Setup	$1,000	50 hours	$500	30 hours	$300	20 hours	$200
Run	$30,000	500 hours	$20,000	300 hours	$12,000	200 hours	$8,000
Tooling	$10,000		$2,000		$5,000		$3,000
Handle	$10,000	60	$5,455	30	$2,727	20	$1,818
Maintain	$1,000	500 hours	$455	400 hours	$364	200 hours	$182
Support	$3,000	1	$1,000	1	$1,000	1	$1000
Total	$55,000		$24,409		$18,391		$12,200
Unit Costs	$183.20		$24.40		$36.80		$122.00

This cost information is valuable if it can eliminate uncertainty related to decisions but is of little value if it is difficult to identify what causes the uncertainty. In particular, product variety makes it difficult to construct production costing since it increases production complexity and amplifies uncertainty. Suppose that the handling activity is performed for 10 products instead of three. The current activity cost driver rate is estimated at $90.90 ($10,000/110) per production run. Can designers apply the same cost diver rate when more products are handled? To answer to this question, more extensive activity analysis is necessary, but it is likely to raise the rate if adding more products increases complexity in production and leads to consuming more resources. As a result, ABC for multiple products focuses more on the area where the increased number of products increases the number of activities and resource use (Martin and Ishii, 1996; Anderson, 1997). Therefore, studies on activity behaviors affected by the increased number of products should be conducted and included in developing a cost system for product family and product platform design.

Another problem, when cost information is used for product family and product platform design, is that ABC for multiple products needs to be reexamined to estimate possible cost savings by sharing features, parts, and subassemblies among multiple products. Suppose that Products A, B, and C require a common set of tooling instead. Tooling cost cannot be traced to each product since it is shared by all three products. The cost should be allocated to each product via a cost driver such as machining hours. Suppose that tooling cost is negotiable if three sets of common tooling are purchased and that its cost savings is about $1000. This cost savings from common tooling increases the unit cost of Product A from $24.40 to $ 26.90 and decreases the unit cost of Product B from $36.80 to $32.20 and the unit cost of Product C from $122 to $110; however, total unit costs of the three products are reduced from $183.20 to $169.10. These cost variations are shown in Table 14-9. This example indicates that by sharing resources (e.g., tooling), total indirect cost is reduced, but the indirect cost of individual products can either increase or decease depending on the quantity of the cost driver required for each product. Therefore, individual cost variances from resource sharing in the family should be considered as discussed next.

Table 14-9. Activity costs for a product family with common tooling.

Activities	Costs	Cost Drivers of Product A	Product Costs of A	Cost Drivers of Product B	Product Costs of B	Cost Drivers of Product C	Product Costs of C
Sales Volume		100		500		100	
Setup	$1,000	50 hours	$500	30 hours	$300	20 hours	$200
Run	$30,000	500 hours	$20,000	300 hours	$12,000	200 hours	$8,000
Tooling	$9,000	500 hours	$4,500	300 hours	$2,700	200 hours	$1,800
Handle	$10,000	60	$5,455	30	$2,727	20	$1,818
Maintain	$1,000	500 hours	$455	400 hours	$364	200 hours	$182
Support	$3,000	1	$1,000	1	$1,000	1	$1,000
Total	$54,000		$26,909		$16,091		$11,000
Unit Costs	$169.10		$26.90		$32.20		$110.00

3.   AN ACTIVITY-BASED COSTING METHOD TO SUPPORT PRODUCT FAMILY DESIGN

In the previous section, we reviewed ABC and how to implement ABC for multiple products. However, ABC itself does not always provide cost information for product family design because the way in which costs are to be used defines the way in which ABC should be developed. To facilitate platform design, we need to develop a method for using activity-based cost information to support product family design. Platform leveraging strategies and platform planning provide the context for developing the proposed method. Platform leveraging strategies are already covered in Chapter 5; therefore, we only review platform planning in the next section before introducing the proposed method in Section 3.2.

3.1   Platform planning

Effective planning for a product platform allows a company to deliver distinctive products to different market segments while maintaining the costs of development and production resources. Platform planning helps reduce the increased cost of addressing distinctive market needs while ensuring that they are closely met. According to Robertson and Ulrich (1998), platform planning involves two difficult tasks. First, marketing managers focus on individual product planning, addressing the problems of which market segments to target, what customers in each segment want, and what product attributes will appeal to those customers. Second, designers focus on designing a family of products, addressing the problem of what product architecture should be used to deliver the different products while sharing parts and production processes across the products. These two tasks are often in conflict and confronted when planning the commonality and differentiation underlying the platform. Platform planning seeks to resolve this conflict in term of differentiation value to customers and cost of variety.

In light of platform planning, the following definitions for differentiating attributes and chunks are offered to provide context for platform planning (Robertson and Ulrich, 1998):

1) *Differentiating attributes* (DAs): this term denotes characteristics that customers think are important when distinguishing between products.
2) *Chunks*: this term refers to the major physical elements of a product, its key components, and subassemblies.

Platform planning is comprised of three plans: (1) a product plan, (2) a differentiation plan, and (3) a commonality plan, as shown in Figure 14-3. The *product plan* lays out highly differentiated products changing over time in the distinct market segments and should fit well within the company's overall product portfolio. The product plan describes a top-level layout of products containing the customer needs and a basic business plan such as expected sales, volumes, and selling price range. The *differentiation plan* sets the target values of DAs for each product in the product plan to achieve maximum appeal to customers in the target segments as shown in Figure 14-4. The values of the DAs for competing products serve as a useful benchmark for differentiation. The plan is an explicit accounting of the costs associated with developing and producing each product.

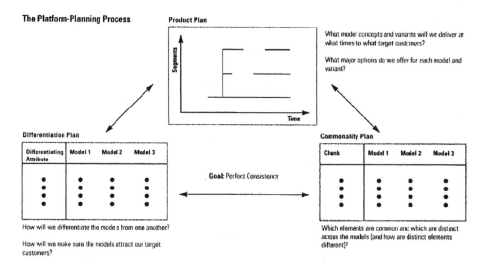

Figure 14-3. The platform planning process; adapted from (Robertson and Ulrich, 1998).

Given the cost information from the differentiation plan, the company next seeks to minimize costs such as development cost, tooling investment, and manufacturing costs by sharing components, subassemblies, and processes. The *commonality plan* describes the extent to which the products in the plan share physical elements. The example commonality plan is illustrated in Figure 14-5. In addition, platform planning involves finding architectural solutions where the differentiation plan brings high value to customers but requires high costs while the commonality plan brings low costs to the company but gives low value to customers. The solutions can be provided by modifying the target level of differentiation for the DAs that are particularly critical drivers of production costs. The cost information of effectively differentiating a product for a particular segment helps determine which part is feasible for the platform. The example of revised commonality plan is illustrated in Figure 14-6.

Differentiating Attributes	Sport Coupe	Family Sedan/Station Wagon	Importance to Customer
Curvature of window glass	More curvature	Straight, vertical	•••
Styling of instrument panel	Evocative of English roadster	Highly functional	•••
Relationship between driver and instrument panel	Driver sits low to ground, distant from steering wheel, with seat reclined.	Driver sits higher, closer, more upright	•••
Front-end styling	Shorter nose; vehicle appears to attack the road.	Longer nose, more substantial look	•••
Colors and textures	Darker colors and mix of leather and textiles	Practical surfaces and colors	••
Suspension stiffness	Stiff, for improved handling	Softer, for improved comfort	•••
Interior noise	Some engine noise desirable, 70 decibels	Noise minimized, 60 decibels	•

Figure 14-4. Example differentiation plan (Robertson and Ulrich, 1998).

Instrument Panel Chunks	Sporty Coupe				Family Sedan/Station Wagon				Comments
	Number of Unique Parts	Development Cost ($ millions)	Tooling Cost ($ millions)	Manufacturing Cost	Number of Unique Parts	Development Cost ($ millions)	Tooling Cost ($ millions)	Manufacturing Cost	
HVAC system	45	$ 4	$ 9	$ 202	35	$ 3.8	$ 7.5	$ 200	Duct work and support structure different Share motors and other components
Dash cover and structure	52	$ 4	$ 7	$ 123	48	$ 3.8	$ 6.5	$ 120	Share some brackets and components
Electrical equipment	115	$ 4	$ 2.2	$ 420	65	$ 2	$ 2.1	$ 430	Share switches, wiring, and central module
Cross-car beam	12	$ 2	$ 2	$ 35	12	$ 2	$ 2	$ 35	Cross-car beam entirely different
Steering system and airbags	26	$ 2	$ 0.1	$ 200	26	$ 2	$ 0.1	$ 195	All components different
Instruments and gauges	16	$ 1	$ 0.2	$ 22	13	$ 0.8	$ 0.2	$ 20	Can share some instruments
Molding and trim	10	$ 0.4	$ 0.2	$ 11	10	$ 0.4	$ 0.2	$ 10	All molding and trim different
Insulation	3	$ 0.2	$ 0.2	$ 8	1	$ 0.1	$ 0	$ 10	Change insulation in coupe to let in more engine noise
Audio and radio	8	$ 0.2	$ 0	$ 300	0	$ 0	$ 0	$ 300	Same radio option in all vehicles
Total	287	$ 17.8	$20.8	$ 1,321	210	$ 14.9	$ 18.5	$ 1,320	

Figure 14-5. Example of commonality plan (Robertson and Ulrich, 1998).

Instrument Panel Chunks	Sporty Coupe				Family Sedan/Station Wagon				Comments
	Number of Unique Parts	Development Cost ($ millions)	Tooling Cost ($ millions)	Manufacturing Cost	Number of Unique Parts	Development Cost ($ millions)	Tooling Cost ($ millions)	Manufacturing Cost	
HVAC system	45	$ 4	$ 9	$ 196	8	$ 0.4	$ 0.5	$ 195	Share all but ends of ducts.
Dash cover and structure	52	$ 4	$ 7	$ 123	48	$ 3.8	$ 6.5	$ 120	All new shape and structure for coupe.
Electrical equipment	115	$ 4	$ 2.2	$ 412	30	$ 0.5	$ 0	$ 415	Share wiring, control module, and combination switch.
Cross-car beam	12	$ 2	$ 2	$ 33	1	$ 0.2	$ 0	$ 33	Change horizontal beam length.
Steering system and airbags	26	$ 2	$ 0.1	$ 196	21	$ 1.0	$ 0	$ 192	Change only steering wheel and cover.
Instruments and gauges	16	$ 1	$ 0.2	$ 22	13	$ 0.8	$ 0.2	$ 20	Share gauge mechanisms.
Molding and trim	10	$ 0.4	$ 0.2	$ 11	10	$ 0.4	$ 0.2	$ 10	All molding and trim must be different.
Insulation	3	$ 0.2	$ 0.1	$ 8	1	$ 0.1	$ 0	$ 10	Change insulation in coupe to let in more engine noise
Audio and radio	8	$ 0.2	$ 0	$ 300	0	$ 0	$ 0	$ 300	Same radio option in all vehicles
Total	287	$ 17.8	$20.8	$ 1,301	132	$ 7.2	$ 7.4	$1,295	

Figure 14-6. Example of revised commonality plan (Robertson and Ulrich, 1998).

Costs relevant to platform planning vary. For example, in some settings, tooling cost may be insignificant and may be omitted from the plan;

however, in the automotive industry, for example, tooling and capital investment costs are primary drivers when planning the platform. In other settings, development time may be important, for example, in complex, technology-intensive products such as elevators, air conditioning units, and jet engines. We consider three costs to be most relevant to product family and product platform development: (1) development, (2) manufacturing, and (3) tooling costs. These costs are estimated because actual costs cannot be determined until the products have been produced. Consequently, *cost information for platform planning relies on the importance of costs affected by providing differentiated products and reliable estimation.* This concept lays the foundation for developing the proposed method for using cost information to support product family design as discussed next.

3.2   Proposed method for using activity-based costing

As stated in Section 1, ABC can estimate production costs more accurately by allocating indirect costs to products that require different levels of activities. It alone is not sufficient to provide cost information for product family design because a product platform for a family of products is designed through implementing a platform plan under a selected platform strategy, which in turn affects costs. In order to identify the costs affected in product family design, the section presents a method for using activity-based cost information to support bottom-up approaches to product family design. The proposed method and its five steps are illustrated in Figure 14-7.

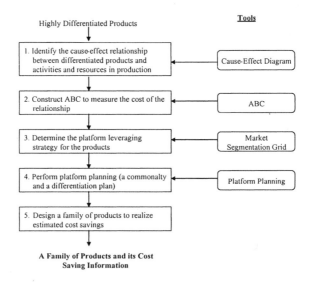

Figure 14-7. Steps and tools associated with the proposed method.

The method starts with a set of highly differentiated products that have been developed one at a time and ends with a redesigned family of products and the estimated cost savings. The tools utilized in each step are shown on the right side of Figure 14-7; these steps are elaborated in Sections 3.2.1 through 3.2.5 wherein the implementation of each step is described. These steps prescribe how to capture and use cost information to redesign a family of products; the actual implementation of each step is problem-specific.

3.2.1   Step 1 - Identify cause-effect relationships for products

Given are a set of highly differentiated products, Step 1 in the proposed method is to identify the cause-effect relationships between differentiated products and activities and resources in production because the method is focused on redesigning or consolidating a group of distinct products to reduce production costs (i.e., a bottom-up approach to platforming). During this step, the designer needs to investigate the product architecture of the set of differentiated products and find their effect on production. To display the possible effects of differentiated products for increased levels of activities and use of different resources in production, the cause-effect diagram is used as an analysis tool. Figure 14-8 illustrates an example cause-effect diagram for Black & Decker's power tools (Meyer and Lehnerd, 1997). For instance, 104 uniquely designed armatures increased the number of tools and thereby contributed to increased resource costs. This diagram helps identify possible activity and resource information affected by differentiated products, which provides cost information relevant to platform planning. By tracking the effects of differentiated products along activities and resources, we can identify a possible platform plan to address each cause.

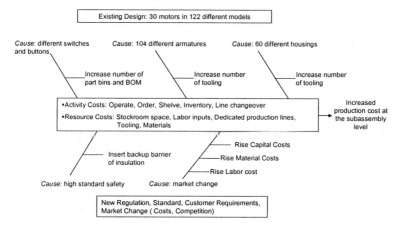

Figure 14-8. Cause-effect diagram for Black & Decker's motors in their power tools.

3.2.2 Step 2 - Use ABC to measure the costs of each relationship

Once the cause-effect relationships have been identified, Step 2 uses ABC to measure the costs of each relationship in such a way that overhead costs are separated into activity costs affected by the differentiated products. To use ABC, past accounting data is analyzed, and aggregate accounting data is allocated into activity costs according to identified the key activities and resources affected by the differentiated products. Once overhead costs are assigned to the activities, the costs are allocated to each product via activity cost drivers. Implementing ABC follows the steps in Section 2:

- Steps 1 and 2: Identify major activities and the resources used by each activity. For example, the following activities and resources can be identified from the example shown in Figure 14-8.
    - Activities: Operate, Order, Shelve, Inventory, Changeover
    - Resources: Stockroom space, dedicated machines, tooling, parts
- Step 3: Collect costs on the activities and resources.
- Step 4: Identify activity cost drivers that drive the activity costs and measure activity cost rates in terms of these cost drivers.
- Step 5: Assign the activity costs to each product via activity cost rates.

This procedure is shown in Figure 14-9. Once ABC is developed, it can be used to measure the effect of differentiated products on the production costs.

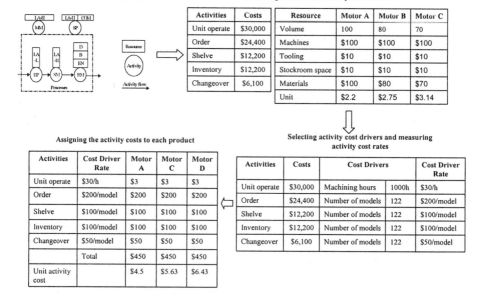

Figure 14-9. Example implementation of ABC.

3.2.3 Step 3 - Determine a suitable platform leveraging strategy

ABC provides relevant cost information as to which part of the product structure leads to the greatest cost savings by having a product platform. The following question then needs to be addressed: what market segments are being targeted when redesigning and consolidating the products into a family? To answer this question, an appropriate platform leveraging strategy needs to be identified, addressing the market opportunities for the products. For the Black & Decker motor example, a beachhead approach was used that vertically scaled the motors within the power tool segment, increasing their power and torque as needed, and then horizontally leveraged the motors into different market segments (e.g., kitchen appliances, lawn and garden, etc.) using standardized interfaces. Once the platform leveraging strategy is identified, a product platform is developed to help realize cost saving opportunities in consideration of the costs affected by the differentiated products, which is elaborated in the next step.

3.2.4 Step 4 - Perform platform planning

To develop the product platform for the selected platform leveraging strategy, the designer needs to develop a platform plan to realize a platform across the market segments identified in the previous step. As mentioned in Section 3.1, platform planning seeks to plan component commonality and differentiation based on the value of variety to customers and the estimated cost of variety. Highly ranked components or subsystems are considered as candidates for the platform with respect to the cost of variety, which are identified in Steps 1 and 2 as part of the differentiation plan. The candidates are redesigned and consolidated to reduce the cost of variety without reduction of the value of variety to the customers identified in Step 3 as part of the initial commonality plan. The following redesign guidelines can be used for the commonality plan to address the cause-effect relationships: 1) modularity, (2) commonality, (3) standardization, (4) consolidation, (5) delayed product differentiation, and (6) reusability.

- *Modularity* assigns required functions to modules and achieve product variety by combining the modules according to product requirements. Each module is created by dedicated processes and is cost-effectively assembled to its platform. When multiple functions are integrated into one module, careful design is necessary to avoid designing too specific to its product in terms of compatibility to the platform. Interface design is the most critical issues (Ishii, et al., 1995; Otto and Wood, 2001; Martin and Ishii, 2002).

- *Commonality* seeks to reduce the number of unique parts, modules, and subassemblies that are used to satisfy product requirements without sacrificing (Kota, et al., 2000).
- *Standardization* promotes standard parts that are commercially available and may reduce the number of unique parts. In most cases, standard parts are much cheaper than custom ones (Perera, et al., 1999).
- *Consolidation* seeks to integrate several parts or materials into one that can be processed in a single machine or may lead to modularity. In the case where tooling costs are high, large cost savings can be expected by integrating them into a single tool if possible (Constantine, et al., 2001).
- *Delayed product differentiation* saves costs by reducing the amount of inventory in assembly line and distribution. Delayed product differentiation provides considerable benefits to the company where multiple products are assembled across the supply chain and when their inventory costs are a big concern (Feitzinger and Lee, 1997).
- *Reusability* tries to reuse current parts/subassemblies over to new products. The parts and subassemblies that are designed robust and interchangeably can be reused over time if their functions still meet customer requirements. Robust interface design is again the most critical issue (Scheidt and Zong, 1994; Blackenfelt and Sellgren, 2000).

Different redesign guidelines for the commonality plan are chosen at different product hierarchical levels:

1) At the feature level, designers try to find ways to reduce resource use by standardizing or sharing features.
2) At the component level, redesign guidelines are chosen to standardize components or increase commonality across products.
3) At the subassembly level, redesign guidelines are chosen to reduce the number of components by designing consolidated, common, or modular subassemblies.
4) At the assembly level, easy assembly and delayed product differentiation are considered as the main redesign guideline.

At all levels, reusability is an important redesign guideline since the parts and subassemblies used over time generate large savings when developing new products. To obtain high reusability, the parts and subassemblies should be robust to changes over time; Martin and Ishii (2002) present a methodology to assess the impact of generational changes on product variety and how the coupling within the architecture affects this. At all levels, the product platform is realized by choosing appropriated redesign guidelines as summarized in Table 14-10.

Table 14-10. Relationships between redesign guidelines and different architectural levels.

Redesign Guideline	Architectural Level			
	Feature	Component	Subassembly	Assembly
Standardization		x	x	
Consolidation	x		x	
Commonality		x	x	
Modularity			x	
Late differentiation				x
Reusability		x	x	

Platform planning provides a framework upon which product families and product platforms can be strategically developed to satisfy the market segments identified by the platform strategy, resulting in cost savings. In many cases, a critical task in platform planning is to find architectural solutions to maintain the value of variety to customers while lowering the cost of variety in conjunction with manufacturing process redesign, which is addressed in the next step.

3.2.5 Step 5 - Revise commonalty plan and estimate cost savings

Steps 1 and 2 reveal the relevant costs affected by the differentiated products using ABC. By reducing the levels of activities and use of different resources through the initial commonality plan and its associated redesign guidelines identified in Step 4, cost savings can be estimated within the market segments identified in Step 3. During the initial commonality plan, only a few components and subassemblies can be shared since some of the initial commonality plan may fail to satisfy the value of variety to customers. Searching for architectural solutions to maintain the value of variety to customers while lowering the cost of variety often results in additional cost savings. During this step, therefore, the redesign guidelines from Step 4 are reapplied, and the associated chunks are redesigned for an alterative architecture that allows for the platform in the revised commonalty plan. In particular, architectural solutions need cost information when developing and implementing the solution (i.e., design and investment costs). The costs (cash outflow) need to be compared with return by increased cash inflows in the future attributable to the costs (Atkinson, et al., 2004). The amount of cost savings or net present value (NPV)[23] can justify the platform plan for a family of products. In this example, by choosing such redesign guidelines as commonality and modularity for switch and button parts in the initial commonality plan, some activity costs and resource costs identified in Step 2 can be saved. The revised commonality plan for Black & Decker shares the motor housing and armatures, and the differentiation plan scales the motor stack length (Meyer and Lehnerd, 1997). This revised commonalty plan

---

[23] Net present value (NPV) is a representative method to compute the sum of the present values of all cash inflows and outflows associated with a project.

enables overall cost savings through reducing the activity costs such as operate, order, shelve, inventory, and changeover and resource costs such as stockroom space, dedicated machines, tooling, and parts. These cost savings are necessary information for developing the product family.

4.   ELECTRIC SCREWDRIVER FAMILY EXAMPLE

To demonstrate the proposed method, we applied it to a set of electric screwdrivers that have not been developed as a product family. The goal is to demonstrate the usefulness of the proposed method for redesigning them based on a common platform. For this example, the five screwdrivers (P1-P5) are shown in Figure 14-10 along with a part diagram; complete details can be found in (Park, 2005).

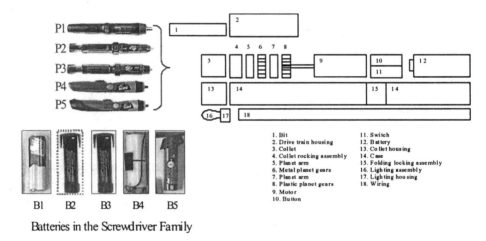

Figure 14-10. Part diagram of the screwdrivers.

Step 1 in the proposed method is to identify the cause-effect relationships between differentiated products and activities and resources in production. We investigated the product architecture of this set of differentiated products and found its effect on production. These cause-effect relationships are constructed for the electrical subassemblies since the subassemblies expect to use more activities and resources and have the highest cost for variety in the family: individually designed parts reduce opportunities to achieve economies in procurement, and the increased number of parts raises activity costs relating to ordering, setup, handling, storage, and support, as shown in Figure 14-11.

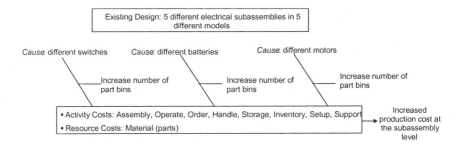

Figure 14-11. Cause-effect diagram for increased production cost at the subassembly level.

Step 2 measures cost variation for the screwdrivers using ABC, focusing on the costs caused by having differentiated products. The products consist of *custom components* manufactured using traditional, in-house processes and *standard components*, which can be purchased from suppliers for use by any number of companies. The specifications of the products are collected through product dissection. Determining all production activities affected by the distinct products is one of the most challenging aspects since production activities vary significantly depending on product specifications. For this example, part costs are estimated through part cost handbooks, Internet searches, and by applying design for manufacturing and assembly (DFMA) (Boothroyd, et al., 2002). Since indirect cost information is not readily available, the following activities and their costs are assumed to be used to produce the products: OR/MA-H/MA-S (order/material handle/material storage), SE (setup), SU (Support), and IV (Inventory). These indirect costs can vary considerably depending on activity characteristics and cost drivers. The part or subassembly costs and the activity costs of the five screwdrivers are listed with the functions of the products in Table 14-11, and the costs and cost drivers of Product P1 are given in Table 14-12. Complete details for all five screwdrivers can be found in (Park, 2005).

Step 3 maps the platform leveraging strategy to a market segmentation grid to help identify potential market opportunities for the products. The platform strategy for this family is chosen as the vertical leveraging strategy since all the products are targeted at a single market with a slightly different range of prices based on performance. The platform strategy consists of a low-end (2.4V) and a high-end (3.6V) platform as shown in Figure 14-12.

Table 14-11. Production costs for the screwdrivers.

Direct Costs						
Function	**Subassemblies**		**A ($)**	**B ($)**	**C ($)**	**D ($)**	**E ($)**
Bit	Bit		1.06	1.06	1.06	2.12	2.12
Storage	Housing		0.99	0.94	0.94	2.67	2.97
Power convert	Gear train		2.92	2.94	2.94	5.5	5.5
	Shaft assembly		1.28	1.27	1.33	1.37	1.37
		Sub total	4.20	4.21	4.27	5.50	6.87
Electricity	Motor		1.13	1.13	1.13	1.13	1.13
	Battery assemblies		4.27	5.97	5.97	5.55	7.25
	Switch assemblies		1.07	1.07	1.07	1.07	1.07
		Sub total	6.47	8.17	8.17	7.75	9.45
Auxiliary	Locking/Positioning/Lighting		0.18	0.04	0.04	0.09	0.09
Subtotal			12.90	14.42	14.48	18.13	21.50
Assembly			1.91	1.91	1.91	1.91	1.91
Indirect Costs							
OR/MA-H/MA-S	-		2.00	2.00	2.30	2.30	2.30
Setup (SE)	-		0.50	0.50	0.50	0.50	0.50
Support (SU)	-		2.50	2.50	2.89	2.89	2.89
Inventory (IV)	-		1.00	1.00	1.13	1.13	1.13
Indirect total	-		6.00	6.00	6.82	6.82	6.82
Total costs							
Unit Costs	-		20.81	22.68	23.21	30.75	34.29
Family Costs			131.74				

Table 14-12. Production costs for Product P1.

Category	Quantity	Part Costs	Material-related activities	Conversion-related activities		Operating-related activities	
Activity	-	-	OR/MA-H/MA-S	Assembly	SE	SU	IV
Cost driver	-	-	# of parts	Assembly hours	# of products	# of parts	# of parts
Bit	1	$1.06	$0.06	-	$0.10	$0.07	$0.03
Storage	6	$0.99	$0.35	$0.495	$0.10	$0.44	$0.18
Power converter	12	$1.28	$0.71	$0.65	$0.10	$0.88	$0.35
Electric	13	$6.47	$0.76	$0.65	$0.10	$0.96	$0.38
Auxiliary	2	$0.18	$0.12	$0.12	$0.10	$0.15	$0.06
Sub-total:	34	$12.90	$2.00	$1.91	$0.50	$2.50	$1.00
Total:				$20.81			

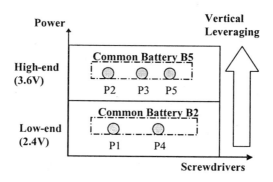

Figure 14-12. Proposed platform strategy using the market segmentation grid.

To develop a family of products reflecting the platform leveraging strategy, the designer needs to investigate a platform plan to realize a platform across the market segments identified by the leveraging strategy. As mentioned in Step 4, platform planning seeks to plan component commonality and differentiation by specifying redesign guidelines to reduce the cost of variety without reduction of the value of variety to customers. The electrical subassemblies identified in the cause-effect diagram can be leveraged in the family using the redesign guidelines of standardization, commonality, and modularity. Cost information for the individual electrical subassemblies is shown in Table 14-13.

Table 14-13. Production costs of the electrical subassemblies.

Category	Qty	Part costs	Material-related activities	Conversion-related activities		Operating-related activities	
Activity	-	-	OR/MA-H/MA-S	Assembly	SE	SU	IV
Cost driver	-	-	# of parts	Assembly hours	# of products	# of parts	# of parts
Bit	13	$6.47	$0.76	$0.653	$0.1	$0.96	$0.38
Storage	11	$8.17	$0.65	$0.653	$0.1	$0.81	$0.32
Power converter	11	$8.17	$0.65	$0.653	$0.1	$0.81	$0.32
Electric	14	$7.75	$0.83	$0.653	$0.1	$1.04	$0.41
Auxiliary	13	$9.45	$0.77	$0.653	$0.1	$0.96	$0.38
Sub-total:	62	$40.01	$3.66	$3.27	$0.5	$4.58	$1.81
Total:				$53.83			

The initial commonalty plan for the batteries is to modularize the batteries of Products P1 and P4 like the battery of Product P2 and replace the batteries of Products P2 and P3 with the battery of Product P5. This plan reduces the number of battery types to two, making the batteries modular with the high-end batteries without losing performance; however, this commonality plan must be justified by its cost savings. The initial commonalty plan for the motors is that all motors are made common across the products using a standard motor. As the revised commonalty plan, the switch subassemblies are redesigned such that the switch subassemblies are interchangeable across the products. Figure 14-13 shows the commonality plan and its associated redesign guidelines. Table 14-14 shows the resulting component cost variation of the electrical subassemblies due to increased purchase volume, and Table 14-15 shows the production cost variation.

Table 14-14. Component cost variation of electrical subassemblies based on the commonality plan due to increased volume (Unit: $, discount rate: 0.95).

Electrical Subassemblies	Production Costs				
	P1	P2	P3	P4	P5
Motor	1.00	1.00	1.00	1.00	1.00
Battery	5.67	6.69	6.69	5.67	6.69
Switch	0.95	0.95	0.95	0.95	0.95
Sub-total:	7.62	8.64	8.64	7.62	8.64

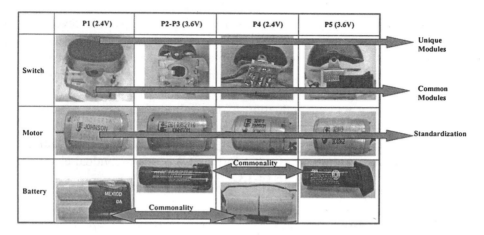

Figure 14-13. Commonality plan and its associated redesign guidelines.

Table 14-15. Production cost variation of electrical subassemblies for commonality plan.

Category	Quantity	Part costs	Material-related activities	Conversion-related activities		Operating-related activities	
Activity	-	-	OR/MA-H/MA-S	Assembly	SE	SU	IV
Cost driver	-	-	# of parts	Assembly hours	# of products	# of parts	# of parts
Bit	3	$7.62	$0.18	$0.10	$0.00	$0.22	$0.09
Storage	3	$8.64	$0.18	$0.10	$0.00	$0.22	$0.09
Power converter	3	$8.64	$0.18	$0.10	$0.00	$0.22	$0.09
Electric	3	$7.62	$0.18	$0.10	$0.00	$0.22	$0.09
Auxiliary	3	$8.64	$0.18	$0.10	$0.00	$0.22	$0.09
Sub-total:	15	$41.16	$0.90	$0.50	$0.00	$1.10	$0.45
Total:				$44.11			

Overall cost variation based on the commonality plan is shown in Table 14-16. The component costs increase since the high-end components (i.e., P2 and P5) are used as common components; however, overhead costs are significantly reduced due to the reduced level of activities. Consequently, this commonalty plan is feasible only if the product information and cost estimation are considered. With this cost advantage, designers propose to construct a new architecture for the family based on the electric subassembly as a platform. The cost method and analyses presented here can be applied to a more complex set of products.

Table 14-16. Cost variance without and with a platform.

Switch Assembly	Part Costs			Overhead Costs			Total Cost Variation
	Individual	With Platform	Variation	Individual	With Platform	Variation	
P1	$6.47	$7.62	($1.15)	$3.66	$0.90	$2.76	$1.61
P2	$8.17	$8.64	($0.47)	$3.27	$0.50	$2.77	$2.30
P3	$8.17	$8.64	($0.47)	$0.50	$0.00	$0.50	$0.03
P4	$7.75	$7.62	$0.13	$4.58	$1.10	$3.48	$3.61
P5	$9.45	$8.64	$0.81	$1.81	$.045	$1.36	$2.17
Total:	$40.01	$41.16	($1.15)	$13.82	$2.95	$10.87	$9.72

5. CLOSING REMARKS

A product platform for a family of products is created by implementing a platform plan to cover target market segments using the selected platform leveraging strategy in consideration of potential cost savings from having a platform. Hence, the most important step in product family design is to plan and develop a platform that actually saves costs. The proposed method can support product family design by providing relevant cost information using ABC, the market segment grid, and platform planning. ABC plays a crucial role in providing the activity and resource cost information caused by product differentiation. We focus mainly on two types of cost savings: (1) cost savings from shared resources and (2) cost savings from reduced levels of activities. The market segmentation grid is used to help identify market opportunities, and platform planning is used to create a platform plan to realize a platform across the targeted market segments. The proposed method is intended to support a bottom-up approach to platform-based product family development by providing relevant cost information and guidelines to support product redesign.

Designers need production costs for new products when the products are designed with new technologies. In many cases, this cost estimation is conducted with an assumption that the products would be produced with a cost structure similar to the current production. This assumption implies the products consume resources similar to existing products, which allows us to extend ABC for multiple products to for product family design. However, innovative product design combined with new production technologies might yield significant estimation errors unless only relevant costs are used. Therefore, it is a difficult and challenging task to design a cost system to find cost information relevant to new technologies. It is also not an easy task for designers to perform ABC and replace their current accounting system; however, when undertaken properly, ABC is very valuable to product family design. It helps designers estimate the production costs of newly designed products more accurately and make informed decisions for product family design in terms of production cost savings. No one can accurately predict future production costs for a product family, but product designers and cost engineers who understand their production including suppliers and distributors will be able to make better decisions and react more quickly to today's fast paced markets.

6.   ACKNOWLEDGMENTS

This work was supported by the National Science Foundation under Career Grant No. DMI-0133923. Any opinions, findings, and conclusions or recommendations presented in this chapter are those of the authors and do not necessarily reflect the views of the National Science Foundation.

Chapter 15

PRODUCT FAMILY REDESIGN USING A PLATFORM APPROACH
Assessing Cost and Time Savings

Zahed Siddique
School of Aerospace and Mechanical Engineering, University of Oklahoma, Norman, OK 73019

1.   SETTING: JUSTIFYING THE MOVE TOWARD A PLATFORM STRATEGY

Product strategy at the platform level simplifies the product development process and encourages a long-term view, because there are fewer platforms than products and major platform decisions are only made every few years. A move towards implementation of a platform strategy, which is significantly different from design and development of each product separately, can be a challenging undertaking. While the move is difficult, potential benefits from product family approach include decrease in development cost and time over a range of products. Consequently, key questions and issues that need to be addressed to justify a company's decision to allocate resources for refocusing their product strategy at the platform level are:

1. What will be the potential decrease in development cost for implementing a product platform strategy?
2. What will be the potential decrease in development time for implementing a product platform strategy?

Design and development cost and time are two of the parameters essential to quantify potential benefits for moving towards a platform strategy. In addition, cost and time associated with design and development of individual products, without a platform strategy, needs to be determined to estimate potential savings for a company. Activity Based Cost (ABC) and Activity Based Time (ABT) models can be developed and simulated to answer the two key questions. ABC is based on the idea: activities consume resources and products consume activities (Cooper, 1989). The cost of a product is then the sum of costs of activities associated with the product. An ABC system gives visibility to how effectively resources are being used and how all activities contribute to the cost of a product. These ABC related can also be utilized to estimate development time for a product.

One of the problems encountered, during development of ABC and ABT models, is that cost and time information related to product platform and family are not readily available. Available information includes cost and time data associated with development of individual product[24]. To complicate problems further, activities involved in design and development of individual products or families of products have inherent uncertainty associated with them, which needs to be included in the ABC and ABT models. Emblemsvag and Bras (1994) addressed this problem by using a combination of ABC and modeling of uncertainty as continuous and discrete probability distributions. In their method fuzzy numbers are used to model the uncertainty. The Monte Carlo simulation technique is then used to solve the model and to determine the effects of uncertainty on cost. In this chapter we employ a similar procedure to develop and solve Activity Based models with uncertainty to approximate financial effects and time savings related to implementing a product platform strategy.

2.   MODEL DEVELOPMENT

Keywords and related activities associated with addressing the key questions are used to concisely present the overall problem formulation (see Table 15-1). The problem formulation starts with information that need to be *Given* to *Identify* design and development activities associated with moving towards a platform strategy. These identified activities and their associated uncertainty can then be utilized to *Formulate* and *Simulate* the ABC and ABT models. Statistical hypothesis testing can be used on the simulation statistics to decide if implementing a platform strategy will be beneficial for the company. A five-step approach (see Figure 15-1) is presented in this chapter to develop and solve ABC and ABT models for the problem formulation shown in Table 15-1. The outputs from the models, after simulation, are the cost and time estimate for implementing a product

---

[24] In this chapter, individual product denotes development of a variety of products with a platform approach, which requires each product variety to be designed and developed separately.

platform strategy for a family of products. It is assumed that (1) the company has knowledge about the current market, which includes market segmentation and requirements for each segment, and (2) the company is looking into employing a platform approach to satisfy multiple market segments. The five steps are detailed in the remainder of this section.

Table 15-1. Problem formulation to decide if cost and time savings justify moving towards a product platform approach for an existing family of products.

Keywords	Tasks
Given	Existing product family approach, activities involved in development and manufacturing of the product family, uncertainty involved with cost and time for each activity, new platform approach for the product family
Identify	Activities involved in development and manufacturing of the new product family approach, uncertainty involved with cost and time for each activity for the new platform approach
Formulate	Activity Based Cost and Time models for existing and new product family approach
Simulate	Activity Based Cost and Time models for statistical data related to existing and new product family approaches
Test	Hypothesis
Select	Approach with better financial and time savings

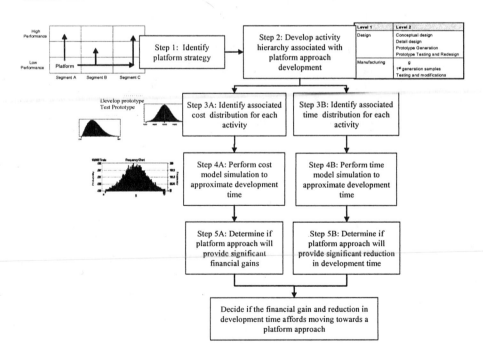

Figure 15-1. Steps for generating activity based cost and time model for product family.

2.1   Step 1: Identify platform strategy

The first step involves determining a platform leverging strategy for the product family. Market segmentation grid (Meyer, 1997) is one appraoch that can be utilized to specify the platform strategy. The market segmentation grid is setup by listing the major market segments serviced by a company's products in the horizontal axis, with different tiers of price and performance within each market segment listed in the vertical axis. Three types of platform leveraging strategies can be identified within the market segmentation grid as discussed in Chapter 5: horizontal leveraging, vertical leveraging, and the beachhead approach. In this chapter, the market segmentation grid is utilized to aggregate cost and development time for platform and family members. Step 1 corresponds to organizing some of the product family information provided in the *Give* of the problem formulation.

2.2   Step 2: Develop activity hierarchy associated with platform approach

Activity hierarchies are created to systematically identify tasks and steps related to design and development of product platform and family. A two stage approach is employed to create the activity hierarchy for the product platform and family members, from design and development activities for individual products. The activity hierarchy of individual product is created in the first stage, since companies usually have well-established procedures and/or are knowledgeable about activities involved in developing individual products. The second stage involves modifying/extending the single product activity hierarchy to separately and explicitly create new activity hierarchies for developing (1) the initial platform and (2) the family members from the platform (see Figure 15-2).

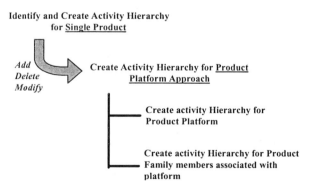

Figure 15-2. Development of activity hierarchy for platform approach.

As an example, consider the activity hierarchy shown in Table 15-2 for design and development of an individual product. This activity hierarchy can be modified for initial product platform and development of family members from the platform. In this example, some of the manufacturing activities (preliminary manufacturing, testing and process modifications) are not required for product family members that are supported by a platform. The reductions of these activities are a direct result of reuse of a platform that has already been tested.

Table 15-2. Design and manufacturing activity hierarchy for single product, product family platform and family members supported by the platform.

Level 1	Level 2 for single product	Level 2 for initial product platform	Level 2 for product family members
Design	Conceptual design	Conceptual design	Detail design
	Detail design	Detail design	Prototype Generation
	Prototype Generation	Prototype Generation	Prototype Testing
	Prototype Testing	Prototype Testing	Product Modification
	Product Modification &	Product Modification	& Redesign
	Redesign	& Redesign	
Manufacturing	Tooling	Tooling	Tooling
	Preliminary manufac.	Preliminary manufac.	Initial production
	Testing	Testing	Process modifications
	Process modifications	Process modifications	Ramp up
	Initial production	Initial production	
	Ramp up	Testing	
		Process modifications	
		Initial production	
		Ramp up	

2.3 Step 3: Identify associated cost and time distribution for each activity

Cost and time associated with each activity has variability associated with it, which needs to be included in the model. Cost and time for each activity in the hierarchy, developed in Step 2, are represented as probability distributions. Uncertainty in cost and time, associated with the activities, is modeled based on experience from engineering and finance, because cost and time data associated with activities related to development of individual product or product family are not available. The type of distribution to use, as well as the mean, the left deviation, and the right deviation, are modeled based on experience. Step 3 corresponds to utilizing information from *Identify* to perform the *Formulate* task of the problem formulation.

With the cost distribution information related to each activity specified, the ABC model can be developed (Step 3A of Figure 15-1). Development of the cost model for the platform approach follows the same overall procedure as the activity hierarchy development (described in Step 2)– (1) estimating cost distribution for single product development activities and (2) modifying these estimations for the initial platform and subsequent products.

Development of the ABT model (Step 3B of Figure 15-1) requires identifying not only the time distribution associated with completing each activity, but also requires identifying the sequential and concurrency of these activities, which can be gathered from the development team. Development time for sequential activities can be estimated by direct addition of time for the activities. Development time for concurrent activities can be estimated in two ways, depending on the situation:

❑ Using percentage calculations, which can be utilized if for a set of concurrent activities, the later activities are started when a certain percentage of former activities have been performed.

❑ Using time gaps, which can be utilized in cases where later activities start after a certain amount of elapsed time for former activities.

2.4   Step 4: Perform simulation to approximate development cost

In this step, cost and development time is estimated by simulating the ABC and ABT models, with uncertainty, developed in Step 3. Design and development cost for entire product family (Step 4A of Figure 15-1), using a platform approach, is estimated as:

$$F_{cost} = P_{cost} + n*M_{cost}, \qquad (1)$$

where:

F_{cost} = Cost for entire product family
P_{cost} = Development cost for initial product platform
M_{cost} = Development cost for product family members from platform
n = Number of family members

Development cost for product varieties (S_{cost}) without using a platform strategy is estimated by first simulating the cost models developed for individual products (Sp_{cost}), then multiplying the cost estimate with the number of family members (n).

$$S_{cost} = n*Sp_{cost} \qquad (2)$$

The cost saving for utilizing a platform strategy can be estimated as:

$$\text{Cost Saving} = F_{cost} - S_{cost}. \qquad (3)$$

If all activities involved in design and development are included in the model then the error related to the simple relationships shown in Eqs. (1)-(3) will be negligible. Otherwise, an additional term to address the error can be included in the model that can be determined from previous project data. A process is utilized to estimate development time for the entire product family, using a platform strategy:

$$F_{time} = P_{time} + n*M_{time}, \qquad (4)$$

where:

F_{time} = Development time for entire product family
P_{time} = Development time for initial product platform
M_{time} = Development time for product family members from platform
n = Number of family members

The estimated time for developing the product family without using platform (S_{time}) strategy and the potential development time saving for utilizing a platform approach is estimated as (F_{time} - S_{time}), which is similar to Eqs. (2)-(3).

Monte Carlo simulation technique is used to determine the effects of the uncertainties in the final cost and time for the product family. To simulate the model using Monte Carlo technique, the Crystal Ball software is used, which adds on to Microsoft Excel. The Monte Carlo simulation provides random samples of numbers from the assumed probability distributions. These random numbers then propagate through relationships/equations in the model to estimate the desired final output, which includes development cost and time. This step corresponds to *Simulate* of the problem formulation.

2.5   Step 5: Determine approach with the better financial prospect

The Monte Carlo simulation output of the ABC and ABT models forms a new statistical distribution, when a considerable number of samples has been generated. Since the assumptions propagated through the model are random, the statistical distribution can be used in ordinary statistical analysis to make decision regarding moving towards a platform strategy. The decision maker is usually concerned if the potential cost and time savings will be more than a specified amount (δ), given a specified confidence level. The question is answered separately for cost and time utilizing hypothesis testing. The null hypothesis of interest is:

$$H_0: \mu_1 - \mu_2 = \delta$$
$$H_1: \mu_1 - \mu_2 > \delta \qquad (5)$$

where:

> μ_1 corresponds to the mean (cost or time) of the existing approach
> μ_2 corresponds to the mean (cost or time) of platform approach

A large and same number of samples are used to simulate both platform strategy and non-platform strategy, hence the test statistic becomes:

$$z = \{(\bar{x}_1 - \bar{x}_2) - \delta\} / \sqrt{(\sigma_1^2 + \sigma_2^2)/N} \tag{6}$$

where N is the number of samples for the simulations.

The null hypothesis in this case will be rejected if $z > z_\alpha$, where α is the confidence level. Step 5 corresponds to *Test* of the problem formulation (see Table 15-1).

The results of hypothesis testing, for both cost and time, are used to select between platform strategy and individual product development approach. Although the hypothesis test results provide guidelines to reach a decision, the final selection should also include opinion of designers, manufacturing, and management.

3.   COMPUTER DISK DRIVE SPINDLE MOTOR FAMILY CASE STUDY

The computer storage Industry has grown rapidly with increase in computer usage and storage demands. With the advent of personal computers and computer applications in various fields, new markets have opened up for data storage. With the constantly changing demands of computer industry and the existing competitions among different manufacturers, time cycle needed for hard disk development is decreasing and has become a never ending challenge for the disk drive manufacturers. In addition, the competition of bringing the products into market at an earlier time than the competitors is creating urgency in every new product release. These needs and challenges in the hard disk drive industry are forcing manufacturers to implement product platform concepts. Manufacturers are trying to implement platform strategy for components and modules of the hard disks. To make rational decisions on modules/components that should utilize a platform strategy, for a set of products, manufacturers need to identify potential investment outcome, which includes reduction in development cost and time.

3.1 Case scenario

One of the hard disk performance measures is the revolution speed of the spindle motor. The spindle motor of the hard disk drive is responsible for rotating the hard disk platters, allowing the hard disk drive to operate. Increasing performance and demand of storage capacity has increased the spinning speeds of the spindle, because with the increased speed the data can be read faster from the recorded media and thus quicken the operations of hard disk drive. Based on the spindle motor speed, the hard disk drive market can be segmented for both consumer (PC) and desktop drives (i.e., Unix-based desktops).

The spindle motors also need to meet certain specifications. First, the motor should be of high quality to run for thousands of hours with start and stop cycles without failures. Second, it must not generate particles, heat, or noise while operating over extended period of time. Third, it must be smooth with minimum vibration. This is needed as the tolerances between the media and head are very low, which if not maintained will affect the data. Finally it should able to run at constant speed. The spindle motor has a base with a vertical cylindrical hub (see Figure 15-3) that holds the platters and rotates it at constant speed, whenever computer is operating. The spindle motor is fixed to the base plate of the hard disk drive during assembly. Most disk drives have several disks that are separated by disk spacers and clamped to the rotating spindle by means of screws. The spindle, and consequently the disks, is rotated at a constant speed, usually disk drives speed range from 4200 RPM up 12000 RPM.

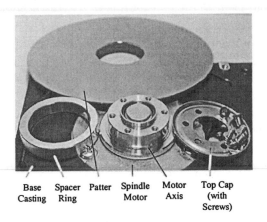

Base	Spacer	Patter	Spindle	Motor	Top Cap
Casting	Ring		Motor	Axis	(with
					Screws)

Figure 15-3. Components of computer hard disk spindle motor.

Based on the market needs and demand, three new hard disks will be introduced with varying speed. As shown in Figure 15-4, the speeds are:

(1) Consumer Drive with 4200 rpm (CD-1)
(2) Desktop Drive with 5400 rpm (DD-1)
(3) Desktop Drive with 7200 rpm (DD-2)

These disk drives will be introduced in the market over a period of time. The spindle motors used in these drives have the potential to be manufactured from the same motor platform, which is being considered by the developer. The consumer drive motor (CD-1) will be first developed, while DD-1 and DD-2 will leverage the spindle motor of CD-1. The manufacturer wants to identify the possible advantages over developing the products separately to make the decision of moving towards a platform strategy.

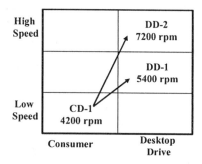

Figure 15- 4. Product platform approach for spindle motor product family.

Questions that need to be answered are: *What will be the financial gain from using one platform for the three spindle motors? What will be the potential decrease in development time from moving towards a platform strategy?* The decision to move toward the platform strategy for the spindle motor will be approved if there will be a cost savings of at least $2.25 Million and a decrease of 25 months in development time for launching the new program. These target values indicate minimum cost and time savings the company must achieve to change the current method and move towards a platform approach. If the cost saving or decrease in development time is less than the specified target then the motors for the three drives will be developed individually. It has been assumed that the technical problems associated with providing different speed for the motor can be solved.

3.2 Problem formulation and model development for spindle motor family

The problem formulation for the motor product family is shown in Table 15-3. The formulation summarizes tasks and required information to determine if implementation of the platform approach will be a beneficial undertaking and increase utilization of resources for the company.

3.2.1 Step 1: Platform strategy for spindle motor family

The platform strategy for the motor is shown in Figure 15-4. The initial platform for the motor family will be the motor used in CD-1 drive. The motor for the desktop drives, DD-1 and DD-2, will be developed from the initial platform.

Table 15-3. Problem formulation for motor product family.

Keywords	Tasks
Given	- Activities involved in development and manufacturing of motors individually. - Uncertainty involved with cost and time for each activity. - New platform approach (see Figure 15-4) for the CD-1, DD-1 and DD-2 motor product family, with CD-1 as the platform.
Identify	- Activities involved in development and manufacturing for the new product family approach (see Figure 15-4) - Uncertainty involved with cost and time with activities for the motor platform approach.
Formulate	Activity Based Cost and Time models for development of the 3 hard disk motors individually and using CD-1 as the platform.
Simulate	Activity Based Cost and Time models for statistical data related to design and development of individual motors and product platform approach.
Test	Hypothesis for cost savings with 90% and 99% confidence level: H_0: $\mu_1 - \mu_2 = 2.25M$ H_1: $\mu_1 - \mu_2 > 2.25M$ Hypothesis for decrease in development time with 90% and 99% confidence level: H_0: $\mu_1 - \mu_2 = 25$ wks H_1: $\mu_1 - \mu_2 > 25$ wks
Select	CD-1 as platform or development of the motors individually based on better financial and time savings.

3.2.2   Step 2: Develop activity hierarchy for individual, platform and product family motors

Activities associated with new motor development include both: (1) component level and (2) drive level activities. Drive level activities include engineering and testing to determine system level compatibility of the motor. Component level activities include cost and time associated with development of the motor excluding drive level activities.

Activity hierarchy for the current individual motor development process, which is gathered from designers and engineers, is shown in Figure 15-5. The individual spindle motor development process activities are then modified by engineers and designers to approximate development activities required for initial platform and subsequent spindle motor family members. The activity hierarchy for the initial motor development process and individual motor development process (without platform) were determined to be same. Activities involved in developing subsequent spindle motors from the platform are shown in Figure 15-6.

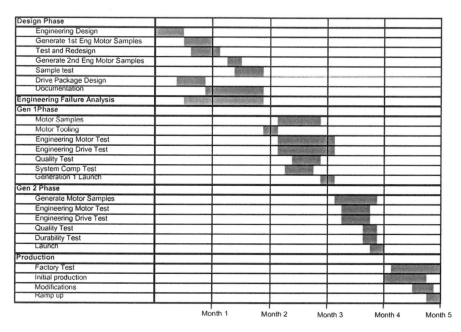

Figure 15-5. Gantt chart for single motor development.

Gantt charts that represent approximate development time for individual motor and family member, using a platform approach, are shown in Figure 15-5 and Figure 15-6, respectively. The Gantt charts will be utilized for summation of each activity time to approximate total time required for development. As an example, for the product family member development, the three main phases are sequential, with most of the activities performed in each phase being concurrent. In the Production phase, Factory test of drives starts one week after Initial production, Modifications start half week after Factory test, as problems arise. Ramp up of production begins half week after Initial production ends.

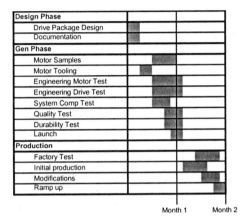

Figure 15-6. Gantt chart for motor product family member development.

3.2.3   Step 3: Identify associated cost distribution for each activity

Cost and development time associated with each activity is added to spindle motor development process for both individual and platform approach to complete the ABC and ABT models. The labor cost has been calculated with an approximate salary of $10,000 per month.

Uncertainty associated with each activity has been included in the activity hierarchy for single product development process. As an example, for the Engineering Design activity, the hours required to perform the activity can vary from 80-120 hours, with the possibility that on average the hours spent will be close to the minimum. The hours associated with Engineering Design activity are distributed among two designers, and the task is completed in approximately 1.5-4 weeks. A Weibull distribution was chosen, by the designers and engineers to reflect the uncertainty involved with the parameter. Scale and Shape parameters for the distribution for the Engineering Design activity are shown in Table 15-4. Distribution parameters associated with uncertainty for different activities for component level activities are shown in Table 15-5. The ranges are given in hours for labor, dollar for cost, number of items for other activities, and weeks to complete the task for completion time.

Table 15-4. Weibull distribution parameters associated with "Engineering Design" activity for labor hours and completion time in weeks.

Weibull distribution with parameters		
Labor hours	**Time in weeks**
Minimum	80.00 hrs	1.5 weeks
Scale	15.00	1
Shape	2	2
Distribution shape	Design Engineering (labor) - Comp Level	Engineering Design
Selected range	80.00 to 120.00 hrs	1.5 to 4.0 weeks

Addition of the uncertainty to each activity completes the ABC and ABT models for single product development, and platform approach. The initial platform model is same as the single platform development approach. Distribution associated with activities involved in product family member development is estimated from single product development data. The probability distributions for activities involved at the component level for spindle motor family members are also presented in Table 15-5.

3.2.4   Step 4: Simulate model to approximate cost and time

The development cost and time for the entire spindle motor family is estimated by simulating ABC and ABT models for the initial motor platform, CD-1, and members of the product family, DD-1 and DD-2. In each case the simulation was performed by gathering data for 10,000 random samples. An approximation for the total development cost for the family of spindle motors, using a platform approach, is determined using Eq. (1) (i.e., estimated cost for developing the initial platform, CD-1, and the two motors of the family, DD-1 and DD-2). The estimated total cost for developing the spindle motors individually is obtained by running three cost models, representing each spindle motor simultaneously and then using Eq. (2).

Simulation is run on the entire model and results are obtained to demonstrate the applicability of the model. Statistical test data for total cost without platform, total cost with platform and total cost savings are shown in Table 15-6. The mean total cost saving for implementing the specific platform approach, instead of developing the spindle motors individually, is almost $2.3 Million. Frequency distribution for total cost savings is shown in Figure 15-7. The simulation data can be used to perform percentile calculations and other statistical analysis to help decide the financial gains in implementing a platform approach for the three spindle motors family.

Table 15-5. Range for motor development parameters at component level.

	Labor Hours and cost		Completion Time	
	Single product & initial platform	Subsequent Family members	Single product & initial platform [wks]	Subsequent Family members [wks]
Engineering Design	80 to 120 Hrs		1.5 to 4.0	
Generate 1st Eng Motor Samples	20 to 32 Samples		1.5 to 3.0	
Test and Redesign	160 to 240 Hrs		1.5 to 4.0	
Generate 2nd Eng Motor Samples	100 to 150 samples		1.0 to 2.0	
Sample test	80 to 120 Hrs		2.0 to 3.0	
Motor Samples	500 to 600 Samples	100 to 200 Samples	2.5 to 5.0	1.5 to 3.0
Motor Tooling	$250K to $300K	$20K to $80K	1.0 to 2.0	1.0 to 2.0
Engineering Motor Test	480 to 520 Hrs	200 to 225 Hrs	3.0 to 5.5	2.0 to 3.5
Engineering Drive Test	480 to 580 Hrs	240 to 340 Hrs	3.0 to 5.5	2.0 to 3.5
System Comp Test	80 to 120 Hrs	40 to 65 Hrs	1.5 to 3.0	1.0 to 2.0
Launch	7500 to 8000 Samples	3000 to 3500 Samples	0.75 to 2.0	0.5 to 2.0
Factory Test	10K to 12.5K Samples	5000 to 6250 Samples	3.0 to 5.0	1.0 to 3.0
Initial production	480 to 520 Hrs	200 to 225 Hrs	2.5 to 4.0	1.0 to 3.0

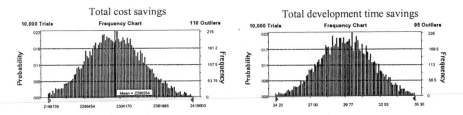

Figure 15-7. Frequency Chart for total cost savings and total time saving.

The development time for the entire family is estimated by simulating the ABT models to approximate the initial motor platform and members of the product family and then using Eq. (4) for total development time. In the case of the spindle motor, the estimated total time for the non-platform approach is calculated by running three models, representing each motor, simultaneously and then adding each model approximation to estimate the total. In the case of the using the platform approach, the total time is approximated by estimating the time for developing the initial platform (CD-1) and the other two motors of the family (DD-1 and DD-2) supported by the platform. Statistical data obtained from simulating the ABT models for total development time without platform, total development time with platform and total development time savings are shown in Table 15-6.

Table 15-6. Simulation results for the cost and development time savings.

	Development Cost (dollars * 1000)			Development Time (weeks)		
	No Platform	With Platform	Total Cost Savings	No Platform	With Platform	Time Savings
Mean	4299	2002	2296	71.5	41.69	29.88
Median	4297	2001	2296	71.45	41.59	29.86
Standard Deviation	37	27	45	1.75	1.22	2.14

3.2.5   Step 5: Determine approach with better financial prospect

The management team of the company want to know if using a platform approach for the three spindle motors will save at least $2.25 Million and 25 weeks in development time for the company. Statistical hypotheses testing is used to determine the outcomes from the ABC and ABT models separately.

The hypothesis for cost saving, for the development of the three spindle motors, is formulated as:

$$H_0: \mu_1-\mu_2=2.25*10^6$$
$$H_1: \mu_1-\mu_2>2.25*10^6 \tag{7}$$

where: μ_1 corresponds to the mean of the existing approach and
           μ_2 corresponds to the mean of platform approach

The hypothesis is tested for both 90 percent and 99 percent confidence level. Using statistical data obtained from ABC model simulation (see Table 15-6) the test statistics is $z = 102.6$. From statistical tables: $z_{0.10} = 1.282$ and $z_{0.01} = 2.326$. For both confidence level, the null hypothesis is rejected because $z>z_\alpha$. Hence it can be stated that with 99 percent confidence, for the cost model developed, the platform approach will yield at least $2.25 Million in savings.

In a similar way, the hypotheses associated with decrease in development time for the three spindle motors can be formulated as:

$$H_0: \mu_1 - \mu_2 = 25$$
$$H_1: \mu_1 - \mu_2 > 25 \qquad (8)$$

where: μ_1 corresponds to the mean of the existing approach and
μ_2 corresponds to the mean of platform approach

The hypothesis is tested for both 90 percent and 99 percent confidence level. Using statistical data from Table 15-6 the test statistics is $z = 223$. From statistical tables: $z_{0.10} = 1.282$ and $z_{0.01} = 2.326$. For both confidence level, the null hypothesis is rejected because $z > z_\alpha$. Hence it can be stated that with 99 percent confidence, for the ABT model developed, the platform approach will yield at least a 25 weeks decrease in development time.

5.   SUMMARY

In the present global market high quality, reduced cost, and development time are some of the challenges facing the manufacturers. Product platforms to support a family of product can reduce cost and development time for a family of products. Manufacturers need to estimate potential development cost and time savings to move toward a platform strategy. The Activity Based Cost and Time model were developed to assist designers/management in making decisions regarding implementation of product platform strategy. Using uncertainty in the model provides managers and designers to include the investment risks in the model. These cost and time estimates for the platform approach were compared with existing single product development approach to determine possible financial gains. The developed ABC and ABT models incorporated uncertainty associated with development cost and time of products. The addition of uncertainty is incorporated in the model using fuzzy numbers and then employing Monte Carlo simulation to simulate the models. The activity hierarchy, developed for the ABC and ABT models, provided information on the process of developing new products and platform approach.

The method of developing the ABC and ABT models for the platform approach was demonstrated using a family of hard disk drive spindle motors. Statistical results, which included frequency chart, quartile calculations and other data, associated with the models were calculated from the simulations. The statistical data, obtained from the simulation, were then used to determine if the platform approach meets a specified cost and time saving target. The statistical data can also be used to better understand the cost and time associated with platform development and be used to identify cost and time drivers associated with the specific product development to reduce cost and time. The current ABC and ABT models only address development cost and time, other life cycle activities associated with developing product platforms need to be added to better estimate the effect of utilizing a platform approach.

Chapter 16

PROCESS PLATFORM AND PRODUCTION CONFIGURATION FOR PRODUCT FAMILIES

Jianxin (Roger) Jiao, Lianfeng Zhang, and Shaligram Pokharel
School of Mechanical and Aerospace Engineering, Nanyang Technological University, Singapore 639798

1. INTRODUCTION

One of the pressing needs faced by manufacturers nowadays is quick response to the requirements of individual customers while achieving high quality and near mass production efficiency, namely mass customization (Pine, 1993a). Due to product proliferation, manufacturing organizations are confronted with difficulties in dealing with frequent design changes and recurrent process variations, which augments the complexity of product and process structures (Westkämper, et al., 2000). Developing multiple products as product families based on common platforms has been well recognized as a successful approach in many industries (Sanderson and Uzumeri, 1997). Current practice in developing product families only encompasses the design domain – dealing with the transformation of diverse customer needs to functional requirements and subsequently the fulfillment of these requirements through a variety of design parameters (Simpson, 2004). It seldom, if not at all, explicitly considers the input from the backend of product realization, viz., production processes. While seeking technical solutions is the major concern in design, it is at the production stage that product costs are actually committed and product quality and lead times are determined *per se*. For a given design, the actual cost depends on how the production is planned and to what extent the economy of scale can be realized within the existing manufacturing capabilities. This implies that the claimed rationale of product family design can only be fulfilled at the production stage (Jiao and Tseng, 2004).

The direct consequence of product customization on production is observed as an exponentially increased number of process variations (referred to as process variety), such as diverse machines, tools, fixtures, setups, cycle times, and labor (Wortmann, et al., 1997). Process variety introduces significant constraints to production planning and control, e.g., preventing make-to-order systems from building up customization capabilities. Regardless of the negative impact of process variety, the common components and the same basic product structure assumed by the customized products in a family introduce similarity in the associated production processes. Similar to a product family, a process family comprises a set of similar production processes that share a common process structure (referred to as a process platform). Consequently, companies are interested in configuring existing operations and processes (termed as production configuration) by exploiting similarities among product and process families so as to take advantage of repetitions (Schierholt, 2001). In addition to leveraging the costs of delivering variety, exploiting process families around process platforms can reduce development risks by reusing proven elements in a firm's activities (Sawhney, 1998).

A process platform entails the conceptual structure and overall logical organization of producing a family of products, thus providing a generic umbrella to capture and utilize commonality, within which each new product fulfillment is instantiated and extended so as to anchor production planning to a common process structure (Martinez, et al., 2000). The rationale of such process platforms lies not only in minimizing the planning of variant forms of the same production process, but also in modeling the production of a class of products that can widely variegate the operations and process sequences in accordance with specific design changes within a coherent framework (Meyer and Lehnerd, 1997).

2.   RELATED WORK

The importance of managing variety has been well recognized (Ho and Tang, 1998). To overcome the limitations of traditional bills-of-materials (BOMs) in handling variants, the generic BOM concept has been developed (van Veen, 1992). Hastings and Yeh (1992) propose to combine routings with traditional BOMs to provide material requirement data for each scheduled operation, resulting in a time-phased material requirement plan derived from a feasible schedule. Blackburn (1985) demonstrates how combining routings and BOMs in one document can support just-in-time manufacturing in a traditional MRP-implemented production environment. The generic bill-of-materials-and-operations (GBOMO) is put forward by (Jiao, et al., 2000) to unify BOMs and routings for the purposes of easing production planning and control as well as accommodating large number of product and process variants.

Focusing on reducing subassembly proliferation and the cost of offering product variety, Gupta and Krishnan (1998b) propose a methodology for designing product family-based assembly sequences. While attempting to create common assembly processes, their method neglects the link between product and process families. De Lit, et al. (2003) put forward the integrated design of product families and the corresponding assembly systems. Their focus is on new product families with little attention to the reuse of existing assembly plans. He and Kusiak (1997) discuss the design of assembly systems for modular products. They suggest to divide an assembly line into the basic and variant subassembly lines, such that the basic subassembly line is used for common and basic operations, whereas the variant ones for variant operations.

The concept of process platforms is introduced by Jiao, et al. (2003) to facilitate coordination in product and process variety management. To support computerized production process derivation, the principle of group technology is adopted to build coding schemes for the set of process variants in relation to product variants (Zhang, et al., 2004). Jiao, et al. (2005) study the modeling of process variety using object-oriented Petri-Nets with changeable structures for supporting production configuration. Shierholt (2001) presents the concept of process configuration that combines the principles of product configuration and process planning and thus process configuration is *de facto* an alternative of computer aided process planning.

3. PROCESS PLATFORM

3.1 Product and process families

In mass customization, the concept of product families has been widely accepted as an efficient means to provide sufficient variety for the market while shortening product development lead times and reducing costs (Halman, et al., 2003; Simpson, 2004). A product family is defined as a group of related products that share common features, components and subsystems remaining constant from product to product and differ in others varying from product to product (Messac, et al., 2002). While a product family typically addresses a market segment, i.e., a group of customers, each specific product within the family, called a product variant, targets a niche within the segment, i.e., a particular customer in the group. In accordance with a given product family, a process family, consisting of a set of process variants, is concerned with the fulfillment of all product variants in the family. Commonality across the variety of product variants leads to a number of same or similar operations, processes, and sequences among process variants (Schierholt, 2001). Therefore, there exist a common product structure and a common process structure within a product family and variety is embodied in different variants (instances) of these common structures.

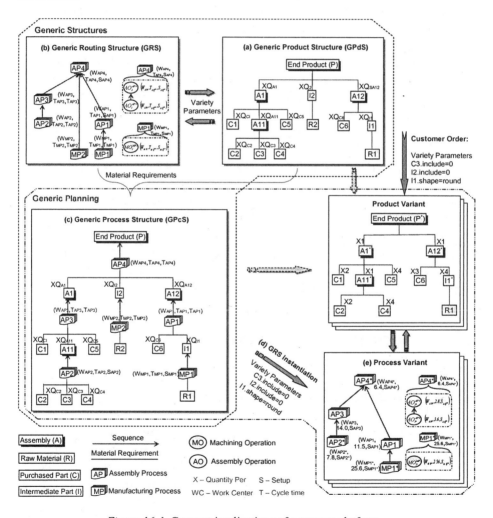

Figure 16-1. Concept implications of a process platform.

3.2 Process platform

The general gist of *platform thinking* (Sawhney, 1998) is to develop product families and the associated process families so as to produce high variety while maintaining economies of scale and scope. A comprehensive review of platform-driven development can be found in both (van Vuuren and Halman, 2001) and (Simpson, 2004).

A process platform involves three main aspects: (1) a common process structure shared by all process variants; (2) derivation of specific process variants from the common structure; and (3) correspondence between product and process variety, which resembles the correlation between the generic product and routing structures. In this research, the above three issues are approached by generic structures, generic planning, and variety parameters, respectively. Figure 16-1 illustrates the concept of a process platform. As noted, each generic or specific process, may it be a manufacturing type or an assembly one, contains one or more than one ordered operations. For example, $AP4$, the generic assembly process for forming the family of end products, involves two generic assembly operations AO_1^{AP4} and AO_2^{AP4}, while $MP1$, a generic manufacturing process, has only one generic machining operation MO_1^{MP1}. The cycle time and setup of a process are the aggregation of these of its operations. If a process contains one operation only, the process can be replaced by the operation.

3.2.1 Generic Variety Representation

The concept of generic representation proposed by (van Veen, 1992) is adopted to describe the large number of specific items with minimal data redundancy. These items can be the product related, including end products, assemblies, intermediate parts, and raw materials, and the process related, including operations, sequences, processes, and manufacturing resources. An item is generic in the sense that it represents a set of similar items (i.e., variants) of the same type (i.e., a family).

Instead of using part numbers (so called direct identification), the identification of individual variants of a generic item is based on variety parameters and their instances (i.e., a list of parameter values). This is referred to as indirect identification (Hegge, 1992). Such indirect identification entails a type of class-member relationships (exhibiting a meta-structure) between a family and its variants (Jiao, et al., 2000). In this way, generic variety representation facilitates the specification of feasible variations of items (both products and processes) with respect to optional and alternative values of variety parameters.

3.2.2   Generic Process Structure

Product data can be represented by a BOM that is used for an end product to state raw materials, intermediate parts and assemblies required for making the product. Production information is concerned with how a product is built, that is, the specification of processes, operations, and their sequences to be performed along with related resources such as workcenters/machines, labors, tools, fixtures and setups. Similar to describing a product structure using a BOM, an operations routing is usually adopted to represent the production process structure for a given product (Jiao, et al., 2000).

A process platform is underpinned by two generic structures. While a generic product structure (GPdS; Du, et al., 2001) shown in Figure 16-1(a) represents the set of product variants in the family, the related production processes can be generalized as a generic structure of standard routings (GRS) shown in Figure 16-1(b). These standard routings form the basis of various process variations matching product variety.

The relationship between the product structure (i.e., BOMs) and the routing structure is embodied in the materials required by particular production operations. The link between BOM and routing data can be established by specifying each component material in the BOM as required by the relevant operation of the routing for making its parent product (Mather, 1987). Through these links, the GPdS and the GRS can be synchronized into a unified generic structure, called generic process structure (GPcS). Therefore, as conceptually described in Figure 16-1(c), the GPcS distinguishes the common structure of the process platform, from which process variants are derived according to given product data.

While the GPdS associates each component material directly with its parent product, a component material in the GRS is associated with the relevant process or operation in the GPcS for producing its parent component. For each manufactured part or intermediate component, its GPcS can be derived by specifying the sequence of operations required for producing that component in connection with materials and resources including workcenters, cycle times, and setups required. The complete GPcS with respect to a GPdS can be composed by linking the GPcSs of lower-level product items through the processes or operations that require them.

For example, in Figure 16-1, assume a variety parameter, shape, and its value set, {normal, special}, are associated with a generic component, I1. The generic identification of I1 family is described as a set, $I1 \equiv \{I1_1^*, I1_2^*\}$. Thus two I1 variants can be identified using this variety parameter, i.e., $I1_1^* = \{I1 \mid I1.shape = "normal"\}$ and $I1_2^* = \{I1 \mid I1.shape = "special"\}$. The corresponding process variation for producing I1 family involves two variants identified from a generic manufacturing process MP1 that has a generic machining operation MO_l^{MP1}. To make II_2^*, a particular workcenter (WMP1*), other than that (WMP1) in the standard routing for making II_1^*, has to be employed, in which the cycle time and the related setup (same as these of the operation variant $MO_l^{MP1'}$ which is identified first) become different from those for making II_1^* (change from 12.5 and FMP1 to 25.6 and FMP1*, respectively).

3.2.3   Generic Planning

Generic planning is introduced to determine specific process variations in standard routings in order to accommodate product variants in a family. Within a process platform, the variation of an operation thus the related process results from the differences in product item variants to be processed by this operation. Therefore, derivation of process variants from the GPcS becomes the major concern in generic planning. Taking advantage of the meta-structure inherent in the generic variety representation, variant derivation can be implemented through the instantiation of a GPcS with respect to the given values of particular variety parameters transformed from customers' individual needs, as shown in Figure 16-1(d).

Under the umbrella of a GPcS, not only the GPdS and the GRS are unified by the material requirement links, but also they employ exactly the same set of variety parameters and their values to handle variety (Jiao, et al., 2000). Thus the class-member relationships between generic items and their variant sets can be consistently used for process variant derivation. In addition, the correspondence between product and process varieties can be maintained throughout the variation of both product and routing structures. As shown in Figure 16-1, the same set of parameters, {C3.include, I2.include, I1.shape}, is used in generic planning to derive the process variant in response to the product of a customer order, that is,

Variety parameters and their values:
{C3.include=0, I2.include=0, I1.shape="special"};
Product variant specification:
$$\{I1 \xrightarrow{\text{I1.shape="special"}} II^*, I2 \xrightarrow{\text{I2.include=0}} \varnothing, C3 \xrightarrow{\text{C3.include=0}} \varnothing\};$$
Process variant derivation:
$$\{MP1 \xrightarrow{\text{I1.shape="special"}} MPI^*, MP2 \xrightarrow{\text{I2.include=0}} \varnothing, AP2 \xrightarrow{\text{C3.include=0}} AP2^*, AP4 \xrightarrow{} AP4^*\}.$$

Prior to deriving the process variant, the product variant is specified based on the product platform of the product family according to the customer order.

During the derivation process, every generic item, more precisely, generic process items, involves an instantiation process, thus giving rise to a coordination issue among different types of instances of multiple generic items. Jiao, et al. (2000) propose a generic variety structure to coordinate multiple variants in regard to parameter values when exploding a GPdS. Similarly, rules and constraints are introduced to define relationships between generic items (in and between product and process types), and between parameter values of child and parent product items for the specification of variants of machines, operations, and processes. The rules and constraints should guarantee for given specifications of each product item, valid variants of related generic operations, desired operation variants, and related process variants are generated through the derivation process.

4.   PRODUCTION CONFIGURATION

A process platform, Ω, contains a set of production process variants, $\{P_i\}_P$, for producing the set of product variants in a family. It is defined as a tuple, $\Omega = \langle SPI, \gg \rangle$. $SPI = \{PI_i\}_{N^{PI}}$ is a set of process classes each of which is for producing a family of product items, may it be a part type or an assembly. Different valid configuration of these items forms product members of the family. $\gg$ is the set of sequence relations between two process classes in SPI such that $PI_i \gg PI_j, \forall i \neq j = 1, \cdots, N^{PI}$ indicates PI_i should be completed before the commencement of PI_j. Furthermore, a transitive closure of $\gg$ is reflexive so that with SPI, $\gg$ forms a tree.

With respect to the associated product item families, SPI can be further classified into two sets, i.e., $SPI = SPI^M \cap SPI^S$, where $SPI^M = \{PI_i^M\}_{N^{PM}}$ is a set of master process classes that are compulsory to all process variants, $SPI^S = \{PI_i^S\}_{N^{PS}}$ is the set of selective process classes that are optional to process variants. A process class, $PI_i, \forall i = 1, \cdots, N^{PI}$, contained in the process platform for producing an item family, can be classified into one of the following three categories: (1) a type of manufacturing process consisting of a series of submanufacturing processes (including machining operations and non-machining operations, e.g., material transfer) for manufacturing a part family, (2) a type of assembly process consisting of a series of subassembly processes (including assembly operations and non-assembly operations) for producing an assembly family; and (3) a mixed process involving the above two for making an assembly family.

Each $PI_i, \forall i = 1, \cdots, N^{PI}$ encompasses a set of ordered operation classes, $SO^{PI_i} = \{O_m^{PI_i}\}_{N^{OPI_i}}$, i.e., $PI_i = \langle SO^{PI_i}, \succ \rangle$, where $\succ$ is the set of precedence relations between two operations classes in SO^{PI_i} such that

$O_m^{PI_i} \succ O_n^{PI_i}, \forall m \neq n = 1, \cdots, N^{O^{PI_i}}$ suggests that $O_m^{PI_i}$ should be performed before $O_n^{PI_i}$. In addition, a transitive closure of $\succ$ is reflexive so that with SO^{PI_i}, $\succ$ forms a tree. In SO^{PI_i} of $PI_i, \forall i = 1, \cdots, N^{PI}$, two sets are distinguished, i.e., $SO^{PI_i} = SMO^{PI_i} \cap SSO^{PI_i}$, where $SMO^{PI_i} = \{MO_m^{PI_i}\}_{N^{MO^{PI_i}}}$ is the set of master operation classes necessary to all $\{P_i\}_P$, $SSO^{PI_i} = \{SO_m^{PI_i}\}_{N^{SO^{PI_i}}}$ is the set of selective operation classes optional to $\{P_i\}_P$.

A process variant, $P_i, \forall i = 1, \cdots, P$, contains a series of sequenced processes, i.e., $P_i = \langle spi_i^*, \gg \rangle$, where $spi_i^* = \{pi_{ij}^*\}_{P \times M}$ is a set of processes, each of which is to produce a specific item of the product variant, $\gg$ is the set of sequence relations between two processes $pi_{ia}^*, pi_{ib}^*, pi_{ia}^* \neq pi_{ib}^* \in PI_i$, $\forall i = 1, \cdots, N^{PI}, a \neq b = 1, \cdots, M$ in spi_i^*, such that $pi_{ia}^* \gg pi_{ib}^*$ indicates the process pi_{ia}^* for an item must be completed before producing another item by pi_{ib}^*. Each $pi_{ij}^*, \forall j = 1, \cdots, M$ of a production process variant, $P_i, \forall i = 1, \cdots, P$, is composed of a set of operations, may it be a machining type or an assembly, i.e., $pi_{ij}^*, \forall j = 1, \cdots, M = \langle SO^{pi_{ij}^*}, \succ \rangle$, where $SO^{pi_{ij}^*} = \{O_k^{pi_{ij}^*}\}_{N^{SO^{pi_{ij}^*}}}$ is a set of specific operations, $\succ$ is the set of precedence relations between two operations in $\{O_k^{pi_{ij}^*}\}_{N^{SO^{pi_{ij}^*}}}$. If $pi_{ij}^*, \forall i = 1, \cdots, P, j = 1, \cdots, M$ is for manufacturing a part, then $\{O_k^{pi_{ij}^*}\}_{N^{SO^{pi_{ij}^*}}}$ is a set of machining operations, if $pi_{ij}^*, \forall i = 1, \cdots, P, j = 1, \cdots, M$ is for producing an assembly, then $O_k^{pi_{ij}^*}, \forall k = 1, \cdots, N^{SO^{pi_{ij}^*}}$ can be a machining operation or an assembly one.

A type of manufacturing or assembly process producing a family of item variants (either a part type or an assembly one) employs a set of machine classes, each of which in turn have a number of similar machines, a number of material handling devices (i.e., material handlers) and a number of buffers. In the transfer of materials, semi-finished items (i.e., WIP), and finished items (including final products), many types of material handlers, such as AGVs, robots, or human operators, are involved. To keep the smoothness of manufacturing and assembly processes, a number of buffers, including input, WIP, and output buffers are used to store materials, WIP, and finished items/products, respectively.

The underlying principle of process platform-based production configuration is to select the set of process concepts first. The selection is accomplished by referring to the hierarchy of the given product. Production knowledge, such as rules and constraints, guides the process selection. Meanwhile, the precedence relations among selected processes are also determined. Then, each assembly item is decomposed recursively to parts. For each assembly at every decomposition level, processes, execution sequences and required resources for its child assemblies and parts are specified at the immediate lower level. Corresponding to parts at the lowest level of the decomposition paths of each assembly, the set of appropriate resources, machining operations and their execution order are provided. Figure 16-2 illustrates the set of elements of a process platform in relation to production configuration.

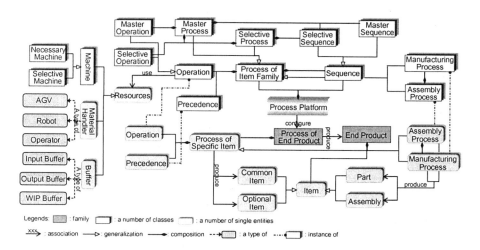

Figure 16-2. Platform elements associated with production configuration.

5.   PRODUCTION CONFIGURATION MODELING

The successful implementation of mass customization calls for the computerization of production configuration. In turn, the development of such automatic systems necessitates the process of configuration to be transparent. Hence, it raises the importance in the formal representation of production configuration, i.e., to model the obtainment of a complete production process for an end product by means of configuration. One important issue in modeling is to understand the characteristics of systems or processes to be modeled so as to design or select proper modeling tools. The fundamental issues associated with production configuration are as follows.

(1) Variety handling. For fulfilling a variety of customer expectations, a high number of product components are designed. Inevitably, their production creates a large variety of operations, precedence relations, and manufacturing resources. Thus, in production configuration, attention should be paid to diversified variety regarding process, product, and resources.

(2) Process changes. The frequent design specification changes to the customized products cause recurrent variations in production. Exhibited by changeovers in operations, sequences and resources, process differences, especially structure changes, must be explicitly considered in the way that a more appropriate production process can be obtained.

(3) Levels of abstraction. To alleviate difficulties in focusing on all details at one time, which is arduous especially in large scale configuration, production configuration adopts the strategy of problem decomposition. The production process to be configured is broken down into a number of process concepts according to the hierarchy of the given product, which are again subdivided, etc. These process concepts are specified for the associated product items at each level of the product hierarchy. Refinement of each process is made at the lower level of decomposition.

(4) Constraint satisfaction. Due to heterogeneous varieties, the compatibility issue is a major concern in production configuration. Three types of constraints are observed. The first type constraints, i.e., inclusion conditions, specify the circumstances under which the processes and operations are to be included in a configuration. Constraints of the second type tackle the interrelations among processes (operations) and determine which processes (operations) to be completed before the commencement of others. Constraints of the third type, i.e., execution rules, specify the operation details, e.g., machines to be used and estimated cycle times.

To tackle the above issues involved in production configuration, a multilevel system of nested colored object-oriented Petri Nets with changeable structures (NOPNs-cs) is introduced here as the modeling formalism. The principles of colored Petri Nets (CPNs; Jensen, 1992), object-oriented PNs (OPNs; Jensen, 1992) and the mechanism for handling structure changes in PN models (Jiang, et al., 1999) are adopted to define the nets in the formalism. The relevant data regarding product item, process elements, and manufacturing resources are attached to colored tokens in CPNs to tackle multiple configuration constraints. Applied to OPNs, the colored tokens also deal with the large and various varieties involved. The change handling mechanism is intended to address the modeling of process variations. A concept of net nesting is introduced for addressing the issues of specifying process details at different levels, that is, lower level nets are nested in the places of higher level nets. In the proposed formalism, a resource net (*RNet*) is defined to reflect the internal behaviors of physical objects (i.e., the set of manufacturing resources). Similarly, a manufacturing net (*MNet*) is defined to reflect both the manufacturing processes of a family of parts and parts themselves when it is nested in a place of the higher level net. Further, an assembly net (*ANet*) is introduced to represent the processes of an assembly family, and similar to the *MNet* , it is also used to indicate the assemblies when it is nested in a place of the higher level PN. Lastly, the process net (*PNet*) is defined to describe the abstract production process of the family of final products. It includes a set of conceptual processes selected for major product items at the first level of the hierarchy, the precedence relations among them, and the required manufacturing resources.

To specify the firing conditions of transitions with respect to firing colored tokens in an *MNet* , *ANet* or *PNet* , each of such transitions with *OR* relations among input arcs is decomposed into several input transitions, a state place and an output transition. In the nets, a single resource object (i.e., the number of such object is one) may have more than one input arc. Thus, conflict may occur when multiple objects or subprocesses require such a single object to perform operations at the same time. To maintain 1-bounded property and the safeness of an object place, the inhibitor arcs (Wang and Wu, 1998) are applied to these resource objects. $Inh(p_i, t_j) = 1$ implies that no operation request can be passed to the object represented by the place, p_i, unless the object is not occupied, i.e., there is no token in the place.

The multilevel nested net system (*NNSys*) is specified to represent the complete production configuration based on a process platform. It provides abstraction mechanisms for process engineers focusing on selected conceptual processes to work out details while without being distracted by other details of the remaining. In an *NNSys* , the highest level is the *PNet* . A number of *RNets* , *MNets* and *ANets* are located at the second level. Each of these nets provides more details for the corresponding places in the *PNet* . Similarly, at all the following levels, nets in the lower levels provide further descriptions of the assembly and manufacturing processes nested in places in immediate higher level nets. At the lowest level of each path originating from the places representing *ANets* in the *PNet* , all nets are *RNets*, whilst a mixture of *RNets* , *MNets* and *ANets* can be found at any arbitrary level in between the highest level and the lowest level. Figure 16-3 gives an example of an *N+2* level nested net system for production configuration of an end product with an *N* level hierarchy. Due to the space constraint, not all of the nested *MNets* , *ANets* and the encapsulated *RNets* are given in the figure.

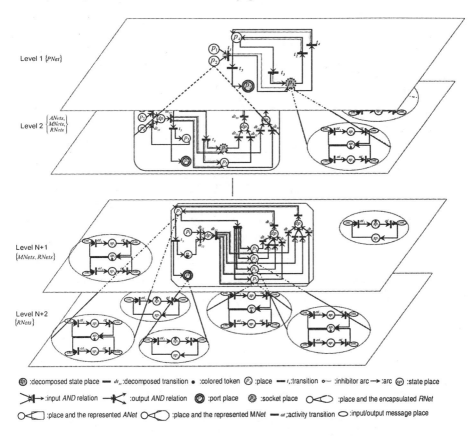

Figure 16-3. NOPNs-cs for production configuration modeling.

To enable the communication through sending and receiving messages between objects at two adjacent levels, *port places* in the lower level nested nets and *socket places* in the higher level nets are introduced. They are only defined for resource objects. The specification of port and socket places attempts to address the connection between lower and higher level nets and thus the continuity of the modeling from the lowest level to the highest level. For example, as shown in Figure 16-4 (two levels in a *NNSys*), when a token representing a part is produced in the *MNet* at level i+1, which is nested in place p_2 in the *ANet* at level i, and loaded into the output buffer represented by place p_6, a token with the same color appears in place p_2 in the *ANet* at level i. Meanwhile, a message requesting machine setup from the output buffer p_6 in the nested *MNet* is sent to the place p_3 representing an assembly machine in the *ANet* at level i.

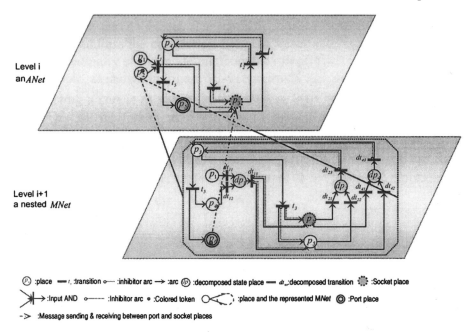

Figure 16-4. Port places and socket places.

6.   PROCESS PLATFORM CONSTRUCTION

In manufacturing practice, a large amount of production information and process data are available in an organization's databases. Besides reducing development risks thus saving time and cost, the knowledge reuse from historical production data also facilitates the handling of process variety and tradeoffs between design changes and process variations. Data mining has been used for knowledge discovery of previously unknown and potentially useful patterns of information from historical data (Chen, et al., 1996). The construction of a process platform, *de facto*, is to identify the underlying GPcS from existing production data. Therefore, data mining techniques is proposed to solve the GPcS identification problem.

The sequenced operations suggest production processes (PROCs) can be represented by tree like precedence graphs, i.e., tree representations (Martinez, et al., 2000). While operations details, such as machines, setups, and cycle times, are embedded in nodes of a representation tree of a PROC, the sequences information is reflected by the tree structure. Thus two types of data, i.e., textual data and structural data, are included in the GPcS identification. For this reason, a systematic data mining methodology is introduced by integrating text mining and tree matching techniques to solve this unique data mining problem. The methodology includes three stages: (1) PROC similarity measure, (2) PROC clustering, and (3) PROC unification.

6.1 PROC similarity measure

This step deals with measuring similarity of the set of PROC variants in a family. The PROC similarity measure can be simplified as node content similarity measure and tree structure similarity measure because of the classification of textual data (i.e., operations details) and structural data (i.e., operations sequences).

6.1.1 Node content similarity measure

In a PROC family with P members, each PROC is composed of a number of instances of a set (or subset) of N operation types. Accordingly, the node content similarity measure of two PROCs is given as the sum of their operation similarity measure. Since an operation is described by three aspects: materials, product and resources, comparison of two operations can be measured as material, product, and resource similarity.

The material similarity of two operation variants of the same type is compared as the weighted sum of similarity of all their material components. The weight assigned to each type of material component is to reflect their different importance levels or contribution to the functionality of their parent item. With respect to two types of components, namely primitive components (i.e., manufactured and purchased parts) and compound components (i.e., assemblies), different approaches are adopted. For the primitive components, text mining techniques are used, while for compound components, weighted bipartite matching is employed.

The major procedure of comparing primitive components is summarized next. Data file preparation and component description deals with pre-processing raw data so that the mining tools can work on. The different types of primitive and compound components extracted from operation nodes are saved in different files. Each file is created for each family. For describing a component, two types of attributes are concerned: nominal type and numerical one. Parts of either purchased or manufactured are described exactly the same way by their attribute value pairs. Subsequent data file analysis concerns the processing of prepared data by mining tools. The result is a list of extracted keywords, i.e., attribute values, along with the occurrence counts. The aim of quantifying nominal attribute values is to convert their text format values to numerical values (ranging from 0 to 1) for similarity comparison. Attribute weight calculation addresses the relative importance of each attribute (i.e., weight) based on the extracted occurrence counts. Attribute similarity measure attempts to calculate similarity of two attributes based on the distance of their value instances in two components. With attribute similarity available, the similarity of two components is measured as their weighted sum. Finally, with the presence of the pairwise comparison of same type components, a *PxP* matrix os established to record the obtained similarity. Repeating the above procedure for other primitive component files, a number of *PxP* matrices are constructed in the same way.

After the construction of the primitive component similarity matrices, weighted bipartite matching (Romanowski and Nagi, 2005) is carried out to measure similarity of compound components of the same type. Similarly, a number of *N* compound component similarity matrices can be constructed. Since products of operations are of compound components, product similarity measure is the same as that of compound components. As described by machines, cycle times, and setups, the resource similarity measure is computed as a weighted sum of similarity measures of the three descriptive attributes. To obtain the pairwise comparisons of machine, setup and cycle time (i.e., three descriptive attributes), the text mining procedure is applied to the extracted variants of the three types again. At last, a total number of *N* resource similarity matrices are constructed for all operations.

With the availability of similarity of materials, products and resources, the similarity of operations is computed as their sum. Subsequently, node content similarity of two processes is calculated as the sum of these of same type operations. For a relative measure (i.e., between 0 and 1), the node content similarity measure is normalized using the max-min normalization method. After the normalization, a *PxP* node content similarity matrix is established to document pairwise comparison among process variants.

6.1.2 Tree structure similarity measure

Tree structure similarity measures the degree of commonality of two PROCs in terms of their operations sequences (i.e., the arcs of precedence graphs). To deal with such structural data, the tree matching technique is applied. The procedure proceeds as follows.

The first step is to determine the base PROC between two PROCs being compared. Owing to the symmetric property of distance measure and cyclic representation of a partial order (a PROC is a partial order; Martinez, et al., 2000), the pairwise comparisons of all tree pairs of two PROCs can be simplified to merely compare an arbitrary tree of one PROC (referred to as the base PROC) with all representation trees of the other one. For reducing the total number of tree comparisons of two PROCs, the one with the higher number of representation trees is specified as the base PROC. A table is established for recording PROCs according to the ascending order of the numbers of their representation trees attempting to ease the pairwise comparison. In the PROC table, except the first one, all of the representation trees of the following *P-1* PROCs are generated for the succeeding PROC comparison (by constructing tree edit graphs), so that during pairwise comparison, PROC only needs to be compared with PROCs that follow it.

The basic principle of tree matching is to compare two trees based on tree transformation – to transform one tree to exactly the same as the other one. For the most accurate tree structure similarity measure between two PROCs, the tree edit graph (Valiente, 2002) is employed to provide an indirect way of tree transformation. To facilitate comparisons based on a consistent common ground, the same cost value is assigned to each tree editing operations represented by arcs in the graph. From the top-left corner to the bottom-right corner in a graph, the shortest path with the minimum number of arcs takes fewest editing operations and thus the minimal transformation cost, i.e., the distance of two trees. The distances of the other tree pairs of two PROCs are obtained by repeating the above process. The tree distance measure of two PROCs is defined as the minimum distance among all obtained comparison of tree pairs. Repeating the process for all PROC pairs, their structure distances are measured. For a consistent comparison, the max-min normalization method is employed to normalize the above absolute tree structure distance values. Subsequently, tree structure similarity can be calculated. Finally, a *PxP* matrix can be established for pairwise tree structure similarity of the PROC family.

6.1.3   PROC similarity measure

As note content similarity and tree structure similarity are two independent measures, the overall PROC similarity is thus suggested to be measured by a Euclidian distance rather than a simple sum. PROC similarity calculation is repeated for all the PROCs in the family. The obtained PROC similarity values are normalized and a $P \times P$ matrix is established for documenting pairwise PROC similarity measure. Figure 16-5 shows the logic of PROC similarity measure. As shown, starting from measuring tree structure and node content similarity using tree matching and text mining procedures, respectively, PROC similarity measure ends with the construction of the $P \times P$ similarity matrix.

6.2   PROC clustering

PROC clustering aims to group a set of individual processes into classes of similar ones. Considering the complex data types involved, this research adopts a fuzzy clustering approach. In comparison with the k-means method, fuzzy clustering partitions PROC instances based on the similarity degree that is derived from the real data of production processes, rather than based on subjectively pre-defined clusters.

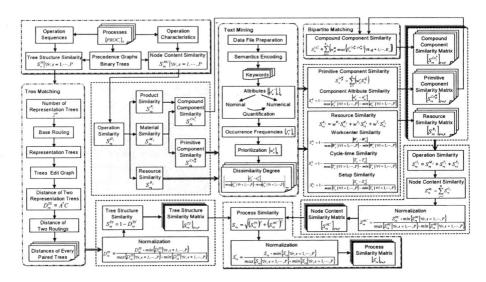

Figure 16-5. Overview of PROC similarity measure.

The first step in PROC clustering procedure is to define a fuzzy compatible relation R as similarity measures for the given PROC set Ω. R is constructed in a matrix form such that it is identical to the PROC similarity matrix, that is, R is a compatible matrix. The second step is to construct a fuzzy equivalence matrix. The fuzzy compatible relation R is a fuzzy equivalence matrix if and only if the transitive condition can be met. For converting a compatible matrix to an equivalence matrix, the continuous multiplication method introduced by (Lin and Lee, 1996) is implemented. Thirdly, a λ-cut of the equivalence matrix should be determined. The λ-cut is a crisp set that contains all the elements of the set Ω, such that the similarity grade of R is no less than λ. Then each λ-cut is an equivalence relation representing the presence of similarity among PROC instances to the degree. Finally, the PROC clusters can be identified based on the equivalence matrix by adopting the netting graph method (Yang and Gao, 1996).

6.3   PROC unification

PROC unification attempts to unify all members of a PROC cluster into a GPcS. The GPcS is formed by maintaining a valid tree structure through a tree growing process. The formation of a GPcS involves four major steps, including assorting basic process elements, identifying master and selective process elements, forming basic trees, and tree growing, as discussed below.

Basic process elements refer to operations and precedence. The assortment of elements by breaking down each PROC leads to a lead node set, an intermediate node set, a lead node arc set and an intermediate node arc set. Accordingly, the associated four types can be distinguished. The second step is to identify the set of master elements, i.e., elements that are common to all PROCs, and the set of selective elements, i.e., elements that are optional to PROCs. The basic trees are defined as the trees with such structures that are assumed by groups of PROC variants in a cluster. The generalization of basic trees can simplify the tree unification, in that fewer numbers of trees need to be unified, that is, the tree unification from individual PROC variants is converted to that of basic trees.

Tree growing aims to form a generic tree by pasting all basic trees one by one. Thus, an initial generic tree, i.e., a seed, is selected for growing. The basic tree with the longest path and the maximal number of intermediate nodes should be specified as the seed, because such a comprehensive tree encompasses most production conditions occurring in the PROC family. Then the initial generic tree starts to grow by unifying with the other basic trees. While all nodes representing operations are to be included in the final GPcS, arcs (i.e., operations precedence) cannot be directly added into the growing tree, because the addition of some arcs may damage the tree structure. Their addition is based on the result of evaluation, which is performed between the arc of the tree being unified and the associated arc in the growing tree.

Upon the completion of the tree growing process, the formed GPcS consists of a generic tree structure and an additional arc set. Repeating the procedure, the GPcS for other clusters are obtained. Treating such formed GPcS of each cluster as member trees and performing the unification process again leads to the formation of the GPcS for the entire process family. Similarly, the final GPcS includes a unified generic process structure and an extended additional arc set resulted from each cluster. Due to the presence of selective arcs, the GPcS is by no means a simple union of all member trees.

7. PRODUCTION CONFIGURATION EVALUATION

For a given product variant, more than one production processes can be configured. Thus it raises the evaluation issue of configured processes, i.e., production configuration evaluation. Production configuration evaluation involves two aspects. Firstly, the evaluation is conducted among the number of process alternatives that are generated for a product variant. Among them, an optimal one for the product is specified. However, the optimal process for each individual product may not be the optimal one when considering the cohort of a product family. Thus, the second aspect deals with the evaluation of all configured process variants with consideration of all product variants in the family as a whole. Then the optimal set of process variants with respect to the product variant set is determined.

To support process variant evaluation, the PN simulation software can be employed. A commercial tool, Petri .NET Simulator 2.0, is adopted in this research to perform the evaluation. To build the simulation model, the process specifications resulted from product specifications, e.g., cycle times and operation names, need to be input to the property fields of places representing resources (and associated operations).

A simulation model and its result are shown in Figure 16-6. In the simulation model, each circle represents a place that is associated with a particular resource, e.g., machines. Red indicates that the corresponding resources are processing, e.g., material items. The three black solid dots in the final place represent the number of products or items produced at the current simulation run time. The center window in the figure shows the set of machines used in a production process and their structure connections. Specifications for some places/machines including P2, P3, P5, and P8 are shown in the small windows around the center one. In the simulation result, the number of tokens is presented as a function of time. Thus, during the simulation time, the number of produced token is clearly stated in the result.

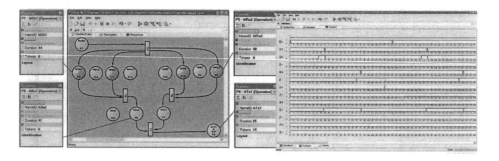

Figure 16-6. Simulation model and evaluation result.

8.   CASE STUDY

The proposed concept of process platforms has been tested on a family of vibration motors for hand phones produced by an electronics company.

Since every hand phone model is unique, the vibration motors matched with these unique models are typical customized products. The differences in motor's design result in the production characterized by a huge number of variations including changes to work centers, machines, tools, fixtures, and setup activities. The main parts of a vibration motor are rubber holder, weight, and mainbody, which further consist of armature assy, bracket assy, and frame assy. The BOM for a vibration motor is shown in Figure 16-7.

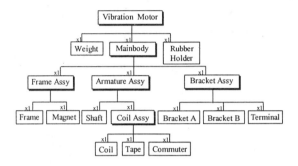

Figure 16-7. The BOM structure for vibration motors.

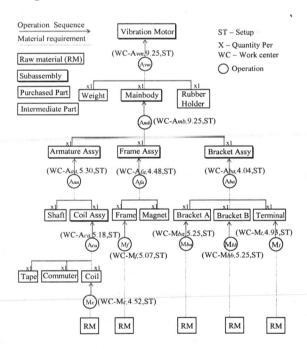

Figure 16-8. The GPcS of the motor family.

The manufacturing process for vibration motors involves six assembly operations (A*vm,* A*mb,* A*aa,* A*ca,* A*fa,* A*ha*) and five machining operations

(M*t*, M*ba*, M*bb*, M*f*, M*c*). Figure 16-8 shows the underlying GPcS of the process platform for the motor family. As shown, each process to form a product item includes only one operation type involving either machining or assembly. So we directly use operations rather than processes in the figure. Table 16-1 lists the generic item bracket assy and its child components to illustrate the concept of generic variety representation.

Table 16-1. Generic item and variety parameters for vibration motors.

Hierarchy Level	Generic Parent Item	Generic Component Item	Variety Parameter	Parameter Value Set
. 2	Bracket Assy (BAssy)	Bracket A (BA)	P_{BA0} Cardinal flag	$P_{BA01}^{\bullet} =$ "*not include*" $P_{BA02}^{\bullet} =$ "*include*"
			P_{BA1} Color	$P_{BA11}^{\bullet} = Blue$ $P_{BA12}^{\bullet} = Red$ $P_{BA13}^{\bullet} = Black$
			P_{BA2} Shape	$P_{BA21}^{\bullet} = "T"$ $P_{BA22}^{\bullet} = "U"$ $P_{BA23}^{\bullet} = "L"$
			P_{BA3} Width	$P_{BA31}^{\bullet} = "5"$ $P_{BA32}^{\bullet} = "6"$ $P_{BA33}^{\bullet} = "7"$
. 2	Bracket Assy (BAssy)	Bracket B (BB)	P_{BB0} Cardinal flag	$P_{BB01}^{\bullet} =$ "*not include*" $P_{BB02}^{\bullet} =$ "*include*"
			P_{BB1} Color	$P_{BB11}^{\bullet} = Blue$ $P_{BB12}^{\bullet} = Red$ $P_{BB13}^{\bullet} = Black$
			P_{BB2} Shape	$P_{BB21}^{\bullet} = "T"$ $P_{BB22}^{\bullet} = "U"$ $P_{BB23}^{\bullet} = "L"$
			P_{BB3} Width	$P_{BB31}^{\bullet} = "5"$ $P_{BB32}^{\bullet} = "6"$ $P_{BB33}^{\bullet} = "7"$
. 2	Bracket Assy (BAssy)	Terminal (TL)	P_{TL0} Cardinal flag	$P_{TL01}^{\bullet} =$ "*not include*" $P_{TL02}^{\bullet} =$ "*include*"
			P_{TL1} Length	$P_{TL11}^{\bullet} = "6"$ $P_{TL12}^{\bullet} = "8"$
			P_{TL2} Width	$P_{TL21}^{\bullet} = "5"$ $P_{TL22}^{\bullet} = "7"$

For a given customer order shown in Table 16-2, the configured production process is illustrated in Table 16-3. The production configuration model for the motor variant specified in Table 16-2 is built using NOPNs-cs. Due to the space constraint, only the first two levels, which include the assembly processes of the final motor and mainbody, and the level including the manufacturing process of coil are shown in Figure 16-9. Applying the deadlock detection algorithm in (Wang and Wu, 1998) to the model, the firing of the sequence of enabled transitions reaches the goal state. Therefore, the model is live and deadlock free.

Table 16-2. An individual customer order.

Order #: xxxxx	Customer Info.: xxxxxxx
Due date: xxxxx	Delivery: xxxxxxx
Volume: xxx	Description: xxxxxxxxxx

Variety parameter and value pairs:
Coil.Length = 4mm
Coil.WindingMode = BFT
Frame.Thickness = 4mm
Frame.Length = 11mm
Bracket A.Shape = O
Bracket A.Color = Red
Bracket A.Diameter = 4.5mm
Bracket B.Shape = U
Bracket B.Color = Red
Bracket B.Width = 5mm
Terminal.Length = 6mm
Terminal.Pitch = 2.5mm
Tape.Width = 3mm
Tape.Color = Red
Commuter.Thickness = 2mm
Shaft.Length = 12mm
Shaft.Material = PVC
Magnet.OutDiameter = 3.5mm
Weight.Radius = 2.5mm

Table 16-3. The production process for the motor variant specified in Table 16-2.

Sequence no.	Operation	Work Center	Cycle Time Sec./item	Setup	Material (Comp. Item)	Product (Comp. Item)	Quantity per
50	Vibration motor assembly	$WA^{\bullet}_{vm2}$	9.00	$S^{\bullet}_2(TF^{\bullet}_2)$	$Weight^{\bullet}_2$ $Rholder^{\bullet}$ $Mbody^{\bullet}_1$	$Wmotor^{\bullet}_2$	1 1 1
40	Mainbody assembly	$WA^{\bullet}_{mb1}$	9.25	$S^{\bullet}_1(TF^{\bullet}_1)$	$AAssy^{\bullet}_1$ $FAssy^{\bullet}_3$ $BAssy^{\bullet}_1$	$Mbody^{\bullet}_1$	1 1 1
30	Armature assembly	$WA^{\bullet}_{aa1}$	5.00	$S^{\bullet}_1(TF^{\bullet}_1)$	$CAssy^{\bullet}_2$ $Shaft^{\bullet}_3$	$AAssy^{\bullet}_1$	1 1
20	Coil assembly	$WA^{\bullet}_{ca2}$	5.12	$S^{\bullet}_2(TF^{\bullet}_2)$	$Coil^{\bullet}_4$ $Tape^{\bullet}_2$ $Commuter^{\bullet}_3$	$CAssy^{\bullet}_2$	1 1 1
30	Coil fabrication	$WM^{\bullet}_{c4}$	4.87	$S^{\bullet}_4(TF^{\bullet}_4)$	Raw material	$Coil^{\bullet}_4$	N/A
20	Frame assembly	$WA^{\bullet}_{fa3}$	4.36	$S^{\bullet}_3(TF^{\bullet}_3)$	$Magnet^{\bullet}_2$ $Frame^{\bullet}_3$	$FAssy^{\bullet}_3$	1 1
10	Frame fabrication	$WM^{\bullet}_{f3}$	5.08	$S^{\bullet}_3(TF^{\bullet}_3)$	Raw material	$Frame^{\bullet}_3$	N/A
20	Bracket assembly	$WA^{\bullet}_{ba1}$	4.04	$S^{\bullet}_1(TF^{\bullet}_1)$	$BA^{\bullet}_1$ $BB^{\bullet}_6$ $TL^{\bullet}_4$	$BAssy^{\bullet}_1$	1 1 1
10	BA fabrication	$WM^{\bullet}_{ba1}$	5.10	$S^{\bullet}_1(TF^{\bullet}_1)$	Raw material	$BA^{\bullet}_1$	N/A
10	BB fabrication	$WM^{\bullet}_{bb6}$	5.12	$S^{\bullet}_6(TF^{\bullet}_6)$	Raw material	$BB^{\bullet}_6$	N/A
10	TL fabrication	$WM^{\bullet}_{TL4}$	5.18	$S^{\bullet}_4(TF^{\bullet}_4)$	Raw material	$TL^{\bullet}_4$	N/A

* 1): Same sequence numbers indicate parallel operations
2): Setup means the preparation of different tools/fixtures, materials for making particular parts at particular work centers.

The data mining methodology for identifying the GPcS is applied to 30 variants in the motor family. The node content similarity matrix, tree structure similarity matrix, and PROC similarity matrix for the 30 motor variants are constructed. For illustrative simplicity, Figure 16-10 shows the final PROC similarity matrix. Based on this PROC similarity matrix, four groups are clustered as shown in Table 16-4. Applying the proposed tree unification approach to the PC1 in Table 16-4, the GPcS is formed and shown in Figure 16-11. The implementation of the process platform model shows its viability in mass customization. This method for high variety management can be used and refined by both practitioners and researchers.

Table 16-4. The identified clusters for a motor family.

PROC Cluster	PROC Variants
PC1	P1, P3, P10, P13, P14, P17, P20, P22, P25
PC2	P2, P4, P5, P6, P7, P8, P9, P11, P16, P18
PC3	P23, P26, P27, P28, P29, P30
PC4	P12, P15, P19, P21, P24

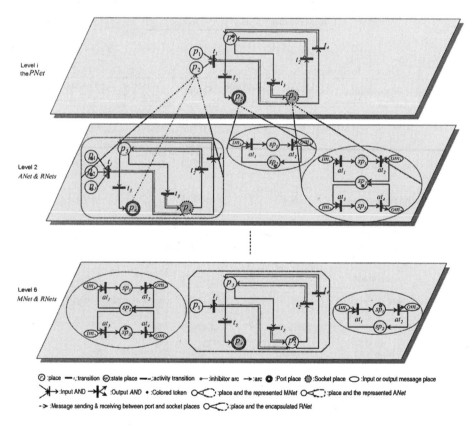

Figure 16-9. The built model for the production configuration using NOPNs-cs.

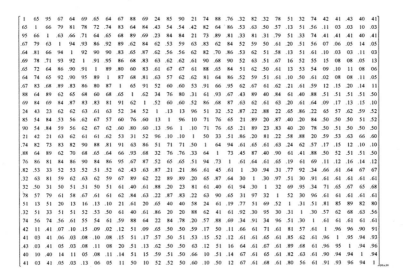

Figure 16-10. PROC similarity matrix.

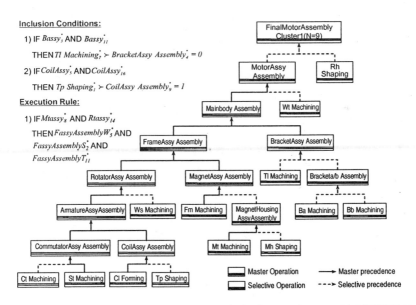

Figure 16-11. The Identified GPcS for process cluster "PC1" in Table 16-4.

9.   SUMMARY

Process platforms have been proposed to support configuration of production processes for new members of product families in mass customization. The implications of a process platform include three aspects: (1) generic variety representation, (2) generic structures, and (3) generic planning. As more companies are required to tackle product customization, the proposed approach of developing structure and obtaining relevant data from the existing company database would be useful in reducing the cost and increasing the efficiency of the production process, thus leading to the cost effectiveness of mass production. As the model proposed here is conceptual yet applicable, we expect further research in refining the model to make it more robust and then applying the general model to a specific product family so that further specialization can be introduced for efficient mass customized production systems.

Chapter 17

MEASURING SHAPE COMMONALITY
Identifying Common Shapes for a Product Family

Zahed Siddique and Manojkumar Natarajan
School of Aerospace and Mechanical Engineering, University of Oklahoma, Norman, OK 73019

1. THE PLATFORM SHAPE COMMONALIZATION PROBLEM

The current market has become customer driven and heterogeneous, and these shifts in the market have caused companies the additional problem of providing greater variety with existing challenges of providing greater quality, competitive pricing, and greater speed to market. Many companies are moving towards a platform approach to address the challenges posed by the market, which requires aggregation of the existing varieties to design and develop common platforms. Product platform aggregation is a bottom-up approach that focuses on development of a common platform for an existing family (see Chapter 1). In a given product family, each product will have a basic/core function in combination with a unique set of functions to appeal to the targeted market segments (Kota, et al., 2000; Kota and Sethuraman, 1998). Consequently, one of the important questions that need to be addressed is "What is common among the different products of the family?" A key factor to answer this question is measuring commonality of components across the product family to identify common components that have the potential to be included in the platform for the family.

Computer-Aided Design (CAD) is used extensively during mechanical product design, which involves creating 3D models of components and then assembling them into modules and systems. These 3D CAD models are very close representation of the physical components and systems, hence they are used for different types of analyses. Methods and tools to compare and identify common platform using these 3D CAD models of components would enable faster commonalization and standardization of components, thus facilitating to specify the architecture of the product family. Common product platform development, in general, has a component perspective and an assembly perspective associated with it. The set of common elements of both the perspectives, across a product-line, makes up the common platform (Siddique, et al., 1998; Siddique, 2000). This chapter focuses on the component perspective.

The Commonalization problem addressed in this chapter can be stated as:

Given a set of geometrical models for n similar components C_1, C_2, C_3,...., C_n

1.   Identify the common and similar features among the components
2.   Measure the geometrical similarity of the components
3.   Establish a common platform for the components based on shape commonality and features

The focus is to develop means for comparing geometry, identifying the common /distinct features, measuring commonality, and identifying a common platform for a set of similar components. The chapter addresses issues related to shape commonality of components, which can be used to identify and develop common platform.

2.   BACKGROUND AND RELATED RESEARCH

2.1   Product family metrics

Commonality and standardization are two primary issues during the development of common platforms for a set of similar products. Wheelwright and Clark (1992) suggested designing "platform projects" that are capable of meeting the needs of a core group of customers and can be easily modified into derivatives through addition, substitution, and removal of features. McGrath (1995) emphasized the requirements for a well-designed product platform for a family of products. Researchers (Barker, 1985; Barker, et al., 1986; Collier, 1981) have shown that parts commonality can help in minimizing inventory investment, while maintaining a desired level of customer service. McDermott and Stock (1994) described how the use of common parts can shorten the product development cycle by saving both time and money in the manufacturing process. De Lit and Delchambre

(2003) presented approaches to develop assembly line layout for a family of products.

Different indices have been developed to measure commonality of product members in a family as discussed in Chapter 7. The Non-Commonality Index (NCI) (Simpson, 1998; Conner, et al., 1999) is a measure of the variability in the design parameter settings across members of the product family. Kota, et al. (2000) illustrated a design strategy that helps minimize non-value added variation across models within a product family without limiting variety. They also introduced the Product Line Commonality Index (PCI) that included shape of components. Erni and Lewerentz (1996) introduced the concept of a multi-metric, which can be used to represent high-level objectives (i.e., commonality) as functions of several low-level product family characteristics.

2.2   Shape similarity measurements

Although shape matching methods have been studied and applied in computer graphics application, it has not received much attention in mechanical design of components. Shape matching methods are classified according to their representations of shape: 2D contours, 3D surfaces, 3D volumes, structural models, or statistics. In design of mechanical components the interest is in 3D models, including 3D surfaces. Several representations and techniques have been developed to match 3D surfaces, some of these approaches include Extended Gaussian Images (Horn, 1984), Spherical Attribute Images (Delingette, et al., 1992), Harmonic Shape Images (Zhang and Hebert, 1999), and Spin Images (Johnson and Hebert, 1999). Model-based approaches, which have also been used for shape matching, first decompose a 3D object into a set of features (or parts), and then compute a dissimilarity measure between objects based on the differences between their features and/or their spatial relationships.

Graph based approaches to determine shape similarity and search databases include feature based graph structures that contain manufacturing information to retrieve similar designs from databases (Elinson, et al., 1997). Cicirello and Regli (2002) developed model dependency graphs to carry out similarity assessments of solid models of mechanical parts based on machining features. McWherter, et al. (2001; 2002) integrated graph based data structures to enable indexing and clustering of CAD models.

Most of the recent approaches in shape-based matching techniques work on polygonal meshes to measure the similarity. Issues that have to be addressed to ensure that the shape similarity measures facilitate product family commonalization are:

4. Support hierarchy, which requires measuring similarity of primitives and combining the measurements for the entire product.
5. Identify dissimilarity measures associated with translation, rotation, scaling, deformation etc. and then incorporate them accordingly.

2.3   Neutral CAD format - IGES

Solid models are developed using a step-by-step process, where each step has a modeling operation and an associated design feature. Almost all CAD software makes use of a history tree to show the sequence of operations carried out to design the model. These history trees are non-unique, i.e., the same tree can be ordered in a different way and yet result in the same geometry. This characteristic of the history tree makes it unsuitable to be used in comparing. Furthermore, the structure of the history tree usually depends on the CAD software being used; hence developing a general approach to compare shape of components requires using a CAD model representation that is not software dependent, such as IGES and Step. In this chapter, the IGES format is used to represent the geometry of components.

The fundamental units of data in the IGES file are the entities, which are categorized into two types: (a) geometry and (b) non-geometry. Geometry entities represent the definition of the physical shape. Points, curves, surfaces, solids, and relations are included in the geometry entities. Non-geometry entities typically serve to enrich the model by providing a viewing perspective in which a planar drawing may be composed and by providing annotation and dimensioning appropriate to the drawing. Non-geometry entities further serve to provide specific attributes or characteristics for individual or groups of entities, and definitions and instances for groupings of entities.

An IGES model file consists of five sections Start, Global, Directory Entry, Parameter Data, and Terminate. A file may include any number of entities of any type as required to represent the product definition. Each entity has an occurrence in the directory entry and the parameter data entry. The directory entry provides an index and includes descriptive attributes about the data. The parameter data provides the specific entity definition. The directory data are organized in fixed fields and are consistent for all entities to provide simple access to frequently used descriptive data. The directory data and parameter data for all entities in the file are organized into separate sections, with pointers providing bi-directional links between the directory entry and parameter data for each entity.

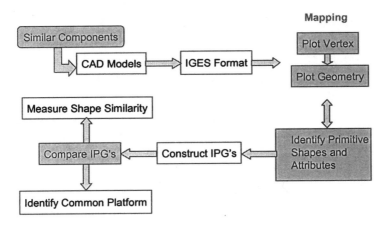

Figure 17-1. Approach for measuring shape commonality and identifying common platform.

3. OVERVIEW OF APPROACH

The approach presented in this chapter primarily focuses on individual components, which serves as the foundation to determine architecture and shape of common components for a product platform. The overall process to identify a common platform for a set of similar components consists of five steps (see Figure 17-1):

Step 1: Design components in a CAD environment
Step 2: Convert CAD models to IGES file format
Step 3: Construct IGES Parametric Graph (IPG)
Step 4: Measure shape commonality
Step 5: Identify common shapes

The common shapes for the sets of similar components can then be combined to determine the overall platform architecture and set of components that can be potentially commonalized.

3.1 Step 1: Design components in a CAD environment

The process starts by developing 3D models of components in the CAD environment. CAD models are widely used in product design, hence in most cases this step, of the overall platform commonalization approach, will not require considerable amount of time or resources.

3.2   Step 2: Convert CAD models to IGES file format

As mentioned earlier, the approach presented here utilizes IGES format, which is a CAD software neutral format, is used and almost all CAD software has the capability to convert their native CAD models to the IGES format. Once the CAD model is created, this step becomes trivial.

3.3   Step 3: Construct IGES parametric graph (IPG)

The 3D shape is represented using IGES Parametric Graph (IPG), which is a Rooted Attribute Labeled Tree. The root of the IPG is the function block, which is used to specify the basic function of the component to ensure that during comparison, components that have similar functions are considered. The root of the tree is attached to the basic building block, which is defined as the shape feature that works as the base for other shapes to be added or subtracted. The leaves of the tree are either features that are added/subtracted from the basic building block, or shape sub-features such as chamfer, rounds, fillets, etc. Each node in the IPG, except for the root, has parameters associated with it, thus making it an attribute tree. Conceptually, the IPG is similar to Constructive Solid Geometry (CSG) Trees, with the exception that in the IPG all primitive shapes are added to the base component (because negative volumes are defined using cuts and holes), thus usually there is only one level under the base feature. The idea is to break the component shape into shape features like rectangles, polygons, cylinders etc. (see Figure 17-2). The set of attributes associated with each leaf of the IPG are distinguished based on the following fields.

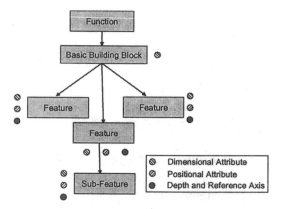

Figure 17-2. IPG Structure.

1. Dimensions and scaling: Dimensional attributes give the overall dimensions of a feature, such as the length of the sides. In complex shape, the dimensions can correspond to the position of control vertices and other defining shape parameters.

2. Position: The position of a feature is represented with respect to the basic building block. The IPG of the basic block consists of only the dimensional feature because it is the first block and is used as a reference for other (sub)features.

3. Reference axis and depth: The third attribute shows the depth of a cut and the axis about which the cut is made.

The steps for converting IGES file to IPG (see Figure 17-3) involves first identifying IGES entities used to represent the component shape to facilitate identify shape and associated attributes. The IGES entities are used first to determine the shape features, and then to determine the values of the attributes associated with the shape features. The structure of the IGES entity representation and the hierarchy of the IGES file are then used to construct the IPG. Since the IPG is created following the IGES, file hierarchy and order of features in the IGES representation, the IPG is independent of the order of solid modeling operations used by a designer and CAD software.

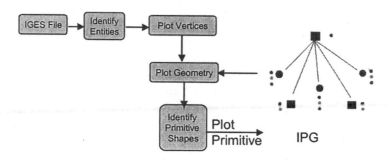

Figure 17-3. Process flow for creating IPG from IGES file.

With the IGES entities identified, the next step is to map the geometric information using the solid model representation entities (Kennicott, 1996, pp. 504-520): Vertex List Entity (Entity 502), Edge Entity (Entity 504), Loop Entity (508) and Face Entity (Entity 510). First, the set of vertices (Entity 502) are plotted and numbered from the coordinate values and information provided in the IGES file. Edges that join these vertices are then plotted (Entity 504) and relationships among the different edges are identified using the Loop Entity (Entity 508) and Face Entity (Entity 510). The Loop Entity shows the bound of a face, i.e., a single collection of face boundaries, seams and poles of a single face (Kennicott, 1996, p. 518). The Face Entity shows which loops are involved in a surface. This information is then used to determine the shape of the feature, followed by determining the dimensional attributes of the shape from the boundary edges and positional attributes from the coordinates of the vertices and the identified shape features. The same process is applied to each set of Vertex List, Edge, Loop and Face Entity to identify all the shape features defining the component geometry. The first feature in the IGES file is the base shape for the component; hence, it is used as the basic building block of the IPG. Subsequent shape features, identified from the IGES file, are added to the basic building block of the IPG as they are identified.

3.4   Step 4: Shape commonality measurement

With the IPGs developed for components being compared, measuring the shape commonality is performed using the following steps:

1. Identify the features present in all the models being compared,
2. Identify the common features,
3. Compare dimensional and positional attributes of common features, and
4. Identify and compare sub-features.

The IPG provides a representation that enables comparison of shape features and attributes, which includes dimension and position, defining the shape of the components. Solid Model Commonality Index is developed to measure the shape commonality among a set of similar components by comparing the IPG's of similar components and measuring shape commonality, taking into account dimensional differences, positional differences, distinct features, and sub-features features. The index is developed based on the assumption that two or more similar components will have a number of common features as well as distinct features and any amount of difference in an attribute has the same effect on shape commonality. The later assumption corresponds to stating that the effect of an attribute value being different has a more significant impact on commonality than how much the attribute value differs by. The Solid Model Commonality Index for component a, denoted by Φ_a, for the entire family, can be calculated using Eq. (1):

$$\Phi_a = \phi_{feat} \times \phi_{attrib}$$

(1)

ϕ_{feat} =   Commonality of Geometrical Features
ϕ_{attrib} =   Commonality of Attributes associated with features that are common for the entire family

Commonality of Geometrical Features is calculated using Eq. (2).

$$\phi_{feat} = \frac{\left(\text{number of total featurs - number of unique features}\right)}{\text{number of total featurs}} \qquad (2)$$

Commonality of Attributes for geometrical features that are common is calculated using Eq. (3). In Eq. (3), the sub-features (e.g., chamfer, round, etc) are treated as an attribute of the feature. Sub-features are considered common if all sub-feature attributes match. In the case when any of the attributes do not match, they are aggregated and considered as uncommon.

$$\phi_{attrib} = \frac{\begin{pmatrix} w_1\left(Tot_{Dim} - Uncom_{Dim}\right) + w_2\left(Tot_{Pos} - Uncom_{Pos}\right) \\ + w_3\left(Tot_{Subfeat} - Uncom_{Subfeat}\right) \end{pmatrix}}{w_1 \times Tot_{Dim} + w_2 \times Tot_{Pos} + w_3 \times Tot_{Subfeat}} \qquad (3)$$

where:

Tot_{Dim}	=	Total number of dimensional attributes in common features
$Uncom_{Dim}$	=	Total number of uncommon dimensional attributes in common features.
Tot_{Pos}	=	Total number of positional attributes in common features
$Uncom_{Pos}$	=	Total number of uncommon positional attributes in common features.
$Tot_{Subfeat}$	=	Total number of sub-features
$Uncom_{Subfeat}$	=	Total number of uncommon sub-features.
w_1, w_2, w_3	=	Relative importance of dimensional attributes, positional attributes, and sub-features on component shape.

$$\sum_{i=1}^{3} w_i = 1$$

The dimensional and positional attributes are compared if the features are similar. The weights for the different shape attributes will be specified by the designer, depending on the product and importance of factors as determined by the designer. The Solid Model Commonality Index is 1, if all shape (sub)features are same with same attribute values for the components being compared.

The Solid Model Commonality Index gives a measure of how similar components are on a geometrical level. It facilitates the decision making whether a common platform could be established from the compared models. If the index is very low for a set of components, then the designer should carefully consider if it should be included in the platform.

3.5   Step 5: Identifying the common shapes

The IPG is a Rooted Attribute Labeled Tree, which has been used to represent the shape components. The root of the tree represents the function of the component, with the other nodes specifying the features and sub-features that define the shape of the component. Each node, other than the root, has attributes associated with it that specifies the parameters of the shape features. The common shape features for the set of components can be determined from their respective IPGs.

Development of a common product platform to support family of products requires identifying not only the components that have common shapes, but also identifying common and similar (sub)features of component shapes to determine the potential for the set of components to be commonalized. This identification of common and similar features correspond to identifying subtrees of the IPGs that are common in the set of components being compared, which is determining the isomorphic Sub-IPGs (IGES Parametric Subgraphs). Two or more sub-IPGs are considered isomorphic if:

(1) The roots of the IPGs are same:
(2) The (sub)feature type for the nodes are same
(3) The set of attribute values that is common in the identified common (sub)features for all components being compared. (At the attribute level, only the common attribute values for the common sub-features are included in the Sub-IPG).

Given two or more IPG's representing CAD models of similar components the common component geometry can be identified using the following procedure:

(1) Search and compare the function node of IPG for all the models. Identify functions that are common for all the models. This step corresponds to determining if all the models perform the basic function.
(2) For each set of IPG's with the same function names, determine the nodes or (sub)feature shapes that are common in all the IPG's. This step corresponds to determining the (sub)features present in the common platform.
(3) For all the common geometrical feature shapes, compare the attributes. This step corresponds to determining the common attributes and the attributes that need to be modified or redesigned to establish a common platform.

3.6 Effects of common component shape on family architecture

The product family architecture, for a set of similar products, needs to specify components that are part of the platform and options, and spatial relationships among these components. As mentioned earlier, this chapter primarily focuses on the identifying and measuring shape commonality for a set of similar components. Since the components are the building blocks to specify a product platform or family architecture, the common component shapes and the Solid Model Commonality Index can be integrated to facilitate platform and product family design. In this section, a brief discussion on how the shape commonality and identified common shapes can be utilized to specify common platform and product family architecture is presented.

The Solid Model Commonality Index for all components in the platform should be 1, which indicates that the shapes of the components in the platform are used in all members of the product family without any modification. The component Solid Model Commonality Index can be used to identify components that have the potential to be commonalized across the entire product family. The isomorphic sub-IPGs indicate shape features that are already common in the components. This information can be used to modify features and attributes that are not common and consequently improve the platform for a set of products.

The Shape Commonality Index (SCI) for the jth product family member is:

$$SCI_j = \frac{n + \sum_{i=1}^{n'} \Phi_i}{n + n'}$$

(4)

where:

n = Number of components in the platform
n' = Number of optional components in the product family member
Φ_i = Solid Model Commonality Index for component i, calculated using Eq. (1).

4. EXAMPLE: END CASING FAMILY

The end casing product family used in this section has four members in the family (see Figure 17-4), and is used to demonstrate applicability of the process to determine Solid Model Commonality Index and to identify the common shape platform. The casing family members have similar shape, hence the manufacturer wants to (i) measure how similar the shapes of the housing are, and (2) identify the (sub)features that are same and constitute the shape platform for the family members. The Solid Model Commonality Index depends on positional, dimensional, distinct shape features, and sub-features of the casings. The steps required to perform the evaluation and to identify the common solid model features are presented next.

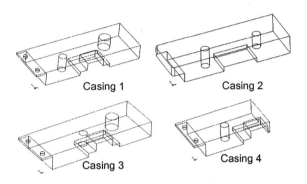

Figure 17-4. Casing product family.

4.1   Generation of IPG

The CAD solid models of the four members of the end casing family are first converted to the IGES format to generate the IPGs. The entities used in the IGES models of the casing family members are then identified from the file and are used to generate the IPG using the steps presented in Section 3.3. These steps involve: (1) plotting the vertices using the co-ordinates given by Entity 502, (2) connecting the vertices by plotting the edges using Entity 504, and (3) determining the faces, type of features and attributes of the features using the Loop Entity (Entity 508) and Face Entity (Entity 510).

The feature recognition and generation of the IPG for Casing 1 is shown in Figure 17-5. The basic building block for Casing 1 is a rectangular block of dimension 15"x5"x2". The next feature, identified from the IGES entities, is a cylindrical cut with a radius of 0.5" with an axis through (5, 2, 0) and is a through hole in the Z-direction. The third feature identified is also a cylindrical cut with radius of 0.75" with axis through (12.16, 3.81, 0) and a depth of 1.5" about the Z-axis. The fourth feature is a rectangular cut with dimensions (4.25"x 3"x 0.5"). The position of the cut is defined using the bottom left edge, specified by the co-ordinate of the end points of the edge

(6.5,0,0) and (6.5, 0, 0.5), of the rectangular cut. The fifth and sixth feature are also rectangular cuts with dimensions (2.25" x 2" x 1") and (1.5" x 5" x 1.75"), with the positional reference edge joined by end-points (7.5,0,0.5) and (7.5,0,1.5) for the fifth, and (0, 0, 0) and (0, 0, 1.75) for the sixth feature. The last two features are cylindrical cuts with radius of 0.25" and axis through points (0.75, 4, 2) and (0.75, 1, 2), respectively, and a depth of 0.5" each. The IPGs for the other three casings are generated in a similar manner and are shown in Figure 17-6.

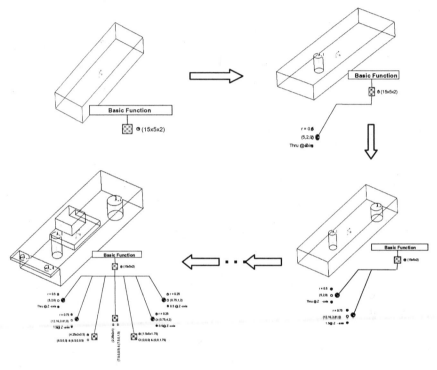

Figure 17-5. Sequence of feature identification and generation of IPG for Casing 1.

4.2   Shape commonality measurement

The Solid Modeling Commonality Index of the four casings can be determined from the constructed IPGs. Although the objective is to measure the Solid Modeling Commonality Index for all members of the casing family, two additional comparisons are first presented in this section to illustrate the process of measuring the index using Eq. (1).

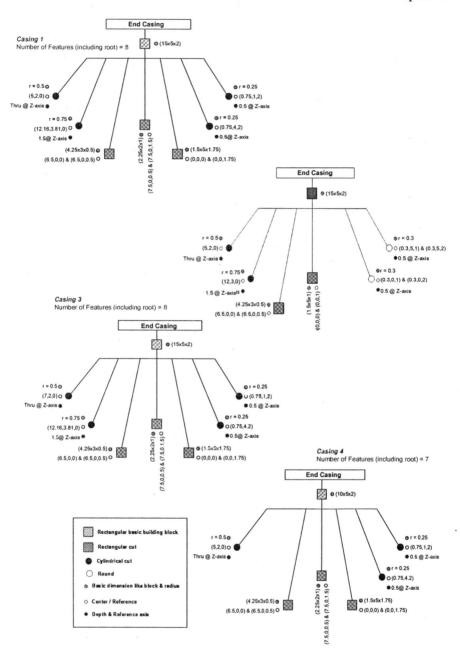

Figure 17-6. IPGs for four casings.

4.2.1 Comparing Casings 1 and 2

The shape of Casing 1 and Casing 2 (see Figure 17-4) differs primarily because of several features that are not present in both casings. Casing 2 has three features missing, i.e., two holes and one rectangular cut, which are present in Casing 1. In addition, the two round edges (sub-features) that are present in Casing 2 are not present in Casing 1. The Solid Model Commonality Index for the two casings can be determined from the IPGs (see Figure 17-6) using Eqs. (1)-(3). The number of total features, which is 15, can be counted from the IPGs of Casing 1 and Casing 2, with three features that are not present in Casing 2. The Commonality of Geometrical Features, ϕ_{feat}, is calculated to be (ϕ_{feat} = (13-3)/13) 0.77 using Eq. (2). The relative importance of dimensional attributes (w_1), positional attributes (w_2), and sub-features (w_3) on component shape is first specified by the designer to determine the Commonality of Attributes (ϕ_{attrib}). Values of different variables, obtained from comparing IPGs of Casing 1 and Casing 2, are shown in Table 17-1. The Commonality of Attributes (ϕ_{attrib}) for Casing 1 and 2 is 0.0.86 (using Eq. 3). Using the calculated values of Commonality of Features and Commonality of Attributes, the Solid Model Commonality Index for Casing 1 and 2 is determined as 0.66 using Eq. (1).

4.2.2 Comparing Casings 1 and 3

Casings 1 and 3 have very similar shapes; the difference is only in positional change of the first cylindrical hole (see Figure 17-6). Since all features are same in both, the Commonality of Features is (ϕ_{feat}=)1 using Eq. (2). The Commonality of Attributes for Casing 1 and 3 is calculated as (ϕ_{attrib}=)0.98 using Eq. (3) and values of variables shown in Table 17-1. The Solid Model Commonality Index, when Casing 1 and 3 are compared is 0.98 (using Eq. 1), which indicates that both casings are very similar in shape.

Table 17-1. Values of variables determined from IPGs to determine the Commonality of Attributes (ϕ_{attrib}) for comparing Casing 1 and 2, Casing 1 and 3, and Casing 1,2,3 and 4.

Variables	Comparison of Casing 1 and 2	Comparison of Casing 1 and 3	Comparison of Casing 1, 2, 3 and 4
w_1	0.5	0.5	0.5
w_2	0.4	0.4	0.4
w_3	0.1	0.1	0.1
Tot_{Dim}	13	20	11
$Uncom_{Dim}$	1	0	2
Tot_{Pos}	18	30	15
$Uncom_{Pos}$	3	1	2
$Tot_{Subfeat}$	2	0	2
$Uncom_{Subfeat}$	2	0	0

4.2.3   Comparing Casings 1, 2, 3, and 4

Casing 1, 2, 3 and 4 have four common features that are present in all four casings. The total number of shape features for the four casings is 28; hence, the Commonality of Features is calculated to be (ϕ_{feat}=(28-16)/28 =0.43), using Eq. (2). Values of different variables, obtained from comparing IPGs of Casing 1, 2, 3 and 4 are shown in Table 17-1. The Commonality of Attributes (ϕ_{attrib}) was calculated as 0.85 (using Eq. 3). Using the calculated values of Commonality of Features and Commonality of Attributes, the Solid Model Commonality Index for Casing 1, 2, 3 and 4 is determined as 0.36 using Eq. (1).

4.3   Identifying the common platform

The common platform shape from the IPGs (see Figure 17-6) is determined using the approach to identify isomorphic Sub-IPGs, presented in Section 3.5. The comparison of the four casings highlighted that even though the basic building block, which is a rectangular block feature, is same, there is a dimensional difference in the X direction (this difference is indicated in the isomorphic sub-IPG shown in Figure 17-7 as X_1). All four casings have one cylindrical hole in common, with a radius of 0.5", which has positional variation, represented by X_2 in Figure 17-7. The third feature common to all the four casings is a rectangular cut of dimension 4.25"x 3" x 0.5". The fourth feature common to all is a rectangular cut with dimensional difference in Z-direction. The common platform IPG for the casing product family is shown in Figure 17-7. From Figure 17-7 the Tot_{Dim}, $Uncom_{Dim}$, Tot_{Pos} and $Uncom_{Pos}$ (see Table 17-1) can be easily determined which are used in the Solid Model Commonality Index calculations. As an example, total number of dimension attributes in the common features for Casing 1, 2, 3 and 4 are 11 with 2 (X_1 and Z_1) that are uncommon.

The casing family consisted of one component; hence, the Solid Model Commonality Index does not require using Eq. (4). Although not the focus of this chapter, in the case of developing product family architecture, the Solid Model Commonality Index value for different components will have to be combined (Eq. 4) to determine the overall index value for the family.

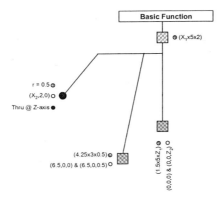

Figure 17-7. IPG of common platform shape for Casing 1, 2, 3, and 4.

5. CONCLUDING REMARKS

In this chapter, an approach to measure and identify common geometries for a set of similar components, from their respective CAD solid models, was presented. A Rooted Attribute Labeled Tree representation, called IPG, is used to convert a 3D CAD model for comparison and measurement of shape commonality and identification of common platform. The Solid Model Commonality Index incorporated comparison of dimensional, positional and sub-features to measure the commonality of similar components from the CAD models. The approach used to determine the common platform also identified the geometric features that need to be redesigned to establish a more effective common platform. The overall approach was then applied to an end casing product family. The IPG's of the family were developed, and a common platform was identified along with the shape commonality index.

One of the primary limitations of the presented approach is related to the use of IGES format, which limits representation of complex shape features, such as spline geometries. The simple shapes and geometries covered in this chapter needs to be expanded. Hence, more shapes and geometries need to be studied to expand the approach discussed in this chapter to ensure that geometries that are more complex can be compared. More IGES entities need to be studied and evaluated to accomplish this task. Future extensions for this approach include: (1) The IPG's were created manually from the IGES files. For complicated geometries, the IGES files can be very long and cannot be read manually. An application that will use an IGES file as input and generate an IPG as output needs to be developed. (2) The present approach has been applied at single component level. However, in the real world, there are very few products with only single components: the approach needs to be expanded to families with multiple components.

Chapter 18

PROCESS PARAMETER PLATFORM DESIGN TO MANAGE WORKSTATION CAPACITY

Christopher B. Williams, Janet K. Allen, David. W. Rosen, and Farrokh Mistree
George W. Woodruff School of Mechanical Engineering, Georgia Institute of Technology, Atlanta, GA 30332

1. OFFERING VARIETY THROUGH PLATFORM DESIGN

Offering product variety affordably is the crux of mass customization. Unfortunately, this is the foremost difficulty that enterprises face in making the transition to this paradigm. Anderson (1997) addresses the problem of offering affordable variety through the identification of the cost of variety. The cost of variety is the sum of all the costs of attempting to offer customers variety with inflexible products that are produced in inflexible factories and sold through inflexible channels. This cost includes the cost of customizing or configuring products, the cost of excessive variety, the cost of excessive procedures, and the cost of excessive processes and operations, among others. The key to mass customization, therefore, is the development of products and production processes that minimize the cost components.

It is neither feasible nor effective to cope with customers' demands for product variety through a simple increase in inventory, a reaction commonly found in mass production. Manufacturing enterprises are recognizing that product design presents the best control over offering such a variety (Anderson, 1997). The core issue of transitioning to mass customization now becomes how to *design* a product and its manufacturing process for affordable customization.

While previous chapters of this text focus on the design of the product, the focus in this chapter is the design of an aspect of the manufacturing process, specifically, the determination of process parameters for a single workstation. In order to provide some context, a brief overview of current manufacturing philosophies is provided in Section 1.1. In Section 1.2, the focus is narrowed to address the problem of workstation process parameter design in the presence of changing capacity requirements. In Sections 1.3 and 1.4, we present the chapter's focus: a methodology for the design of a process parameter platform.

1.1   Agile manufacturing of customized products

As a result of the shift to mass customization, enterprises are forced to manufacture more complex products (multiple features, multiple variants) with reduced product life cycles, reduced time-to-market, and volatile demand (McFarlane and Bussmann, 2000). As such, the complexity of the production process design problem is dramatically increased. Today's manufacturing approach must enable the quick launch of new product models, rapid adjustment of the manufacturing system capacity to market demands, and rapid integration of new process technologies into existing systems (Mehrabi, et al., 2000). Realizing that improving the flexibility and productivity of a manufacturing system is the "crucial challenge of modern industrial management" (Ferrari, et al., 2003), many system-level production process design approaches have been developed to enable manufacturing enterprises to affordably produce customized products.

Using the fundamental concepts of Group Technology - grouping parts with similar production processes together (Rolstadas, 2001) - cellular manufacturing involves processing a collection of similar parts (part families) on a dedicated cluster (or cell) of machines or manufacturing processes (McAuley, 1972). This strategy has the potential to reduce setup times, reduce in-process inventory, improve part quality, shorten lead-time, reduce tool requirements, and improve productivity. Cellular manufacturing systems have a high level of flexibility that allow organizations to "quickly respond to changes in the market demand or product structure with minimum disruption to their prior manufacturing commitments" (Malakooti, et al., 2004).

Flexible Manufacturing Systems (FMS) is a manufacturing technology and system-level philosophy that focuses on designing a production system that is capable of producing several families of parts, with shortened changeover time, and without major retooling. FMS is a programmable machining system configuration that incorporates software to handle changes in work orders, production schedules, part programs, and tooling for several families of parts (Hopp and Spearman, 2001).

Reconfigurable Manufacturing Systems (RMS) take this concept further by striving to create production processes that are capable of not only adapting to producing a variety of parts, but also changing the system itself easily (Mehrabi, et al., 2000). Modular machines and open architecture controllers are the key enabling technologies for RMS, and have the ability to integrate/remove new software/hardware modules in response to changing market demands or technologies without affecting the rest of the system. "This offers RMS the ability to be converted quickly to the production of new models, to be adjusted to exact capacity requirements quickly as market grows and product changes, and to be able to integrate new technology" (Mehrabi, et al., 2002). The objective of a RMS is to provide exactly the functionality and capacity that is needed, precisely when it is needed.

These system-level philosophies and their related implementation technologies provide general strategic direction for the planning of the production process. These ideologies have generated research towards the planning and design of various aspects of the production process - from sequencing and synchronization of multiple machining and assembly operations, to line balancing and capacity planning. In this chapter, however, we focus on improving the agility of individual workstations.

1.2   Need for the agile definition of process parameters

Consider a manufacturer of customized widgets. Due to the volatile demand of the different widget variants, and the manufacture-to-order nature of the process, the capacity requirement of each of the workstations in the production line changes daily. In the context of a single workstation, this change forces the manufacturing engineer to reconfigure the process parameters of the workstation (e.g., turning speed, tool size, laser power, temperature, etc.) in order to maintain the best compromise between three conflicting objectives: minimization of cost, maximization of throughput, and maximization of quality. This reconfiguration not only requires a new evaluation of the process parameters, but also entails a costly and lengthy setup of the workstation. The engineer is in need of a means of making the setup of this specific workstation more efficient and effective at adapting to changing capacity requirements.

This scenario describes a problem of process parameter design. While there are parameter design techniques in the literature, they do not address defining process parameters so as to improve the agility of the workstation in the face of varying capacity requirements. Robust parameter design, inspired by Taguchi's robust design principles, is focused on choosing the values of controllable parameters so as to improve a defined quality characteristic, while minimizing the variation imposed on the process via uncontrollable (noise) parameters/factors (Robinson, et al., 2004). While typically used to maintain quality in the presence of uncertainty, robust parameter design has also been used to make decisions regarding capacity planning and machine investment (Paraskevopoulos, et al., 1991).

In order to direct the reader's focus towards the process parameter design problem in the context of mass customization, we pose the following question: "How can a designer determine a workstation's process parameters so as to efficiently and effectively handle fluctuating capacity requirements?" In order to maximize a workstation's efficiency in the presence of different production capacity requirements, one must identify a means to make efficient transitions between process parameters (i.e., minimize the cost and time of the workstation setups). We look to the development of product platforms as inspiration of achieving this goal.

1.3   Product platform development as inspiration

Although they are two separate domains, product design and process parameter design share similarities in the context of producing customized goods. In the realm of product design, a designer must find an affordable manner in which to offer variety in product specification and/or product function. In the realm of process parameter design for individual workstations, a manufacturer must find an efficient means to offer variety in production capacity requirements.

In the context of designing customized products, variety is efficiently offered through the development of product platforms – a set of common components, modules or parts from which a stream of derivative products can be created (Meyer and Lehnerd, 1997). The design of product platforms for customized products enables the manufacturer to maintain the economic benefits of having common parts and processes while still being able to offer product variety to customers.

We assert that the core concept of platform design – offering variety efficiently through commonality and/or modularity – can be applied to the design of the process parameters for a workstation involved in the manufacture of customized goods. As such, the concept of a *process parameter platform* is introduced: "A *process parameter platform* is defined as a set of common process parameters from which a stream of derivative process parameters can generate a customized product efficiently despite changes in required capacity."

The concept of a process parameter platform is very similar to a product platform. Product platforms are a set of design parameters that are commonalized across various intervals in the design space in order to offer product variety. Process parameter platforms comprise a set of process parameters that are commonalized across various intervals in order to offer variety in the workstation's production capacity requirement. Just as the commonality of design parameters in a product platform lowers the cost of offering product variety, the commonality in a process parameter platform lowers the cost of the setups encountered with the reconfiguration of a production workstation for different capacity requirements.

The concept of grouping similar system-level manufacturing/assembly processes into a family is the crux of agile manufacturing. We introduce the concept of the process parameter platform in order to extend this philosophy to the lower-end of the production hierarchy – the design of a family of process parameters for individual workstations.

1.4   Context

The goal in creating a platform of process parameters is to reduce the cost and time of workstation setups and thus create an efficient manner of offering variety in production capacity. We address two main issues in this chapter: how one should design a process parameter platform and, more importantly, whether or not the development of a process parameter platform is an advantageous venture. In order to answer these questions, we look to different product platform design techniques as potential foundations for the design of this new type of platform.

Simpson provides a thorough review of 32 existing optimization-based product platform design approaches wherein their different characteristics are compared and contrasted (Simpson, 2004). The following limitations are identified in Simpson's review: (1) Two-thirds of the techniques require *a priori* specification of the platform to optimization; (2) Half of those techniques surveyed assume that maximizing product performance maximizes demand, maximizing commonality minimizes production costs, and that resolving the tradeoff between the two yields the most profitable product family; (3) Only half of the methods integrate manufacturing costs directly; (4) Less than one-third incorporate market demand or sales into the problem - those that do assume that demand is uniform, and use single objective optimization (with the goal of either minimizing cost or maximizing profit); and (5) Only two methods are capable of handling multiple methods of managing variety (modularity and product scaling).

These limitations are significant, as the design of a process parameter platform requires a methodology that enables a designer to handle multiple design objectives, synthesize multiple manners in which to offer variety, model the manufacturing process and the non-uniform demand of the market effectively, and handle the inherent tradeoffs between commonality and platform performance. Of those product platform techniques surveyed by Simpson, only one technique is capable of satisfying all of the above listed requirements - the Product Platform Constructal Theory Method.

The Product Platform Constructal Theory Method (PPCTM) is a top-down product platform design approach for developing product platforms that facilitates the realization of a stream of customized product variants, and which accommodates the issue of multiple levels of commonality and multiple customizable specifications (Hernandez, et al., 2003; Williams, et al., 2004). The result of the use of the PPCTM is a hierarchical organization of multiple approaches for achieving commonality, as well as the specification of their range of application across the product platform.

In this chapter, we present a methodology for the development of a process parameter platform using the PPCTM. In Section 2, the application of the PPCTM to process parameter design is described in detail. The methodology is presented in Section 3 and is explained and validated through its application to an example problem. Results are presented in Section 4, and closing remarks are offered in Section 5.

2. HIERARCHICAL PROCESS PARAMETER PLATFORM DESIGN AS A PROBLEM OF ACCESS IN GEOMETRIC SPACE

The Product Platform Constructal Theory Method (PPCTM) serves as the theoretical foundation for the methodology of designing process parameter platforms. The fundamental problem addressed in the PPCTM is how to determine and organize different methods of offering variety systematically when creating a platform (Hernandez, et al., 2002; 2003).

2.1 The product platform constructal theory method

In the Product Platform Constructal Theory Method, the approach for organizing common components for a very large number of product variants is anchored in the thesis of Herbert Simon (1996), who observed that complex structures adapt and evolve more efficiently when they are organized hierarchically. Considering this, Hernandez and coauthors (2003) propose to determine and organize commonality of product parameters in a hierarchic manner. With the formulation of the PPCTM, platform design is represented as a problem of optimization of access in a geometric space.

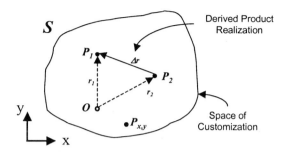

Figure 18-1. Illustration of product customization as an optimal access problem.

An optimal access problem is characterized by the need to determine the optimal "bouquet of paths" that link all points, $P_{x,y}$, of a geometric space, S, with a common destination, O (see Figure 18-1). Adrian Bejan (1996) initiated constructal theory as a result of studying problems of optimal access. Bejan's constructal theory embodies the notion that the hierarchic organization we observe in Nature is the result of a sequential process of optimization that works towards improving the "access" of elementary geometric space elements, which are then assembled into larger space elements until the entire relevant space is connected (Bejan, 1997). Constructal theory has been applied to several different types of engineering problems ranging from thermodynamics, fluid analysis, heat transfer, and the design of product platforms. Fundamentally, the crux of constructal theory is that access in a geometric space can be made most efficient through (i) a hierarchic organization of the several means of achieving access, and (ii) the use of a multistage decision process to determine the range of application of each technique (Bejan, 2000).

In order to abstract constructal theory and problems of optimal access to product platform development, Hernandez and coauthors (2002) introduce the concept of *space of customization* as the geometric space set of all feasible combinations of values of product specifications that a manufacturing enterprise is willing to satisfy (i.e., space S in Figure 18-1).

Mathematically, let N be the number of quantitative parameters that define the requirements of a product. Let r_i represent these parameters, where $i=1,...,N$. Then the space of customization, M^N, is the set:

$$M^N \equiv \{(r_1, r_2, ..., r_N)\}$$

$$(1)$$

It should be noted that a space of customization is not limited to continuous variables; it can be formed by continuous, discrete or mixed-valued requirements.

Based on this definition, a product i can be represented by a unique specification of product requirements in an N-dimensional space of customization, i.e., a vector $r_i(r_{i1},\ldots,r_{iN})$:

$$\boldsymbol{r}_i = r_{i1}\hat{e}_1 + r_{i2}\hat{e}_2 + \ldots + r_{iN}\hat{e}_N,\qquad (2)$$

where $\hat{e}_k$ is the unit vector in each direction k of the space of customization. Using this representation of a product, the derivation of a new product, r_j, based on an existing product, r_i, is referred to as a *"product customization,"* represented by a vector in the space of customization:

$$\Delta\boldsymbol{r}_{ji} = \sum_{k=1}^{N}(r_{jk} - r_{ik})\hat{e}_k = \sum_{k=1}^{N}\Delta r_{jik}\hat{e}_k.\qquad (3)$$

This representation of product customization is illustrated in Figure 18-1. Generic approaches to "access" points in the space of customization, i.e., to achieve product customizations from a baseline design, are referred to as *modes for managing product variety* (as shown in Figure 18-1 as Δr).

With the introduction of these definitions, the problem of designing a platform for customizable products becomes an effort to define a baseline set of components (the product platform) from which all the points of a space of customization can be accessed through the systematic use of a series of modes for managing product customization, and improving a (set of) given objective(s) (e.g., cost, profit, product performance, etc). The fundamental problem addressed in the application of the PPCTM to product platform design is how to organize and determine the extent of application of modes for managing product variety systematically in order to create a product platform for customized products. Through the application of the tenets of constructal theory, this optimal access problem is formulated as a multi-stage decision wherein the ranges of application of each mode for managing product variety are the decision variables. The goal of each decision is to improve the objective functions in order to provide the most efficient manner of offering product variety. More detailed information can be found in (Hernandez, et al., 2003).

2.2 Abstracting the PPCTM to process parameter platform design

In order to abstract constructal theory and problems of optimal access to the design of a process parameter platform, one must first define a geometric space that captures the essence of the problem. Since the production capacity requirement of the workstation is the only specification being varied, it serves as the lone dimension of the resulting geometric space. This geometric space is defined as the space of capacity: "A *space of capacity* is the range of workstation production capacity that a manufacturing enterprise is willing to satisfy with a process parameter platform."

The bounds of this single dimension are determined by the manufacturing enterprise as the amount of production capacity that their manufacturing system should satisfy. Each point in this geometric space represents a unique required level of production capacity (see Figure 18-2).

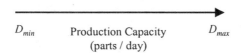

D_{min}　　　Production Capacity　　　D_{max}
(parts / day)

Figure 18-2. Visualization of the space of capacity.

With this geometric space defined, the crux of the application of the PPCTM to process parameter platform design is the synthesis of multiple methods of offering variety in order to provide any variant within the geometric space. These methods are called modes for managing capacity variety, and are defined as: "A *mode of managing capacity variety* is any generic approach in the design of the process parameters of a workstation for achieving a change in the required manufacturing capacity."

Examples of modes of managing capacity variety include, but are not limited to, process parameter standardization, batch size commonalization, machine type commonalization, and the modular combination of machine capacity (i.e., adding a new machine to the manufacturing system). These modes serve as the linking mechanism between the individual production capacity requirements that compose a family of process parameters.

With a means for linking the geometric space established, the problem of designing a platform for customizable workstation setups becomes an effort to define a set of parameters from which one can access all the points of a space of capacity through the systematic use of a series of modes for managing capacity variety, and improving some given objective(s) (e.g., cost, throughput, quality, etc.). The fundamental problem addressed in the application of the PPCTM in this realm is how to organize and determine the extent of application of the modes for managing capacity variety systematically in order to create a process parameter platform that will enable the efficient reconfiguration of a workstation.

The end result of the application of the PPCTM to the design of process parameters is the synthesis, hierarchic organization, and determination of the ranges of multiple modes of managing capacity variety in order to construct a process parameter platform. With these fundamental theoretical constructs presented, the Process Parameter Platform Constructal Theory Method is presented in detail in Section 3. An example problem is presented alongside each of the six steps of the PPCTM to serve as a tutorial for the reader.

3. THE PROCESS PARAMETER PLATFORM CONSTRUCTAL THEORY METHOD

Using constructal theory, the development of a process parameter platform for a workstation is formulated as an optimal access problem and solved as multi-stage decision wherein the ranges of application of each mode for managing capacity variety are the decision variables. The goal of each decision is to improve the given objective function(s) in order to provide the most efficient manner of offering production capacity variety. The six steps of the methodology are shown in Figure 18-3.

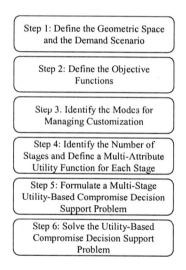

Figure 18-3. The product platform constructal theory method.

The first step of the PPCTM involves abstracting the development of a process parameter platform as a problem of access in a geometric space by identifying the space of capacity. In the second step, the objective functions are defined. Typical objective functions include production performance metrics such as the minimization of average cost, or the maximization of throughput and/or product quality. The modes for managing variety are identified in the third step and are hierarchically organized in Step 4. The determination of the range of application of each mode for managing variety is done through the formulation and solution of a multi-stage utility-based compromise Decision Support Problem. With the extent of application of each mode known, a designer is capable of fully defining the process parameter family that offers the best compromise to the objective functions. An example problem, the design of a process parameter platform for the manufacture of customizable hearing aid shells, is presented as a means to illustrate the method.

3.1 Example: Process parameter platform design for customized hearing aid shells

Consider a manufacturer of hearing aids that seeks a competitive advantage by offering personalized hearing aids (see Figure 18-4a). The manufacturer whishes to create these personalized hearing aids through the use of an additive fabrication technology (i.e., rapid prototyping). Specifically, it can create a personalized hearing aid shell from a 3D CAD model of a patient's ear canal (obtained by laser scanning a clay impression of the ear), using Stratasys' Fused Deposition Modeling (FDM). As shown in Figure 18-4b, FDM is a rapid prototyping technology that creates objects by extruding a heated filament through a nozzle that lays down the part's cross-section one layer at a time (Stratasys, 2003). The manufacturer believes that the use of the FDM rapid prototyping technology will provide the manufacturing enterprise the agility and flexibility to offer a personalized hearing aid to each customer at a competitive price. Due to the nature of the technology, a rapid prototyping machine is capable of creating multiple, different geometries in a single build without having to change the primary machine tool, and is thus capable of creating thousands of unique hearing aids more efficiently than traditional hard tooling alternatives. This example is based on an actual product line resulting from collaboration between Siemens and Phonak (Masters, 2002).

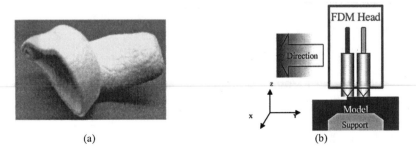

Figure 18-4. (a) Customized hearing aid shell, and (b) the fused deposition modeling process (Williams, et al., 2003).

While the manner in which product variety will be offered has been determined, the specific manner in which the manufacturing process will be configured has not. The manufacturing engineer faces a dilemma, as the selection of process parameters for Fused Deposition Modeling is a very difficult and complex problem, especially when confronted with the need to satisfy three conflicting objectives: the minimization of production cost, the minimization of production time, and the maximization of the quality of each part. Specifically, the manufacturing engineer must determine the appropriate batch size, process parameters for FDM, as well as the type and number of machines to be used for different levels of production capacity.

The problem is further obfuscated by a production constraint: the parts must be completed no more than one week from the date they were ordered. Furthermore, as is typical with most mass-customized parts, the demand for this product is highly non-uniform – the demand for customized hearing aid shells ranges from an arrival rate of 120 parts per day to 1000 parts per day. As a result, each time the capacity requirement changes, the manufacturer is required to re-evaluate all of the process parameters in order to maintain maximum production efficiency, as well as pay the costly setup penalty of changing the process parameters.

In this example problem, the manufacturing engineer will benefit from the use of the PPCTM to generate a process parameter platform for the range of capacity requirements presented by this scenario. Through the commonalization of process parameters, the setup and reconfiguration of the FDM will be more efficient when faced with capacity requirement changes. There are two key assumptions in the formulation of this example: (1) this manufacturing enterprise seeks to offer variety only through hearing aid shell geometry. Offering variety through a change in product material, product color, or in functionality is not considered in this example problem; and (2) all relevant information and models needed to apply the PPCTM are available, complete, and certain. The role of uncertainty and risk is not taken into account in the problem formulation. The reader is directed to (Williams, et al., 2003) for more details on the modeling for the problem.

3.2   Step 1: Define the geometric space

The first step of the PPCTM is the abstraction of a geometric space from the problem. For the development of a process parameter platform, this geometric space is the space of capacity (defined in Section 2.2). The definition of an appropriate space of capacity involves the identification of the range of production capacity that the manufacturing enterprise wishes to offer. The resulting space of capacity is a one-dimensional space that is bounded by the range of required production capacity.

As stated in the problem definition, the production capacity will fluctuate from 120 parts per day to 1000 parts per day. The resulting space of capacity is provided in Figure 18-5. Each point along this space of capacity represents a different level of production capacity.

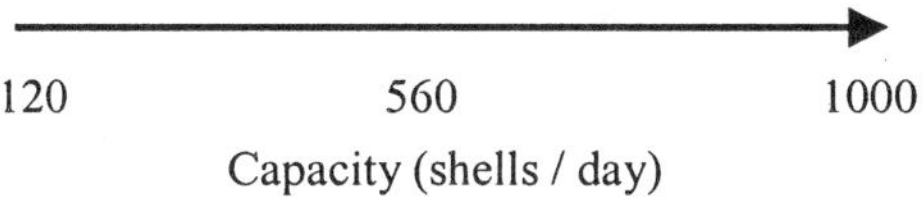

Figure 18-5. Space of capacity for the hearing aid shell example problem.

3.3 Step 2: Define the objective functions

As stated in Section 3.1, the manufacturing engineer wishes to find the best compromise between three conflicting objectives: to minimize the cost of the production process, to minimize the amount of time to build a batch of parts, and to maximize the quality of each part. The focus in this particular step of the PPCTM is to define the necessary objective functions.

The calculation of average time for the entire process family is based on the amount of time to build one batch, t_{batch}, of a particular capacity requirement from Eq. (4). The average build time of the process family is:

$$t_{avg} = \left[\left(\sum_{i=D_{min}}^{D_{max}} t_{batch,i} N_{batch,i} \right) + \sum_{j=1}^{N_{setup}} t_{setup,j} \right] \Big/ \left(D_{max} - D_{min} \right) \qquad (4)$$

where D_{max} and D_{min} represent the upper and lower bounds of the capacity space respectively. It is important to note that this build time metric includes a setup time penalty, t_{setup}, of 30 minutes that is accrued for each different arrangement of process parameters, N_{setup}, across the production family. The time to build a single batch, t_{batch}, is calculated as:

$$t_{batch} = t_{warm} + \left(t_l N_l + t_{base} \right) N_{pb} \qquad (5)$$

where:

t_{warm} = setup time; warming the machine for a build (0.5 hr);
t_l = time to build each layer;
N_l = number of layers;
t_{base} = time to build the base of the part; and
N_{pb} = number of parts per batch.

The time to build each layer is directly dependent on the road width parameter that defines the width of the material deposited by the extrusion nozzle. As this width increases, the amount of material deposited in a single pass is increased; thus the amount of time spent depositing a layer decreases. The number of layers (N_l) needed to complete a part is the quotient of the height (h) of the part and the layer thickness (t_{layer}).

$$N_l = h/t_{layer} . \tag{6}$$

While it is naïve to assume that the time to build each layer will be the same for each layer of each hearing aid shell, this will not significantly influence the overall results or interfere with the validation of the PPCTM.

The calculation of the average cost of the process family is performed in a similar fashion. The cost per batch of a single capacity requirement, C_{batch}, is calculated as:

$$C_{batch} = (V_{material}C_{material})N_{pb} + (C_{labor} + C_{operation}) t_{batch}$$
$$+ [(C_{maint} + C_{machine})N_{machine}/N_{ppy}] N_{pb} \tag{7}$$

where:

$V_{material}$ = volume of material used to build one part (~229 mm^3);

$C_{material}$ = cost of material; includes build and support materials (0.26 \$/cm^3 and 0.23 \$/cm^3, respectively);

C_{labor} = cost of labor (\$25 / hour);

$C_{operation}$ = hourly cost of machine operation (\$80 / hour);

C_{maint} = annual machine maintenance cost (\$50,000 / year);

$C_{machine}$ = cost of purchasing machine, to be paid through one year of production; and

N_{ppy} = number of products produced in one year.

The hourly and annual fees for labor, operation, and maintenance are estimates of actual costs. The cost of purchasing the machine is presented in Table 18-1 along with other machine specifications. Their values do not change the fundamental validation strategy of applying the PPCTM to this example problem. The average cost is calculated as:

$$C_{avg} = \left(\sum_{i=D_{min}}^{D_{max}} C_{batch,i}N_{batch,i} \right) + \left(\sum_{j=1}^{N_{setup}} C_{setup,j} \right) \bigg/ D_{max} - D_{min} . \tag{8}$$

Table 18-1. Machine characteristics (Stratasys, 2003).

Machine	Max. Parts	Scan Speed	Layer Thickness	Road Width	Cost
Prodigy Plus	234	64 mm/s	0.178 – 0.33 mm	0.19 – 0.21 mm	$70,000
Titan	818	127 mm/s	0.24 – 0.26 mm	0.19 – 0.21 mm	$210,000
Maxum	1703	254 mm/s	0.127 – 0.25 mm	0.193 – 0.965 mm	$260,000

A penalty for the setup of a different arrangement of process parameters is added to this metric with C_{setup}, which is equal to $50.

For many products, quality is a metric with a subjective nature. When manufacturing with rapid prototyping, quality can be quantified by modeling the difference between the desired geometry and the actual, produced geometry. This error occurs because of the nature of this layer-based, additive manufacturing technology (known as the "stair stepping" effect, see Figure 18-6). The quality of the part is directly related to two process parameters: road width (w_{road}) and layer thickness (t_{layer}) of the deposition.

As can be observed in Figure 18-6, the best quality is achieved with a minimal layer thickness and minimal road width. The average quality of the process family is calculated as:

$$Q_{avg} = \left(\sum_{i=D_{min}}^{D_{max}} Q_i \right) \bigg/ \left(D_{max} - D_{min} \right), \qquad (9)$$

where Q_i is the quality of a single part, estimated as the sum of the road width and the layer thickness. These objectives are inherently contradictory. For example, to maximize the quality, one would set the layer thickness and road width parameters to a minimum. This however, would increase material costs as well as production time.

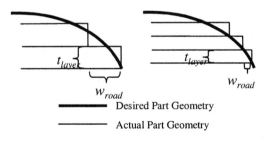

Figure 18-6. Quality metric diagram.

3.4 Step 3: Identify the modes for managing variety

For this step, a designer identifies appropriate methods for providing variety in production capacity for the manufacturing system. Modes of managing variety are the linking mechanism between the different production capacity requirements. For this third step in the PPCTM, the designer identifies those modes that will be used to offer variety in the production capacity of the manufacturing process. Since the space for capacity is one-dimensional, each mode will be directed towards achieving variety in capacity only. For this example problem, five different modes for managing process customization are implemented.

3.4.1   Mode D1: Customization of the batch size

One mode of managing capacity variety for this example problem is the standardization of the batch size (i.e., number of parts per build). With the use of this mode, changes in capacity are achieved by choosing a different batch size. The determination of batch size is a very important parameter decision. Larger batch sizes are preferred because of the associated decrease in costs. The batch size cannot be too large as the process is constrained both by the physical capacity limitation of each machine type (see Table 18-1), and by the weeklong production-time constraint.

3.4.2   Modes D2 and D3: Standardization of the process parameters

For these two modes, changes in capacity are achieved by standardizing the process parameters of the manufacturing process: layer thickness (the height of each individual layer deposited by the process) and road width. The determination of these parameters is very important to the overall process, as each of the parameters plays a large role in the calculation of each of the three objective functions. In comparison to modifying the road width of the process (Mode D3), production capacity changes of higher fidelity can be achieved by slightly changing the layer thickness (Mode D2).

3.4.3   Mode D4: Commonalization of machine type

Changes in capacity are also achieved by changing the type of machine used in the production process. Changing the machine type changes the maximum batch size constraint, and also affects the speed at which the layers of the parts can be drawn. Of course, different machines cost different amounts of money, so the objective of minimizing cost is severely altered with each different machine type. Similar to all forms of technology, larger, faster machines cost more money and typically offer better production quality (see Table 18-1). As such, the selection of machine type is crucial. The use of this mode for managing variety commonalizes the scan speed and maximum parts allowed per batch over a range of capacities.

3.4.4 Mode D5: Modular combination of machines

This fifth and final mode of managing capacity variety for this example problem is the selection of the number of machines used for the production of the shells. This mode involves the notion of modular combination to achieve variety. For this example, "modular combination" refers to the addition of machines to compensate for large demands of capacity. Since it is assumed that only similar machines can be added to the production process, the addition of a machine simply doubles the capacity of the manufacturing process. This doubling comes at a large cost however.

These five modes of managing process customization are the approaches considered for accessing all the points of the space of capacity generated in Step 1 (Section 3.2) of this example. Next, in Steps 4 through 6 (Sections 3.5 - 3.7), a hierarchic organization of these modes is synthesized in order to offer a variety of production capacity in the space of customization.

3.5 Step 4: Identify the number of hierarchy levels and allocate the modes for managing variety to the levels

In the fourth step of the PPCTM, it is established how and when each mode of managing variety is used. Modes that are capable of achieving the smallest variations in production capacity are typically used at the lower levels of the hierarchy (i.e., before modes that can only achieve large variations in capacity). Economical and technological considerations play an important role in mode hierarchicy for managing variety.

Following the tenets of constructal theory, each level of the hierarchy represents a geometric "sub-space" of the entire space of customization. The sizes of each sub-space represent the extent of application of each mode for managing variety and are the decision variables of this multi-stage problem.

3.5.1 The first stage and the first space element

For this example problem, of all the process parameters to be varied, altering the batch size provides the best "control" over changes in the required production capacity. Customizing the batch size provides the simplest (changing the batch size does not require additional operator input) and most cost effective manner (there are no setups costs associated with changing the batch size) for a designer to offer a specific production capacity requirement. For these reasons, this mode is placed at the bottom of the mode hierarchy. The mode D1, "Customization of the Batch Size," is used to define a common batch size for a range of capacities that are bounded in the first space element, ΔD_1, as shown in Figure 18-7.

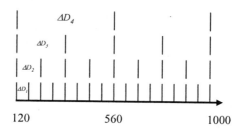

Figure 18-7. The space elements of the hearing aid example space of capacity.

3.5.2 The second stage and the second space element

The mode D2, "Standardization of Layer Thickness," is chosen as the mode for managing variety for the second stage in the hierarchy. Of the modes remaining for selection, altering this process parameter provides a designer with the ability to make smaller adjustments in the level of production capacity. Mode D2 is used to define a common layer thickness for a range of capacities that are bounded by the second space element, ΔD_2 (see Figure 18-7). Each second space element is composed of a number of first space elements.

3.5.3 The third stage and the third space element

This third stage follows the same formulation as found in the previous stages. Mode D3, "Standardization of Road Width," is used to define a common road width for a range of capacities that are bounded by the third space element, ΔD_3 (see Figure 18-7). Mode D3 is placed at a higher level of the hierarchy than mode D2 (changing the layer thickness) because changing the value of road width does not provide as such high fidelity changes in production capacity. The build time metric is not as sensitive to changes in road width as it is with changes in layer thickness. Similar to layer thickness, increasing the road width of a part decreases the build time and cost, as well as lowers the average part quality.

3.5.4 The fourth stage and the fourth space element

The fourth and final space element is composed of a number of assemblies of the third space element as shown in Figure 18-7. In this final space the remaining two modes of managing variety, Mode D4 ("Commonalization of Machine Type") and Mode D5 ("Altering the Number of Machines") are used.

The application of these modes is different than the application of the other modes in the previous stages. Mode D4 is based on the concept that certain capacity requirements are suited for different types of machines. Whether limited by maximum build size or by a slow scan speed, lower-end machines simply cannot produce parts at a sufficient rate to meet larger demands. This applies to the other end of the capacity space spectrum as well; higher-end machines may be too expensive to justify their use for lower capacity needs.

Mode D5, the combination of similar machines, is investigated in this space as an opportunity to offer more capacity. Increasing the number of machines produces an interesting tradeoff: it not only increases the amount of capacity of the production process by decreasing the build time, but it also increases the costs associated with purchasing and maintaining the machines. Since these two modes are only capable of expensive, large, discrete changes in production capacity, they are used at the highest level of the hierarchy.

The focus in this decision stage is the assignment of different machine types and quantities to specific ranges of the capacity space. This is achieved by identifying six cutoff points along the space (each cutoff representing each of the discrete upgrade of production capacity). As can be seen in Figure 18-8, each cutoff point represents a different combination of machine type and quantity. This specific ordering of each cutoff point is based on the maximum number of parts per build of each machine type/quantity combination (see Table 18-1).

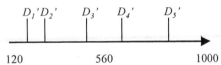

$D \leq D_1'$ corresponds to one Prodigy Plus; $D_1' < D \leq D_2'$ corresponds to two Prodigy Plus; $D_2' < D \leq D_3'$ corresponds to one FDM Titan; $D_3' < D \leq D_4'$ corresponds to two FDM Titans; $D_4' < D \leq D_5'$ corresponds to one FDM Maxum; $D_5' < D$ corresponds to two FDM Maxums

Figure 18-8. Visualization of the fourth stage of the hearing aid example problem.

With all of the modes for managing variety successfully organized (see Figure 18-9), the formulation of the multi-stage utility-based compromise Decision Support Problem begins.

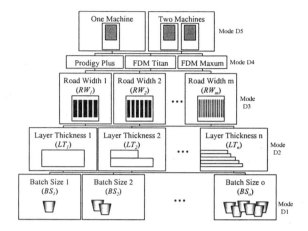

Figure 18-9. Hierarchic organization of the modes for managing capacity variety.

3.6   Step 5: Formulate a multi-stage utility-based compromise decision support problem

Following the tenets of constructal theory, the determination of the range of application for each mode for managing variety that composes a level of the hierarchy (or sub-space) represents one stage in a multi-stage decision. With the order of the use of the modes established, a designer proceeds by formulating a proper multi-stage decision problem.

In this work, each decision stage is formulated with a utility-based compromise Decision Support Problem. The utility-based compromise DSP (u-cDSP) is a decision support construct that is based on utility theory and permits mathematically rigorous modeling of designer preferences such that decisions can be guided by expected utility in the context of risk or uncertainty associated with the outcome of a decision.   While any appropriate decision formulation technique is serviceable, we prefer to use the u-cDSP because its use "provides structure and support for including human judgment in engineering decisions involving multiple attributes, while simultaneously providing an axiomatic basis for accurately reflecting the preferences of a designer with regard to feasible tradeoffs among these attributes under conditions of uncertainty" (Fernández, et al., 2001). Furthermore, the u-cDSP has proven useful in previous product platform techniques as it provides a decision construct in which a designer can model multiple, conflicting objectives (Seepersad, et al., 2002). The formulation of each utility-based compromise Decision Support Problem follows the four steps listed in (Seepersad, et al., 2002) and shown in Figure 18-10.

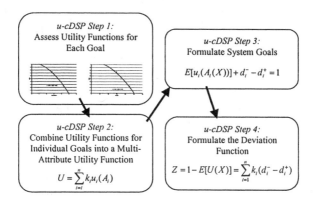

Figure 18-10. The formulation of the utility-based compromise decision support problem.

First a utility function for each of the objectives is formulated by qualitatively and quantitatively assessing the preferences of the designers (all designers' preferences are modeled as risk averse in this work). These individual utility functions are then combined into a multi-attribute utility function as a weighted average of the individual utilities. Finally, goal and deviation functions are developed for each stage. The deviation function of the u-cDSP is formulated to minimize deviation from the target expected utility (i.e., 1, the most preferable value), which is mathematically equivalent to maximizing expected utility. The goal and deviation functions formulated for each u-cDSP inherently consider the compromise of the tradeoffs between the each objective function. With the goal of minimizing the deviation of the expected utility from the ideal value, parameters that provide the best values for this overall objective are chosen while maintaining consistency with the designer's preferences. With the presence of the u-cDSP, designers are given the ability to model multiple objectives in each decision stage of the PPCTM.

Table 18-2. Utility functions for cost, time, and quality.

Utility Value	Cost (\$), $k_c = 0.25$ Metric Value	Time (hrs), $k_t = 0.25$ Metric Value	Quality (mm), $k_q = 0.5$ Metric Value
1	500	13	0.35
0.75	8500	63	0.65
0.5	15250	105	0.9
0.25	21000	139	1.12
0	25000	168	1.3
Utility Function[a]	$u(C) = 2 - e^{\left(\frac{0.685(C-C_{min})}{(C_{max}-C_{min})}\right)}$	$u(t) = 2 - e^{\left(\frac{0.685(t-t_{min})}{(t_{max}-t_{min})}\right)}$	$u(q) = 2 - e^{\left(\frac{0.6861(q-q_{min})}{(q_{max}-q_{min})}\right)}$

[a] the *min* and *max* subscripts refer to the values of each parameter with utilities of 0 and 1 respectively.

In this example, the multi-attribute utility function for each stage is based on the individual utility functions of cost, time, and quality. The values to be used in the utility assessment are calculated using the average values of the metrics calculated in each decision stage. Each utility function is presented in Table 18-2.

The k-values presented in the table above are calculated by solving a system of equations wherein a designer establishes quantitative preferences for each scaling constant. The k-values are used in concert with the individual utility functions for the determination of the overall expected utility, $E[U(D_d)]$.

$$U(D_d) = k_c u(C) + k_t u(t) + k_q u(Q) \qquad (10)$$

The goal of each objective is to maximize the value of each individual utility function (i.e., for the value to reach the ideal value, 1).

$$E[u(C_i)] + d_{C_i}^- + d_{C_i}^+ = 1 \qquad (11)$$

$$E[u(t_i)] + d_{t_i}^- + d_{t_i}^+ = 1 \qquad (12)$$

$$E[u(q_i)] + d_{q_i}^- + d_{q_i}^+ = 1 \qquad (13)$$

The deviation function for the multi-attribute utility function of each decision stage, therefore, is to minimize the deviation of the expected utility function from the ideal value:

$$Z_i = 1 - E[U_i(D_d)] = \sum_{j}^{3} k_j (d_j^- + d_j^+). \qquad (14)$$

In order to evaluate the values of the three objective functions for each subspace, the values of parameters to be commonalized across the space must first be determined. In order to satisfy the property of near-decomposability of hierarchic systems, the choice of these design variables for each space element must be independent of the choice for the other space elements. The decision of appropriate values to be commonalized is based on the largest values of capacity requirement of the subspace, or geometrically the rightmost point of each space element, since its solution will be sufficient for all variants within that space.

Due to the multiple objectives involved with this example, one cannot identify a specific value of the process parameter that satisfies production constraints and also minimizes cost and time, and maximizes quality without a detailed analysis. In order to determine the appropriate values of the parameters to be commonalized for each subspace a designer must employ a utility-based compromise DSP. This initial decision is presented in Figure 18-11 as "Decision 0." The *max* and *min* subscripts in Figure 18-11 refer to the imposed bounds of the machine for each process parameter (see Table 18-1). The individual utility functions of Decision 0 ($E[u(C)]$, $E[u(t)]$, $E[u(q)]$) are evaluated using Eqs. (4), (8), and (9) respectively. Decision 0 will be used inherently in each of the decision stages in order to determine the value of the parameter that will be commonalized across each subspace.

3.6.1 The first stage and the first space element

The first space element is defined by the range of application of Mode D1: "Customization of the Batch Size". ΔD_l determines the extent of the commonality of the batch size, and is therefore the decision variable for this first stage. The focus in this decision is the determination of ΔD_l that provides the best value for the overall objective – the maximization of the expected utility of all the objectives.

In this first stage, the calculation of cost, time, and quality at each node begins with the determination of the value of the batch size that should be commonalized. As stated previously, this value is determined at the maximum capacity value of the current sub-space.

Given the layer thickness and road width, the batch size that provides the largest expected utility within the space is solved for using the u-cDSP outline in "Decision 0" (see Figure 18-11).

Given: Capacity Requirement; D_d (parts /day)
Machine Type; Prodigy Plus, FDM Titan, or FDM Maxum
Machine Number; N_{mach}

Find: Batch Size; N_{pb}
Layer Thickness; t_{layer}
Road Width; w_{road}
Deviation variables; d_i^- and d_i^+

Satisfy: Bounds: $0 \leq N_{pb} \leq N_{pb,max}$

$t_{layer,min} \leq t_{layer} \leq t_{layer,max}$

$w_{road,min} \leq w_{road} \leq w_{road,max}$

Constraints: $t_{cycletime} \leq 1$ week

$d_i^-, d_i^+ \geq 0$

$d_i^- \Box d_i^+ = 0$

Goals: $E[u(C)]+d_C^- -d_C^+ =1$
$E[u(t)]+d_t^- -d_t^+ =1$
$E[u(Q)]+d_Q^- -d_Q^+ =1$

Minimize: $Z = 1 - E[U(D)] = 1 - \left[k_c(d_c^- - d_c^+) + k_t(d_t^- - d_t^+) + k_q(d_q^- - d_q^+) \right]$

Figure 18-11. Decision formulation for "Decision 0".

This common batch size is then applied to each capacity requirement of the first space. The resulting time, cost, and quality metric values are then averaged across the first space. This is done by first evaluating the number of first space elements in the space of capacity:

$$N_1 = (D_{max} - D_{min})/(D_{max,1} - D_{min,1}).$$ (15)

The values of D_{max} and D_{min} are total range of capacity offered; $D_{max,1}$ and $D_{min,1}$ are based on the corresponding value of ΔD_1:

$$D_{max,1} = D_{min,1} + \Delta D_1.$$ (16)

The average time of all of the first space elements is calculated as:

$$t_{avg} = \left(\sum_{i=1}^{N_1} t_{avg,1,i} + \sum_{j=1}^{N_{setup}} t_{setup,j} \right) \Big/ (D_{max} - D_{min}),$$ (17)

where $t_{avg,1}$ is the average time of a single first space element. Similarly the average cost of the first space elements is calculated with:

$$C_{avg} = \left(\frac{1}{D_{max} - D_{min}} \right) \left[\left(\sum_{i=1}^{N_1} C_{avg,1,i} \right) + \left(\sum_{j=1}^{N_{setup}} C_{setup,j} \right) \right].$$ (18)

The third objective, the maximization of average quality, is averaged across the space of capacity via:

$$Q_{avg} = \left(\frac{1}{D_{max} - D_{min}} \right) \left(\sum_{i=1}^{N_1} Q_{avg,1,i} \right).$$ (19)

These averaged values are then used to calculate an expected utility of the first space. The individual utility functions ($u(C)$, $u(t)$, and $u(Q)$), and the resulting expected utility function ($E[U(D)]$), are calculated as presented in Table 18-2. The resulting decision formulation for the first stage is shown in Figure 18-12.

Inherent in this first stage decision is the value of layer thickness and road width process parameters, as well as the type and number of machines used in the production process. The individual utility functions cannot be evaluated without these values. These crucial details are determined in the following decisions.

Given: The one-dimensional capacity space
          Mode D1: Customization of the Batch Size
Find: The value of the decision variable ΔD_I
          Deviation variables, d_C^-, d_C^+, d_I^-, d_I^+, d_Q^-, d_Q^+
Satisfy: Bounds: $0 \leq \Delta D_I \leq 880$
          Constraints: $t_{cycletime} \leq 1$ week
                         $d_i^-, d_i^+ \geq 0$
                         $d_i^- \Box d_i^+ = 0$
          Goals: $E[u(C)]+d_C^- - d_C^+ = 1$
                         $E[u(t)]+d_i^- - d_i^+ = 1$
                         $E[u(Q)]+d_Q^- - d_Q^+ = 1$
Minimize: $Z_1 = 1 - E[U(D_d)] = 1 - \left[k_c(d_c^- - d_c^+) + k_t(d_t^- - d_t^+) + k_q(d_q^- - d_q^+) \right]$

Figure 18-12. Decision formulation for the first space element.

3.6.2   The second stage and the second space element

The range of application of Mode D2, "standardization of the layer thickness," defines the size of the second space element. Each second space element is composed of a number of first space elements. The number of first space elements that compose a second space element, $N_{1,2}$, is:

$$ N_{1,2} = \Delta D_2 / \Delta D_1 = \left(D_{max,2} - D_{min,2} \right) / \left(D_{max,1} - D_{min,1} \right). \qquad (20) $$

The range of commonality of the layer thickness process parameter is defined by the decision variable ΔD_2. Similar to the previous decision stage, "Decision 0" is used to identify the layer thickness that provides the best compromise between the three conflicting objectives for the largest capacity requirement of each second sub-space. This value is then commonalized over a series of capacity requirements as dictated by the value of ΔD_2.

Similar to the first stage, the focus of this second decision stage is the selection of a value of ΔD_2 that maximizes the expected utility of the space of customization defined in Step 1. The calculation of the average time, cost, and quality of each second space element is dependent on the number of first space elements that compose the second space element:

$$ t_{avg,2} = \left(\sum_{i=1}^{N_{1,2}} t_{avg,1,i} \middle/ N_{1,2} \right), \qquad (21) $$

$$ C_{avg,2} = \left(\sum_{i=1}^{N_{1,2}} C_{avg,1,i} \right) \middle/ N_{1,2}, \qquad (22) $$

$$Q_{avg,2} = \left(\sum_{i=1}^{N_{1,2}} Q_{avg,1,i} \right) \Big/ N_{1,2} . \qquad (23)$$

The average time, cost, and quality of all of the second space elements is calculated in a similar fashion as is done for the first space elements (see Eqs. 17-19). Following the format of Stage 1, the averaged values are then used to calculate the individual utility functions (see Table 18-2). Finally, these individual utility functions are combined into an overall expected utility function. The focus in the decision of this second stage (see Figure 18-13) is the minimization of the deviation of this expected utility from 1.

Given: The one-dimensional capacity space
Mode D2: Standardization of Layer Thickness
The value of ΔD_1

Find: The value of the decision variable ΔD_2
Deviation variables, d_C^-, d_C^+, d_t^-, d_t^+, d_Q^-, d_Q^+

Satisfy: Bounds: $0 \le \Delta D_2 \le 880$

Constraints: $\Delta D_1 \le \Delta D_2 \le 880$

$t_{cycletime} \le 1$ week

$d_i^-, d_i^+ \ge 0$

$d_i^- \Box d_i^+ = 0$

Goals: $E[u(C)] + d_C^- - d_C^+ = 1$
$E[u(t)] + d_t^- - d_t^+ = 1$
$E[u(Q)] + d_Q^- - d_Q^+ = 1$

Minimize: $Z_2 = 1 - E[U(D)] = 1 - \left[k_c (d_c^- - d_c^+) + k_t (d_t^- - d_t^+) + k_q (d_q^- - d_q^+) \right]$

Figure 18-13. Decision formulation for the second space element.

3.6.3 The third stage and the third space element

For the third space Mode D3, standardization of road width, is used to offer variety in the production capacity of the manufacturing process. The range of commonality of road width, ΔD_3, determines the size of each third space element. Each third space element is composed of a number of second space elements. The number of second space elements that compose a third space element, $N_{2,3}$, is calculated as:

$$N_{2,3} = \Delta D_3 / \Delta D_2 = \left(D_{max,3} - D_{min,3} \right) \Big/ \left(D_{max,2} - D_{min,2} \right). \qquad (24)$$

The formulation of the decision third for the third stage is identical to that of the previous two. There are two decision variables for this third stage, the value of road width to be commonalized, and its range of commonality, ΔD_3. "Decision 0" is used to identify the road width that provides the best compromise between the three conflicting objectives for the largest capacity requirement of each second sub-space. This value is then commonalized over a series of capacity requirements as dictated by the value of ΔD_3.

Since the formulation of the objective functions and the decision of a third space element is extremely similar to that of a second space element (see Eqs. 21-23 and Figure 18-13), their explicit forms are not presented here for the sake of brevity.

3.6.4 The fourth stage and the fourth space element

The size of the fourth and final space element is determined by the range of commonality of each machine type / quantity combination. The range of application of Modes D4 and D5 are defined by the placement of the "cut-off" points – the capacity requirements that require a different machine type and/or quantity. As with the previous decision stages, the focus of the implementation of Modes D4 and D5 in this fourth stage is the maximization of expected utility for the entire capacity space. Similar to the previous steps, average time, cost, and quality are all calculated as a sum of the spaces of the previous stage. The analysis begins with the calculation of the number of third space elements in each fourth stage cutoff point, $4,i$.

$$N_{3,4_i} = \Delta D_{4,i} / \Delta D_3 = \left(D_{\max,4,i} - D_{\min,4,i} \right) / \left(D_{\max,3} - D_{\min,3} \right) \qquad (25)$$

The average time, cost, and quality of a fourth space element is evaluated using Eqs. (26)-(28).

$$t_{avg,4_i} = \left(\sum_{j=1}^{N_{3,4_i}} t_{avg,3,j} \right) / N_{3,4_i}, \qquad (26)$$

$$C_{avg,4_i} = \left(\sum_{j=1}^{N_{3,4_i}} C_{avg,3,j} \right) / N_{3,4_i}, \qquad (27)$$

$$Q_{avg,4_i} = \left(\sum_{j=1}^{N_{3,4_i}} Q_{avg,3,j} \right) / N_{3,4_i}. \qquad (28)$$

The average time, cost, and quality of all of the fourth space elements is calculated in a similar fashion as is done for the first space elements (see Eqs. 17-19). Following the format of the previous stages, these averaged values are then used to calculate the individual utility functions as shown in Table 18-2. Finally, these individual utility functions are combined into an overall expected utility function (see Eq. 10). The focus in the decision of this fourth stage (see Figure 18-14) is the minimization of the deviation of this expected utility from 1.

Given: The one-dimensional capacity space
             Mode C4: Commonalization of Machine Type
             Mode C5: Altering the Number of Machines
             The value of ΔD_1
             The value of ΔD_2
             The value of ΔD_3
Find: The value of the decision variable ΔD_4
             The location of each cutoff point, D_1', D_2', D_3', D_4', D_5'
             Deviation variables, d_C^-, d_C^+, d_t^-, d_t^+, d_Q^-, d_Q^+
Satisfy: Bounds: $0 \leq \Delta D_4 \leq 880$

             Constraints: $\Delta D_3 \leq \Delta D_4 \leq 880$

                            $t_{cycletime} \leq 1$ week

                            $d_i^-, d_i^+ \geq 0$

                            $d_i^- \square d_i^+ = 0$

             Goals: $E[u(C)]+d_C^- -d_C^+=1$
                            $E[u(t)]+d_t^- -d_t^+=1$
                            $E[u(Q)]+d_Q^- -d_Q^+=1$
Minimize: $Z_4 = 1 - E[U(D)] = 1 - \left[k_c(d_c^- - d_c^+) + k_t(d_t^- - d_t^+) + k_q(d_q^- - d_q^+) \right]$

Figure 18-14. Decision formulation for the fourth space element.

The result of the solution process is the determination of the ranges of the modes of managing process customization that will produce the best compromise between the three objectives (minimize cost, minimize build time, and maximize quality).

3.7   Step 6: Solve the multi-stage utility-based compromise decision support problem

The final step of the PPCTM is the formulation of an appropriate solution algorithm. Any appropriate solution technique can be used; the primary goal in the solution is the determination of the ranges of the modes that provide the largest expected utility (i.e., the best compromise between the conflicting objectives of minimizing cost, minimizing build time, and maximizing part quality). A graphical representation of the solution method for this problem is presented in Figure 18-15.

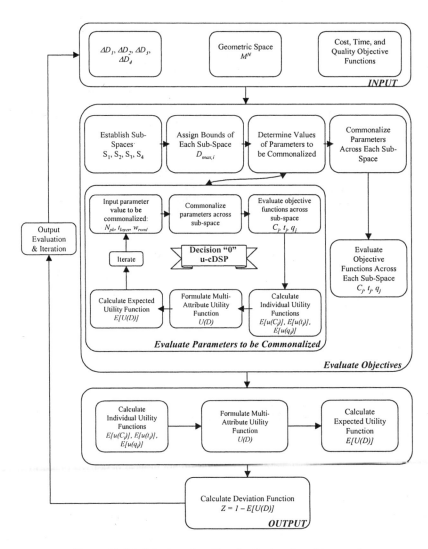

Figure 18-15. Solution algorithm for the hearing aid example.

This solution method involves iterating through values of the modes of managing variety (ΔD_1, ΔD_2, ΔD_3, and ΔD_4), establishing the dimensions of the sub-spaces, commonalizing the design parameters (N_{pb}, t_{layer}, and w_{road}) across each sub-space, evaluating the objective functions, and comparing the resulting overall utility of each iteration.

4.   RESULTS

The result of the application of the PPCTM to this example problem is the hierarchic organization of the five modes of managing production variety (see Section 3.4), and the determination of their range of application across the process parameter platform that provides the best compromise among the three objectives. The extent of application of each mode and the resulting average cost, time, and quality for the platform is shown in Table 18-3.

Table 18-3. Range of each mode of managing capacity variety for hearing aid example.

Mode of Managing Customization	Parameter Commonalized	Range of Commonalization (parts / day)
ΔD_1 (customization of batch size)	Batch Size	5
ΔD_2 (standardization of layer thickness)	Layer Thickness	21
ΔD_3 (commonalization of road width)	Road Width	21
ΔD_4 (commonalization of machine type; modular combination of machines)	One FDM Maxum	880
$t_{avg} = 47.42$ hrs	$C_{avg} = \$11,659.32$	$Q_{avg} = 0.509$ mm
Expected Utility = 0.8360		

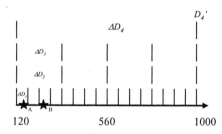

Figure 18-16. Graphical representation of resultant process parameter platform.

This table of results provides a manufacturing engineer with the range of commonalization for each mode of managing process customization. This is presented graphically in Figure 18-16. These results inform the manufacturing enterprise that, in order to achieve the best compromise between all three objectives, the best configuration of the modes of managing variety is to use one FDM Maxum, to commonalize the batch size for every interval of 5 parts per day, and to commonalize road width and layer thickness parameters for every interval of 21 parts per day. From these ranges of application for each mode of managing variety, the specific values of the design variables are derived by the use of the PPCTM (see Table 18-4).

Table 18-4. A sample of the mapping of the process parameters.

Capacity (parts/day)	Batch Size (parts)	Road Width (mm)	Layer Thickness (mm)	Machine Type	Number of Machines
120	727				
125	727				
130	761		0.14668		
135	777				
140	811				
145	845				
150	862		0.15652		
155	878				
160	912	0.193			
165	946				
170	962				
175	979				
180	1013				
185	1047			FDM	1 Machine
190	1063		0.1762	Maxum	
195	1080				
200	1097				
205	1164				
210	1181				
215	1215		0.17128		
220	1232				
225	1249				
230	1282	0.22388			
235	1299		0.1762		
240	1316				
245	1333				
250	1383				
255	1400		0.18604		
260	1417				

To illustrate the significance of this result, the following tutorial example is presented. Imagine that the manufacturing enterprise reports an arrival rate of demand of 234 parts per day (point A in Figure 18-16). Looking at Table 18-4, the manufacturing engineer, using one FDM Maxum, selects a batch size of 1299 parts/batch, a road width of 0.22 mm, and a layer thickness of 0.18 mm. If the manufacturing capacity requirement changes to 183 parts per day the following week (point B in Figure 18-16), the engineer is only required to change the batch size to 1047 parts/batch and road width to 0.19mm without changing the value of the layer thickness parameter.

A thorough investigation of these results provides interesting lessons about the methodology and the example problem itself. For instance, it can be questioned why only one machine type and quantity was chosen to be commonalized across the entire space of capacity. Specifically, how can the most expensive machine be the best choice for low capacity requirements? After careful observation it is concluded that while the less expensive machine does offer a lower total cost, the difference in cost is not enough to compete with the higher-end machine which offers a faster build time and a better part quality. As a result, the higher-end machine is the best choice for the entire space of capacity; it simply offers more machine for its price. This observation provides a great opportunity to witness the ability of a designer to consider multiple objectives in the design of a production platform with the PPCTM.

Another interesting observation is that, from watching the iteration history, the value of expected utility does not change drastically with different ranges of application of each mode for managing variety. From this observation one can witness a major limitation of the use of an expected utility function for this specific example problem. The formulation of the expected utility function (see Table 18-2) is broad enough to capture the extremely different capabilities and properties of each machine. This results, however, in a function that is insensitive to small changes in decision parameters. It is therefore recommended that a designer's preferences be clearly identified, and be as scoped as possible for the application of the PPCTM to this type of problem. It is noted that the use of the u-cDSP is not a limitation of the PPCTM; the expected utility function's insensitivity to small parameter changes is a result of the model itself, as documented in (Williams, et al., 2003).

Through the solution of this example problem, it is evident that the PPCTM is an effective means of designing process parameter platforms. The issue of the usefulness of production process itself has yet to be addressed, however. In order to answer this issue, the values of each objective are compared between the results from the development of a production platform (using the PPCTM), with the result of having no commonality in process parameters for each capacity requirement (i.e, parameters are reevaluated and reconfigured for each change), and with the result of making all process parameters common across the space of capacity (i.e., there is one parameter configuration for all capacity requirements). This comparison is provided in Table 18-5.

Table 18-5. Comparison of results of a) PPCTM, b) No commonality, and c) Strict commonality of process parameters.

	t_{avg} (hrs)	C_{avg} ($)	Q_{avg} (mm)	$E[U(D)]$
(a) PPCTM Result	47.42	11,659.32	0.509	0.8360
(b) No Commonality ($\Delta C_X = 1$)	48.16	11,734.33	0.507	0.835
(c) Pure Commonality ($\Delta C_X = 880$)	49.52	12,681.23	0.756	0.720

As can be observed from the table, the concept of developing a process parameter platform for this example problem provides an improvement to all three objective functions (and thus provides the largest expected utility) when compared to having "pure" and "no" commonality. Although there is only a small quantitative benefit for this specific problem, it is evident that the development of a process parameter platform improves the workstation's ability to adapt efficiently to changes in its required capacity.

Although it is evident that the development of a process parameter platform is of some benefit for this example problem, it is important to note that this success is not necessarily universal. It is important to note that this methodology is most beneficial when the cost and time required for the changeover of a workstation is very costly. Furthermore, if the manufacturer expects the workstation to encounter wildly different required capacity from day to day (e.g., a change of over 200 parts/day for this example problem), the benefits of the process parameter will be lost. Since commonality between the process parameter variants is dependent upon the order of arrival, it is possible that no commonality between different setups will be encountered if the change in production capacity is continually larger than the range of application of each mode for managing variety. This limitation can be alleviated by accounting for the frequency of different capacity requirements with a probability distribution function.

While this specific example problem provides an appropriate means for validating the proposed methodology, it is important to note its unique characteristics. Rapid manufacturing is a special class of production process because it is capable of producing multiple products in one batch. It is also capable of producing an entire part with a single workstation without an assembly step, and thus, there is no differentiation between a "part" and the end product. Since it is possible to manufacture a product with a single rapid manufacturing machine, decisions regarding batch size, machine selection, and machine quantity are able to be made during the determination of process parameters. This is not a valid assumption when designing a production process with the more traditional and less flexible means of manufacturing. Despite the uniqueness of this example problem, we are confident in the adaptability of the methodology to a wide range of applications. We are aware that more traditional means of manufacturing involve complex decisions regarding sequencing and capacity planning. For this reason, this methodology is scoped to be applicable to the process parameter design of single workstations.

5. CLOSURE

In this chapter we present the concept of process parameter platforms. Process parameter platforms are a set of common process parameters from which a stream of derivate process parameters can generate a customized product efficiently despite changes in capacity requirement. The goal in creating a platform of process parameters is to create an efficient manner of offering variety in production capacity. Similar to product platforms, process parameter platforms achieve variety efficiently through commonality and/or modularity.

Along with this concept, we present a design methodology for realizing process parameter platforms. Leveraging from previous work with the Product Platform Constructal Theory Method, a powerful product platform design technique, the design of process parameter platforms is treated as a problem of access in a geometric space. The use of the PPCTM provides a designer the ability to accommodate the issues of:

(1) Multiple design objectives: As shown in Section 3.3, the development of the platform required the compromise of three conflicting objectives: the minimization of production cost, the maximization of quality, and the maximization of throughput.

(2) Multiple modes of offering variety: As illustrated in Section 3.4, a designer must synthesize multiple modes of offering variety (standardizing process parameters, □ommunalizing batch size, standardizing machine type, and modularly combining machines) in order to provide a means of achieving all variants within the space of capacity. The use of the PPCTM synthesizes multiple modes of offering variety through hierarchic organization in order to offer variety efficiently.

(3) Volatile markets: Through the definition of the space of capacity in Section 3.2, a designer is able to develop platforms in the presence of changing capacity requirements for workstations – a feature inherent in the manufacture of customized goods.

(4) The inherent tradeoffs between platform extent and performance: As described in Sections 3.6 and 3.7, the determination of the range of application of each mode for managing variety is achieved systematically through the rigorous formulation of a multi-stage utility-based compromise Decision Support Problem.

An example problem, the design of a process parameter platform for the manufacture of a line of customizable hearing aid shells, is presented as an example problem to aid in the description of the methodology. Through the application of the methodology to this example problem, it is shown that the design of a process parameter platform for a specific workstation minimizes successfully the necessary setup and changeover encountered with changes in capacity requirement; however, it is noted that this benefit is only seen when the changes in capacity are not widely distributed along the range of required capacity.

6. ACKNOWLEDGEMENTS

We gratefully acknowledge the support of NSF Grants DMI-0085136 and DMI-9900259. Christopher Williams is a Georgia Tech President's Fellow and a NSF IGERT Research Fellow through the Georgia Tech TI:GER program. The cost of computer time was underwritten by the Systems Realization Laboratory at the Georgia Institute of Technology.

7. NOMENCLATURE

C_{avg}	Average cost of family of processes; $
C_{batch}	Cost of building a single batch of hearing aid shells; $
C_{labor}	Cost of labor to operate FDM machines; $/hr
$C_{machine}$	Cost of purchasing FDM machine; $
C_{maint}	Annual cost of maintaining FDM machine; $
$C_{material}$	Cost of FDM material; $/mm3
$C_{operation}$	Annual cost of operating FDM machines; $
C_{setup}	Cost of changing process parameters due to setup; $
D	Production capacity; parts/day
d_i^-, d_i^+	Deviation variables
h	Height of hearing aid shell; mm
N_{batch}	Number of batches
N_i	Number of ith space elements
N_l	Number of layers
$N_{machine}$	Number of FDM machines
N_{pb}	Number of parts per batch
N_{ppy}	Number of parts per year
N_{setup}	Number of production setups
Q_{avg}	Average part quality; mm
t_{avg}	Average production time; sec
t_{base}	Time required to build the base of each part; sec
t_{batch}	Time require to process a batch of parts; sec
$t_{cycletime}$	Production lead time constraint; 1 week
t_l	Time required to build each part layer; sec
t_{layer}	Layer thickness; mm
t_{setup}	Machine setup time; sec
t_w	Machine warm-up time; sec
$U(X)$	Overall, multi-attribute utility function
$V_{material}$	Volume of material used to build one hearing aid shell; mm^3
w_{road}	Road width of FDM deposit; mm
Z	Deviation function
ΔD	Range of capacity in hearing aid shell space of capacity
ΔD_i	The dimension of ith space element in direction D of market space; a decision variable for a stage i.

PART IV: APPLICATIONS OF PLATFORM-BASED PRODUCT FAMILY DEVELOPMENT

Chapter 19

ICE SCRAPER PRODUCT FAMILY DEVELOPMENT AT INNOVATION FACTORY

Steven B. Shooter
Department of Mechanical Engineering, Bucknell University, Lewisburg, PA 17837

1.   INTRODUCTION

Many companies have developed product families based on common platforms with varying degrees of success. Many studies of these platforms are based on product dissection. It can be challenging to gain complete information from the companies about the product development for a number of reasons including intellectual property protection. Also, many new products are developed by teams, making it difficult to get the complete picture from any individual. The case study in this chapter comes from a small company of only two primary people, so it was possible to gain insight about the complete process. Of particular interest is that the company started their design with full intent of using platform strategies for developing their product family. The following is a description of their top-down approach to platform-based product development.

2.   INNOVATION FACTORY

On a cold March morning, two business partners were flying from Philadelphia to San Francisco to pitch a new product idea. On the flight they lamented their frustrations with cleaning ice from their windshields that morning. They then discussed the problems associated with current ice scrapers, and by the time they landed they had developed design criteria and preliminary sketches for an innovative ice scraper. This led to the founding of Innovation Factory, a company whose design philosophy is to "take a high-tech approach to solving the mundane but often life-endangering challenges of daily life."

Innovation Factory is a "virtual company" with only two employees who manage the complete product life cycle process. That means they have built effective partnerships with other companies and organizations to support their product development efforts. These include marketing research firms, government technical support agencies, an industrial design company, production companies, sales and distribution networks. Tucker Marion, the partner who acted as the project manager, has a B.S. in Mechanical Engineering and an MSE in Technology Management. In graduate school he took a class on product platforms and was eager to incorporate the techniques for the development of the ice scraper family. Their efforts led to the IceDozer products shown in Figure 19-1.

Figure 19-1. Innovation Factory product family of ice scrapers.

Figure 19-2 shows some of the common ice scrapers on the market, the majority of which are low cost and low performance. Additionally, little engineering thought or intellectual property has been added to the market in decades. Innovation Factory wanted to take advantage of this stagnant market and address the performance issue common to the current offerings. One main problem with current scrapers is that the straight blades do not conform to curved windows and windshields. They do not maximize the scraping pressure and do not have ergonomic handles. Innovation Factory performed a national survey and discovered that 96% of the responders did not like the current ice scrapers on the market, and they did not know the brand of ice scrapers that they used. Over 50% of the responders said that they would pay several times more than the average current price of $3 or $4 for a better ice scraper. Innovation Factory therefore focused on developing a scraper that would actively deform with the windshield and fix the problems of existing scrapers. They would also focus on a higher end market and establish brand identity.

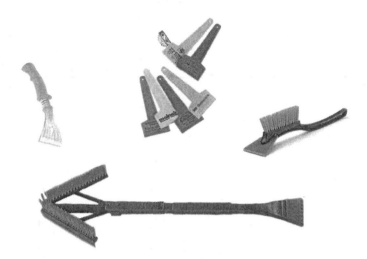

Figure 19-2. Common ice scrapers available on the market.

3. MARKET-DRIVEN PRODUCT MANAGEMENT PROCESS

Effective platform management maintains product innovation across multiple generations. Meyer and Lehnerd (1997, p. 37) state that "effectively managing the evolution of a product family requires that management consider in collective fashion three essential elements of the enterprise: (1) derivative products made for various customer groups; (2) the company's product platforms; and (3) the common technical and organizational building blocks that are the basis of product platforms."

Figure 19-3 shows the Power Tower indicating the market segment opportunities for the ice scraper family. The fundamental building blocks at the bottom of the figure and the market segment applications drive the derivative products and successive generations of the product platform. Figure 19-4 illustrates the intended price points for the IceDozer product family. The core product is the IceDozer Classic with an intended sales price of $15. The variant MiniDozer targets the lower market with a more compact model priced at $8-$10. The IceDozer Extreme would include additional features for the higher-end specialty market.

Figure 19-5 shows the resulting product family based on the platform, which is the scraper blade. Because the blade was the distinguishing element of the platform and at the center of their patent claims, its technical development was a large focus.

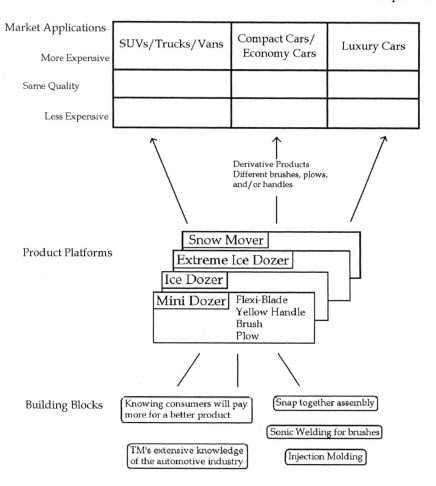

Figure 19-3. Power tower for ice scraper product family.

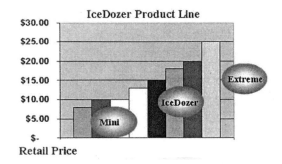

Figure 19-4. Price segmentation of the IceDozer family.

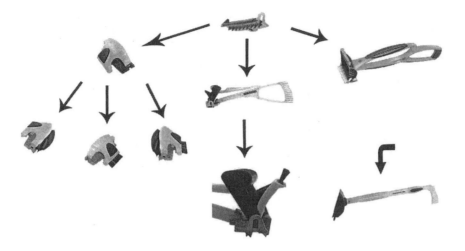

Figure 19-5. IceDozer product family and the accompanying SnowMover.

Innovation Factory followed a beachhead strategy for their product family development. At the center of the platform is the IceDozer which was the first product launched. The platform was then extended for customers in different market segments. The MiniDozer is the second stage of evolution including its extensions to provide step-up functions required by mid and high-end users in other segments. The IceDozer Extreme is at the top end. The design drivers for the product line were that they would be more rugged than competitors (and have a rugged look), work effectively in multiple conditions such as frost, thin ice and thick ice, include a brush, and allow different attachments for variants. The SnowMover does not contain the blade platform. Its function is to brush snow off a car rather than scrape ice. It was launched with the IceDozer to add breadth to the product line.

4. PRODUCT ARCHITECTURE AND CONCEPT INSTANTIATION

When establishing a product family, it is often helpful to establish the architecture of the individual products and the platform. This architecture follows the guidelines described in Otto and Wood (2001). Figure 19-6 shows the function structure for the IceDozer based on its primary functions and the flow of material, energy and information. The material elements are the human hand, the windshield, ice, and snow; the energy is the human force, and the information tests cleanliness of the windshield. The function architecture is useful for understanding the needed elements for the design.

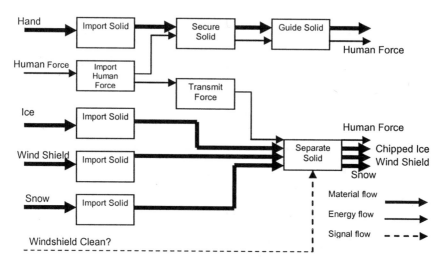

Figure 19-6. Function architecture for IceDozer.

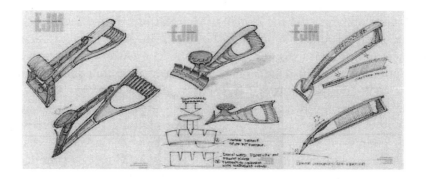

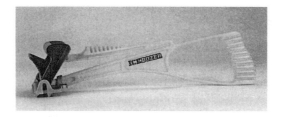

Figure 19-8. Final IceDozer.

Innovation Factory worked closely with an industrial design company to develop seventeen different concepts based on a flexible blade and two-handed operation. The top three concepts are shown in Figure 19-7. This evolved into the final IceDozer shown in Figure 19-8, which has three components. The assembly structure is shown in Figure 19-9. This shows how the blade is secured by the primary handle and the plow handle. We will examine the connections in more detail later; however, it is significant that the outside of the blade is secured by the plow, and the inside of the blade is secured by the primary handle. The force from the hand on the plow is then applied at the edges and the force from the primary handle is applied at the middle. The primary handle is designed such that compliance in the material allows the blade to flex to the contour of the windshield. The instantiation of the IceDozer elements with the functional architecture is shown in Figure 19-10.

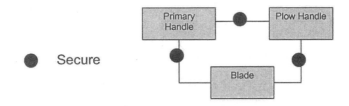

Figure 19-9. IceDozer assembly architecture.

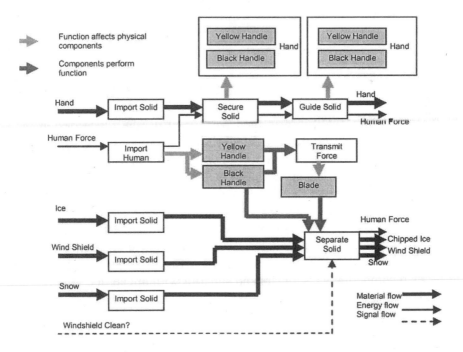

Figure 19-10. Components and function architecture for IceDozer.

5.   FOCUS ON THE BLADE DESIGN

The FlexiBlade was identified as the fundamental platform element and would therefore be the focus of the technical development. Innovation Factory researched properties of ice during its different phases of formation and thicknesses to gain a better understanding of scraping ice. They found that as ice freezes and gets colder, it becomes more brittle with a crystal-like structure resembling quartz. Instead of scraping, it was discovered that fracturing the ice proved more effective for thicker ice. This insight led to the final blade design (see Figure 19-11) to include three methods for varying ice conditions allowing scraping against thin frost layers and breaking or cracking of thicker ice. The blade was designed with corrugations to be rigid in the lateral direction for scraping the ice yet flexible in the horizontal direction to contour to the shape of the windshield. It has a pronounced front edge for scraping ice and pointed teeth on the bottom for cracking thick ice.

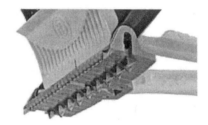

Figure 19-11. Features of the FlexiBlade.

Finite element analysis was then performed on the blade design with reaction forces from the windshield as shown in Figure 19-12. These first months of development led to an overall size and shape with a well-analyzed blade. The team then began testing prototypes. Three scrapers were made using stereolithography for the handles and computer numerical control (CNC) machined blades. In August they tested the scraper in a rented ice garage facility. The goal was to perform testing scenarios with different ice thicknesses and determine the time required to clear a path leaving no remaining frost or ice. The tests indicated that the IceDozer was many times faster than competing ice scrapers and superior with thicker ice.

Although the IceDozer outperformed other scrapers, failures occurred during testing. The inside of the handle failed on one of the prototypes. This prompted reinforcement with the addition of ribs. One of the tabs on the blades broke from an impact force with the windshield. They rounded out this stress concentration feature to remedy the problem. The testing also indicated the need for larger and pointier teeth on the blade. They did additional finite element analysis to help make improvements on features to reduce stress concentrations while maintaining the desired flex of the blade.

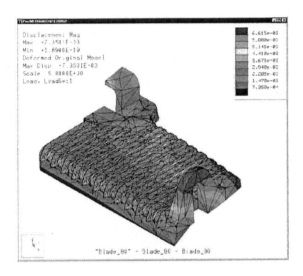

Figure 19-12. Finite element analysis of the FlexiBlade.

6. PRODUCTION PARTNERSHIPS

Because Innovation Factory does not have any of its own production facilities; all production needed to be outsourced. They used the Internet to obtain multiple quotes from molders that placed bids on the tooling job. Close collaboration between Innovation Factory, the technical consultants, the industrial design consultants and the production facility occurred. The focus was placed on inexpensive injection-molded parts; consequently, the elements were designed so that the tooling required no pulls or other expensive tool action. Durability and a rugged look were consumer-focused elements that were maintained. The coordination activities through this stage of development proved the most challenging to Innovation Factory. Through repeated iterations between design and manufacturing, the final IceDozer was produced. Total time from concept to market was about 11 months. The IceDozer was launched on January 15 with the support of an article in Popular Science (February, 2002). In order to expand the core IceDozer line, it was decided through customer feedback to attach a brush feature. Using the existing front handle, a multifunction brush extension was added to make the IceDozer Plus. This brush was designed in early 2003, and offered for the 2003/4 sales cycle.

7.   REUSING THE PLATFORM ON MINI-DOZER

The MiniDozer was a planned element of the product family from the beginning. The timeline shown in Figure 19-13 illustrates the product family development process. Note that the MiniDozer conceptual design began at the same time as the IceDozer. Innovation Factory then decided upon the FlexiBlade as the platform and focused development efforts on the base IceDozer product. The MiniDozer was then brought to production very rapidly because of product platform techniques.

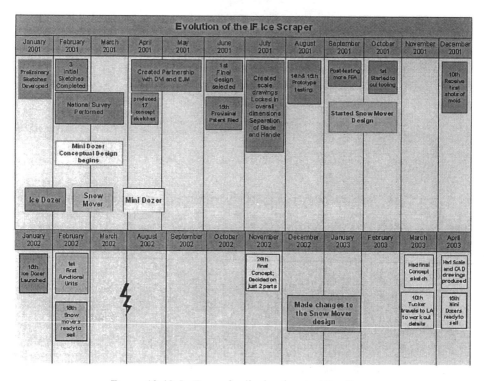

Figure 19-13. IceDozer family development timeline.

The MiniDozer development focused on trying to maintain all of the design drivers. The original concept shown in Figure 19-14 tried to maintain the same architecture with three elements. This concept reused both the blade and the plow but would use a reduced-sized handle. When the MiniDozer design was revisited several months later, Innovation Factory understood better the production cost involved in the three-element design. Meeting the desired price-point for the product family forced a reduction to two elements: the platform blade and the handle. Because the FlexiBlade was being reused, they focused on the handle design as another platform for the MiniDozer line of variants. Market feedback from the IceDozer indicated an interest in a brush by some users, and others liked the plow.

Figure 19-14. MiniDozer concept similar to IceDozer with three components.

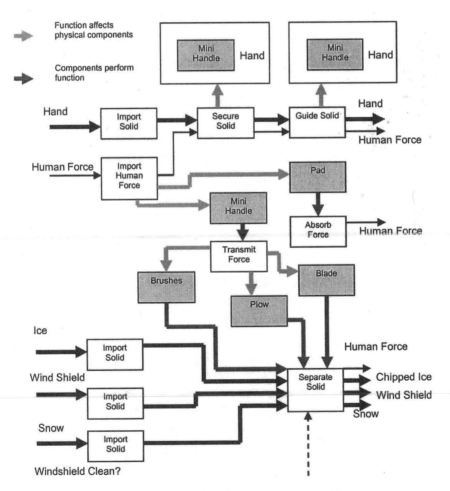

Figure 19-15. MiniDozer function-to-component architecture.

Figure 19-15 shows the architecture for the MiniDozer with the components instantiated. Prototype testing indicated the need for a pad at the placement of the palm whose function is to absorb some of the force from the hand and provide comfort. Innovation Factory debated strongly over the addition of this element because of the added cost. Ultimately it was added to the MiniDozer because of the intended consumer view of a high-end product compared to the competition. Figure 19-16 shows the assembly architecture for the MiniDozer with the base and desired variants. The base unit contains the blade, handle and pad. The variants allow for a brush and/or a plow. The challenge was then to design the handle so that it accommodated the variants and attached well to the blade platform. Figure 19-17 shows the MiniDozer Basic, the Classic with the brush, and Deluxe with the plow and brush.

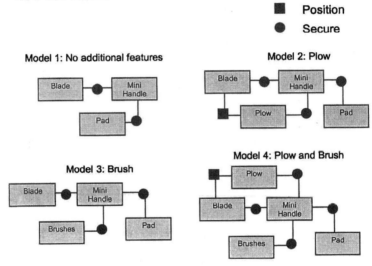

Figure 19-16. MiniDozer assembly architectures for variants.

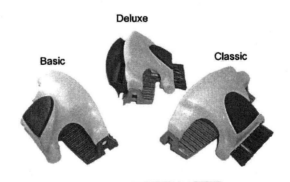

Figure 19-17. MiniDozer and variants.

8. FOCUS ON HARDPOINTS FOR THE PLATFORM AND EXTENSIONS

A modular approach to product family design requires special attention to the interfaces among physical components often referred to as *hardpoints*. Careful design of hardpoints facilitates reuse of the platform in a product line. The FlexiBlade needed to connect differently for the IceDozer and MiniDozer as shown in Figure 19-18. The hardpoint highlighted in the center of the picture connects the blade to two different handle configurations in the same manner. These outside connections gave the blade its edge stiffness. Notice that there is more material surrounding the connection on the MiniDozer to add some additional lateral support at the connection. The hardpoint highlighted on the left and right of the picture uses the same blade feature to connect to the handle in a different manner. The IceDozer handle inserts into the loop. The design of the handle then allows the blade to flex. For the MiniDozer the loop inserts into a pocket. The loop serves as a guide in that pocket to allow the blade to deflect. In all of the hardpoints the design focus was on strength and ease of assembly.

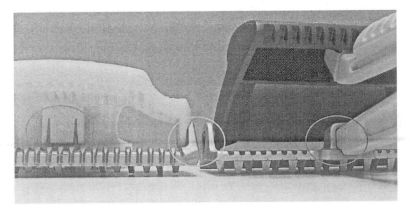

Figure 19-18. Hardpoints on Blade-Handle Connection –connection in center is used in the same manner on both, connection on left and right uses same blade feature differently.

The handle of the MiniDozer was also designed with hardpoints for including extensions. The left picture in Figure 19-19 shows the inside of the handle with the slots for connecting elements to the front of the scraper such as the plow extension. The rear surface was intentionally designed flat for the inclusion of additional features such as the brushes that are sonically welded to the handle.

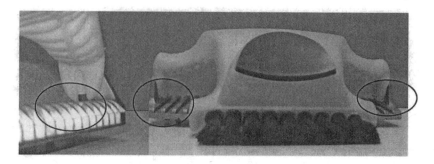

Figure 19-19. MiniDozer handle hardpoints.

9.   OTHER PRODUCTS IN THE FAMILY

Innovation Factory also has developed a third line of ice scrapers in the IceDozer Extreme which has yet to make it to production. The Extreme uses the same platform FlexiBlade and plow but incorporates a longer handle for reach on large vehicles such as SUVs, vans, and trucks. Figure 19-20 shows concept sketches of the IceDozer Extreme, which has gone through prototype testing but has not appeared on the market as of this writing.

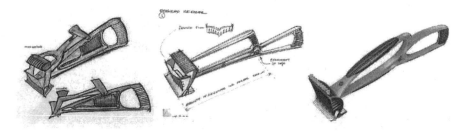

Figure 19-20. Three concept sketches of the IceDozer Extreme.

The SnowMover is sold as a complement to the IceDozer or Minidozer. The large plow supplies the user with a tool that clears deep snow from the surface of the car without damage. Customer surveys indicated that this was important to drivers because many would operate their vehicles after cleaning only a small section of the windshield, creating a driving hazard if a section of snow would dislodge from the car. It was originally hoped that aspects of the IceDozer could be reused for the SnowMover; however, the functional differences of the products did not support the reuse of components. Innovation Factory decided to develop this as a separate product with materials and colors that identify it as part of the Innovation Factory family. However, the main body, telescoping handle, and clips of the SnowMover were designed with the hopes of reuse. For example, several mock-ups of the IceDozer Extreme reused components of the SnowMover. It was the intention of Innovation Factory to reuse as much of these components in future products as possible.

10.   CONCLUSIONS AND LESSONS LEARNED

The development of the IceDozer product family is a good example of a top-down approach to product family planning. It proved an especially good approach for Innovation Factory because they were able to leverage their resources while adding to variety in a product line. They were able to focus their research and development on the signature FlexiBlade platform that appears across the family. They were also able to incorporate similar production techniques for the variant elements and extensions. This resulted in lower tooling costs, and shorter developmental lead times. These proved essential to the early success of the Innovation Factory, particularly when increasing volume in season three. The common FlexiBlade was able to be produced in bulk in advance, allowing very quick order to delivery lead times. Without common components, this may not have been possible.

The ability to offer multiple products is important to small companies for facilitating adoption by vendors and customers. Offering a more complete product line adds legitimacy to a small company as they work with vendors who might be hesitant to work with one-off product producers. The Innovation Factory, through the use of common components, was able to quickly develop and introduce line extensions. Not only did the use of platforming techniques increase product breadth, but it allowed a signature feature to be co-marketed among different models and price points. Product families are not only beneficial to large firms; they can aid new companies as well. The Innovation Factory is a case in point of a start-up company successfully translating platform theory into real-world cost and product development cycle time reduction.

11.   ACKNOWLEDGEMENTS

This work was supported by the National Science Foundation under Grant No. IIS-0325321. Any opinions, findings, and conclusions or recommendations presented in this chapter are those of the authors and do not necessarily reflect the views of the National Science Foundation.

Chapter 20

ARCHITECTING AND IMPLEMENTING PROFITABLE PRODUCT FAMILIES AND SHARED ENGINEERING PLATFORMS
Strategies for Overcoming Organizational Constraints

Srinivas Nidamarthi and Harshavardhan Karandikar
ABB Corporate Research Center, Ladenburg, Germany 68526

1. BUSINESS OBJECTIVES

Every company has the business objectives of maximizing customer choice as well as its profitability. Typically, companies address maximum customer choice through a large spectrum of variants in their products and complete flexibility in creating engineered solutions to satisfy varying customer needs. For example, a camera manufacturer may wish to offer various choices such as fixed focus, auto-focus, variable zoom, different zoom ranges, SLR, APS, and digital cameras, and in different combinations, to satisfy customers with different demands (including the price that they wish to pay). The business goal, therefore, is to design a family of products or systems that satisfy many customers but at a minimum cost. These goals, customer choices and profit margin, are not as contradictory as they seem.

We have developed a set of systematic methods through which a product business can identify the essential design elements of profitable product families. We have successfully applied this method in a number of product families in ABB ranging from commodity-like products (that are mass produced) to custom manufactured ones. In this chapter, we explain these methods and our experience in applying them for product families such as fluid flow meters, air-handling fans, electrical drives, synchronous motors, instrument transformers, high voltage cable accessories, robot controllers and secondary substations. We present a platform approach to two very distinctive businesses within ABB – product business, where our customers can buy products off the catalogues, and engineering business, where customers place a turnkey order for a system including design, build and commissioning. We show how platforms can be applied to both businesses and how organizational constraints can be overcome during implementation.

Unlike a product business, where product development is an independent dimension to manufacturing, a systems or solutions business achieves success through efficient and effective project management, where the final solution is designed, built, tested and commissioned for customers. In this case, engineering platforms can be built by sharing knowledge, competence, services, and sales-, design- and supply management-processes across countries and projects. In ABB, we are designing such platforms for systems businesses developing discrete manufacturing lines for the automotive industry. These are engineering platforms for Body-In-White, Press, Paint, and Powertrain Assembly Automation businesses that are distributed over many locations worldwide. In this chapter we discuss our approach, the challenges we have faced, and results.

2.   ABB: THE COMPANY

ABB (http://www.abb.com/) is a global automation and power technology company that delivers products, systems and services to customers worldwide. Its products range from household circuit breakers to industrial robots, systems ranging from simple plant automation applications to substations installation and commissioning, and services from breakdown repairs to life cycle and complete plant maintenance. ABB Corporate Research introduces product technology as well as business process innovation for all ABB companies. One initiative within the R&D project portfolio has been to introduce and sustain product platform methodologies focusing mainly at cost reduction and profitability. In this work, we have collaborated with MIT and Stanford, as well as developed our own methods and tools. We have successfully applied these methods and tools for ABB companies, and transferred them for regular use in these companies.

As is typical for any such large multinational business the responsibility for profits and strategy is split between the local business unit and country organization versus a global management structure. This creates specific and complex hurdles in the execution of a product/solution platform strategy. This aspect has not been addressed in existing literature.

This chapter is based on our experience over the last 5 years in implementing platform strategies for twelve product businesses and for four systems businesses where engineering is distributed over fourteen locations.

3.   RESEARCH PROBLEMS

Platforms optimize assets that are shared by a set of end results (products or engineered systems) produced from them. These assets are (Roberston and Ulrich, 1998): components, processes, knowledge, and people and relationships. At the same time the end results must enable a company to be profitable by capturing market share. That is, products or systems produced over a platform must achieve maximum profit at minimum cost for a company to be competitive and efficient. Because markets are dynamic, platforms must also enable flexibility in products or systems to satisfy changing demands from customers. Therefore platform development must consider customers, market position, and the way the assets are leveraged to produce maximum profit over a time.

Another unique problem in a global and mature company like ABB is that decisions for platform development or improvement are never from a clean slate. Existing assets play a very complex role in determining optimum product architecture, processes and organization.

Given these influences, we have organized platform development as a collaborative effort between corporate research and the respective business units. We, from R&D, guided and supported the businesses, while they themselves re-organized and implemented our platforms recommendations. This mutual understanding of scope between R&D and ABB businesses worked well from introducing state-of-the-art platforms methods (that are brought by us) to realizing the improved platforms (the end results that the units must themselves adapt and realize). Thus our experience is based on both outside knowledge of platforms research, as well as inside challenges in implementing them. Within this scope of our work, platforms research problems can be broadly classified into: (1) Designing profitable product architecture for product businesses, and (2) Developing work processes, shared and structured knowledge, tools and organizational culture for systems engineering businesses.

3.1   Profitable product architecture

Product architecture is one of the most important parameters influencing revenue and cost levels and thus profitability of a company. In order to maximize market share, companies attempt to offer large product variety. This variety must be designed in such a way that costs of providing it should be minimal. We call such a spectrum of variety as *profitable product architecture*. We have developed methods to determine most profitable product architecture given its current sales, revenues, and costs. Two examples where we have determined profitable product architectures follow.

ABB Robotics wanted to design a new family of robot controllers satisfying its customer demands as of today and for the future, as well as reducing the product cost by at least 25%. In this project, the product family needed to be defined in its variety (customer options like IO interfaces, electrical drives; and choices in each option like analog IO or digital IO) that meets customer demands as well as maximizes the product's profitability. In another project, ABB Instrumentation wanted to streamline its range of magnetic flow meters (to measure any fluid flow capable of carrying electro-magnetic flux) produced at three production units, in three different countries, that have little sharing of components across their products although all serve similar functional purposes. In this project, the flow meter variety (customer options like meter size, electronic display; and choices like meters varying from 3 mm to 2.5 m in diameter) needed to be determined for maximum profitability and customer satisfaction.

While flow instruments are like standard and commodity products, where customers can quickly replace a failed product by a new one, the robot controllers are very often built to custom specifications. Nevertheless, both these problems are alike and typical of any product business. We have successfully applied our methods to architect a number of such product families. Sections 4 and 5 describe our research methods, and how they were applied to such product businesses.

3.2   Shared engineering platforms

In a global systems business cost and delivery time are key competitive factors to maintain profitability and growth. Both cost and time can be reduced if engineering activity can be shared across countries, based on standard solutions and a systematic process. In this case, we are building engineering platforms for globally distributed and relatively independent business units within Manufacturing Automation business in ABB.

ABB Manufacturing Automation consists of four units that develop and deliver custom lines for automotive manufacturers in a project-based business. The engineered deliverables from these four business units is illustrated in Figure 20-1. In Press automation, sheets of aluminum are handled and pressed to make parts like car doors and roofs. Engineering here is to deliver robots and machines to automate this manufacturing process. Similarly in Body-In-White (BIW) where the car body is assembled, in Paint Automation where it is painted according to the consumer's order, and in Powertrain where its gearbox, engine and axles are assembled.

Each type of automation line is completed according to customers' requirements via project management. Typical customers include automotive OEM (Original Equipment Manufacturers) like DaimlerChrysler and BMW, and Tier-1 suppliers for OEMs like gearbox, axel and car-body components (sheet metal) manufacturers. In each project, customer requirements such as gearbox parts that need to be assembled, degree of automation required, throughput rate (number of gearboxes per day), and the floor layout of the assembly line are identified. Solutions are then designed, manufactured, assembled internally for testing and debugging, disassembled, shipped, and then finally installed and commissioned at customer site. No two engineering projects are alike, and the engineering is often distributed over multiple locations. Therefore, it is a complex and very challenging problem to implement engineering platforms for such business units. Sections 6 and 7 describe how we have implemented these platforms using the Powertrain business unit as an example, and how we are currently improving them.

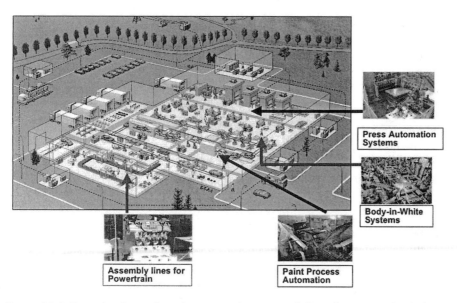

Figure 20-1. Example of manufacturing automation system delivered to automotive industry.

4. DESIGNING PROFITABLE PRODUCT ARCHITECTURE

The product architecture to achieve maximum profit and optimum customer coverage can be obtained through a systematic method containing three major steps:

(1) *Determine functional variety*: Functional requirements that vary from customer to customer are determined, and customers and sales personnel are interviewed to analyze existing functional variety and future trends.

(2) *Determine profitable product variety*: Using statistical analysis and optimization methods profit patterns from the existing product choices are analyzed. This method is based on the principle that a particular variety offered to customers is only satisfying to them if a) it functionally meets their requirements and b) the price they wish to pay is proportional to this functional satisfaction. The method is also based on the fact that the company must meet its cost targets to be profitable, and remain in business. This method uses cluster algorithms to compute changes in profit margins with changes in product family. Here we also present methods to model revenues and costs as a function of variety. These models enable designers to consider business rationale (i.e., profit/variety) when designing a family.

(3) *Design for Commonality*: Using results from the above two steps, design changes are determined to maximize the product family's scope of meeting customer functionality at minimum cost.

4.1   Determining functional variety

Functional variety serves to satisfy various needs of customers. For example, customers choose various power ratings for electrical drives in a robot controller according to their robot's operational needs (a robot carrying large weights requires higher torque, which in turn requires a high power electrical drive). From a company's point of view, these needs also define customer-segments – groups of customers needing specific functions. For example, customer needs for robots operating in a foundry are different from those operating for welding. Such segmentation is a very important factor in architecting a product family – to design its functional variety and offer it at right price. Therefore, determining right functional variety is essential to:

(1) *Differentiate customer segments and thereby determine market capturing product variety and its pricing*: For example, magnetic flow meters need to be corrosive and abrasive resistant as per the fluid flow they need to measure. These requirements are of high importance for chemical and pharmaceutical customers (customer segments) who can pay a good price for the value they get (high quality, resistant design), but are not so important for customers in wastewater segment who need a cheaper product.

(2) *Design the product variety at lowest possible cost*: A functional variety that can serve several customer segments and little physical variety (parts, components, etc.) that can serve for several functional varieties will always have high potential to reduce product cost. For example, Hastelloy-based flow meters can serve corrosive as well as abrasive needs, thus serving both chemical and pharmaceutical customers. This alloy can also serve for wastewater flow measurement; however, it is a high-priced solution for this

market segment, which requires a cheaper solution. Therefore, additional variety in the portfolio must remain in order to ensure market share.

We determine functional variety for a product family through the following steps: (1) Understand existing functional variety for the product family in business; (2) Identify customer demands as of today and as expected in future; and (3) Evaluate the identified variety with design, production and business logic (i.e., technical feasibilities, estimated costs and profit potential, etc.).

In this process, we interview several stakeholders of the product family – customers, sales engineers, product managers, designers, business managers and service units. For example, while re-designing the robot controller family we found out that many customers would like to have an IO board, which can read both analog and digital inputs at certain configurations.

4.2   Determining profitable variety

After determining functional variety, we determine the product variety – customer options and choices (as illustrated previously) – that can maximize a company's profit. An overview of this method is shown in Figure 20-2.

4.2.1   Product variety model

The functional variety, determined as described in the previous section, is categorized according to factors that clearly distinguish customer needs, revenue creators, and cost drivers. For example, variety in flow meters are categorized according to their sizes (in mm), materials (for corrosive resistance), regional standards (CE, US and Canadian regulatory standards), and display/electronic options, because this variety varies from customer to customer, and revenues and costs of the company can be broken down to these variety categories.

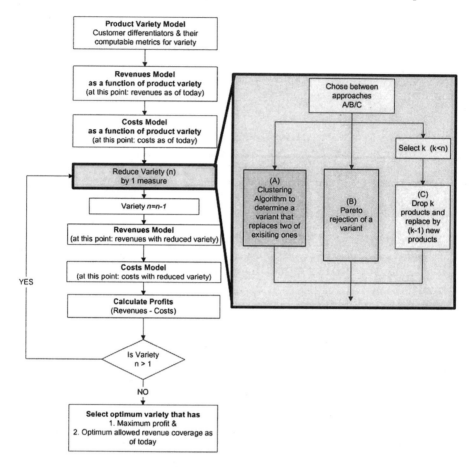

Figure 20-2. Method for determining product variety maximizing profit.

4.2.2   Revenue model

For example, each customer j wants their ideal target τ_j but we offer him x. The price P_j that a customer is willing to pay for that offer could be:

$$P_j = P_{0j} - w_j\left(x - \tau_j\right)^2 \tag{1}$$

where P_{0j} is the ideal revenue we can get, and w is a sensitivity factor that reflects how fast revenue drops when we change variety (e.g., price needs to be adjusted steeply low for high priority functions). Thus, if we were to offer x to all customers then the revenue we get is:

$$R = \sum_{\text{customers } j} \left\{ P_{0j} - w_j \left(x - \tau_j \right)^2 \right\} = \overline{R}_0 - \overline{W} \sigma^2 \qquad (2)$$

where R_0 is the ideal revenue if τ_j is offered, and σ is the deviation from meeting the τ_j. In other words, revenue reduces proportional to the standard deviation – or the dissatisfaction of customers for not getting exactly what they want (this model also includes customers who will not buy because of the difference). This σ is obtained from existing customer preferences, and based on changes in product variety meets these preferences (see Eqs. 4 and 5). Therefore, the redesigned variety's revenue potential is estimated from σ, which is obtained from the product family's sales performance as of today.

4.2.3   Costs model

Costs are also modeled according to product variety and using Activity Based Costing (Hicks, 1999) principles. However, because accurate costing is time consuming and such detail is not required at this stage of analysis, we use a simplified costs model containing Fixed Costs and Variable Costs of Variety. Fixed Costs of Variety (FCV) are costs independent to the extent of variety in a product family (i.e., costs that occur even if no products are sold). These are usually building rent, machine depreciation, etc. Variable Costs of Variety (VCV) depend on extent and volume of product variety. These are usually material costs, assembly work hours, etc. We use the following steps to model Fixed Costs of Variety (FCV):

(1) *Determine fixed cost components of a company* – Administration, Depreciation, Rent and Utilities, Office expenses, etc.

(2) *Determine share of those fixed costs for the product variety in consideration* – this is usually done in consultation with sales engineers, product owners and management (for example, if certain products are sold directly over the Internet, they carry little or no sales overhead).

(3) *Determine how these fixed costs change if we change variety* – for example, it takes large investments (building, etc.) to expand certain product variety, and only a small change in assembly process using existing machines in case of another variety (hence little or no added fixed costs, see Figure 20-3).

We use the following steps to model Variable Costs:

(1) *Determine what costs vary along with volume of products produced in that variety category* – material, number of machining/assembly activities, number of people, etc.

(2) *Fit a mathematical model of varying costs as a function of volume of variety that is produced, and the above cost drivers* – in order to simplify the

model, one can model according to selected (major) cost drivers rather than to fit for all of them.

(3) *Determine how these variable costs change if we change variety* – the algorithm (see Figure 20-2) will estimate volume of products that a company can sell with a changed variety. Given that volume and material and other cost driver information, this costs model will estimate the variable cost component. Figure 20-3 illustrates these costs.

Total cost of product variety is the sum of Fixed and Variable Costs, and profit is obtained by subtracting this cost from revenues.

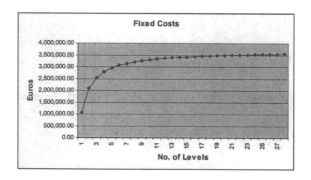

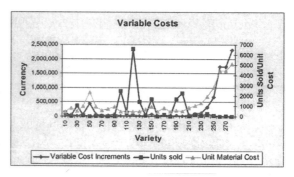

Figure 20-3. Illustration of a simplified cost model: fixed costs of variety (FCV) increase rapidly during initial levels of variety, and not so much later because the same infrastructure is assumed to be used to offer increased variety. Variable costs of variety (VCV) vary with the specific costs of variety and with quantities produced.

4.2.4   Variety reduction

Consider the revenue distribution in Figure 20-4 from a flow meter variety of sizes from 1 cm to 10 cm sold in the last 3 years (an illustration only). If all this variety is to be replaced by just two, what could be those sizes? From the revenue distribution, one can visualize two clusters – one at

3, 4 and 5 cm, and the other at 7, 8, 9 and 10 cm. Using cluster analysis, weighted with revenues, these two sizes could be computed as 4.1 and 8.2 cm (μ_{21} and μ_{22} in Figure 20-4, where subscripts indicate level and cluster).

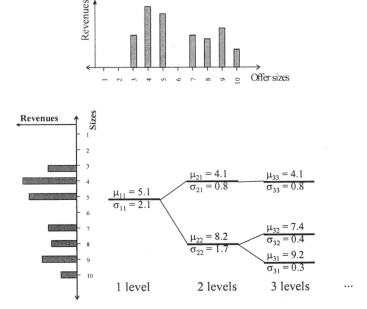

Figure 20-4. Illustration of revenues in a product variety category (left), and computation of new variety values given the revenue distribution (right).

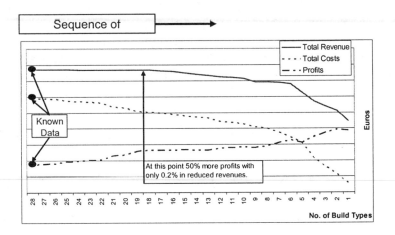

Figure 20-5. Using the revenue, cost and cluster algorithms, we found that 18 types of build and casing designs will result in 50% more profit, without compromising existing customer revenues, as of today's profit with a variety of 28 types. Our method also determines design values (like sizes) for these 18 types.

The customer dissatisfaction (σ) with these two sizes (4.1 and 8.2 cm) can be computed from the number of customers and revenues from the known data (see Eqs. 4 and 5). Such clusters are sequentially determined from the known data of 10 levels up to single choice as shown in Figure 20-4. Using σ, the number of customers in each cluster is estimated.

Using the revenues and costs model, and given new variety values and number of customers from cluster algorithm, we determine changed revenues and costs at each level. Figure 20-5 shows results in case of variety in flow meter build and casing types (actual revenue and profit values have been edited to preserve confidentiality of our client).

At each step in the iteration (see Figure 20-2), the cluster algorithm computes variety from n to (n-1) levels by replacing two of the weakest candidates (those that have lowest revenues and farthest from rest of variety) with a substitute $x_j^{(n-1)}$ using the formula:

$$x_j^{(n-1)} = \frac{\sum\limits_{i=1}^{N} w_{i,j}^{(n)} \cdot x_i^{(n)} + \sum\limits_{i=1}^{N} w_{i,j+1}^{(n)} \cdot x_i^{(n)}}{\sum\limits_{i=1}^{N} w_{i,j}^{(n)} + \sum\limits_{i=1}^{N} w_{i,j+1}^{(n)}}, \tag{3}$$

where w_{ij} are weights that help determine $x_j^{(n-1)}$ proportional to revenues. Note that the distance scale represents customer differentiation. For example, wastewater segment usually requires large flow meters compared to pharmaceutical customers who need small and precise meters. Here, size can be used as a differentiator. In other cases (e.g., non-numeric scale such as color as a customer preference), the differentiation scale is arrived at from customer interviews (e.g., priorities, overlapping needs, etc.).

The corresponding standard deviation (σ), customer dissatisfaction, due to the newly computed product variety is calculated as follows:

$$\sigma_j^{(n-1)} = \sqrt{\frac{\sum\limits_{i=1}^{n} \left(\sigma_{i,j}^{(n-1)}\right)^2 \cdot w_{i,j}^{(n)} + \sum\limits_{i=1}^{n} \left(\sigma_{i,j+1}^{(n-1)}\right)^2 \cdot w_{i,j+1}^{(n)}}{\sum\limits_{i=1}^{N} w_{i,j}^{(n)} + \sum\limits_{i=1}^{N} w_{i,j+1}^{(n)}}}, \tag{4}$$

where:

$$\sigma_{i,j}^{(n-1)} = \sqrt{\left(x_i^{(n)} - x_j^{(n-1)}\right)^2 + \left(\sigma_i^{(n)}\right)^2}.$$

(5)

At the very beginning of this algorithm, that is when product variety is at the same level as it is offered today, the initial sigma could be estimated according to customer dissatisfaction with existing product variety.

We also use two other algorithms for sequential reduction of variety: Pareto Rejection and K-Cluster approach – with or without manual overriding of variety values. In Pareto Rejection, the weakest variety values – those that have lowest revenues and are farthest from the rest of variety – are rejected using multi-criteria optimization (a farthest loner represents an exclusive customer preference, and low revenues means not many customers). Here, revenue and the distance are the criteria. In K-cluster approach, one can specify the number of clusters to extract from the known data. Detailed discussion of this algorithm is out of scope of this chapter. Often business logic proclaims the new variety and how many of them to be introduced in a product family (e.g., new customer demands for new business opportunities). In such cases, the values found by algorithms need to be overwritten manually. In those cases, the σs will be recomputed, and revenues and costs are estimated in view of new business opportunities. According to various business contexts, we also use various revenue formulae:

(1) Revenue with one offer (R_{1offer}) as value between minimum (R_{2min}) and maximum (R_{2max}) of revenues with two offers:

$$R_{1\,offer} = R_{2\,max} + R_{2\,min}\frac{\sigma_{2\,max}^2 + \sigma_{2\,min}^2}{\sigma_1^2}.$$

(6)

This model is most often used because it predicts reasonable drop in revenues with reduced variety. The results in Figure 20-4 use this formula.

(2) Revenue with one offer as a proportion of sum of revenues with two offers:

$$R_{1\,offer} = (R_1 + R_2)\frac{\left(\sigma_1^2 - (\sigma_{21}^2 + \sigma_{22}^2)\right)}{\sigma_1^2}.$$

(7)

This model is used when revenue drops will be high or low, often influenced by market or other external factors (a multiplier as a weight can be introduced), according to the variety in consideration.

(3) Revenue with one offer as sum of revenues with two offers (no loss of revenue):

$$R_{1\,offer} = R_1 + R_2. \qquad (8)$$

This model is used when reduced variety is likely to keep or even increase revenues (usually by meeting customer choices better).

Because the design process is an iterative process, the results from these algorithms must be crosschecked with relevant people as described in Section 3.1. For example, the design of IO board for robot controllers (introduced in Section 3.1) resulted in a variety reduction of 40% with this algorithm in its initial iteration. During the cross verification, designers could realize a better revenue potential if they could create an IO board combining three existing designs. A new design combining all three attributes was technically possible. We immediately verified market attraction for such new design through sales channels and customer contacts. We received a very positive response from our customers. Based on this feedback, we have repeated the algorithm now using the additive revenue formula (third one above), and manually overwriting the new variety values for the new IO board. The end result showed even higher potential for profit increase because of cost reduction (the new design turned out to be 20% cheaper) and increased customer reach even though variety has been dropped by 40%.

4.3   Design for commonality

The analysis method described so far derives business logic (revenues and costs per variety) and key design decisions (what variety values to change, and to which values) using the product variety performance as of today. These results are used to realize the product design using the concepts of Design for Commonality (Fisher, et al., 1997). Here, the primary goal is to increase commonality in physical components to meet various customer needs, thereby reducing costs due to physical variety. Supplementing this research, we have developed a method to further bring in business logic in decision making during the design process. In this method we have mapped physical variety to the functional variety along with the business results (profit and volume) from that variety as shown in Figure 20-6.

Figure 20-6 shows various alloys satisfying requirements in four different customer segments for flow meters. Each alloy has a purpose, and meets specific customer demands in its corresponding customer segment. In this figure, we have simplified these differences to convey how the design process works. Using this matrix, we realized that Alloys C and G could be replaced by Alloy A (see Item 1 in Figure 20-6), because from our analysis the costs saving from variety reduction was more than the loss of profit (as not all customers will be satisfied with the replacement). Moreover, we found that Alloy A can be better served for Alloy B (see item 2 in Figure 20-6), and made cheaper because of increased volume in manufacturing.

| | Profits | | Unit Cost (if left blank, then the material does not satisfy the customers) | | | |
Number sold	(Euro)	Electrode Material	Water / wastewater	Chemical	Pharmac eutical	Food
2988	1608167	Alloy A	8	8	8	
2787	1274754	Alloy B	5			
11	8518	Alloy C		14		
371	336937	Alloy D		76		
487	533451	Alloy E		501		
74	49891	Alloy F				7
26	12950	Alloy G		21		

Figure 20-6. Illustration of decision-making using business logic (profits and volume) and design for commonality reasoning.

5. KEY EXPERIENCES

The one-dimensional metric used in the clustering algorithms is a simplification in light of the full complexity of a product. Several functional attributes can be analyzed simultaneously, in an n-dimensional space, using a suitable metric that maps these dimensions into distances according to the functional differences (as perceived by the customers). However, in practice, we found that combining several dimensions complicates both business decision-making and the design process. For the sake of transparency, especially to understand what variety influences profits, we believe it is best to consider one dimension after another, and conclude at the end with all results in consideration.

More important to product businesses is to expand customer choices without increasing, in fact reducing, product costs. The methods presented here could be extended from both algorithmic and the design methods points of view. Our algorithm could be extended to identify increase in variety from high and densely distributed revenue regions. The design methods can incorporate more customer specific data such as Quality Function Deployment (QFD) with conjoint analysis to estimate initial distribution of customers to identify increase in choices. These methods must be supplemented by design for commonality that controls the costs of variety, which can find substitute designs that fulfill more customer choices at reduced costs.

5.1   Implementing new platforms

Once the new profitable variety has been identified, the issue becomes one of how to introduce it into the production mix and the sales process. From our experience, this process can take anywhere up to 3 years. The time lag is the shortest if the variety analysis is conducted as part of the planning process for the planned new generation of products.

As a bridging measure, we recommend pursuing a strategy of price increases or cost reduction measures for specific product features and options. A pure focus on variety reduction, often espoused, is misplaced. To reemphasize, the goal is not to offer less variety in the market but to maximize customer choice at minimum cost to the manufacturer. This distinction is subtle and crucial.

5.2   Sustaining product platforms

Because customer demands are dynamic a product family evolves over time. We need to add products, accessories and features to meet the needs of new markets. In an industry where product life cycles span from 5 up to 40 years, we cannot discontinue production, service and support of old products. Thus our product portfolio invariably grows over time. Therefore, the efficient management of product families over time is a major challenge. We are addressing this issue through periodic application of platforms related design methods such as the method presented here and with information tools and organizational change. Organizational change is discussed in detail within engineering platforms development (see Section 6.2). The need for IT systems to manage product variety is discussed next.

5.2.1   IT systems for managing variety

Managing variety in a product family is an enormous information management problem. Before a new product can be introduced to the market, all elements of company's information infrastructure have to be prepared to handle the new product. This issue emerges in situations of product redesigns or changes in portfolio. Product variety generates choices at customer level with rules to determine feasible mix of these choices in various options (e.g., a high voltage transformer requires withstanding cables and expensive insulating accessories), components and parts at manufacturing level also with rules for feasible and efficient manufacturing (e.g., high voltage cables need more time to cure), and product support and service information specific to the variety.

Figure 20-7 depicts necessary data flows of product definition data among company subunits and information systems. The Design Department creates a physical design of a product that fulfils functional specifications received from Product Manager. As the design process is mainly supported by PDM (Product Data Management) system, after a design has been completed this system contains all the latest information about product. Before updated product can be offered to customers, its data has to be synchronized in both ERP (Engineering Resource Planning) and Sales Configurator (SC). As these systems cover different aspects of information about a product, different data has to be provided. Synchronization processes depend strongly on specific information systems implemented in a company and on the level of their integration. In particular cases, portions of the system may constitute one software suite and use common data repository. In these ideal situations cost of information synchronization is minimized. Currently, the very common situation in many manufacturing companies is coexistence of various systems with separate databases, and therefore additional effort is required for the data transfers. The first direction of updated data propagation is synchronization of the SC and the Configuration Model. In the studied empirical example at ABB Robotics just the maintenance of configuration rules in the Sales Configurator costs 300k USD per year.

The manufacturing and accounting subsystems also have to be simultaneously updated. As changes in the product component structure generate need for new components, this fact has to be introduced into appropriate modules of ERP system. New objects need to be inserted into Material Master and supply chains need to be established for them what requires involvement of procurement department into design process. "As-designed" product structure has to be translated into "as-manufactured" structures to enable the production. Many of today's PDM system support multiple viewpoints to product structure. In that case, "as-manufactured" structure is prepared within the PDM, and should be linked with the data in ERP. In practice, integration of these systems to the level that would enable automatic information exchange is very difficult and highly expensive. In the studied case of magnetic flow meters manufacturing company, creation of manufacturing rules for the "as-manufactured" product structure is performed manually. For a product possessing 203 different choices in 17 option types, more than 12,000 manufacturing rules are required, which requires more than 1.5 man-months of effort. These rules have to be fed manually, which dramatically increases the probability of errors. Figure 20-8 shows the increase of data objects processed in consecutive stages in the IT systems for a family of industrial robots.

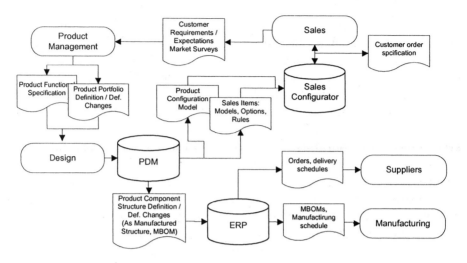

Figure 20-7. Product information flow.

An efficient product variety management has to deal with this information from its identification, representation, storage, and usage for various activities such as product configuration, manufacturing, service, and product development. We are currently working on an integrated system that connects sales configurators, PDM and ERP systems to improve information management for product variety.

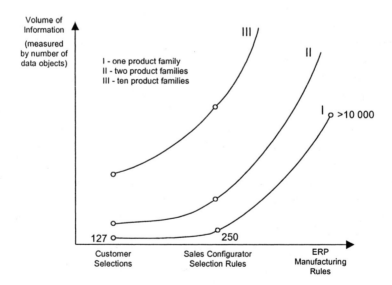

Figure 20-8. Variety dependent volume of information.

6.   DESIGNING SHARED ENGINEERING PLATFORMS

Sections 4 and 5 discussed platform approaches for product businesses. In case of engineering business, introduced in Section 3.2, organizational assets (people and competence), project management, and supply chain play central role in delivering a complete system. For example, before we developed engineering platforms for the Powertrain systems business, the engineering for the unit took place in three countries, US, Germany and Sweden, as shown in Figure 20-9. Each country organization was independently responsible for approaching customers, gathering requirements for the assembly lines, and selling the concept. When the customers purchase the line, it is then designed (layout and simulation), detailed, tested internally before commissioning at customer sites. For example, in a typical gear-box assembly line, about 500 parts such as gears, shafts, pins, springs, nuts, washers and casings, needs to be assembled. Such a line consists of several pallets moving past manual, semi-automatic, and fully-automatic stations performing various operations such as inserting, pressing, turning, tightening, testing, and sealing. Design of these lines must consider customer requirements such as various parts, throughput (number of gearboxes to be produced in a given duration), layout of the line, logistics (parts feeding), life cycle cost of the line and safety.

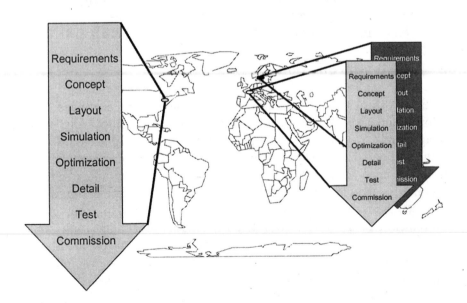

Figure 20-9. Independent engineering locations before developing platforms.

In order to achieve customer requirements, each location had a common organizational structure consisting of sales, engineering and purchasing (supply management) departments. The sales department is responsible for gathering the requirements for a proposal, and winning the contract. If this line has to be designed, the engineering will detail the requirements, complete conceptual and detail design, simulate the line for optimization, detail, assemble and test internally. It will then be disassembled for transport, and commissioned at customer location. This engineering process is also shown in Figure 20-9. Each customer line is accomplished through turnkey project management.

In order to be competitive, the group of three units needed to drastically reduce cost and delivery time. This could only be achieved if the engineering could be shared across the three locations, and with reduced engineering time by reusing standardized solutions rather than designing from scratch for each project. We approached this challenge with three strategic initiatives:

(1) *Standardization*: Develop modular and pre-engineered solutions that can be readily used, or as starting point for designing rather than from scratch each time.

(2) *Organizational Improvement*: Develop engineering teams across countries to share standards, and re-distribute engineering both on standards and in projects with an aim of overall reduction in engineering effort.

(3) *Global Supply Strategy*: Develop supply management based on standards, and common supplier base that could supply to any engineering location with globally negotiated prices.

The following sections describe these initiatives in more detail.

6.1   Standardization

Standardization is about developing efficient work processes to fully exploit standards from sales, engineering to purchasing (supply management). We define standards as modular, pre-engineered and long-living solutions that satisfy customer functionality. Standards can be used to define or benchmark customer functionality when gathering requirements. By doing so, the engineering department can complete the design either by using standards as they were, or by adapting them to meet customer requirements. This saves considerable amount of engineering time as compared to designing from scratch. Moreover, for customers, standards give a choice to choose repeatedly used solutions (i.e., proven solutions) at a better price and faster delivery time.

In order for this strategy to work well, standards must be well communicated and understood within the organization in sales, engineering and purchasing departments. It is important to have easy access to standards within the organization, and to maintain them for their integrity and long-life. In other words, standards are a collection of information necessary for:

(1) *Sales*: All information necessary to sell standards – brochures and technical datasheets – that clearly show how to meet customer functionality, and benefits from buying standards to customers.

(2) *Engineering*: All information necessary to complete engineering faster than before – details such as 3D models and 2D drawings – that can be readily used, or adapted to produce solutions.

(3) *Supply Management (Purchasing)*: All information necessary to buy standards from suppliers – Bill of Materials (BOM), past suppliers, and bulk of standards bought from them in the past – that can be used globally, and leverage prices for bulk purchase.

For example, before standardization, the engineering teams used to detail requirements for a conveyor by determining about 15 parameters such as weight and width of pallet, and speed of their transportation. Such a method produces long engineering times until all customer discussions are complete.

After standardization, pre-engineered instances of conveyor that meet various parameters are produced and are type-coded. Thus sales can select them very early based on their customer discussions and avoiding repetitive discussions. This enables engineering to use the corresponding details (3D model) of selected instances to complete the project design very quickly. This is depicted in Figure 20-10.

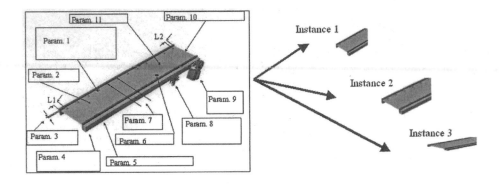

Figure 20-10. Conveyor solutions before (left) and after (right) standardization.

6.2   Organizational improvement

After standardized solutions have been achieved, it was easier to create the powertrain assembly line with these as building blocks. Standards enabled us to establish globally shared engineering resources. For example, an assembly line can now be made with conveyor standards from the US, station standards from Germany, sealing systems from Sweden and so on. The organizational structure shown in Figure 20-11 is optimized and shared to implement engineering platforms.

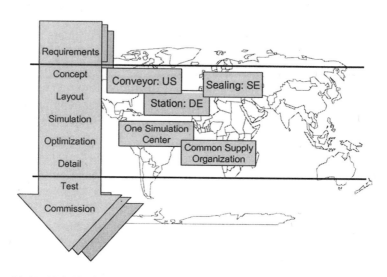

Figure 20-11. Globally sharing organization based on standards for engineering platforms.

Requirements gathering and their detailing are kept local to countries because of customer and regional specific demands. Testing and commissioning was also kept local because this work needed to be done on site at customers' plants. All engineering in between is now shared and organized based on standards. This platform also enables having only one center of expertise for simulation and optimization and a common supply organization.

6.3   Global supply strategy

Shared standards and engineering activity increases amount of items or services that an organization buys globally. For example, this company can now by three times the amount of conveyors that it used to buy before developing these platforms. This will save additional costs by negotiating bulk price discounts with suppliers.

7. KEY EXPERIENCES

Establishing engineering standards is a technical as well as a "political" problem. It requires top-down management push to get started and a bottom-up lift to sustain and succeed. In order for engineering platforms strategy to be effective, people – salespeople, engineers, supply managers – must share and understand priorities, strategies and common decisions. For example, standards could only be managed well if clear ownership and responsibilities are assigned to maintain their quality, completeness, accuracy and integrity. Often, new standards are added or existing ones are modified without fully justifying why we need to change or add them. Such actions destabilize sharing and supplier strategy to fully leverage engineering platforms. Standards' owners are central and technical authority to manage them while the others use them directly or by adapting (only if necessary) to complete lines. New or modified standards are approved only in complete agreement with the owners and with a formal review process.

Such responsibilities, processes, and clear rules for maintaining standards could only be developed with systematic change in organizational culture. We are driving this change by clearly showing benefits for all, management commitment, frequent communication and openly resolving conflicts. Here, IT-based and engineering systems for managing standards play a crucial role especially in communication and conflict resolution. Changes at organizational and processes level, and in implementing IT systems are always challenging tasks but are essential steps for efficient management of product and engineering platforms, and keeping them profitable.

8. SUMMARY

We have presented how platforms can be implemented for two very different businesses. In both cases, respective assets are optimized to increase profitability.

In a products business, the existing product portfolio can be mapped to customer satisfaction, and based on its existing market performance it can be architected for maximum profits. This chapter presents detailed steps for computing the optimum product family architecture. The organizational challenge here is to sustain the optimum architecture. In this process, a major constraint is the complexity, and subsequent dependency, of the information related to product variety. This complexity can only be handled by transparent integration of various information systems such as Sales Configurators (SC), Product Data Management (PDM), and Engineering Resource Planning (ERP).

In a systems engineering business, engineering content, resources and supply chain need to be optimized and better synchronized to implement platforms. In order for the engineering platform to work effectively, people must be aware of individual responsibility, and how overall benefit could be maximized by using standardized solutions that are synchronized with a global supply chain. This requires a systematic change management approach to align organizational culture with global strategy. This chapter presented how such engineering platforms are implemented while simultaneously addressing the cultural constraints.

Chapter 21

A CASE STUDY OF THE PRODUCT DESIGN GENERATOR[*]
A Methodology for Web-Based Product Platform Customization

Gregory M. Roach[1] and Jordan J. Cox[2]

[1]*Department of Mechanical Engineering, Brigham Young University, Rexburg, ID 84330;*
[2]*Department of Mechanical Engineering, Brigham Young University, Provo, UT 84602*

1.   PRODUCT DESIGN GENERATOR METHODOLOGY

A Product Design Generator is a web-based tool, developed for a specific product platform, for automatically creating all of the design artifacts and supporting information necessary for the design of a particular product. The PDG is modeled as a transformation function where a set of customer requirements is transformed into finished designs that will meet those requirements. Several methods have been presented for configuring and defining a product platform and are not reviewed here. Once the concept and embodiment have been selected, scaling, reconfiguration, artifact creation, and testing must occur to complete the design. Variants of the product platform are achieved by modifying the customer requirements. The development of the transformation function must account for the envelope of variation desired to encompass the range of product family members. The development of the PDG demonstrates how this is accomplished.

The PDG approach to systematic design is different from many other systematic approaches to design. Rather than giving a recommended top-level design process that can apply to any design project, the PDG approach

---

[*] This chapter is a modified version of the paper: Roach, G. M., Cox, J. J., and Young, J. M., 2003, "A New Strategy for Automating the Generation of Product Family Members and Artifacts Applied to an Aerospace Application," *ASME Design Engineering Technical Conferences,* Chicago, IL, ASME, Paper No. DETC2003/CIE-48185. Reprinted with the permission of the ASME.

focuses on a specific product platform. Best-practice steps for designing that product are identified and captured in a web application. As best practices are captured and automated in a web-based PDG, the emphasis in designing custom products changes. Rather than focusing on creating design artifacts and completing engineering analyses, the engineer can focus on interpreting the analyses automatically completed by the PDG and making appropriate tradeoffs by considering a large number of detailed candidate designs.

New design skills will be necessary for developing PDGs rather than detailed designs for individual products. In addition to thoroughly understanding the best practices in design of the product class, PDG developers must be able to generalize the process so that it will apply to any member of the class. Further, they must be able to understand principles of reuse and modularity that are common to software engineering. Finally, they must understand how the design process is interrelated with the other business processes of the company. The methodology for constructing a web-based PDG is presented in the context of a case study application of a product platform for an axial turbine disk of a gas turbine engine (Roach, et al., 2003). An additional application involving a flow valve product platform for large industrial plants is presented in (Roach, et al., 2005).

2.   CASE STUDY: AXIAL TURBINE DISK

A turbine disk is a component in a gas-turbine engine. The basic function of the turbine is to transform a portion of the kinetic energy and heat energy in the exhaust gases to mechanical work, thereby driving the compressor and other accessories (Otis and Vosbury, 2001). The product platform chosen for this case study is that of an axial flow turbine (see Figure 21-1). A typical axial flow turbine is made up of a number of rotating airfoils that are typically inserted into slots in an otherwise solid disk. The disk functions to maintain the circular motion of the airfoils and couples them to one of the rotating engine shafts.

The turbine section of an aircraft engine is located immediately after the combustor section and absorbs most of the energy from the combustion process. Consequently, the turbine is the most highly stressed component in the engine. The stresses on the turbine disk come from the extremely high temperatures of the combustion gases, the enormous inertial loads due to rotation at tens of thousands of rpm, and thermal cycling during flight (Otis and Vosbury, 2001). The objective is to design a disk to withstand the operating stresses, fit within a specified spatial envelope, and weigh as little as possible. These conflicting objectives make turbine disk design an inherently iterative process, sometimes taking months to execute.

Figure 21-1. An example axial turbine disk for a modern gas turbine engine.

2.1 The product transformation function

The PDG leverages a product platform beyond simple combinations of family members and develops a process for defining a quasi-continuous range of members of the product family. This is accomplished by defining a transformation function to transform a specific set of customer requirements to a member of the product family. This includes the generation of all the artifacts associated with the product family member so that new members can be combinations of derivatives of existing members and not exclusively fixed configurations of existing members. The development of this transformation function requires that the concept for the product family remain fixed and predefined. No concept development is implied in the PDG.

For a PDG, the process follows a similar sequence. All the terms or elements of the PDG must be defined that map the customer requirements through to the definition of the product family member and the definition of all the associated artifacts. Once these definitions have been completed a transformation function can be identified and implemented to generate product artifacts and variant designs within the product family.

The resulting PDG application is an automated web-based implementation of the transformation function for a product platform capable of producing a quasi-continuous range of product family members and their associated design artifacts and information. The complex transformation function is decomposed to intermediate transformations to deal with the complexity of the process. The intermediate transformations account for behavior predictions, company rules and best practices, the generation of design artifacts, data and artifact vaulting strategies, testing procedures and design artifact delivery procedures (Cox, et al., 2001).

2.2   PDG construction method

The PDG application is constructed in three steps: selection of the product concept, development of the product generation schematic (PGS), a blueprint for constructing the web application, derived from the definition of the intermediate functions and their workflow execution sequence, and construction of the reusable intermediate functions and integration of them into the automated PDG application.

2.2.1   Concept selection

The first step in building a PDG is to select the product concept or platform. Methods for developing the product platform are not reviewed here. However, along with the chosen concept and embodiment, the best-practice steps for designing the chosen product must be identified so that they can be captured in the PDG. This also includes the identification of all the design artifacts, performance predictions, knowledge, and other outputs from the design process.

The product concept for the axial turbine disk involves a solid disk with inserted turbine blades. The inserts will allow one, two, and three lobe attachments. The product platform consists of variations to the shape of the disk which provide scalability, a modular attachment that varies in the number of lobes and their location, modular disk to disk coupling features, and additional modular features for secondary flow guidance and testing procedures that are not required for all disk design variants. In order to completely define the embodiment of the concept, the company best-practice steps for turbine disk design were identified by looking at the current process for disk design and speaking with the designers, analysts and other potential users of the turbine disk PDG. It is important here to gain as much knowledge of the entire process as possible. It is not sufficient to know the role of a single designer and leave out the analyst if major organizational improvements are to be made. The entire process, rather, from start to finish must be understood.

The current design process for this component was identified by observing the process and by speaking to those that participate in the process. In the current preliminary mechanical design cycle, a specification of flow path is initially provided but the airfoil geometry does not yet exist. The mechanical designer takes this flow path as an input and designs the airfoil from the standpoint of stress without determining the actual aerodynamic shape. A material is selected for the airfoil and assumptions are made regarding the taper from the tip of the airfoil to the hub. This preliminary shape is then analyzed and iterated on until the stresses are within acceptable limits. The airfoil itself then becomes the input load to the attachment portion of the turbine disk. An initial shape for the attachment is selected based on experience and analyzed to determine stresses. The attachment is then modified until it meets acceptable criteria. The method for coupling the disks together is selected based on the overall engine configuration. If secondary flow guidance is required, additional features are added to the definition of the disk geometry. A taper is then selected for the disk, the disk is analyzed, and iterations occur until stresses are found to be below a specified threshold value. If a material can be found and stresses are found to be within acceptable limits, the preliminary design is complete. If the criteria cannot be met, the flow path is changed, preliminary aerodynamic analysis revisited and the mechanical disk design begins anew.

2.2.2   Develop the Product Generation Schematic (PGS)

The PGS is a schematic representation of the overall transformation function that will be used as a blueprint for the construction of the PDG. In it, the members of the domain and range sets for the intermediate transformations that comprise the overall product transformation are enumerated and their dependencies are identified and defined in detail. Also, the sequencing of the execution of the intermediate transformation functions, which will control the workflow of the web application is defined.

The PGS is constructed in four phases: (1) identification and classification of product elements and intermediate transformations, (2) layout of plans for constructing the intermediate transformations, (3) rectification of the master parameter list, and (4) layout of the design and release workflows.

In the first phase, the results of the best-practice process for designing the product are classified as members of various domain and range sets for the intermediate transformations. The classifications are divided into eight major sets: customer specifications (C), product behavior predictions (B), company rules and best practices (K), governing master parameters (M), test results (T), product artifacts (A), vaulted artifacts (V), and final product deliverables (U). Intermediate transformations are defined for all of the domain and range sets. The intermediate transformations include predictive models, parametric CAD/CAE/CAM models, testing processes, delivery procedures, data vaulting procedures, parametric document models, etc. A graphical representation of these sets and the intermediate transformations is provided in Figure 21-2.

The arrows between the sets represent the intermediate transformation functions. For example, the set M, the governing master parameters is the domain set for the intermediate transformation which is constituted of

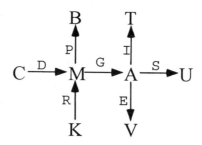

Figure 21-2. A graphical representation of the intermediate transformations with their domain and range sets.

artifact models (i.e., CAD solid and drawing models, NC code models, etc.) that produce the range set A, the set of product design artifacts. The other arrows on the graphical schematic represent similar types of transformations.

Next, parametric models and procedures are identified to transform the various domain sets to the range sets. For example, a CAD drawing is defined as an output design artifact. Thus, a parametric CAD solid model and drawing model represent the reusable transformation function that can be instantiated to produce the desired drawing artifact.

The second phase is to design and define the parametric intermediate transformations that were identified as part of phase 1 that will produce each of the desired outputs. The reusable transformation functions are not created in any specific design tool at this point. Instead, they are abstract definitions of the strategies that will be used when constructing the models. Typically, these intermediate transformations are defined in terms of feature structures, parameter schemes, and parameter relationships. The feature structure is a broad definition of the features of the specific intermediate transformation. Parameters are used to establish specific sizes or quantities associated with the features. Relationships tie values for one parameter to another or establish constraints on the parameters. Careful planning in the selection of features, parameters and relationships is required as these have a significant influence on the variability of the product platform.

In the third phase, the governing parameters for all of the various intermediate transformations are gathered and rectified into a single independent list of governing parameters called the master parameter list. The result of this phase is a set of unique and independent parameters.

The final phase in the construction of the PGS is the layout of the design and release workflows that must be instantiated with values and executed to generate a specific design. This process is referred to as storyboarding. The storyboard is divided into two parts, corresponding to a design workflow and a release workflow. The goal of the storyboarding effort is to control the process by which values for the parameters in the master parameter list are determined and to control the sequence of execution of the intermediate transformation functions. The storyboard also depicts the look and feel of the actual web application by showing the intended design of the user interface.

The goal of the design workflow is to find acceptable values for every parameter in the master parameter list. This necessitates an iterative process which is constructed by identifying the appropriate sequence for evaluating the intermediate transformations for predicting product behavior, developing a strategy for making changes to the values in the master parameter list and defining the criteria that determine when the values have been chosen appropriately. The workflow of the web application must be precisely controlled and synchronized for data persistence and access by simultaneous users. This not only includes the predictive models and applications but also additional integrations such as writing to and retrieving data from a database.

Initially the storyboard for the release workflow is quite complicated due to the large number of design artifact creations that must be sequenced. However, once the sequence is established and coded into the PDG the final user interface for the release workflow is very simple because the creation of the artifacts is entirely automatic and the user is only required to review the resulting artifacts. The result of these four phases is a blueprint for the construction of the PDG application.

The four-phase approach for constructing a PGS was followed in the development of the turbine disk PDG. First, the product elements and intermediate transformations were organized into their appropriate categories. The organizational categories for the product elements are customer specifications, product behavior predictions, test results, product artifacts, vaulted artifacts, and product deliverables. The customer specifications were identified along with the design artifacts and analyses necessary to produce the physical turbine disk. The customer specifications identified for the turbine disk include but are not limited to the following: spatial constraints, flow path definition, inlet flow conditions, temperatures, pressures, flow rates, material properties, mission data, thermodynamic cycle data, specification of high pressure turbine or low pressure turbine, engine stage to be designed, cooling air specifications, minimum number of hours for low cycle fatigue, number of lobes in the attachment, number of airfoils, geometric constraints, bore diameter, and the attachment broach angle.

Next, the product artifacts specific to the turbine disk product platform were identified. These include: a CAD solid model of the disk, CAD drawings of the disk, a technical report, and the manufacturing operation sheets. Creation of these artifacts depends on the engineering predictions that verify that customer specifications can be met. These product behavior predictions consist of the following: disk stress, disk deflection, disk life, disk loading due to rotating airfoils, attachment stresses, and cost.

The generation of product behavior predictions and product artifacts require significant company knowledge. It is this experience and knowledge that gives one company a competitive advantage over another. Typically this knowledge is under-documented and exists only as an intellectual asset possessed by an individual employee. For this reason, the knowledge and best practices are the most challenging to identify and capture. Much of the turbine disk design process documentation in this particular aerospace company was out of date or missing, which is not uncommon. The knowledge, therefore, had to be captured in detailed interviews with all those involved in the process of disk design and manufacture. Because of the proprietary nature of the company knowledge, only a sample of the parameters is provided: material properties of proprietary materials, combustor profile, design for manufacture standards, loss coefficients, element type and size in the FEA model, manufacturing feed rates, and tooling diameters are representative elements of this knowledge. The final product deliverables are defined as the following: turbine disk hardware, CAD solid model, CAD drawings, manufacturing process sheets, and the technical documentation.

Many companies are required, for legal and other reasons, to store some of the artifacts generated during the product development process for future reference. In this case study the vaulted artifacts are the CAD solid model and drawings, the master parameter list, and the technical documentation.

After manufacture, the disk is tested according to the approved testing procedure and the results are stored in the test result category. The test that is performed is a spin test with the result being pass or fail.

The membership of the various sets is used to identify the parametric models required for the PDG. These parametric models represent the intermediate transformations. The easiest parametric models to identify are those needed to create the previously identified product artifacts. For the axial turbine disk product platform the following parametric models are required: a CAD solid model (to map master parameters to weight and volume and to generate the engineering drawings), a CAD drawing model (to map master parameters to physical turbine disk), a manufacturing process sheet model (to map master parameters to the process steps required to manufacture a physical turbine disk), and the technical documentation model (to map master parameters to various approval documents and reports).

Similarly, parametric models are required to map the customer requirements to the product behavior predictions. The parametric models identified for the turbine disk PDG are the following: disk loading model (to map temperature and airfoil weight to thermal and inertial loading conditions), combustor profile model (to map input temperatures to disk temperatures), disk deflection model (to map master parameters and attachment deflection to disk deflection), disk stress model (to map master parameters to stress), attachment deflection model (to map master parameters to attachment deflection), and the attachment stress model (to map master parameters to attachment stress). These intermediate transformations generate the set of product artifacts and values for the product behavior predictions.

This information is then organized into a schematic framework. Figure 21-3 shows one possible graphical representation of the sets and the mappings for the turbine disk Product Generation Schematic.

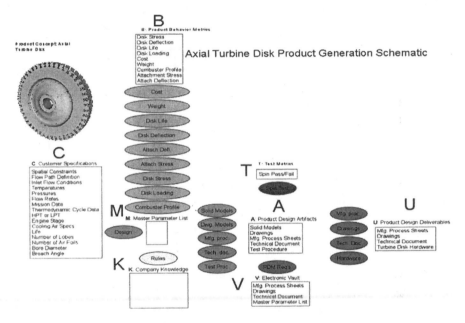

Figure 21-3. A graphical representation of the Product Generation Schematic.

The second phase is to design the parametric intermediate transformations that were identified as part of phase 1. Recall that these transformations are defined in terms of a feature structure, parameter scheme, and any necessary parameter relationships. One of the features in the CAD solid model of the disk, for example, is the cross section of the disk. It is defined by the parameters and relationship shown in Figure 21-4.

For the turbine disk, a combination of modularity and scalability was used to achieve the desired variability in the turbine disk platform. The disk cross section was created to allow scaling while the disk to disk coupling module and the attachment module were created to allow both scaling and modularity based on the choice of coupling connections and the number of blade attachment lobes. For the drawing model, features are defined as the various views of the solid model, the notes contained in the drawing, and text in the title block. The parameters are part number, date, surface finish, etc. Features of the technical report are defined as figures, graphs, tables, and text fields. The parameters for the report are images, stress values, name of the designer, etc. One of the predictive models, the combustor profile model, is a closed form mathematical model that predicts the disk temperature at a given radial location. In this case only the radial location is necessary to calculate temperature and radius is the only parameter. All of the remaining intermediate transformation functions were defined in a similar manner.

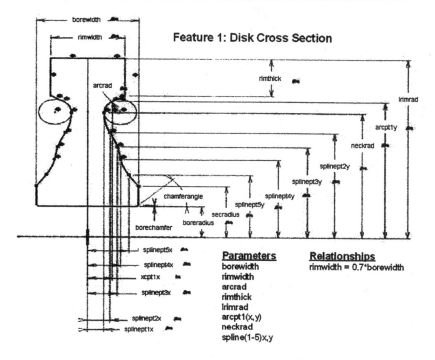

Figure 21-4. The feature definition for the disk cross-section for the CAD solid model.

Once all of the intermediate transformations were defined and their parameters identified, the master parameter list was rectified and reconciled into a single list of independent parameters. Representative members of the list of master parameters includes: disk RPM, bore radius, number of lobes, part number, number of blades, rim displacement, fatigue life, broach angle, and disk material yield strength. The master parameter list contains all of the necessary information to create instances of any artifact or performance prediction model. The master parameter list for the turbine disk PDG contained several hundred members.

The last step in planning the PDG is storyboarding the design and release workflows. Storyboarding provides a way to plan the look and feel of the

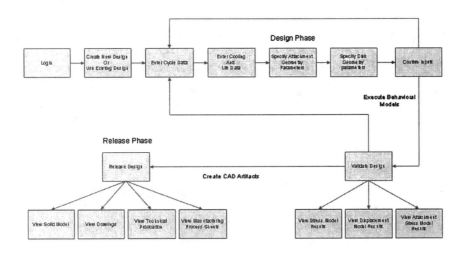

Figure 21-5. The storyboard layout for the turbine disk application.

web application and aids in determining the sequencing of the individual models and the requisite data flows. The storyboard for the turbine disk PDG is shown in Figure 21-5.

During the design workflow, the user provides values for the customer requirements and the web application executes the analysis models to determine stress, deflection, life, weight, etc. When the analyses are complete, the behavior predictions, in terms of values, images, plots, etc. are returned to the user for evaluation. At this point the user can accept these values and move on in the process or redefine the requirements and re-execute the analyses again. Acceptance of the values constitutes completion of the design workflow and initiates the release workflow. During the release workflow, the artifacts are created and vaulted. One of the main benefits of storyboarding the process this way is an understanding of the execution sequence of the parametric models and the data flow that must occur. The disk loading model, for example, must be executed before the finite element model because the disk load becomes a boundary condition to the finite element model. Time invested storyboarding the workflows greatly improves the process and consequently, the PDG. Planning the sequence of parametric models greatly improved the process of turbine disk design. This planning helped to eliminate duplicate calculations, and eliminated entire models that were found to be obsolete or unnecessary. Other models had to be restructured to permit the correct sequencing of models and data flow. Through the planning process, weaknesses were identified in the disk design process and areas of improvement and potential research became apparent. The potential to implement multi-disciplinary optimization, for example, became more realistic.

2.2.3   Construct the reusable intermediate functions and integrate them into the web-based automated PDG application

The final step of construction of the PDG consists of two parts. First, the intermediate transformations identified and defined in the PGS are implemented as executable models in the specific tools selected to produce the desired design artifacts. Second, the storyboard and executable models are integrated into an automated web application. The final results are an easy to use web-based application (the PDG) that walks the company employees through the design process while it simultaneously optimizes reconfigurability, keeps track of formats, manages information, etc.

The first step in constructing the PDG is to implement the intermediate transformations as executable models in the appropriate tools. In this case the CAD solid model and drawing model were implemented in Catia, the disk stress and deflection models were implemented in Ansys, and the technical document was implemented in Microsoft Word. Other models were implemented in the C++ programming language using standard object oriented techniques.

In the second step, the storyboard was implemented as a web interface and the parametric models of the intermediate transformations became server processes. The web-based interface was built using the Common Gateway Interface architecture (CGI) and a MySQL database was used for data management and vaulting of the designs. An example web page from the PDG application is provided in Figure 21-6.

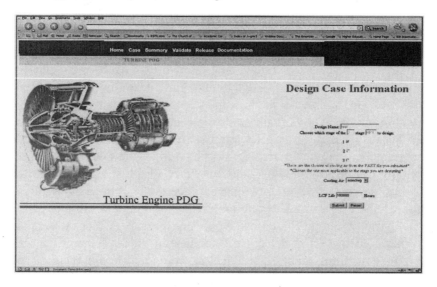

Figure 21-6. A web page from the turbine disk PDG.

The resulting turbine disk PDG can be used to create a quasi-continuous envelope of designs. Figure 21-7 shows sample artifacts produced by an instantiation of the turbine disk platform. This product family member satisfies a specific set of customer specifications. After entering the customer specifications into a web browser, the engineer submits the design to the analysis server. The preliminary behavioral models are executed and the finite element models of the disk and attachment are updated and prepared for execution. An XML message is passed by the application to a daemon running on the analysis server. The daemon processes the message and initiates the execution of the finite element models in Ansys. The results, including numerical calculations and images of the finite element model, are then collected by the analysis server and returned to the user's web browser for review. If the engineer finds the results to be acceptable, the entire set of design artifacts is created at the touch of a button by submitting a request to the artifact server. Again an XML message is passed to another daemon running on the server which processes the message and initiates the execution of Catia to update the Catia solid model, Catia drawing model, and the Catia manufacturing process sheets. Finally, a Visual Basic executable is launched to create the technical documentation in Microsoft Word. These results are then returned to the engineer's web browser in the form of VRML images, JPEG files, and PDF documents so that the user is not required to have specific licenses of software to view the artifacts. Application persistence and the support of simultaneous users is accomplished through a session manager daemon (manages the simultaneous users) and a MySQL database (manages data persistence).

The entire process from start to finish required approximately 35 minutes to execute for each design. Approximately 25 minutes was required by Catia to generate the CAD associated artifacts. This is a significant improvement in design cycle time considering that the design cycle in the traditional

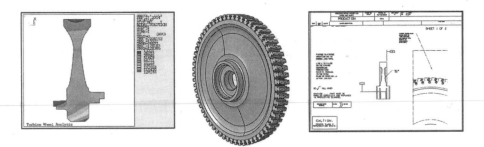

Figure 21-7. Sample artifacts of a typical turbine disk product family member generated by the turbine disk PDG.

product development process takes 3-4 weeks to execute and does not include the creation of the design artifacts which typically take another 5-6 weeks to create.

3.   CONCLUSIONS

The PDG approach provides a methodology for constructing a web-based product platform customization application capable of producing a quasi-continuous range of product family variants. Its implementation yielded significant reductions in design cycle time. A complete cycle of the turbine disk PDG can be completed in 35-45 minutes as compared to approximately 500 man-hours in the conventional process. The time savings allows the engineer to focus on value added activities rather than repetitive tasks.

Besides the high level of customization and automation provided by the PDG several other attractive results are produced: Issues formerly dealt with in detailed design phases are now identifiable in preliminary design. The opportunity to introduce errors in the product development process is reduced because of the standardized, structured process. Each designer now has access to the collective knowledge of the group instead of being forced to re-learn problems that have been previously solved. Finally, the PDG aids in global business strategies by providing 24 hour, 7 days a week design capability accessible by simultaneous users anywhere in the world.

4.   ACKNOWLEDGEMENTS

The authors gratefully acknowledge the help of Jared Young and Honeywell International in completing this work.

Chapter 22

PRODUCT PLATFORM MANAGEMENT PRACTICE AT CETETHERM

Tobias Holmqvist[1], Magnus Persson[1], and Karin Uller[2]
[1]*Department of Operations Management and Center for Modularization Research, Chalmers University of Technology, Göteburg, Sweden;* [2]*Infotiv, Göteburg, Sweden*

1.   CETETHERM AND ITS MARKET SITUATION

Cetetherm[25] is a company developing and manufacturing different types of heat exchanging systems (HES). It has two different product lines, one for small HES and one for large HES. The case example being described in this chapter is about the implementation of a product platform for the large HES, systems that are used by professional users in buildings connected to district heating system. The other product line consists of smaller HES that are mainly used in family houses. The market for large HES is exceptionally heterogeneous, meaning that there are many difficulties involved for individual firms trying to increase their market shares. In the large HES business, there are different rules and regulations in each country, and there are even often several different regions with specific technical demands on products within each country. That is why it is nearly impossible for an individual manufacturer to cover all these policies with a narrow set of standard products and thereby becoming a superior player.

The specific requirements of individual customers, the region-unique products, and the market division contribute to a market with demands difficult to meet. A few years ago, Cetetherm realized the need for improving several areas of the company to achieve a reduction in production time and costs, a more efficient use of the capacity in their different

---

[25] In the beginning of 2005 Cetetherm became fully incorporated in Alfa Laval and is no longer a separate company.

factories, and an increase the time available for the sale personnel. With these goals, Cetetherm started a long journey toward a more modularized product platform (a platform in which components could be more standardized) while still providing customers with a vast variety of products to meet exactly their demand.

This chapter provides a general description of the product platform practice now in use at Cetetherm. This will be clarified by first giving an introduction of the company. The characteristics of the products and its variety are then described, followed by a section explaining the reasons for implementing a product platform development and a description of the implementation process. Cetetherm decided to concentrate their efforts on a specific part of the market where the same range of products could be used instead of trying to cover the whole market at once. The efforts were also structured in a shared sales configuration tool that could also be used to configure individual products. The results for Cetetherm and a description of product platform management conclude this chapter.

2.   ABOUT THE COMPANY CETETHERM

Cetetherm is one of Europe's leading manufacturers of HES. Additionally, it is a world leader in the field of compact district heating installations. The headquarters of Cetetherm are located in the south east part of Sweden, in the small town of Ronneby. Cetetherm was bought in 1987 by Alfa Laval. Alfa Laval is a leading, global supplier of specialized products and engineered solutions, based on optimizing customers' processes in many key areas, such as oil, water, chemicals, and proteins. The product range is brought on the market through its many subsidiaries and distributors worldwide and the four Cetetherm production sites are located in Sweden, Finland, France, and Czech Republic.

3.   HEAT EXCHANGING SYSTEM – A HIGHLY CUSTOMIZED PRODUCT

To understand the challenges faced by Cetetherm and its product platform efforts, there is a need for understanding the product and its market situation. The various actors in the heat exchange market and the versatile standard range of HES products are described in this section.

3.1 A Cetetherm Heat Exchanging System (HES)

The product discussed in this chapter is a complete HES used for large buildings that are connected to a district heating system. In this case, complete means that the HES can be plugged into building heat systems and to the district heat system without the need for any another equipment. The main functionality of the HES is to transfer heat, which is done by exchanging hot water from the district heating network to the local heating system in the building. The energy in the hot water from the district heating network is also exchanged by the HES into hot drinking water for the people in the building.

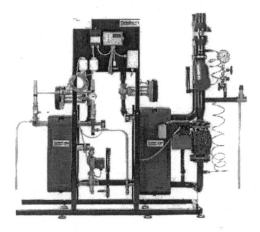

Figure 22-1. An example of a HES built from the product platform at Cetetherm.

As can be seen in Figure 22-1, the HES consists of a number of brazed plate heat exchangers, pipes, pumps, valves, different types of gauges and a digital control system. Cetetherm buys all components from suppliers and assembles them in their production plants as they are system integrators.

3.2 Extensive product variety

The HES can be apprehended as a simple product with a standardized product assortment. The main functionality of this product is to transfer heat from one system to another. Customers buying the HES are professionals and have a lot of opinions about the configuration and the choice of the components. At the same time, customers often have several HE (Heat Exchanger)-systems and therefore would like to have the same type of components in their different HES in order to facilitate product maintenance and keeping the stock of spare parts at a minimum level. Therefore there and keeping the stock of spare parts at a minimum level. Therefore there may be two customers would like to have different components though having exactly the same heating need, and they can also be attached to the same district heating system. All these different customer demands result in a highly customized product, and hence a larger product and component assortment for Cetetherm to manage.

Another challenge for Cetetherm's operations is that the variety of the HES products is not consistent with the functionality. The exact same functionality and performance can actually be achieved by using different combinations of pipes, valves, control systems and heat exchangers. The customer will not notice any functionality difference between the two systems, but the systems will look different from each other. There are many ways to create the same product functionality with different combinations. First, customers can have two different configurations of products with the same product specification. For Cetetherm, this possibility of product configuration leads to increased expenses in all different organizational functions of the company (sales, market, purchasing, engineering, production and logistics), as product variety increases. Second, the demands from customers are very different from country to country, both in terms of what type of configurations they would like, and also in terms of functionality. The characteristics of the district heating system are often unique in different countries. The difference in systems leads to differences between products. In addition, the technical solutions preferred by customers are different as well. In Sweden for example, most customers uses two heat exchangers, one for heating and one for hot drinking water. In Germany, the customer circulates the hot water from the district heating system in the building and would like to have a hot water tank system for the hot drinking water. The legislations in the different countries also vary. In some cases though, the product can be common for two countries from a functional perspective, but must be different due to legislations. As a consequence of the customer and legislations demands, product assortment at Cetetherm was unique from country to country before the platform and modularization project. Many performers in the HES business act exclusively locally and sometimes regional heating-plants even employ their own manufacturers of heating systems. In some countries, the district heating system is different from town to town and sometimes several systems in one single town. The pressure, temperature and chemicals in the water can differ largely from system to system. The consequence for Cetetherm is that customers have very different demands, in addition to the previous mentioned demands, on pipe diameters and welding quality.

4.   REASONS FOR IMPLEMENTING PRODUCT PLATFORM DEVELOPMENT

Before the platform and modularization implementation, Cetetherm did not have a structured offer to their customer. Cetetherm instead designed and assembled what the customer wanted in an order-to-order manner, resulting in highly unique and customized products. The customer had the possibility to specify all the product details. But a few years ago, some employees at Cetetherm got the idea of implementing a platform and modularization based product. These employees got the idea that many of the problems and challenges facing Cetetherm could be reduced by developing a platform based product. One of the major problems was the time-consuming sales procedure. The sales personnel had to spend a lot of time with each customer due to the different and very specific demands from every customer. A simpler and more structured specification of the product would shorten the sale procedure. Instead of spending time on configuring specific customized products, the time could be used for prospecting new potential customers.

Another problem in the sales order process was the large need for engineering. 80% of the sales orders had to be engineered before they could be manufactured. Engineering means development of an already developed product because the existing product needed to be adjusted to fulfill the very specific customer needs and demands. Such use of engineering leads to high costs, and also long order lead-time. The product platform that Cetetherm aimed at developing at that time could simplify, or even take away the need for engineering, since the product would be more standardized.

The manufacturing process and the supply chain at Cetetherm were designed to produce unique products. The manufacturing employees were skilled workers that could build everything that the sales department asked for. The products that were manufactured did not even have any drawings or bill of materials. The manufacturing department simply translated the functionality of the product into components needed, and they then produced the product. Such manufacturing process is very flexible, but also very costly. The time for manufacturing a product was very long. The cost of purchasing materials was high since many of the components were bought in low volume. Both of these problems would be less if the products were more similar to each other and were using more common components.

In addition to the problem for Cetetherm with highly customized products, the HES have a high seasonal demand fluctuation, with a peak of manufactured units from August to September. Before the implementation of the product platform, the possibility to produce HES to store, in order to smooth down the peak in demand, was limited, due to the high customization of the product. Almost all the components were unique as well, resulting in difficulties to pre-assemble components in advance. The result was a need for capacity adjustment in the plants. The employees in the plants worked less time in low seasons or were out on loan to other companies. In high season, everybody in the plant needed to work overtime. The product platform should facilitate the possibility to both pre-assemble components and almost complete HES. The season fluctuations hereby could be managed in an easier way, and hence decreasing the manufacturing cost.

5.   THE PLATFORM DEVELOPMENT PROCESS AT CETETHERM

5.1   The product platform

The idea of implementing a platform and modular-based product had been around Cetetherm for a few years. In the late summer 2001, there were some attempts to kick-start an initial launch of a modularization process of the total product range and outside help was brought in. Representatives from all plants of the Cetetherm group and from many different countries were present: sales managers, production managers, development people, and design engineers to name a few. This process, however, was interrupted when a new standardization appeared: the CE- and PED-certifications, that the products needed to meet to be legal. The development efforts therefore had to be aimed at this certification for the following year, which postponed the development of a product platform for large heating systems.

In the late fall 2002, there was a restart of efforts to develop a product platform for large HES, and the product development department (P&T) at the headquarters was put in charge of the project. The aim was to create an 'International Cetetherm standard' – in other words an assortment of HES that should be market, sold and produced by different plants. This was going to be facilitated by creating a base for an international product platform for the HES. Furthermore, there was a need for developing an international sizing and configuration computer program where the input/output were fitted to all markets and production sites. Initially, the goal was to modularize all German, Finnish, Swedish, and Norwegian HES and to develop a product platform that can be use by other countries within the Cetetherm group. The aim in 2002 was to start out with the German market as their products were frequently engineered and customized. There was less focus on design and more on configuration and documentation of sales support than previously.

The development of product platform resulted in changes for the customers: the market had to accept the change in look of the HES. Production sites also needed to be modified somewhat both regarding layout issues and routines to prepare for the product platform. Valid HES types for different markets had to be defined. A standard assortment of components and sizing rules needed to fulfil both international and national demands.

To facilitate the implementation of a product platform, the project, managed by P&T, was split into four subprojects: design, configuration, production, and market. This was in December 2002, and later on a fifth subproject evolved which handled the product and was headed by the product manager of large heating systems. This way, the project was alive in more than one part of the firm at a time, enabling a parallel development of the product platform.

One of the expected results of the entire project was a platform that could be sized and configured in a software tool – a product platform that could also be built at every production site, and sold to every market. To realize the firm's visions regarding the product platform for HES, key individuals were hired in the P&T department. These were individuals with wide technical expertise, both regarding computer programming as well as HES. The person who came to be the project manager had been with the firm for 10 years and had worked in several different departments making this person very suitable to run the project. The project manager along with some other involved key people also had a profound interest in trying to make the products easier to manufacture and to sell.

In late 2002, the project started to investigate what a shared basic design for the Cetetherm HES should look like. The basic design project was carried out mainly by production site technicians and by the German Engineering division and it was completed in March 2003. The project started to investigate what variety the market needed and how the marked need could be offered with as few components as possible. In spring 2003, it was understood that in order to succeed, one had to switch focus from what was initially stated to a smaller scope, and focus was then set on the Swedish and the Finnish markets. These markets had similar legislation regarding district heating and they were similar in both product and production facilities thereby enabling a much easier merge of the two markets and the development of a new product platform.

The development of the sizing and configuration tool started in the early fall 2003, after a finished prototype had been showed to the end users and agreed upon, and was then completed in May 2004. A goal was for everyone within the Cetetherm group to use the same tool with an international standard for production documentation and printouts, and national standards for customer documentation with demands for quality and delivery documentation predefined.

The product platform project for the HE-systems was completed in June 2004. It enabled Cetetherm to sort out the range of products that the firm wanted to put on the market, and thereby creating a product platform. The carry-over from the product platform project is now focusing on four other markets and the goal is to continue to aim the efforts made so far toward four additional markets; Russia, Poland, Hungary, and Czech Republic.

5.2   Platform development and modularization

Product modularization in its essence is usually to group components into different modules and then to define simple interfaces between the modules. The aim should be that these interfaces stay stable, and unchanged in the forthcoming development of new products. Defining proper interfaces will then give the company a lot of different benefits from modularization, for example increased possibility for efficient management of product variety, the creation of strategic flexibility and reduced task complexity (Persson, 2004). The modularization done at Cetetherm was in some respects a little different from the above described more 'traditional' product modularization. One can argue that Cetetherm's product, the HES, in fact already were naturally modular. A HES typically consists of a number of heat exchangers, valves, pumps, pipes, control system etc. All of these different major components in the system could actually be defined as, and called, product modules. Even though there were already a number of so-called 'natural' modules in the HES, the module interfaces were not the best and the most stable ones. Instead, there was a huge variation of the different modules interfaces for different variants due to the market situation for Cetetherm, the customers' very specific needs and also a lot of other requirements and legislations affecting the product design.

Due to the modularity inherent in Cetetherm's HES, the modularization did not become so much a question of choosing/defining interfaces between the different modules. Instead, it became more of a question to decide how to structure the whole product, modules and component assortment.

When the modularization project started, a lot of highly customization products were sold. The customization for example resulted in that Cetetherm had a lot of different variants of each specific component, but notable most of them fulfilling exactly the same functionality. This could for example be pumps or valves from different suppliers, all having exactly the same functionality. But, the pumps coming from different suppliers and having different brands were due to the very specific and detailed requirements from different markets and customers.

The modularization done was to a large extent a question of sorting the existing component assortment, and also to choose and take decisions about what gives to the customers the best product performance as well as the lowest price possible.This because the annual sales volume of each of the different variants of a specific component are low. Shifting to only one supplier, for example pumps, instead of having many different suppliers for the pumps would increase the yearly purchased and sold volume per year, thus increasing the economy of scale effects.

5.3   The sale and production support software

The aim of the configuration subproject was specified in September 2003, and it was to deliver a software tool that support sales and production of HES included in the product platform range. By installing this tool and making sure that all relevant personnel was instructed on how to use it, a number of old software tools were successfully phased out and the average time to place an order was shortened from hours to less than 10 minutes. In January 2004, the name of the sizing and configuration program was set to Cetetherm Platform System, CPS.

The new tool was an important step in providing the same HES in all the countries where Cetetherm operates. With this tool, volumes in both main components and pipe/frame modules increased and the production cost for each HES would decrease. CPS included more of the total heat exchange system, and therefore let the sales personnel include more in each offer with less effort. Another important benefit was the fact that this tool reduced the mistakes made during calculation and order handling, which would decrease the overall time spent per unit. Easier handling of quotation and production documents also gives salesmen more time to visit potential new customers.

The CPS program was mainly a sales configuration tool, but it was also a production configuration tool. By using it, the HES could be visualized in a way that had not previously been possible. As the program grew, pipe modules gradually had to be defined as system types, sizing rules, and components were also essential. Systems and components were all put into the database that CPS used.

Initially, the sales person configures a product while discussing the appropriate functionality needs with the customer. The program provides possible solutions and choices of modules step-by-step as the system is built up in the program, and there is no possibility to skip one step and go to the next before fully completing the form. Each module represents a certain functionality, which is chosen depending on the costumer needs and choices. When the customer has chosen all essential modules, the result of the run is a comprehensive technical documentation.

The technical documentation is used to form a quotation and to form a blue-print for the production unit. When the documentation is presented for quotation, the data is presented as follows: product summary, technical description, measure sketches, flowcharts (international symbol standard with written explanations), and finally a price list. The information that is transmitted directly to the production unit is the same plus the following: sign data (the sign that is put on the physical product), main component and module list (international article/module numbers), and PED-categorization.

Thus, the CPS sizing and configuration tool works both as a sales configuration tool and as a product configuration tool. The main issue is that the data is stored and transmitted within the company's system and that way there is a minimum information loss.

6.   EFFECTS FOR CETETHERM

The effects of the platform and modularization project started to appear for Cetetherm no more than a few months after project accomplishment. Previously, a large part of the products sold by Cetetherm were highly customized. Most companies, if not all, aims to have the process from customer order to delivery as short and quick as possible. But, in the previous situation for Cetetherm, a large part of the products also had to pass through the company's engineering function. One of the major effects from the newly developed product platform is that this need for engineering work has decreased a lot. Before, around 80% of the total number of sold products had to be engineered. The situation today is that as many as 80% of the ordered products can go directly to the production function and can be started to be assembled immediately, only 20% of the number of sold products has to go via the engineering function before being manufactured and assembled. Due to this decreased need for engineering work, a lot of money can be saved. The development cost for each product decreases; there is also a quite significant effect on the order-to-delivery lead-time. In addition, as less parts of the sold product are engineered, the number of special product solutions decreases, resulting in a better quality of the products delivered to customers. Cetetherm have had problems concerning that a specific customer could have ordered a product a few years ago, and some years later order a new product having the same specification, aimed to fulfill and deliver exactly the same functionality as the previous product. But, many customers become surprised when the design of the product differed a lot from the previous product they already had. This happened simply because of the low degree of standardization and because of the large degree of engineering work on the products before delivered to customers.

The previous situation for the production at Cetetherm can be described as having been a lot of customization and special solutions. This was explained by the market and customers rather specific and detailed needs of how their product should be designed. In certain markets, there have also been needs for special pipes and material for the pipes, making the previous production rather chaotic and unpredictable. Cetetherm had also suffered from very large seasonal variations in the market demands for HES. The situation has changed a lot for Cetetherm's production after the introduction of the new developed product platform. Now that the share for special products has decreased down to 20% of the total sales volume, the production at Cetetherm has become more standardized. Due to the new product platform, it has also become easier for the ones working in the production function to see what the standard products are, built up by modules and components in the platform, and what is not standard product. It has also become much more evident that, from a production point of view, it takes a lot of more time to manufacture and assemble a special product compared to a standard product.

On the basis of these results of the introduction of the new product platform in the production process, Cetetherm has decided to divide its assembly of large HES into three different assembly lines. One assembly line is for standard products consisting of only modules and components that are part of the product platform and that can designed and configured from its computer based sales and configuration IT-system (the CPS-system). There is another assembly line for products that are almost standard yet still require special operations before being delivered to the customer. Finally, the third assembly line is for the products that are highly customized HES. This change is done because of their wish to increase the visibility in the production process even more. With three different separate assembly lines, one will be able to see that the standard products are assembled quickly and without any problems and special operations needed in order to be accomplished. Furthermore, one can easy notice that the special products take more time to assemble and also have a need for a number of different specific operations, operations that may also be of a 'one-time event' character, meaning that they are 'invented' by the assembly operators themselves. These are operations are not easy for the sales and engineering people to predict in advance, which will drive the work of more standardization and less customization even more. It will also be easier to understand that the manufacturing and assembly of special products takes a lot of time and hence cost a lot of money, maybe more than could be measured and estimated before the new developed product platform.

7.   PRODUCT PLATFORM MANAGEMENT

The situation for Cetetherm has changed a lot during the last years, and this due to its initiatives to develop a product platform consisting of a different modules and components to configure, develop and assemble different HES. Due to this, the company's total number of different components in its assortment has decreased a lot. Many of the products have become more standard and less specially customized. But, even after this standardization, there is still a need for some special products due to the market situation, with for example a lot of different legislations in different countries, and even for specific parts of a country. Therefore, there is an appearing risk that the product assortment can start to increase as time goes, and hence the product platform will be including a lot of more variants of the modules and components in some years from now. Due to this risk, the company has taken some decisions to secure that this will not be likely to happen, and the benefits from a product platform strategy and approach will still be there in the future. For example, Cetetherm has a product manager for these large HES, and he is taking a very active role in trying to conserve the product platform. Cetetherm is also about to change its existing product development process to better support that the existing product platform is not changed too much when new products are developed by the company.

Another important thing is the computer-based sales and product configuration system that was developed parallel with the new product platform. This is a product configuration system that is used by Cetetherm's sales people world-wide. By using this system, the salesmen can configure an HES fulfilling customer's different demands, and this can be done in a short time at the same time as the meeting with the customer. The system consists in a very user-friendly interface than is used by the sales people in their customer contacts. When the different product functionalities have been typed into the system, a visualization of the specific HES will appear on the screen in a very visible, and also, printable format. But what is more important from a product platform management perspective is the information and configurations rules lying behind this user interface. There are a lot of rules about how different HES should be built up in order to give the customer the exact product functionality needed. The rules also restrict to some extent what is actually possible to assemble using only the standard modules and components in this product platform. In order to try to conserve the product platform, Cetetherm has implemented restricted rules in what is allowed to be changed in the configuration rules in this computer-based configuration system. They have also restricted how it is allowed to make these changes in order to conserve the benefits from the product platform and a more reduced product, module and components assortment.

8.   CONCLUSIONS

The product platform and the modularization of the product at Cetetherm was not a case of new product development. Instead, the platform and the modules were created by analysing how the market offer could be achieved by using a smaller assortment of products, modules and components. The project first identified what variety was needed in order to cover 80% of the total market demand. They then started to analyse how this variety could be created with as few different and varying components as possible. To be able to do this, a company must understand the potential of using a standardized assortment. Cetetherm reflected their way of doing business and found out that their competitiveness could be improved by giving the customer exactly the desired performance, but not necessarily by using exactly the components the customer wanted.

The process of finding out what components that should be included in the platform is more a decision of which components should be standard; not to invent a totally new product development process. To decide what components to use might seem as an easy task, but it is actually not simply to pick one component of the present. The designer must know how the different components in the product together fulfil the desired product functionality and performance. A certain component can have one type of performance integrated with another component, but different performance when integrated with another component. In the case of Cetetherm, there are many possible ways to have the same performance and most of them were used. Therefore, in order for the company to select, it requires a lot of innovativeness, product knowledge and customer knowledge.

In terms of modularization, the work at Cetetherm is the opposite of what is mostly discussed as modularization. In this case the purchased components, like valves and heat exchanger, accomplish the function. Hence, in terms of modularity, the purchased components are modules that are industrial standard components and cannot be changed to a greater extent. Therefore, the pipes between the components must absorb the variety created by the components. The only possible way to get a modular HES is to design the pipes connecting the components. In some cases, the pipes and components could be that standardized so they could be turned into a module. The benefit of turning components and pipes into modules is that the pipes and components are integrated to one unit, a module, that exists both in the sales system as a function and as a physical part in the production. In other words, the modularization at Cetetherm consists of two different points: first, the selection of which components that can give the performance demanded by the market and the customers; second, the design of pipes that in the smartest way integrate the different components. In other words, the interfaces between the modules should be able to handle the variety of the modules. Normally the interfaces between the modules are argued to be standardized. In this case, the most cost efficient way is instead to adapt the interfaces to the module. Of course, the variants of each modules were selected with as small difference as possible. The smaller differences between the modules the smaller differences between the interfaces are needed.

To exclude the German market from the project, it was necessary to be able to modularize the product. The difference, both in terms of modules and concepts, used in Germany was too large. This fact demonstrates the necessity to consider the limits of the extent of each variety dimension and how many variety dimensions a product can handle without losing platform efficiency and effectiveness. A platform can only cover a limited range of variety (Muffatto, 1999). Considering where the limits are is crucial for the success of the product platform implementation.

The structuring of the total component assortment and the implementation of a standard platform demands acceptance from the sales organization. If the sales organization continue to sell the components that they used to produce, the whole idea of a product platform would gone down. To get the sales personal to choose the components that are best for the company, the activity to select these components must be facilitated. The software facilitating the sales situation and guiding the sales personal hereby became the key to success in this product platform. The platform effect on the company hereby depends on the quality and acceptance of the offer and the sales support system by the sales organization. In those cases where the sales employees do not use the support system the platform, there is no potential benefits in engineering and production. Sales orders of products included in the product platform but not configured in the sales support system include a lot of additional work for both engineering and production. A conclusion from the case at Cetetherm is the importance of having a clear view of who enables and who can harvest the positive effect of the platform. In this case, the sales organization enabled and the P&T (R&D) and production function of the company could harvest the benefit from the product platform.

To turn an organization from offering products in a traditional way without sharing components and resources to offering products that have a clear structure is a demanding task. In a company the organization, processes and IS/IT should support the present product. When a company turns from a less structured product to a product based on a product platform, the organization processes and IS/IT must be adapted to the new platform based product. Such change process needs a real enthusiastic and dedicated employee with both management and technical skills. Without the enthusiastic project leader at Cetetherm, the change to a platform based product would have been almost impossible. The change process had several hard phases where both technical and managerial tasks had to be taken care of. Without a large portion of enthusiastic project, leading those task would have made the project to come to nothing.

The project at Cetetherm included participants from all different organizational functions in the company. The participation from all functions in the company should lead to a cross-functional project, which is necessary for all platform projects (Meyer and Lehnerd, 1997). But, the project leader at Cetetherm would like to have seen more participation and energy from some of the functions in the company. There are many different activities that have to be done when implementing a product platform and several of them are very time consuming. At Cetetherm, as well as other companies, some of the employees that should participate in the platform project were busy with their day-to-day jobs. The daily tasks that had to be solved in order to keep the business running in many cases obstruct the progress of the project. In some cases, the input that was requested from the project from some of the functions was not given. Instead, the project leader had to make assumptions. The lack of input in many cases led to less beneficial solutions resulting in less profit for the company. On the other hand, the project was more or less free from resistance among the employees. The vast majority supported the project and contributed to it.

9. ACKNOWLEDGMENTS

We would like to say thanks to the employees at Alfa Laval Lund/Ronneby, especially Mats R. Nilsson, Tony Svensson, and Mats Persson, making this study possible. This book chapter is based on results of a research project performed under the auspices of the Institute for Management of Innovation and Technology, IMIT.

References

Aboulafia, R., 2000, Airbus pulls closer to Boeing, *Aerospace America,* **38**(4): 16-18.

Ahmadi, R., Roemer, T.A., and Wang, R.H., 2001, Structuring product development processes, *European Journal of Operational Research*, **130**(3): 539-558.

Akundi, S., Simpson, T.W., and Reed, P.M., 2005, Multi-objective design optimization for product platform and product family design using genetic algorithms, ASME Design Engineering Technical Conferences - Design Automation Conference, Gea, H. C. (Ed.), Long Beach, CA, ASME, Paper No. DETC2005/DAC-84905.

Alford, D., Sackett, P., and Nelder, G., 2000, Mass customization - An automotive perspective, *International Journal of Production* Economics, **65**(1): 99-110.

Allada, V., and Jiang, L., 2002, New modules launch planning for evolving modular product families, ASME Design Engineering Technical Conferences - Design for Manufacturing Conference, Montreal, Quebec, Canada, ASME, Paper No. DETC2002/DFM-34190.

Allen, K. R., and Carlson-Skalak, S., 1998, Defining product architecture during conceptual design, ASME Design Engineering Technical Conferences - Design Theory and Methodology, Atlanta, GA, ASME, Paper No. DETC98/DTM-5650.

Almgren, H., 2000, Pilot production and manufacturing start-up: the case of Volvo S80, *International Journal of Production Research*, **38**(17): 4577-4588.

Anderson, D.M., 1997, *Agile Product Development for Mass Customization: How to Develop and Deliver Products for Mass Customization, Niche Markets, JIT, Build-to-Order and Flexible Manufacturing,* Irwin, Chicago, IL.

Anderson, S.W., 1995, Measuring the impact of product mix heterogeneity on manufacturing overhead cost, *The Accounting Review*, **70**(3): 363-387.

Anderson, S.W., and Sedatole, K.L., 1998, Designing quality into products: the use of accounting data in new product development, *Accounting Horizons*, **12**(3): 213-233.

Andrade, M.C., Pessanha Filho, R.C., Espozel, A.M., Maia, L.O.A., and Qassim, R.Y., 1999, Activity-based costing for production learning, *International Journal of Production Economics,* **62**(3): 175-180.

Anonymous, 2002, Volkswagen campaigns for more distinct brand identities, *Professional Engineering,* **15**(5): 13.

Anonymous, 2005, Aircraft families / Commonality: Maximum benefits from today's most modern aircraft, http://www.airbus.com/en/aircraftfamilies/commonality.html, Retrieved on July 16, 2005.

Asiedu, Y., and Gu, P., 1998, Product life cycle cost analysis: State of the art review, *International Journal of Production Research*, **36**(4): 883-908.

Atkinson, A.A., Kaplan, R.S., and Young, S.M., 2004, *Management Accounting,* Pearson Prentice Hall, Upper Saddle River, NJ.

Baker, K.R., 1985, Safety stocks and component commonality, *Journal of Operations Management*, **6**(1): 13-22.

Baker, K.R., Magazine M.J., and Nuttle, H.L.W., 1986, The effect of commonality on safety stock in a simple inventory model, *Management Science*, **32**(8): 982-988.

Balakrishnan, P.V.S., and Jacob, V.S., 1996, Genetic algorithms for product design, *Management Science*, **42**(1): 1105-1117.

Baldwin, C.Y., and Clark, K.B., 1997, Managing in an age of modularity, *Harvard Business Review*, **75**(5): 84-93.

Baldwin, C.Y., and Clark, K.B., 2000, *Design rules: The Power of Modularity*, MIT Press, Cambridge, MA.

Banker, R.D., Datar, S.M., Kekre, S., and Mukhopadhyay, T., 1990, Cost of product and process complexity, in *Measures for Manufacturing Excellence*, Kaplan, R.S. (Ed.), Harvard Business School Press, Boston, MA, pp. 269-290.

Banker, R.D., Potter, G., and Schroeder, R.G., 1995, An empirical analysis of manufacturing overhead cost drivers, *Journal of Accounting and Economics*, **19**(1): 115-137.

Bass, L., Clements, P., and Kazman, R., 2003, *Software Architecture in Practice*, Addison-Wesley, Boston, MA.

Bejan, A., 1996, Street network theory of organization in nature, *Journal of Advanced Transportation*, **255**(7): 85-107.

Bejan, A., 1997, *Advanced Engineering Thermodynamics*, John Wiley & Sons, New York.

Bejan, A., 2000, *Shape and Structure: From Engineering to Nature*, Cambridge University Press, Cambridge, UK.

Ben-Akiva, M., and Lerman, S., 1985, *Discrete Choice Analysis: Theory and Application to Travel Demand*, The MIT Press, Cambridge, MA.

Bhimani, A., and Muelder, P.S., 2001, Managing processes, quality, and costs: A case study, *Journal of Cost Management*, **15**(2): 28-32.

Bielefeld, J.R.J., and Rucklos, G.D., 1992, Cost scaling factors: how accurate are they? *Cost Engineering*, **34**(10): 15-20.

Biersdorfer, J.D., 2001, Module lineup gives a little organizer big ambitions, *The New York Times*, New York, NY: E7.

Blackburn, J.H., 1985, Improve MRP and JIT compatibility by combining routings and bills of material, Proceedings of the American Production and Inventory Control Society, pp. 444-447.

Blackenfelt, M., 2000, Profit maximisation while considering uncertainty by balancing commonality and variety using robust design - The redesign of a family of lift tables, ASME Design Engineering Technical Conferences - Design for Manufacturing, Baltimore, MD, ASME, Paper No. DETC2000/DFM-14013.

Blackenfelt, M., and Sellgren, U., 2000, Design of robust interfaces in modular products, ASME Design Engineering Technical Conferences - Design Automation Conference, Baltimore, MD, ASME, Paper No. DETC00/DAC-14486.

Blanchard, B.S., and Fabrycky, W.J., 1998, *Systems Engineering and Analysis*, Prentice-Hall, Upper Saddle River, NJ.

Blischke, W.R., and Murthy, D.N.P., 1994, *Warranty Cost Analysis*, Marcel Dekker, Inc., New York.

Bode, J., 2000, Neural networks for cost estimation: Simulations and pilot application, *International Journal of Production Research*, **38**(6): 1231-1254.

Boer, M., and Logendran, R., 1999, A methodology for quantifying the effects of product development on cost and time, *IIE Transactions*, **31**(4): 365-378.

Boothroyd, G., Dewhurst, P., and Knight, W.A., 2002, *Product Design for Manufacture and Assembly*, Marcel Dekker Inc., New York.

Bremmer, R., 1999, Cutting-Edge Platforms, *Financial Times Automotive World*, September, pp. 30-38.

Bremmer, R., 2000, Big, bigger, biggest, *Automotive World*, June, pp. 36-44.

Brimson, J.A., 1998, Feature costing: beyond ABC, *Journal of Cost Management*, **12**(1): 6-12.

Bruns, W.J., 1989, Destin Brass Products Co., *Harvard Business Review*, Case No. 9-190-089: 1-10.

Buede, D.M., 2000, *The Engineering Design of Systems - Models and Methods*, John Wiley & Sons, New York.

Caffrey, R.T., Simpson, T.W., Henderson, R., and Crawley, E., 2002, The technical issues with implementing open avionics platforms for spacecraft, the 40th AIAA Aerospace Sciences Meeting and Exhibit, Reno, NV, AIAA, AIAA-2002-0319.

Carney, D., 2004, Platform flexibility, *Automotive Engineering International*, **112**: 147-149.

Cetin, O.L., and Saitou, K., 2004, Decomposition-based assembly synthesis for structural modularity, *ASME Journal of Mechanical Design*, **126**(2): 234-243.

Chakravarty, A.K., and Balakrishnan, N., 2001, Achieving product variety though optimal choice of module variations, *IIE Transactions*, **33**(7): 587-598.

Chandler, C., and Williams, M., 1993, Strategic shift: A slump in car sales forces nissan to start cutting swollen costs, *Wall Street Journal*, New York, NY, A1.

Chang, T.-S., and Ward, A.C., 1995, Design-in-modularity with conceptual robustness, Advances in Design Automation, Azarm, S., et al., (Eds.), Boston, MA, ASME, **82**(1): 493-500.

Chapman, S.J., 1991, *Electric Machinery Fundamentals*, McGraw-Hill, New York, NY.

Chen, M., Han, J., and Yu, P., 1996, Data mining: An overview from database perspective, *IEEE Transactions on Knowledge and Data Engineering*, **8**(6): 866-883.

Child, P., Diederichs, R., Sanders, F.-H., and Wisniowski, S., 1991, The management of complexity, *Sloan Management Review*, **33**(1): 73-80.

Chinnaiah, P.S.S., Kamarthi, S.V., and Cullinane, T.P., 1998, Characterization and analysis of mass-customized production systems, *International Journal of Agile Manufacturing*, **2**(1): 93-118.

Choi, S.C., and DeSarbo, W.S., 1994, A conjoint-based product designing procedure incorporating price competition, *Journal of Product Innovation Management*, **11**(5): 451-459.

Cicirello, V.A., and Regli, W., 2002, An approach to a feature-based comparison of solid models of machined parts, *AIEDAM*, **16**(5): 385-399.

Clancy, D.K., 1998, Strategic design cost analysis, *Cost Engineering*, **40**(8): 25-30.

Clark, J.P., Roth, R., and Field, F.R., 1997, Techno-economic issues in materials selection, in *ASM Handbook Volume 20: Materials Selection and Design*, Dieter. G.E. (Ed.), ASM International, Materials Park, OH, pp. 255-265.

Clausing, D., 1994, *Total Quality Development — a Step-by-step Guide to World-class Concurrent Engineering*, ASME Press, New York, NY.

Dixon, J.R., 1987, On research methodology towards a scientific theory of engineering design, *Artificial Intelligence in Engineering Design, Analysis, and Manufacturing*, **1**(3): 145-157.

Cogdell, J.R., 1990, *Foundations of Electrical Engineering*, Prentice Hall, Upper Saddle River, NJ.

Cohen, L., 1995, *Quality Function Deployment: How to Make QFD Work for You*, Addison-Wesley, Reading, MA.

Cokins, G., 2000, The changing face of cost management in the auto industry, *Journal of Cost Management*, **14**(5): 13-15.

Collier, D.A., 1979, Planned work center load in a material requirements planning system, the Eleventh National AIDS Meeting, New Orleans, LA.

Collier, D.A., 1980, Justifying component part standardization, Twelfth National AIDS Conference, Las Vegas, NV.

Collier, D.A., 1981, The measurement and operating benefits of component part commonality, *Decision Sciences*, **12**(1): 85-96.

Collier, D.A., 1982, Aggregate safety stock levels and component part commonality, *Management Science*, **28**(11): 1297-1303.

Conner, C.G., De Kroon, J.P., and Mistree, F., 1999, A product variety tradeoff evaluation method for a family of cordless drill transmissions, ASME DETC, Las Vegas, Nevada, DETC99/DAC-8625.

Constantine, B., Roth, R., and Clark, J.P., 2001, Substituting tube-hydroformed parts for automotive stampings: An economic model, *Journal of Operations Management*, **53**(8): 33-38.

Cooper, R., and Kaplan, R.S., 1988, How cost accounting distorts product costs, *Management Accounting*, April, pp. 20-27.

Cooper, R., and Kaplan, R.S., 1992, Activity-based systems: measuring the costs of resource usage, *Accounting Horizons*, **6**(3): 1-12.

Cooper, R., and Kaplan, R.S., 1998, The promise - and peril - on integrated cost systems, *Harvard Business Review*, July-August, pp. 109-119.

Cox, J.J., Roach, G. M., and Teare, S., 2001, Reconfigurable models and product templates as a means to increasing productivity in the product development process, Proceedings of the 2001 World Congress on Mass Customization and Personalization, Hong Kong, China, October 1-2, A3C004.

Cusumano, M.A., and Nobeoka, K., 1998, *Thinking Beyond Lean*, The Free Press, New York, NY.

Cusumano, M.A., and Selby, R.W., 1995, *Microsoft secrets: How the World's Most Powerful Software Company Creates Technology, Shapes Markets, and Manages People*, The Free Press, New York, NY.

Dahmus, J.B., Gonzalez-Zugasti, J.P., and Otto, K.N., 2001, Modular product architecture, *Design Studies*, **22**(5): 409-424.

Dahmus, J.B., and Otto, K.N., 2001, Incorporating lifecycle costs into product architecture decisions, ASME Design Engineering Technical Conferences, Pittsburgh, PA, ASME, Paper No. DETC2001/DAC-21110.

Dai, Z., and Scott, M.J., 2004a, Effective product family design using preference aggregation, ASME Design Engineering Technical Conferences - Design Theory and Methodology Conference, Salt Lake City, UT, ASME, Paper No. DETC2004/DTM-57419.

Dai, Z., and Scott, M. J., 2004b, Product platform design through sensitivity analysis and cluster analysis, ASME Design Engineering Technical Conferences - Design Automation Conference, Chen, W. (Ed.), Salt Lake City, UT, ASME, Paper No. DETC2004/DAC-57464.

Das, S.K., Yedlarajiah, P., and Narendra, R., 2000, An approach for estimating the end-of-life product disassembly effort and cost, *International Journal of Production Research*, **38**(3): 657-673.

De Lit, P., and Delchambre, A., 2003, *Integrated Design of A Product Family and Its Assembly System*, Kulwer Academic Publishers, Norwell, MA.

De Lit, P., Delchambre, A., and Henrioud, J.M., 2003, An integrated approach for product family and assembly system design, *IEEE Transactions on Robotics and Automation*, **19**(2): 324-333.

De Weck, O., Suh, E.S., and Chang, D., 2003, Product family and platform portfolio optimization, ASME Design Engineering Technical Conferences - Design Automation Conference, Shimada, K. (Ed.), Chicago, IL, ASME, Paper No. DETC2003/DAC-48721.

Delingette, H., Hebert, M., and Ikeuchi, K., 1992, Shape representation and image segmentation using deformable surfaces, *Image Vision Computing*, **10**(3): 132-144.

Dertouzos, M.L., 1989, *Made in America: Regaining the Productive Edge*, MIT Press, Cambridge, MA.

Dobson, G., and Kalish, S., 1988, Positioning and pricing a product line, *Marketing Science*, **7**(2): 107-125.

Dobson, G., and Kalish, S., 1993, Heuristics for pricing and positioning a product-line using conjoint and cost data, *Management Science*, **39**(2): 160-175.

Doran, D.T., and Dowd, J.E., 1999, Depreciation and amortization cost in activity based costing systems, *Journal of Cost Management*, **13**(5): 34-38.

D'Souza, B., and Simpson, T.W., 2003, A genetic algorithm based method for product family design optimization, *Engineering Optimization*, **35**(1): 1-18.

Du, X., Jiao, J., and Tseng, M.M., 2001, Architecture of product family: Fundamentals and methodology, *Concurrent Engineering: Research and Application*, **9**(4): 309-325.

Duray, R., Ward, P.T., Milligan, G.W., and Berry, W.L., 2000, Approaches to mass customization: configurations and empirical validation, *Journal of Operations Management*, **18**(6): 605-625.

Eisenhardt, K., 1989, Building theories from case study research, *Academy of Management review*, **14**(4): 532-550.

Elinson, A., Nau, D.S., and Regli, W.C., 1997, Feature-based similarity assessment of solid models, Fourth Symposium on Solid Modeling and Applications, Hoffman, C., and Bronsvoort, W. (Eds.), Atlanta, GA, pp. 297-310.

Emblemsvag, J., and Bras, B., 1994, Activity-based costing in design for product retirement, ASME Design Engineering Technical Conferences - Design Theory and Methodology Conference - Advances in Design Automation Conference, Minneapolis, DE- 69(2): 351-362.

Eppinger, S.D., Whitney, D.E., Smith, R.P., and Gebala, D.A., 1994, A model-based method for organizing tasks in product development, *Research in Engineering Design*, **6**(1): 1-13.

Erens, F., 1997, Synthesis of Variety: Developing Product Families, *Ph.D. Dissertation*, University of Technology, Eindhoven, The Netherlands.

Ericsson, A., and Erixon, G., 1999, *Controlling Design Variants: Modular Product Platforms*, ASME, New York.

Erixon, G., 1996, Design for modularity, Design for X - Concurrent Engineering Imperatives, Huang, G.Q. (Ed.), Chapman & Hall, New York, pp. 356-379.

Erni, K., and Lewerentz, C., 1996, Applying design metrics to object oriented frameworks, The 3rd International Software Metrics Symposium, Berlin, Germany, IEEE 96TB100034: 64-74.

Esawi, A.M.K., and Ashby, M.F., 2003, Cost estimates to guide pre-selection of processes, *Materials & Design*, **24**(8): 605-616.

Eynan, A., and Rosenblatt, M.J., 1996, Component commonality effects on inventory costs, *IIE Transactions*, **28**(2): 93-104.

Farrell, R., and Simpson, T.W., 2003, Product platform design to improve commonality in custom products, *Journal of Intelligent Manufacturing*, **14**(6): 541-556.

Feitzinger, E., and Lee, H.L., 1997, Mass customization at Hewlett-Packard: The power of postponement, *Harvard Business Review*, **75**(1): 116-121.

Fellini, R., Kokkolaras, M., Michelena, N., Papalambros, P., Saitou, K., Perez-Duarte, A., and Fenyes, P.A., 2002a, A sensitivity-based commonality strategy for family products of mild variation, with application to automotive body structures, 9th AIAA/ISSMO Symposium on Multidisciplinary Analysis and Optimization, Atlanta, GA, AIAA, AIAA-2002-5610.

Fellini, R., Kokkolaras, M., Papalambros, P., and Perez-Duarte, A., 2002b, Platform selection under performance loss constraints in optimal design of product families, ASME Design Engineering Technical Conferences - Design Automation Conference, Fadel, G. (Ed.), Montreal, Quebec, Canada, ASME, Paper No. DETC2002/DAC-34099.

Fellini, R., Papalambros, P., and Weber, T., 2000, Application of product platform design process to automotive powertrains, The 8th AIAA/NASA/USAF/ISSMO Symposium on Multidisciplinary Analysis and Optimization, Long Beach, CA, AIAA, AIAA-2000-4849.

Fenyes, P.A., 2000, Multidiscplinary design and optimization of automotive structures – A parametric approach, Proceedings of the 8th AIAA/NASA/USAF/ISSMO Symposium on Multidisciplinary Analysis and Optimization, AIAA-2000-4706.

Ferdinand, A., 1993, *Systems, Software, and Quality Engineering*, Van Nostrand Reinhold.

Fernández, M.G., Seepersad, C.C., Rosen, D.W., Allen, J.K., and Mistree, F., 2001, Utility-based decision, support for selection in engineering design, The 13th International Conference on Design Theory and Methodology, Pittsburgh, PA, DETC2001/DAC-21106.

Ferrari, E., Gamberi, M., Manzini, R., and Pareschi, A., 2003, An Integrated optimization process for the production planning and control of a flexible manufacturing system, Advanced Simulation Technologies Conference, Orlando, FL.

Fine, C.H., Vardan, R., Pethick, R., and El-Hout, J., 2002, Rapid response capability in value-chain design, *Sloan Management Review*, 43(2): 69-75.

Fisher, M.L., and Ittner, C.D., 1999, The impact of product variety on automobile assembly operations: empirical evidence and simulation analysis, *Management Science*, 45(6): 771-786.

Fisher, M.L., Ramdas, K., and Ulrich, K.T., 1999, Component sharing in the management of product variety: A study of automotive braking systems, *Management Science*, 45(3): 297-315.

Fixson, S.K., 2005, Product architecture assessment: A tool to link product, process, and supply chain design decisions, *Journal of Operations Management*, 23(3/4): 345-269.

Foster, G., and Gupta, M., 1990, Manufacturing overhead cost driver analysis, *Journal of Accounting and Economics*, 12(1-3); 309-337.

Fredriksson, P., and Araujo, L., 2003, The evaluation of supplier performance: A case study of Volvo cars and its module suppliers, *Journal of Customer Behavior*, 2(3): 365-384.

Fritzsch, R.B., 1997, Activity-based costing and the theory of constraints: using time horizons to resolve two alternative concepts of product cost, *Journal of Applied Business Research*, 14(1): 83-89.

Fujita, K., 2002, Product variety optimization under modular architecture, *Computer-Aided Design,* 34(12): 953-965.

Fujita, K., Akagi, S., and Hirokawa, N., 1993, Hybrid approach for optimal nesting using a genetic algorithm and a local minimization algorithm, Proceedings of the 1993 ASME Design Automation Conference, **DE-65**(1), pp. 477-484.

Fujita, K., Akagi, S., Yoneda, T., and Ishikawa, M., 1998, Simultaneous optimization of product family sharing system structure and configuration, ASME Design Engineering Technical Conferences - Design for Manufacturing, Atlanta, GA, ASME, Paper No. DETC98/DFM-5722.

Fujita, K., Akagi, S., Yoshida, K., and Hirokawa, N., 1996, Genetic algorithm based optimal planning method of energy plant configurations, Proceedings of the 1996 ASME Design Engineering Technical Conferences, Paper No. 96-DETC/DAC-1464.

Fujita, K., and Ishii, K., 1997, Task structuring toward computational approaches to product variety design, Proceedings of the 1997 ASME Design Engineering Technical Conferences, Paper No. 97DETC/DAC-3766.

Fujita, K., Sakaguchi, H., and Akagi, S., 1999, Product variety deployment and its optimization under modular architecture and module commonalization, ASME Design Engineering Technical Conferences - Design for Manufacturing, Las Vegas, NV, ASME, Paper No. DETC99/DFM-8923.

Fujita, K., Takagi, H., and Nakayama, T., 2003, Assessment method of value distribution for product family deployment, Proceedings of 14th International Conference on Engineering Design (ICED 03), Paper No. 1484.

Fujita, K., and Yoshida, H., 2001, Product variety optimization: simultaneous optimization of module combination and module attributes, ASME Design Engineering Technical Conferences - Design Automation Conference, Diaz, A. (Ed.), Pittsburgh, PA, ASME, Paper No. DETC2001/DAC-21058.

Fujita, K., and Yoshida, H., 2004, Product variety optimization simultaneously designing module combination and module attributes, *Concurrent Engineering: Research and Applications*, 12(2): 105-118.

Fujita, K., and Yoshioka, S., 2003, Optimal design methodology of common components for a class of products: its foundations and promise, ASME Design Engineering Technical Conferences - Design Automation Conference, Shimada, K. (Ed.), Chicago, IL, ASME, Paper No. DETC2003/DAC-48718.

G.S. Electric, 1997, Why Universal Motors Turn on the Appliance Industry, http://www.gselectric.com/electric/univers4.htm.

Galsworth, G.D., 1994, *Smart, Simple Design: Using Variety Effectiveness to Reduce Total Cost and Maximize Customer Selection*, Omneo, Essex Junction, VT.

Gerchak, Y., Magazine, M.J., and Gamble, A.B., 1988, Component commonality with service level requirements, *Management Science*, **34**(6): 753-760.

Gershenson, J.K., Prasad, G.J., and Allamneni, S., 1999, Modular product design: A life-cycle view, *Journal of Integrated Design and Process Science*, **3**(4): 13-26.

Gershenson, J.K., Prasad, G.J. and Zhang, Y., 2003a, Product Modularity: Definition and Benefits, *Journal of Engineering Design*, **14**(3): 295-313.

Gershenson, J.K., Prasad, G.J. and Zhang, Y., 2003b, Product Modularity: Measures and Design Methods, *Journal of Engineering Design*, **15**(1): 33-51.

Goldberg, D.E., 1989, *Genetic Algorithms in Search, Optimization, and Machine Learning*, Addison-Wesley Publishing Company, Inc., New York.

Goldratt, E.M., and Cox, J., 1984, *The Goal - A Process of ongoing Improvement*, North River Press, Great Barrington, MA.

Gonzalez-Zugasti, J.P., and Otto, K.N., 2000, Modular Platform-Based Product Family Design, ASME Design Engineering Technical Conferences - Design Automation Conference, Renaud, J.E. (Ed.), Baltimore, MD, ASME, Paper No. DETC-2000/DAC-14238.

Gonzalez-Zugasti, J.P., Otto, K.N., and Baker, J.D., 2000, A method for architecting product platforms, *Research in Engineering Design*, **12**(2): 61-72.

Gonzalez-Zugasti, J.P., Otto, K.N., and Baker, J.D., 2001, Assessing value for platformed product family design, *Research in Engineering Design*, **13**(1): 30-41.

Gordon, P., 2004, Tapping the full potential of product platforms, Platform Management for Continued Growth Conference, Atlanta, GA, PDMA.

Graedel, T., Allenby, H., and Comrie, P., 1995, Matrix approaches to abridged life cycle assessment, *Environmental Science and Technology*, **29**(3): 134A-139A.

Green, P.E., and Krieger, A.M., 1985, Models and heuristics for product line selection, *Marketing Science*, **4**(1): 1-19.

Green, P.E., and Krieger, A.M., 1989, Recent contributions to optimal product positioning and buyer segmentation, *European Journal of Operational Research*, **41**(2): 127-141.

Green, P.E., and Krieger, A.M., 1996, Individualized hybrid models for conjoint analysis, *Management Science*, **42**(6): 850-867.

Grob, B., 1975, *Basic Television - Principles and Servicing, 4th Edition*, McGraw-Hill, New York, NY.

GTI, 2001, GT-Power User's Manual and Tutorial, GT-Suite Version 5.2. Gamma Technologies.

Gulati, R.K., and Eppinger, S.D., 1996, the coupling of product architecture and organizational structure decisions, Working Paper, Cambridge, MA, MIT Sloan School of Management, pp. 31.

Guo, F., and Gershenson, J.K., 2003, Comparison of modular measurement methods based on consistency analysis and sensitivity analysis, ASME Design Engineering Technical Conferences - Design Theory & Methodology, Schmidt, L. (Ed.), Chicago, IL, ASME, Paper No. DETC2003/DTM-48634.

Guo, F., and Gershenson, J.K., 2004, A comparison of modular product design methods based on improvement and iteration, ASME Design Engineering Technical Conferences - Design Theory & Methodology, Salt Lake City, UT, ASME, Paper No. DETC2004/DTM-57396.

Gupta, A.K., and Souder, W.E., 1998, Key drivers of reduced cycle time, *Research Technology Management*, **41**(4): 38-43.

Gupta, S., and Krishnan, V., 1998a, Integrated component and supplier selection for a product family, *Production and Operations Management*, **8**(2): 163-182.

Gupta, S., and Krishnan, V., 1998b, Product family-based assembly sequence design methodology, *IIE Transactions*, **30**(10): 933-945.

Halman, J.I.M., Hofer, A.P., and van Vuuren, W., 2003, Platform driven development of product families: Theory versus practice, *Journal of Product Innovation Management*, **20**(2): 149-162.

Hassan, R., de Weck, O., and Springmann, P., 2004, Architecting a communication satellite product line, The 22nd AIAA International Communications Satellite Systems Conference & Exhibit 2004 (ICSSC), Monterey, CA, AIAA, AIAA-2004-3150.

Hastings, N.A.J., and Yeh, C.H., 1992, Bill of manufacture, *Journal of Production and Inventory Management*, **33**(4): 27-31.

Hauser, J.R., 2001, Metrics thermostat, *Journal of Product Innovation Management*, **18**(3): 134-153.

Hazelrigg, G.A., 1998, A framework for decision-based engineering design, *ASME Journal of Mechanical Design*, **120**(4): 653-658.

He, D.W., and Kusiak, A., 1997, Design of assembly systems for modular products, *IEEE Transactions and Automation,* **13**(5): 646-655.

Hegge, H.M.H., 1992, A generic bill-of-material processor using indirect identification of products, *Production Planning & Control*, **3**(3): 336-342.

Henderson, R.M., and Clark, K.B., 1990, Architectural innovation: the reconfiguration of existing product technologies and the failure of established firms, *Administrative Science Quarterly,* **35**(1): 9-30.

Hernandez, G., Allen, J.K., and Mistree, F., 2002, Design of hierarchic platforms for customizable products, ASME Design Engineering Technical Conferences - Design Automation Conference, Fadel, G. (Ed.), Montreal, Quebec, Canada, ASME, Paper No. DETC2002/DAC-34095.

Hernandez, G., Allen, J.K., and Mistree, F., 2003, Platform design for customizable products as a problem of access in a geometric space, *Engineering Optimization,* **35**(3): 229-254.

Hernandez, G., Simpson, T.W., Allen, J.K., Bascaran, E., Avila, L.F., and Salinas, F., 2001, Robust design of families of products with production modeling and evaluation, *ASME Journal of Mechanical Design,* **123**(2): 183-190.

Heywood, J.B., 1988, *Internal Combustion Engine Fundamentals*, McGraw-Hill, New York, NY.

Hicks, D.T., 1999, *Activity Based Costing: Making It Work for Small and Mid-Sized Companies*, John-Wiley & Sons.

Ho, T.H., and Tang, C.S., 1998, *Product Variety Management: Research Advances,* Kluwer Academic Publishers, Boston, MA.

Hofer, A.P., 2001, *Management von Produktfamilien: Wettbewerbsvorteile durch Plattformen*, DUV Gabler, Wiesbaden.

Höltta, K., and Otto, K., 2005, Incorporating design complexity measures in architectural assessment, *Design Studies*, in press.

Hopp, W.J., and Spearman, M.L., 2001, *Factory Physics*, McGraw-Hill, Boston, MA.

Horn, B., 1984, Extended Gaussian images, *Proceeding of the IEEE*, **72**(12): 1671–1686.

Horngren, C.T., and Foster, G., 1991, *Cost Accounting: A Managerial Emphasis*, Prentice Hall, Englewood Cliffs, NJ.

Horngren, C.T., Foster, G., and Datar, S.M., 2000, *Cost Accounting: A Managerial Emphasis,* Prentice Hall, Upper Saddle River, NJ.

Hu, W., and Poli, C., 1997, To injection mold, to stamp, or to assemble? - Part II: A time-to-market perspective, ASME Design Engineering Technical Conferences, Sacramento, CA, ASME, Paper No. DETC97/DTM-3896.

Huang, C.-C., and Kusiak, A., 1998, Modularity in design of products and systems, *IEEE Transactions on Systems, Man and Cybernetics-Part A: Systems and Humans,* **28**(1): 66-77.

Huffman, C., and Kahn, B.E., 1998, Variety for sale: mass customization or mass confusion, *Journal of Retailing,* **74**(4): 491-513.

Hundal, M.S., 1997, Product costing: A comparison of conventional and activity-based costing methods, *Journal of Engineering Design*, **8**(1): 91-103.

Hundal, M.S., 1997, *Systematic Mechanical Designing: A Cost and Management Perspective*, ASME Press, New York, NY.

Ishii, K., Juengel, C., and Eubanks, C.F., 1995a, Design for product variety: Key to product line structuring, ASME Design Engineering Technical Conferences - Design Theory and Methodology, Boston, MA, ASME: **83**(2): 499-506.

Ishii, K., Lee, B.H., and Eubanks, C.F., 1995b, Design for product retirement and modularity based on technology life-cycle, ASME Manufacturing Science and Engineering, Kannatey-Asibu, E. (Ed.), San Francisco, CA, ASME, MED 2-2/MII-3-2: 921-933.

Ittner, C.D., and MacDuffie, J.P., 1994, *Exploring the Sources of International Differences in Manufacturing Overhead*, The Wharton School, University of Philadelphia, Philadelphia, PA.

Jensen, K., 1992, *Colored Petri Nets: Basic Concepts, Analysis Methods and Practical Use,* **1**, Springer-Verlag Berlin Heidelberg, Germany.

Jiang, L., and Allada, V., 2001, Design for robustness of modular product families for current and future markets, ASME Design Engineering Technical Conferences - Design for Manufacturing Conference, Pittsburgh, PA, ASME, Paper No. DETC2001/DFM-21177.

Jiang, Z., Zuo, M.J., Tu, P.Y., and Fung, R.Y.K., 1999, Object-oriented Petri Nets with changeable structure (OPNs-CS) for production system modeling, *International Journal of Advanced Manufacturing Technology,* 15(6): 445-458.

Jiao, J., and Tseng, M.M., 1999, A pragmatic approach to product costing based on standard time estimation, *International Journal of Operations & Production Management,* 19(7): 738-755.

Jiao, J., and Tseng, M.M., 2000, Understanding product family for mass customization by developing commonality indices, *Journal of Engineering Design,* 11(3): 225-243.

Jiao, J., and Tseng, M.M., 2004, Customizability analysis in design for mass customization, *Computer-Aided Design,* 36(8): 745-757.

Jiao, J., Tseng, M.M., Ma, Q., and Zou, Y., 2000, Generic bill of materials and operations for high-variety production management, *Concurrent Engineering: Research and Application,* 8(4): 297-322.

Jiao, J., Zhang, L., and Pokharel, S., 2003, Process platform planning for mass customization, 2nd Interdisciplinary World Congress on Mass Customization and Personalization, CD-ROM Proceedings, Munich, Germany.

Jiao, J., Zhang, L., and Prasanna, K., 2005, Process variety modeling for process configuration in mass customization: An approach based on object-oriented Petri-Nets with changeable structures, *International Journal of Flexible Manufacturing Systems,* in press.

Johnson, A.E., and Hebert, M., 1999, Using spin-images for efficient multiple model recognition in cluttered, 3-D scenes, *IEEE PAMI,* 21(5): 433–449.

Kahn, B.E., 1998, Dynamic relationships with customers: High-variety strategies, *Journal of the Academy of Marketing Science,* 26(1): 45-53.

Kaiser, M., 2004, *Platform Design for Multi-faceted Application Products using a Customer Driven Product Definition Process,* CNH Global.

Kaplan, R.S., 1991, New systems for measurement and control, *The Engineering Economist,* 36(3): 201-218.

Kaplan, R.S., and Cooper, R., 1998, *Cost and Effect: Using Integrated Cost Systems to Drive Profitability and Performance,* Harvard Business School Press, Boston, MA.

Kassarjian, H.H., 1977, Content analysis, *Journal of Consumer Research,* 4(1): 8-18.

Kaul, A., and Rao, V.R., 1995, Research for product positioning and design decisions: An integrative review, *International Journal of Research in Marketing,* 12(4): 293-320.

Kee, R., 1998, Integrating ABC and the theory of constraints to evaluate outsourcing decisions, *Journal of Cost Management,* 12(1): 24-36.

Kennicott, P.R., 1996, *Initial Graphics Exchange Specification – ANS US PRO/IPO-100-1996,* US Product Data Association, N. Charlston, NC.

Kiley, D., 2005, Can VW find its beetle juice? *Business Week Online,* January 31, http://www.businessweek.com/magazine/content/05_05/b3918132.htm.

Kim, H.M., 2001, Target Cascading in Optimal System Design, *Ph.D. Dissertation,* University of Michigan, Ann Arbor, MI.

Kim, H.M., Kokkolaras, M., Louca, L.S., Delagrammatikas, G.J., Michelena, N.F., Filipi, Z.S., Papalambros, P.Y., Stein, J.L., and Assanis, D.N., 2002, Target cascading in vehicle redesign: A class VI truck study, *International Journal of Vehicle Design,* 29(3): 199-225.

Kimberly, W., 1999, Back to the future, *Automotive Engineer,* 24(5): 62-64.

Kimura, F., Kato, S., Hata, T., and Masuda, T., 2001, Product modularization for parts reuse in inverse manufacturing, *CIRP Annals,* 50(1): 89-92.

Kirchain, R.E., 2001, Cost modeling of materials and manufacturing processes, *Encyclopedia of Materials: Science and Technology,* pp. 1718-1727.

Kirkpatrick, S., Gelatt Jr., C.D., and Vecchi, M.P., 1983, Optimization by simulated annealing, *Science,* 220(4598): 671-680.

Knight, W.A., 1998, Group technology, concurrent engineering and design for manufacture and assembly, Group Technology and Cellular Manufacturing: A State-of-the-Art Synthesis of Research and Practice, Suresh, N.C., Kay. J.M. (Eds.), Kluwer Academic Publishers, Boston, pp. 15-36.

Kobe, G., 1997, Platforms - GM's seven platform global strategy, *Automotive Industries,* 177: 50.

Kogut, B., and Kulatilaka, N., 1994, Options thinking and platform investments: investing in opportunity, *California Management Review,* 36(2): 52-71.

Kohli, R., and Krishnamurti, R., 1987, A heuristic approach to product design, *Management Science,* 33(12): 1523-1533.

Kohli, R., and Sukumar, R., 1990, Heuristics for product-line design using conjoint analysis, *Management Science,* 36(12): 1464-1478.

Kokkolaras, M., Fellini, R., Kim, H.M., Michelena, N.F., and Papalambros, P.Y., 2002, Extension of the target cascading formulation to the design of product families, *Structural and Multidisciplinary Optimization*, **24**(4): 293-301.

Kokkolaras, M., Louca, L.S., Delagrammatikas, G.J., Michelena, N.F., Filipi, Z.S., Papalambros, P.Y., Stein, J.L., and Assanis, D.N., 2004, Simulation-based optimal design of heavy trucks by model-based decomposition: An extensive analytical target cascading case study, *International Journal of Heavy Vehicle Systems*, **11**(3-4): 402-432.

Koltai, T., Lonzano, S., Guerrero, and F., Onieva, L., 2000, A flexible costing system for flexible manufacturing systems using activity based costing, *International Journal of Production Research*, **38**(7): 1615-1630.

Kota, S., and Sethuraman, K., 1998, Managing variety in product families through design for commonality, Proceeding of the 1998 ASME DETC, Atlanta, GA, DETC/DTM-5651.

Kota, S., Sethuraman, K., and Miller, R., 2000, A metric for evaluating design commonality in product families, *Journal of Mechanical Design*, **122**(4): 403-410.

Krishnan, V., and Gupta, S., 2001, Appropriateness and impact of platform-based product development, *Management Science*, **47**(1): 52-68.

Krishnan, V., and Ulrich, K., 2001, Product development decisions: A review of the literature, *Management Science*, **47**(1): 1-21.

Kuhfeld, W.F., 2004, Conjoint analysis, SAS Technical Support Resources, TS-689G, http://support.sas.com/techsup/technote/ts689g.pdf.

Kumar, R., Allada, V., and Ramakrishnan, S., 2004, Ant colony optimization methods for product platform formation, ASME Design Engineering Technical Conferences - Design Automation Conference, Chen, W. (Ed.), Salt Lake City, UT, ASME, Paper No. DETC2004/DAC-57195.

Labro, E., 2004, The cost effects of component commonality: A literature review through a management-accounting lens, *Manufacturing & Service Operations Management*, **6**(4): 358-367.

Lancaster, K., 1990, The economics of product variety, *Marketing Science*, **9**(3): 189-206.

Landers, R.G., Min, B.-K., and Koren, Y., 2001, Reconfigurable machine tools, *CIRP Annals*, **48**(2): 471-495.

Layer, A., Ten Brinke, E., van Houten, F., Kals, H.J.J., and Haasis, S., 2002, Recent and future trends in cost estimation, *International Journal of Computer Integrated Manufacturing*, **15**(6): 499-510.

Lee, H.L., and Tang, C.S., 1998, Variability reduction through operations reversal, *Management Science*, **44**(2): 162-172.

Lehnerd, A.P., 1987, Revitalizing the manufacture and design of mature global products, in *Technology and Global Industry: Companies and Nations in the World Economy*, Guile, B.R., Brooks, H. (Eds.), National Academy Press, Washington, D.C., pp. 49-64.

Leibl, P., Hundal, M.S., and Hoehne, G., 1999, Cost calculation with a feature-based CAD system using modules for calculation, comparison and forecast, *Journal of Engineering Design*, **10**(1): 93-102.

Levy, D.L., 1994, Chaos theory and strategy: Theory, application, and managerial implications, *Strategic Management Journal*, **15**(Summer): 167-178.

Li, H., and Azarm, S., 2002, An approach for product line design selection under uncertainty and competition, *ASME Journal of Mechanical Design*, **124**(3): 385-392.

Liebers, A., and Kals, H.J.J., 1997, Cost decision support in product design, *CIRP Annals*, **46**(1): 107-112.

Lin, C.T., and Lee, C.S.G., 1996, *Neural Fuzzy Systems: A Neuro-Fuzzy Synergism to Intelligent Systems*, Prentice Hall, NJ.

Lingnau, V., 1999, Management accounting and product variety, in *Optimal Bundling*, Fuerderer, R., Herrmann, A., Wuebker, G. (Eds.), Springer, Berlin, pp. 133-155.

Locascio, A., 1999, Design economics for electronics assembly, *The Engineering Economist*, **44**(1): 64-77.

Locascio, A., 2000, Manufacturing cost modeling for product design, *International Journal of Flexible Manufacturing Systems*, **12**(2-3): 207-217.

Lutz, R.A., 1998, *Guts: The Seven Laws of Business that Made Chrysler the World's Hottest Car Company*, John Wiley, New York.

MacCormack, A., Rusnak, J., and Baldwin, C.Y., 2004, Exploring the structure of complex software designs: An empirical study of open source and proprietary code, Harvard Business School Working Paper, Boston, MA.

MacDuffie, J.P., Sethuraman, K., and Fisher, M.L., 1996, Product Variety and manufacturing performance: evidence from the international automotive assembly plant study, *Management Science*, **42**(3): 350-369.

Malakooti, B., Malakooti, N.R., and Yang, Z., 2004, Integrated group technology, cell formation, process planning, and production planning with application to the emergency room, *International Journal of Production Research,* **42**(9): 1769-1786.

Martin, M.V., and Ishii, K., 1996, Design for variety: A methodology for understanding the costs of product proliferation, ASME Design Engineering Technical Conferences and Computers in Engineering Conference, Irvine, California, August 18-22, 1999, ASME, Paper No. 96DETC/DTM-1610.

Martin, M.V., and Ishii, K., 1997, Design for variety: Development of complexity indices and design charts, ASME Design Engineering Technical Conferences - Design for Manufacturability, Sacramento, CA.

Martin, M.V., and Ishii, K., 2002, Design for variety: Developing standardized and modularized product platform architectures, *Research in Engineering Design,* **13**(4): 213-235.

Martinez, M.T., Favrel, J., and Ghodous, P., 2000, Product family manufacturing plan generation and classification, *Concurrent Engineering: Research and Applications,* **8**(1): 12-22.

Masters, M., 2002, Direct manufacturing of custom-made hearing instruments, SME Rapid Prototyping Conference and Exhibition, Cincinnati, OH.

Mather, H., 1987, *Bills of Materials,* Dow Jones-Irwin, Homewood, IL.

Mather, H., 1995, Product variety -- Friend or foe?, Proceedings of the 38th American Production & Inventory Control Society International Conference and Exhibition, Orlando, FL, APICS, pp. 378-381.

Mattson, C.A., and Magleby, S.P., 2001, The influence of product modularity during concept selection of consumer products, ASME Design Engineering Technical Conferences - Design Theory and Methodology Conference, Pittsburgh, PA, ASME, Paper No. DETC2001/DTM-21712.

Maupin, A.J., and Stauffer, L.A., 2000, A design tool to help small manufacturers reengineer a product family, ASME Design Engineering Technical Conferences and Computers and Information in Engineering Conference, Baltimore, MD, Paper No. DETC2000/DTM-14568

McAdams, D.A., Stone, R.B., and Wood, K.L., 1999, Functional independence and product similarity based on customer needs, *Research in Engineering Design,* **11**(1): 1-19.

McAdams, D.A., and Wood, K.L., 2002, A quantitative similarity metric for design-by-analogy, *ASME Journal of Mechanical Design,* **124**(2): 173-182.

McAuley, J., 1972, Machine grouping for efficient production, *Production Engineer,* February, pp. 53-57.

McBride, R.D., and Zufryden, F.S., 1988, An integer programming approach to the optimal product line selection problem, *Marketing Science,* **7**(2): 126-140.

McDermott, C.M., and Stock, G.N., 1994, The use of common parts and designs in high-tech industries: A strategic approach, *Production and Inventory Management Journal,* **35**(3): 65-68.

McFarlane, D.C., and Bussmann, S., 2000, Developments in holonic production planning and control, *Production Planning and Control,* **11**(6): 522-536.

McGrath, M.E., 1995, *Product Strategy for High-Technology Companies,* Irwin Professional Publishing, New York, NY.

McWherter, D., Peabody, M., Regli, W., and Shokoufandeh, A., 2001, Transformation invariant shape similarity comparison of solid models, Proceeding of the ASME DETC 2001, Pittsburgh, PA.

McWherter, D., Peabody, M., Regli, W., and Shokoufandeh, A., 2002, Database techniques for indexing and clustering of solid models, The Sixth ACM/SIGGRAPH Symposium on Solid Modeling and Applications, Dutta, D., Seidel, H. (Eds.), Ann Arbor, MI, pp. 78-87.

Mehrabi, M.G., Ulsoy, A.G., and Koren, Y., 2000, Reconfigurable manufacturing systems: key to future manufacturing, *Journal of Intelligent Manufacturing,* **11**(4): 403-419.

Mehrabi, M.G., Ulsoy, A.G., Koren, Y., and Heytler, P., 2002, Trends and perspectives in flexible and reconfigurable manufacturing systems, *Journal of Intelligent Manufacturing,* **13**(2): 135-146.

Messac, A., 1996, Physical programming: Effective optimization for computational design, *AIAA Journal,* **34**(1): 149-158.

Messac, A., Martinez, M.P., and Simpson, T.W., 2002a, Effective product family design using physical programming, *Engineering Optimization,* **34**(3): 245-261.

Messac, A., Martinez, M.P., and Simpson, T.W., 2002b, A penalty function for product family design using physical programming, *ASME Journal of Mechanical Design,* **124**(2): 164-172.

Meyer, M.H., 1997, Revitalize your product lines through continuous platform renewal, *Research Technology Management,* **40**(2): 17-28.

Meyer, M.H., and Lehnerd, A.P., 1997, *The Power of Product Platforms: Building Value and Cost Leadership,* Free Press, New York, NY.

Meyer, M.H., Tertzakian, P., and Utterback, J.M., 1997, Metrics for managing research and development in the context of the product family, *Management Science*, **43**(1): 88-111.

Meyer, M.H., and Utterback, J.M., 1993, The product family and the dynamics of core capability, *Sloan Management Review*, **34**(3): 29-47.

Michaels, J.V., and Woods, W.P., 1989, *Design to Cost*, John Wiley & Sons, New York, NY.

Michelena, N.F., Park, H., and Papalambros, P.Y., 2003, Convergence properties of analytical target cascading, *AIAA Journal*, **41**(5): 897-905.

Miller, J.G., and Vollmann, T.E., 1985, The hidden factory, *Harvard Business Review*, **63**(5): 142-150.

Miller, S., 1999, VW sows confusion with common pattern for models - investors worry profits may suffer as lines compete, *Wall Street Journal*, New York, A.25.

Miller, S., 2002, Volkswagen to overhaul audi brand --- Effort is first step by European car maker to shed conservative image, *Wall Street Journal*, New York, D.8.

Monroe, K., Sunder, S., Wells, W.A., and Zoltners, A.A., 1976, A multi-period integer programming approach to the product mix problem, Proceedings of the American Marketing Association Meeting, Bernhardt, K. (Ed.), pp. 493-497.

Moore, W.L., Louviere, J.J., and Verma, R., 1999, Using conjoint analysis to help design product platforms, *Journal of Product Innovation Management*, **16**(1): 27-39.

Muffatto, M., 1999, Introducing a platform strategy in product development, *International Journal of Production Economics*, **60-61**: 145-153.

Nair, S.K., Thakur, L.S., and Wen, K., 1995, Near optimal solutions for product line design and selection: beam search heuristics, *Management Science*, **41**(5): 767-785.

Naughton, K., Thornton, E., Kerwin, K., and Dawley, H., 1997, Can Honda build a world car?, *Business Week*, September, **8**: 100-107.

Nayak, R.U., Chen, W., and Simpson, T.W., 2002, A variation-based method for product family design, *Engineering Optimization*, **34**(1): 65-81.

Nelson, S.A., II, Parkinson, M.B., and Papalambros, P.Y., 2001, Multicriteria optimization in product platform design, *ASME Journal of Mechanical Design*, **123**(2): 199-204.

Nobeoka, K., and Cusumano, M.A., 1995, Multiproject strategy, design transfer, and project performance: A survey of automobile development projects in the US and Japan, *IEEE Transactions on Engineering Management*, **42**(4): 397-409.

O'Banion, J., 2004, The C-130 product family, Platform Management for Continued Growth Conference, Atlanta, GA, PDMA.

O'Grady, P., 1999, *The Age of Modularity*, Adams and Steele Publishers, Iowa City, IA.

Oosterwal, D., 2004, Driving company growth and profitability through product platforms, Platform Management for Continued Growth Conference, Atlanta, GA, PDMA.

Ortega, R., Kalyan-Seshu, U., and Bras, B., 1999, A decision support model for the life-cycle design of a family of oil filters, ASME Design Engineering Technical Conferences - Design Automation Conference, Las Vegas, NV, ASME, Paper No. DETC99/DAC-8612.

Ostwald, P.F., and McLaren, T.S., 2004, *Cost Analysis and Estimating for Engineering and Management*, Pearson Prentice Hall, Upper Saddle River, NJ.

Otis, C.E., and Vosbury, P.A., 2001, *Aircraft Gas Turbine Powerplants*, Jeppesen Sanderson, Inc., Englewood, CO.

Otto, K., and Wood, K., 2001, *Product Design: Techniques in Reverse Engineering and New Product Development*, Prentice Hall, Inc., Upper Saddle River, NJ.

Pahl, G., and Beitz, W., 1996, *Engineering Design - A Systematic Approach (Second edition)*, (Translated by Wallace, K., and Pomerans, A.), Springer, London.

Papalambros, P.Y., and Wilde, D.J., 2000, *The Principles of Optimal Design — Modeling and Computation*, Cambridge University Press, Cambridge, UK.

Paraskevopoulos, D., Karakitsos, E., and Rustem, B., 1991, Robust capacity planning under uncertainty, *Management Science*, **37**(7): 787-800.

Park, J., 2005, Methods for Incorporating Cost Information into Platform-Based Product Development, *Ph.D. Dissertation,* Department of Industrial & Manufacturing Engineering, Penn State University, University Park, PA.

Park, J., and Simpson, T.W., 2003, Production cost modeling to support product family design, ASME 2003 Design Engineering Technical Conferences, Chicago, IL, ASME, Paper No. DETC2003/DAC-48720.

Park, J., and Simpson, T.W., 2004, Development of a production cost estimation framework to support product family design, *International Journal of Production Research,* **43**(4): 731-772.

Parkinson, A.R., and Balling, R.J., 2002, The OptdesX design optimization software, *Structural and Multidisciplinary Optimization,* **23**(2): 127-139.

Perera, H.S.C., Nagarur, N., and Tabucanon, M.T., 1999, Component part standardization: A way to reduce the life-cycle costs of products, *International Journal of Production Economics,* **60-61**: 109-116.

Pessina, M.W., and Renner, J.R., 1998, Mass customization at Lutron Electronics - a total company process, *Agility & Global Competition,* **2**(2): 50-57.

Pimmler, T.U., and Eppinger, S.D., 1994, Integration analysis of product decompositions, Proceedings of the ASME Design Engineering Technical Conferences - Design Theory and Methodology, Hight, T. K. and Mistree, F. (Eds.), Minneapolis, MN, ASME, **68**, pp. 343-351

Pine, B.J., 1993a, *Mass Customization: The New Frontier in Business Competition,* Harvard Business School Press, Boston, MA.

Pine, J.B., 1993b, Standard modules allow mass customization at Bally Engineering Structures, *Planning Review,* **21**(4): 20-22.

Pullmana, M.E., Mooreb, W.L., and Wardellb, D.G., 2002, A comparison of quality function deployment and conjoint analysis in new product design, *Journal of Product Innovation Management,* **19**(5): 354-364.

Quiroga, T., 2005, *21st Century Muscle Cars, Car and Driver,* January 2005, http://www.caranddriver.com/article.asp?section_id=15&article_id=8908.

Rai, R., and Allada, V., 2003, Modular product family design: Agent-based pareto-optimization and quality loss function-based post-optimal analysis, *International Journal of Production Research,* **41**(17): 4075-4098.

Raiffa, H., and Keeney, R., 1993, *Decisions with Multiple Objectives: Preferences and Value Tradeoffs,* Wiley, New York, NY.

Rajan, P., Van Wie, M., Campbell, M., Otto, K., and Wood, K., 2003, Design for flexibility – Measures and guidelines, *Design Studies,* to appear.

Ramdas, K., and Sawhney, M.S., 2001, A cross-functional approach to evaluating multiple line extensions for assembled products, *Management Science,* **47**(1): 22-36.

Raymer, D.P., 1989, *Aircraft Design: A Conceptual Approach,* AIAA, Washington, D.C.

Rehman, S., and Guenov, M.D., 1998, A methodology for modeling manufacturing costs at conceptual design, *Computers & Industrial Engineering,* **35**(3-4): 623-626.

Reinertsen, D.G., 1997, *Managing the Design Factory,* The Free Press, New York, NY.

Renders, J.M., and Plasse, S.P., 1996, Hybrid methods using genetic algorithms for global optimization, *IEEE Transactions on Systems, Man and Cybernetics — Part B: Cybernetics,* **26**(2): 243-258.

Roach, G.M., Cox, J.J., and Sorensen C.D., 2005, The product design generator: A system for producing design variants, *International Journal of Mass Customization,* **1**(1): 83-106.

Roach, G.M., Cox, J.J., and Young, J.M., 2003, A new strategy for automating the generation of product family members and artifacts applied to an aerospace application, Proceedings of the 2003 Design Engineering Technical Conferences, Chicago, IL, DETC2003/CIE-48185.

Robertson, D., and Ulrich, K., 1998, Planning for product platforms, *Sloan Management Review,* **39**(4): 19-31.

Robinson, T.J., Borror, C.M., and Myers, R.H., 2004, Robust parameter design: A review, *Quality and Reliability Engineering International,* **20**: 81-101.

Robinson, W.T., 1988, Marketing mix reactions to entry, *Marketing Science,* **7**(4): 368-385.

Roemer, T.A., Ahmadi, R., and Wang, R.H., 2000, Time-cost trade-offs in overlapped product development, *Operations Research,* **48**(6): 858-865.

Rolstadas, A., 1991, Editorial: group technology and design of production systems, *Production Planning and Control,* **2**(4): 297.

Romanowski, C.J., and Nagi, R., 2005, On comparing bills of materials: A similarity/distance measure for unordered trees, *IEEE Transactions on Systems, Man, and Cybernetics, Part A,* in press.

Rothwell, R., and Gardiner, P., 1990, Robustness and product design families, in *Design Management: A Handbook of Issues and Methods,* Oakley, M. (Ed.), Basil Blackwell Inc., Cambridge, MA, pp. 279-292.

Saaty, T., 1980, *The Analytic Hierarchy Process,* McGraw-Hill, New York, NY.

Sabbagh, K., 1996, *Twenty First Century Jet — The Making and Marketing of The Boeing 777,* Scribner, New York, NY.

Sand, J.C., Gu, P., and Watson, G., 2002, HOME: house of modular enhancement - a tool for modular product redesign, *Concurrent Engineering: Research and Applications,* **10**(2): 153-164.

Sanderson, S., and Uzumeri, M., 1995, Managing product families: The case of the Sony Walkman, *Research Policy,* **24**(5): 761-782.

Sanderson, S.W., and Uzumeri, M., 1997, *Managing Product Families,* Irwin, Chicago, IL.

Sands, J., Loughlin, W., and Lu, F., 1998, Equipment Standardization under Acquisition Reform, *Journal of Ship Production*, **14**(2): 110-123.

Sawhney, M.S., 1998, Leveraged high-variety strategies: from portfolio thinking to platform thinking, *Journal of the Academy of Marketing Science*, **26**(1): 54-61.

Scheidt, L.G., and Zong, S., 1994, Approach to achieve reusability of electronic modules, Proceedings of the 1994 IEEE International Symposium on Electronics & the Environment, San Francisco, CA, USA, Paper No. 331-336.

Schlueter, J., 2004, Prepaid wireless: Launching a new platform for the burgeoning market of pay-as-you-go services, Platform Management for Continued Growth Conference, Atlanta, GA, PDMA.

Schonberger, R.J., 1986, *World Class Manufacturing - The Lesson of Simplicity Applied*, The Free Press, New York.

Seepersad, C.C., Hernandez, G., and Allen, J. K., 2000, A quantitative approach to determining product platform extent, ASME Design Engineering Technical Conferences - Design Automation Conference, Renaud, J. E. (Ed.), Baltimore, MD, ASME, Paper No. DETC2000/DAC-14288.

Seepersad, C.C., Mistree, F., and Allen, J.K., 2002, A quantitative approach for designing multiple product platforms for an evolving portfolio of products, ASME Design Engineering Technical Conferences - Design Automation Conference, Fadel, G. (Ed.), Montreal, Quebec, Canada, ASME, Paper No. DETC2002/DAC-34096.

Sella, M., 1998, It cuts both ways, *Worth*, September, pp. 82-93.

Shaukat, M.M., 2001, Modularity, Platforms, and Customization in the Automotive Industry, *M.S. Thesis*, Department of Industrial & Manufacturing Engineering, Penn State University, University Park, PA.

Shierholt, K., 2001, Process configuration: Combining the principles of product configuration and process planning, *Artificial Intelligence in Engineering Design, Analysis, and Manufacturing*, **15**(5): 411-424.

Shimokawa, K., Jurgens, U., and Fujimoto, T., 1997, *Transforming Automobile Assembly: Experience in Automation and Work Organization*, Springer, New York.

Shirley, G.V., 1990, Models for managing the redesign and manufacture of product sets, *Journal of Manufacturing and Operations Management*, **3**(2): 85-104.

Siddique, Z., 2000, *Product Platform Development: Designing for Product Variety, Ph.D. Dissertation*, G. W. Woodruff School of Mechanical Engineering, Georgia Institute of Technology, Atlanta, GA.

Siddique, Z., 2001, Estimating reduction in development time for implementing a product platform approach, ASME 2001 Design Engineering Technical Conferences, Pittsburgh, PA, ASME, Paper No. DETC2001/CIE-21238.

Siddique, Z., and Repphun, B., 2001, Estimating cost savings when implementing a product platform approach, *Concurrent Engineering: Research and Application*, **9**(4): 285-294.

Siddique, Z., Rosen, D.W., and Wang, N., 1998, On the applicability of product variety design concepts to automotive platform commonality, ASME Design Engineering Technical Conferences - Design Theory and Methodology, Atlanta, GA, ASME, Paper No. DETC98/DTM-5661.

Simon, H.A., 1996, *Sciences of the Artificial*, The MIT Press, Cambridge, MA.

Simpson, T.W., 1998, A Concept Exploration Method for Product Family Design, *Ph.D. Dissertation*, G.W. Woodruff School of Mechanical Engineering, Georgia Institute of Technology, Atlanta, GA.

Simpson, T.W., 2004, Product platform design and customization: Status and promise, *AIEDAM*, **18**(1): 3-20.

Simpson, T.W., Chen, W., Allen, J.K., and Mistree, F., 1999, Use of the robust concept exploration method to facilitate the design of a family of products, in *Simultaneous Engineering: Methodologies and Applications*, Roy, U., Usher, J.M., and Parsaei, H.R. (Eds.), Gordon and Breach Science Publishers, Amsterdam, The Netherlands, pp. 247-278.

Simpson, T.W., and D'Souza, B., 2004, Assessing variable levels of platform commonality within a product family using a multiobjective genetic algorithm, *Concurrent Engineering: Research and Applications*, **12**(2): 119-130.

Simpson, T.W., Maier, J.R.A., and Mistree, F., 2001a, Product platform design: method and application, *Research in Engineering Design*, **13**(1): 2-22.

Simpson, T.W., Seepersad, C.C., and Mistree, F., 2001b, Balancing commonality and performance within the concurrent design of multiple products in a product family, *Concurrent Engineering: Research and Applications*, **9**(3): 177-190.

Smith, J., and Duffy, A., 2001, Modularity in support of design for re-use, International Conference on Engineering Design, Glasgow.

Smith, P.G., and Reinertsen, D.G., 1991, *Developing Products In Half The Time*, Van Nostrand Reinhold, New York.

Sodhi, M., and Knight, W.A., 1998, Product design for disassembly and bulk recycling, *CIRP Annals*, **47**(1): 115-118.

Sosa, M., Eppinger, S., and Rowles, G., 2003, Identifying modular and integrative systems and their impact on design team interactions, *Journal of Mechanical Design*, **125**(10): 240-252.

Srinivas, N., and Deb, K., 1995, Multiobjective function optimization using nondominated sorting genetic algorithms, *Evolutionary Computation Journal*, **2**(3): 221-248.

Stalk, G., Jr., and Hout, T., 1990, *Competing Against Time*, The Free Press, New York.

Stalk, G., Jr., and Webber, A.M., 1993, Japan's dark side of time, *Harvard Business Review*, **71**(4): 93-102.

Steiner, W.J., and Hruschka, H., 2002, A probabilistic one-step approach to the optimal product line design problem using conjoint and cost data, Review of Marketing Science Working Papers, 1(4): Working Paper 4, http://www.bepress.com/roms/vol1/iss4/paper4.

Stone, R.B., Wood, K.L., and Crawford, R.H., 1998, A heuristic method to identify modules from a functional description of a product, ASME Design Engineering Technical Conference, Atlanta, Georgia, Sep. 13-16, Paper No. DETC98/DTM-5642.

Stone, R.B., Wood, K.L., and Crawford, R.H., 2000a, A heuristic method to identify modules from a functional description of a product, *Design Studies*, **21**(1): 5-31.

Stone, R.B., Wood, K.L., and Crawford, R.H., 2000b, Using quantitative functional models to develop product architectures, *Design Studies*, **21**(3): 239-260.

Stratasys FDM Technologies, 2003, http://www.stratasys.com.

Sudharshan, D., May, J.H., and Shocker, A.D., 1987, A simulation comparison of methods for new product location, *Marketing Science*, **6**(2): 182-201.

Sudjianto, A., and Otto, K., 2001, Modularization to support multiple brand platforms, ASME Design Engineering Technical Conferences, Pittsburgh, DETC2001/DTM-21695.

Suh, N., 2001, *Axiomatic Design: Advances and Applications*, Oxford University Press, New York, NY.

Sundgren, N., 1999, Introducing interface management in new product family development, *Journal of Product Innovation Management*, 16(1): 40-51.

Tanaka, M., 1989, Cost planning in the design phase of a new product, in *Japanese Management Accounting*, Mondem, Y., Sakurai, M. (Eds.), Productivity Press, Cambridge, MA.

Ten Brinke, E., Lutters, E., Streppel, T., and Kals, H.J.J., 2000, Variant-based cost estimation based on information management, *International Journal of Production Research*, **38**(17): 4467-4479.

Terwiesch, C., and Loch, C. H., 1999, Managing the process of engineering change orders: the case of the climate control system in automobile development, *Journal of Product Innovation Management*, 16(2): 160-172.

Thevenot, H.J., Nanda, J., and Simpson, T.W., 2005, A methodology to support product family redesign using a genetic algorithm and commonality indices, ASME International Design Engineering Technical Conferences - Design Automation Conference, Long Beach, CA, Paper No. DETC2005/DAC-84927.

Thevenot, H.J., and Simpson, T.W., 2004, A comparison of commonality indices for product family design, ASME International Design Engineering Technical Conferences - Design Automation Conference. Salt Lake City, UT, Paper No. DETC2004/DAC-57141.

Thevenot, H.J., and Simpson, T.W., 2006, Commonality indices for product family design: A detailed comparison, *Journal of Engineering Design*, in press.

Thomke, S.H., 1997, The role of flexibility in the development of new products: An empirical study, *Research Policy*, **26**(1): 105-119.

Thonemann, U.W., and Brandeau, M.L., 2000, Optimal commonality in component design, *Operations Research*, **48**(1): 1-19.

Torenbeek, E., 1976, *Synthesis of Subsonic Airplane Design*, Delft University Press, Delft, The Netherlands.

Train, K.E., 2003, *Discrete Choice Methods with Simulation*, Cambridge University Press, UK.

Ulrich, K.T., 1995, The role of product architecture in the manufacturing firm, *Research Policy*, **24**(3): 419-440.

Ulrich, K.T., and Eppinger, S.D., 2004, *Product Design and Development*, McGraw-Hill/Irwin, New York.

Ulrich, K.T., and Pearson, S.A., 1993, Does product design really determine 80% of manufacturing cost? MIT Sloan School of Management, Cambridge, MA, pp. 31.

Ulrich, K.T., and Pearson, S.A., 1998, Assessing the importance of design through product archaeology, *Management Science*, **44**(3): 352-369.

Ulrich, K.T., and Tung, K., 1991, Fundamentals of product modularity, ASME Winter Annual Meeting, Atlanta, GA, ASME, **39**: 73-80.

Umeda, Y., Shimomura, Y., Yoshioka, M., and Tomiyama, T., 1999, A proposal of design methodology for upgradeable products, ASME Design Engineering Technical Conferences - Design for Manufacturing, Las Vegas, NV, ASME, Paper No. DETC99/DFM-8969.

Uppal, K.B., 1996, Estimating engineered equipment costs, *AACE Transactions* **40**: EST.10.1-EST.10.6.

Uzumeri, M., and Sanderson, S., 1995, A framework for model and product family competition, *Research Policy*, **24**(4): 583-607.

Valiente, G., 2002, *Algorithms on Trees and Graphs*, Springer-Verlag, Berlin.

van Veen, E.A., 1992, *Modeling Product Structures by Generic Bills-of-Materials*, Elsevier, New York.

Veinott, C.G., and Martin, J.E., 1986, *Fractional and Subfractional Horsepower Electric Motors*, McGraw-Hill, New York.

Vollrath, K., 2001, Flugzeug der Zukunft fliegt mit Feinguss, *VDI nachrichten, Duesseldorf*, pp. 19.

von Hippel, E., 1990, Task partitioning: An innovation process variable, *Research Policy*, **19**(5): 407-418.

Wacker, J.G., and Trelevan, M., 1986, Component part standardization: An analysis of commonality sources and indices, *Journal of Operations Management*, **6**(2): 219-244.

Wang, L.C., and Wu, S.Y., 1998, Modeling with colored timed object-oriented Petri nets for automated manufacturing systems, *Computers and Industrial Engineering*, **34**(2): 463-480.

Westkämper, E., Schmidt, T., and Wiendahl, H.H., 2000, Production planning and control with learning technologies: Simulation and optimization of complex production processes, Knowledge-based Systems, Leondes, G., (Ed.).

Weustink, I.F., Ten Brinke, E., Streppel, A.H., and Kals, H.J.J., 2000, A generic framework for cost estimation and cost control in product design, *Journal of Materials Processing Technology*, **103**(1): 141-148.

Wheelwright, S.C., and Clark, K.B., 1992, *Revolutionizing Product Development: Quantum Leaps in Speed, Efficiency, and Quality*, The Free Press, New York, NY.

Wheelwright, S.C., and Clark, K.B., 1995, *Leading Product Development*, Free Press, New York, NY.

Wheelwright, S.C., and Sasser, W.E. Jr., 1989, The new product development map, *Harvard Business Review*, **67**(3): 112-125.

Whitney, D.E., 1993, Nippondenso Co. Ltd.: A case study of strategic product design, *Research in Engineering Design*, **5**(1): 1-20.

Wilhelm, B., 1997, Platform and modular concepts at Volkswagen - their effect on the assembly process, in *Transforming Automobile Assembly: Experience in Automation and Work Organization*, Shimokawa, K., Jürgens, U., and Fujimoto, T. (Eds.), Springer, New York, pp. 146-156.

Wilkins, D., 2005, Classic Cars: Cadillac Cimarron, *The Independent*, March 22, Online Edition, http://motoring.independent.co.uk/features/article7325.ece, 9 pgs.

Willcox, K., and Wakayama, S., 2003, Simultaneous optimization of a multiple-aircraft family, *Journal of Aircraft*, **40**(4): 616-622.

Williams, C.B., Allen, J.K., Rosen, D.W., and Mistree, F., 2004, Designing platforms for customizable products in markets with non-uniform demand, Proceedings of ASME Design Theory Methodology Conference, Salt Lake City, UT, DETC2004/DTM-57469.

Williams, C.B., Panchal, J.H., and Rosen, D.W., 2003, A general decision-making method for the rapid manufacturing of customized parts, Proceedings of ASME Computer and Information in Engineering Conference, Chicago, Ill., DETC2003/CIE-48198.

Womack, J.P., Jones, D.T., and Roos, D., 1990, *The Machine that Changed the World*, Rawson Associates, New York, NY.

Wortmann, J.C., Muntslag, D.R., and Timmermans, P.J.M., 1997, *Customer-Driven Manufacturing*, Chapman and Hall, London.

Yang, L., and Gao, Y., 1996, *Fuzzy Mathematics: Theory and Applications*, Springer, New York, NY.

Yano, C., and Dobson, G., 1998, Profit optimizing product line design, selection and pricing with manufacturing cost considerations, in *Product Variety Management: Research Advances*, Ho, T.-H., Tang, C.S. (Eds.), Kluwer Academic Publisher, pp. 145-176.

Yin, R.K., 1994, *Case Study Research: Design and Methods*, Sage publications, Thousand Oaks, CA.

Zamirowksi, E.J., and Otto, K.N., 1999, Identifying product portfolio architecture modularity using function and variety heuristics, ASME Design Engineering Technical Conferences - Design Theory and Methodology, Las Vegas, NV, ASME, Paper No. DETC99/DTM-8760.

Zhang, D., and Hebert, M., 1999, Harmonic maps and their applications in surface matching, IEEE Conference on Computer Vision and Pattern Recognition, **2**, 7 pgs.

Zhang, L., Jiao, J., and Pokharel, S., 2004, Internet-enabled information management for configure-to-order product fulfillment, *Asia Pacific Management Review,* 9(5): 851-876.

Zhang, Y., and Gershenson, J.K., 2003, An initial study of direct relationships between life-cycle modularity and life-cycle cost, *Concurrent Engineering: Research and Application,* 11(2): 121-128.

Zhang, Y., Gershenson, J.K., and Allamneni, S., 2001, Determining relationships between modularity and cost in product retirement, ASME Design Engineering Technical Conferences - Design Theory and Methodology Conference, Pittsburgh, PA, ASME, Paper No. DETC2001/DTM-21686.

Zillober, C., 2001, Software manual for SCPIP 2.2, Informatik, Universitaet Bayreuth, Bayreuth.

Index